P9-CBQ-874
REGISTERED
PROFESSIONAL ENGINEER

Handbook of Temporary Structures in Construction

Handbook of Temporary Structures in Construction

Engineering Standards, Designs, Practices & Procedures

Robert T. Ratay
EDITOR IN CHIEF

McGraw-Hill Book Company

New York St. Louis San Francisco Auckland
Bogotá Hamburg Johannesburg London Madrid
Mexico Montreal New Delhi Panama Paris
São Paulo Singapore Sydney Tokyo Toronto

Library of Congress Cataloging in Publication Data
Main entry under title:

Handbook of temporary structures in construction.

Includes index.
1. Building sites—Handbooks, manuals, etc.
2. Building—Handbooks, manuals, etc. I. Ratay, Robert T.
TH375.H36 1984 624.1 83-23905
ISBN 0-07-051211-6

1234567890 HAL/HAL 8987654

ISBN 0-07-051211-6

The editors for this book were Patricia Allen-Browne and Geraldine Fahey, the designer was Mark E. Safran, and the production supervisor was Teresa F. Leaden. It was set in Century Schoolbook by University Graphics.

Printed and bound by Halliday Lithograph.

To Bizso

Contents

Contributors

Colin P. Bennett, P.E., Manager, Special Projects, Patent Scaffolding Company—A Division of Harsco Corporation, Fort Lee, N.J. (Chapters 13 and 16).

Robert F. Borg, P.E., President, Kreisler Borg Florman Construction Company, Scarsdale, N.Y. (Chapter 1).

J. F. Camellerie, P.E., Manager of Construction, Advanced Technology and Industrial Projects, EBASCO Constructors Inc., Lyndhurst, N.J.; formerly, President, Camellerie & Bialkowski Associates (Chapter 15).

Melvin Febesh, P.E., Vice President, Urban Foundation Company, Inc., East Elmhurst, N.Y. (Chapter 8).

Ben C. Gerwick, Jr., P.E., Professor of Civil Engineering (Construction Engineering and Management), University of California, Berkeley; formerly, Executive Vice President, Santa Fe-Pomeroy Constructors (Chapter 7).

R. Kirk Gregory, M.S., P.E., Consultant; formerly, Director of Engineering, Symons Corporation (Chapter 14).

Julian J. Karp, P.E., Associate Professor of Civil Engineering, Rutgers The State University of New Jersey, Piscataway, N.J. (Chapter 3).

Helmut G. Kobler, P.E., Vice President, Special Projects, Harrison Western Corporation, Healdsburg, Calif. (Chapter 10).

Enno Koehn, Ph.D., P.E., Associate Professor of Civil Engineering, Purdue University, West Lafayette, Ind. (Chapter 5).

Robert G. Lenz, P.E., President, Moretrench American Corporation, Rockaway, N.J. (Chapter 6).

J. Mark Loizeaux, P.E., President, Loizeaux Group International; Vice President, Controlled Demolition, Inc. and Controlled Demolition International, Inc., Phoenix, Md. (Chapter 19).

Bernard Monahan, P.E., Vice President, William L. Crow Construction Company—A subsidiary of J. A. Jones Construction Company, New York, N.Y.; formerly associated with URS-Madigan Praeger Consulting Engineers and Thomas Crimmins Contracting Company (Chapters 11 and 12).

Richard C. Mugler, Jr., MBA, Secretary-Treasurer, Richard C. Mugler Company, Inc., Bronx, N.Y. (Chapter 20).

Louis J. Porcello, Civil Engineer and Construction Consultant, Seattle, Wash.; formerly, Chief Estimator, Harrison Western Corporation, Denver, Colo. (Chapter 10).

Allan G. Provost, P.E., President, Harrison Western Corporation, Denver, Colo. (Chapter 10).

Robert T. Ratay, Ph.D., P.E., Consulting Engineer and Professor of Civil/Structural Engineering, Pratt Institute, New York, N.Y.; formerly, Partner, Severud Perrone Szegezdy Sturm, Consulting Engineers, P.C., New York, N.Y. (Chapter 13).

Robert A. Rubin, Esq., P.E., Partner, Postner & Rubin Attorneys at Law, New York, N.Y. (Chapter 2).

Wheeler H. Rucker, P.E., President, Wheeler H. Rucker & Associates, Engineering Consultants, Colts Neck, N.J. (Chapter 17).

Fred N. Severud, P.E., Partner, Severud Perrone Szegezdy Sturm, Consulting Engineers, P.C., New York, N.Y. (Chapter 9).

Man-Chung Tang, Dr.-Ing., P.E., President, DRC Consultants, Inc., New York, N.Y. (Chapter 18).

Rudi J. van Leeuwen, P.E., Technical Advisor, Slattery Associates, Inc., Maspeth, N.Y.; formerly, Chief Engineer, Spencer, White, and Prentis, Inc. (Chapter 9).

Stuart Wood, Jr., P.E., Regional Manager, J. A. Jones Construction Company, Englewood, Colo. (Chapter 4).

Robert S. Woolworth, P.E., Vice President, Converse Consultants, Caldwell, N.J. (Chapter 8).

Preface

This book presents temporary structures, facilities, and procedures used in construction operations. It is *a first* for it contains the related topics of this field under one cover. It bridges the gap between books on design and books on construction, in that it deals with those *systems that are used to construct the designed projects.* The book presents much-needed practical information on the requirements, design, erection, and maintenance of temporary structures and on the performance of temporary procedures that are used in the execution of construction work.

These systems have a primary influence on the quality, speed and financial success of all construction projects.

Both the *business* and the *practice* of temporary structures and procedures are treated, including legal responsibilities, codes and regulations, design philosophies, construction methods, hardware, inspection, safety, and cost.

Engineers, technicians, contractors, inspectors, building officials, and others concerned with the success of construction projects will find this book valuable as an authoritative reference. Teachers and students of construction engineering will also find it useful as a practical textbook.

Technical information on many types of temporary structures and procedures is not as extensive, well-developed, standardized, and well-documented as on most other subjects of engineering design and construction. Consequently, many parts of this book are original with little or no reliance on previously published works. On topics where there are no standard methods, the more widely used practices are presented. Design methods ranging from rule-of-thumb to engineering analysis are given according to what the generally accepted and warranted degree of sophistication is for the particular problem.

The opening chapter, "Technical and Business Practices," explains how the business of procuring, furnishing, installing, and maintaining temporary structures and procedures should be conducted in order to attain technical and financial success. The second chapter, "The Legal Position," deals with the rights, responsibilities, and liabilities of the involved parties as defined by common law, statute, and the signed contract. The chapter on "Codes and Regulations" introduces the applicable manuals, codes, and regulations, and discusses what we must, may, and may not do according to government regulations and technical code requirements. These first three chapters, as a group, are unique in that they are directed specifically to *temporary* structures and procedures rather than to contracting in general.

Under "Erection and Earthwork Equipment" the reader will learn not only about the types and uses of machinery but also how to select the proper equipment and model, whether to buy or lease it, and what the costs of maintenance, depreciation, and replacement are. A most valuable and heretofore unavailable compilation of operating weights, horizontal breaking, and other forces of over 130 dozers, loaders, trucks, scrapers, graders, rollers, excavators, etc., is given in "Loads Created by Construction Equipment."

"Construction Dewatering" is presented in Chapter 6 as a temporary procedure, with emphasis on the advantages and disadvantages of methods, how to choose the right one, how to estimate costs, as well as the phases of mobilization, installation, operation, and removal. Chapter 7, "Cofferdams," is confined to structural cofferdams and cribs as temporary installations, explaining the proper and safe methods and materials to be used. Chapters 8 ("Temporary Earth-Retaining Structures"), 9 ("Underpinning and Shoring"), and 10 ("Underground Construction Supports") deal with the temporary supports for below-grade construction with emphasis on how to choose, install, and maintain the proper system and ensure safety. Chapters 11 and 12, "Roadway Decking" and "Construction Ramps, Runways, and Platforms," respectively, are not known to be in other books; they present the often neglected topic of various ways of providing access into, out of, over, and around the work area; the discussions are amply illustrated by photographs. Shoring and scaffolding are the most important temporary structures for building construction and are treated in detail in Chapter 13, "Falsework/Shoring," and in Chapter 16, "Access Scaffolding." This may be the most complete treatment of these two subjects in a book and includes design, hardware, installation, purchase and rental costs, safety requirements and numerous illustrations. Perhaps the most often used temporary structure is "Concrete Formwork," discussed in Chapter 14 with emphasis on the contractor's concerns, i.e., types of forms, hardware, details and dimensions, and ACI code requirements. The intricacies of design, installation, and operation techniques of slipforming are discussed in Chapter 15, "Slipforms." The basics of providing lateral stability of structures during construction are discussed in Chapter 17, "Temporary Bracing and Guying." The various construction methods and their required temporary supports are described and illustrated through photographs in Chapter 18, "Temporary Structures in Concrete-Bridge Construction." Although it is not a temporary structure, but is, indeed, a very final procedure, "Demolition Operations and Equipment" are introduced in Chapter

19 including hand, mechanical, and explosive operations; handling and disposal of materials; insurance; and other matters. The final chapter of the book deals with the critical subject of "Protection of Site, Adjacent Areas and Utilities," including signs, barricades, sidewalk sheds, and other protective installations, as well as pointers on avoiding damage to adjacent facilities, and references to OSHA and other safety regulations.

The 20 chapters have been written by 23 authors who were selected for their eminence in their fields of expertise; they prepared manuscripts of high quality and value to the profession. I am grateful to them for their efforts and cooperation.

I owe special thanks to some of the McGraw-Hill people: Harold Crawford for his early encouragement, and Patricia Allen-Browne and Geraldine Fahey for their continuing cooperation and valuable assistance in producing the book.

I thank McGraw-Hill for offering me the opportunity to create this book, the contributing authors for preparing the manuscripts, my family and friends for bearing with me during the years it took to complete the book and I thank you, the reader, for rewarding us all by using the book.

Robert T. Ratay

1

Technical and Business Practices

by Robert F. Borg

A glance at the table of contents of this volume makes abundantly clear that a comprehensive treatment of this subject belongs on the shelf of every practicing engineer, contractor, and building construction subcontractor. After reading this handbook one may well ask, how did the profession get along so long without it?

This opening chapter outlines the broad nontechnical relationships and responsibilities between the temporary construction contractor, the general contractor, the owner, and the public. A temporary structure in construction affects many people. First, and perhaps foremost, it affects the safety of the workers on the job and of the general public. There is also the relationship of the temporary structure to the finished structure. Whether it is finally incorporated into the finished work or is removed at the conclusion of its usefulness, the temporary structure contractor will have supervision work, code requirements, contract and legal requirements, insurance requirements, and perhaps disputes with others over the work being performed.

In this chapter some of these problems and potential problems are touched on. It is possible to point out only some of the general areas of concern. Adequate treatment of most of the topics is given in the chapters that follow. For more extensive treatment of some of the other topics, the reader may be required to review books or treatises on those subjects. Some of them are mentioned in this chapter. Others can be found in, and obtained from, the Engineering Society's Library at the United Engineering Center, 345 East 47th Street, New York, NY 10017.

TECHNICAL PRACTICES

Design

The answer as to who does the design, who makes the drawings, and how general or detailed should the specifications be varies, depending on the temporary structure under consideration. In an extremely complex job involving such temporary work as a cofferdam for bridge piers or a method of driving shafts for a tunnel, the design of the temporary structures will often be done by the designer of the completed project. It will then be the responsibility of each contractor to carry out the details of the design and translate them into work. On the other hand, for simpler types of temporary structures such as temporary ramps that might be used by an excavation contractor for a building, the contractor will do the designing.

Between those two extremes is the type of temporary structure in which specialty contractors, who make a business of doing a specific type of temporary structure, will be employed. Examples of this include concrete forms using patented panels.

Responsibility for Design Adequacy

It is well established that the contractor who builds in accordance with the design that is provided by another is not responsible for the adequacy of the design or for the bringing about of the result for which the design was intended, providing the design is followed. On the other hand, the contractor who provides the design is responsible for both the adequacy of the design and its fitness for the intended use.

In the provision of a temporary structure for sheeting and shoring of an embankment, for example, the design engineer or architect-engineer for the completed structure might indicate that sheeting or shoring is required for a particular embankment. This general directive may be incorporated on the plans or required by specification or code. This may also require that the contractor submit the design for the proposed sheeting and shoring to the engineer. On the other hand, the engineer may decide on the type of design which is required and indicate this design on the contract drawings.

In the first instance the contractor will be asked to submit a detailed plan for the proposed sheeting and shoring and in all likelihood will be required to have the detailed plan checked and signed by a licensed professional engineer. The adequacy of the sheeting and shoring will be the responsibility of the contractor together with his licensed professional engineer.

If, on the other hand, the designer of the completed structure shows the details of the sheeting and shoring and the contractor follows this design in the execution of the work, then the responsibility for adequacy will be that of the original designer.

Should the contractor deviate from this design, and should it be shown that the deviation from the design has resulted in a failure of the sheeting and shoring to properly perform its function, then the contractor has the responsibility of accepting the consequences of the damages which have resulted from this deviation.

Additional criteria that the designer must use in designing temporary structures are the requirements of the Occupational Safety and Health Act (OSHA) of the U. S. government (Title 29—Labor Code, Federal Regulations Chapter XVII, Part 1926, U. S. Government Printing Office).

The Occupational Safety and Health Administration can impose fines for on-the-spot violations, and regulations are enforced not only by federal inspectors but can be enforced as a result of complaints to the federal inspectors by employees, or third persons not connected with the construction job.

Another instance of the failure of design to conform to engineering and construction standards was cited in a report of the South Carolina Department of Labor. The department blamed faulty design as the cause of the collapse of a steel sheet pile cofferdam in about 40 ft (12 m) of water in Lake Keowee, where the water intake structure and a pumping station were built for the City of Greenville, South Carolina. According to the report of *Engineering News-Record* on April 19, 1979, the State Labor Department citation specifically condemned "incorrect basis of design of the cofferdam, in that the method of calculating the requirement for elastic stability and calculating actual stress levels . . . was not adapted to the function of a cofferdam required to retain free water." An independent review of the cofferdam design and of the accident (by Law Engineering Testing Co., Marietta, Ga.) found that "the probable cause of the cofferdam failure was an incorrect basis of design. It appears that the cofferdam was designed as though it were an excavation in earth or rock rather than a cofferdam required to retain free water."

Because of the importance of safety in overall conduct of the business of a temporary construction contractor, the contractor should appoint an individual in his office as the safety officer who is well versed in all requirements of safety regulations and of good construction practices relating to safety. This person should be given authority to order changes in methods or in construction procedures to bring about safe work practices.

Specifications

The specifications for the temporary structure are usually drawn up by the temporary-structure contractor. In most cases architects or engineers designing a complete structure will not provide a design or specifications for the temporary structure work. For example, a temporary-structure contractor working on concrete formwork will have to specify the type of plywood, framing, form ties, snap ties, braces, and other form accessories that are required for the work. These materials are often ordered from a catalog or from a supplier that has supplied materials to the contractor in the past. The responsibility for the adequacy of the specified material and for its quality will be the temporary-structure contractor's.

In the case of a design which is prepared by a temporary-structure contractor for the approval of the architect or engineer, the specifications which are part of that design must also be the responsibility of the temporary-structure contractor. This specification may have to be reviewed by the engineer or architect for its adequacy and for durability and strength. Should there then be a question about any of these qualities, the temporary-structure contractor will have to justify the specified material as being proper.

Permits

The temporary-structure contractor is required to obtain permits for any work done. These permits may consist of approval of the plans for the temporary work by the Department of Housing and Buildings, when this type of approval is required. Such approvals, as an example, might be for shoring or sheet-piling work, or for equipment that the contractor may be required to use in performing the work. The equipment may consist of cranes or hoists to be placed in the street or on the sidewalk. Other permits may be required for sidewalk crossings and for storage of materials in the street. In general, the cost of the permits must also be borne by the temporary-structures contractor. When a permit must be obtained that requires a submission of a drawing prepared by a licensed professional engineer, the drawing is the responsibility of the temporary-structure contractor. If the temporary-structure contracting company does not have a licensed professional engineer as one of its principals, then it is necessary to retain such a qualified individual for the preparation of its plans.

Shop Drawings

In a temporary structure, the shop drawing and its preparation are somewhat different from a shop drawing prepared by a subcontractor or supplier for a permanent part of the finished work. The shop drawing for a temporary structure may, in fact, be a completely original design which does not appear in any way on the general contractor's contract drawings. An example of this might be a shop drawing prepared for the dewatering system using well points. In this case, it is the responsibility of the subcontractor for the dewatering work not only to prepare the details of where the dewatering system will be located and the type of piping and pumping equipment that would be used, but also to incorporate in the shop drawing or its specifications calculations made by the subcontractor to ensure the adequacy of the system being installed.

This then brings up the question of the architect's or engineer's checking and approval of the shop drawings. Since the temporary-structure subcontractor has provided the calculations for the work to be done, it is therefore a question of who will be responsible for those calculations. Normally an engineer or architect will check the shop drawings only for the compliance with the original contract drawings. In this case, since there are no original contract drawings of the material or equipment being shown on the shop drawings, it is necessary for the engineer or architect to decide whether the shop drawings have to be inspected and approved at all. In most cases the engineer or architect will merely review the shop drawings for interferences or for conditions which might affect the finished design.

In the case of a dewatering system, for example, the removal of fine material from a sandy substrate during pumping could cause later problems in the foundations. But in the end it would have to be the temporary-structure subcontractor's responsibility not only to make the calculations for the design of the temporary structure, but also to be responsible for the adequacy of the calculations.

The difficulty sometimes encountered is when a shop drawing is submitted for approval by a temporary-structure subcontractor and is, in fact, checked and approved by the architect or engineer. Is the temporary-structure subcontractor then relieved of any further responsibility for the adequacy of this temporary structure?

The answer to this type of problem will be found in Chap. 2, "The Legal Positions." In any event, it is clear that shop drawings for temporary structures have an important difference in the legal sense from shop drawings for portions of the work that are to be left in the final structure. It is also evident that shop drawings for temporary structures sometimes require a great deal more design work than is normally required for shop drawings for work that will be left in the finished structure.

Supervision

Supervision of a temporary-structure installation includes not only supervision of its initial installation, but supervision of the adequacy of the structure during the entire period that it will be performing its function. The contractor must maintain adequately trained superintendents and foremen who are on the site at all times when work is underway.

A qualified supervisor is one who has had training and experience in the type of work which is being done. Obviously, a foreman who has only a general background in construction but who has never been directly involved in the supervision of a temporary structure such as the one which is being built is not as qualified to do this work as one who has had many years of experience, directly or indirectly related to the work on hand.

Concrete formwork for multistoried high-rise reinforced concrete buildings, for example, requires many intricate details of bracing, posting, struts, and ties so that they will not only be adequate to sustain the load of the wet concrete and the embedded reinforcing bars, but also will be able to undergo the impact of motorized buggies for distributing the concrete and of temporary hoppers, sustain the weight of workers for placement, and have the necessary clearances required for cranes and hoists.

Even after the concrete is poured, it is necessary to observe proper supervision for such activities as winter heating, time constraints before stripping, and reposting of the fresh concrete so that it can attain its required strength for the support of loads of succeeding floors.

Improperly trained or inexperienced supervisors can endanger workers' lives, and the neglect of good practices can cause structural failures.

Inspection

The responsibility for inspection of the work of the contractor for temporary structures is shared by three entities:

The contractor whose own supervisors or engineers must inspect the work as it is being installed, as, for example, when a temporary-structure contractor is installing shoring. As the work is being placed, the foreman and supervisory personnel of the shoring contractor must be alert to all of the requirements of safe and adequate construction practice and be certain that shop drawings or permits which specified the methods of performing the work are being adhered to.

The general contractor who is responsible for the entire finished structure, both temporary and permanent. The second required inspection is to see that the installation of the temporary-structures contractor is being performed safely and adequately, that it is not interfering with other portions of the work that may be a part of the finished structure, and that it is not interfering with other contractors or subcontractors who may be working simultaneously or subsequent to the work of the temporary-structure contractor. The inspection will also cover such things as quality of the materials being installed and compliance with plans and specifications and shop drawings to the extent that they pertain to the temporary-structure contractor.

The party whose responsibility it is to perform the ultimate inspection of the work. This third required inspection may be made by either an architect or an engineer who has the inspection responsibility. It may also be made by a government official such as a member of a building department or government financing agency. This third inspector is the one whose inspection must be the most comprehensive, painstaking, and flawless. This inspection ultimately determines whether, if there is a failure of the temporary structure, the work has been truly reviewed and found to be "approved" by the inspecting party. It is the responsibility of this inspector to be absolutely sure that all necessary compliance is being taken with building codes, safe structural practices, OSHA requirements, and any other procedure for good construction which is made a part of the contract, as well as with the plans, specifications, and shop drawings for the temporary structure.

Codes

An extended treatment of code requirements for the construction of temporary structures is included in Chap. 2 of this book. The code will, of course, outline whose responsibility it is for the design, what permits are necessary, which shop drawings are needed, and how supervision and inspection are to be performed.

What happens, however, if there is no code in the particular location where the structure is to be erected, or if the code is silent on the particular type of structure? In the absence of a code, the contractor for the temporary structure has even more responsibility for the observance of good practices. If inexperienced in building the type of structure being provided or in doubt as to the adequacy of the design, the contractor should get supplemental help from those who have expertise in these areas. And the general contractor should be more alert than otherwise to the possibility of a design or construction failure.

Collapse and Safety

Failures of temporary structures often cause more serious tragedies than collapses of permanent structures, because temporary structures do not always receive the same intense design scrutiny and protection of codes as do permanent structures. Often, casual treatment of temporary-structures design has been the consequence of human inclination to concentrate on the permanent and lasting to the detriment of the impermanent.

The worst construction accident in American history provides an example. The spectacular tragedy occurred on April 27, 1978, when the temporary formwork of tower No. 2 on the new Monongahela Power Company Plant in Willow Island, West Virginia, collapsed, leaving 51 dead. Good practice in this instance was not being observed. After an exhaustive investigation, the National Bureau of Standards, which was retained by the National Occupational Safety and Health Administration, published a detailed analysis with conclusions and recommendations.

The proximate cause that triggered the collapse was failure to permit previously made concrete pours to attain adequate strength. This led to a sequence of events that resulted in the collapse of a jumpform used to pour concrete, thus causing the collapse not only of the form, but also of the temporary scaffolding and a hoisting system as well. ("Investigation of Construction Failure of Reinforced Concrete Cooling Tower at Willow Island, W. Va.," NBS, Rept. 78-1578.)

Preventing such events is, in a sense, what this handbook is all about. By careful planning, training, and learning from our experiences, we all hope that we will be dealing with safe, economical, and adequate temporary structures in the future.

BUSINESS PRACTICES

Estimating

Estimates for the cost of temporary structures entail a three-step process:

1. A layout or design of the structure must be prepared.
2. An estimate of the cost of construction, using standard estimating practices, must be made.
3. An estimate of the cost for maintenance of the temporary structures, such as the labor involved in keeping the temporary structure operative, must be made.

The first part of the estimate, the design phase, is best handled by consultation between the temporary-structure contractor, the general contractor or the general contractor's field forces (if general contractors propose to erect same for themselves with their own forces), and the engineer.

In a typical example involving dewatering by well points, the design would first have to be made showing the location of all of the well points, piping, pumps, settling basins, and disposal facilities. Once the layout is completed, step 2, the estimate, using standard methods, comes next. This, of course, will involve quantity takeoffs of materials, equipment, and labor, and then pricing, using unit prices which have been produced from past experience or from an analysis of the probable size of the work gangs and the probable output per day or per shift in the production of the gangs.

Step 3, which would cover the operation of the system, would estimate the labor and fuel required to maintain the system after it is installed and for the length of time that it is expected to be in operation.

Contract Negotiations

The negotiation of the contract between the temporary-structure contractor and the general contractor, or the owner, is closely related to the concepts set forth in estimating above. One of the important considerations is the type of design which has been prepared by the temporary-structure contractor for the proposed work. In the case where the temporary-structure contractor is competing with others for the same or similar services, differences in the design could, of course, have a marked effect on the estimate and greatly affect the price. If the owner is not aware of the possible advantages and disadvantages of each of the proposed methods of design being presented with various prices, then the temporary-structure contractors should endeavor to make clear the salient features of their designs in the negotiations. In the example given for the estimating section above, the spacing of the well points might vary between two proposals. While the temporary-structure contractors in each proposal may guarantee to dewater the site, a difference in the spacing, which could depend on factors of judgment, could also affect the adequacy of the dewatering system. These factors must be made apparent to the buyer.

Once the buyer and the temporary-structure contractor agree in concept to the design, the hard bargaining begins. Here all the factors that go into the purchase of any commodity come into play. Such matters as wage rates required to be paid, sales tax requirements, insurance requirements, the matter of scheduling and timing, method of payment, and all the other intangible and tangible factors that affect the price will have to be discussed.

The final, agreed-upon price will reflect all these factors.

Forms of Contracts

The temporary-structure contractor may enter into a contract directly with the owner or ultimate user of the facility, or may enter into a subcontract with a general contractor or construction manager who subcontracts on behalf of the owner. If the contract is to be entered into with the owner, one of several forms of contract is most likely to be used.

1. A lump-sum contract between the owner and the temporary-structure contractor. For example, AIA Document A101, "Standard Form of Agreement Between Owner and Contractor For a Stipulated Sum." (All of the referenced AIA forms can be obtained from the American Institute of Architects, 1735 New York Avenue, N.W., Washington, DC 20006.)
2. A cost-plus-a-fee type of arrangement, such as the AIA Form AIII, "Standard Form of Agreement Between Owner and Contractor, Cost of the Work Plus a Fee." The fee could be either as a percentage of the cost of the work or as a lump sum amount added to the cost of the work.
3. A unit-price form of contract between the temporary-structure contractor and the owner. This type of contract would be based on an assumed estimated quantity for each of the classes or types of work to be performed, such as a unit price for steel, a unit price for concrete, and a unit price for lumber. When the work is completed, an accurate measurement can be made of each class or type of work that was actually performed and the unit prices previously agreed to can be applied to the quantities for a calculation of the amount of money payable to the contractor. The AIA does not have a standard form for this type of agreement. An example of a bid form for one, however, can be found in *Construction Contracting,* 4th ed., by Richard H. Clough, Wiley Interscience, New York, pp. 435–437.

On the other hand, the temporary-structure contractor, rather than being in a contractual relationship directly with the owner, may enter into a subcontract with a general contractor under an arrangement similar to the AIA Form A401, "Standard Form of Subcontract Between General Contractor and Subcontractor." In this arrangement, which is most commonly the case, the temporary-structure contractor will have no direct relationship with the owner, but will look to the general contractor for whatever obligations are required to be performed and for payment of any money due. Subcontracts can be entered into between the temporary-structures contractor and the general contractor on a fixed-fee, cost-plus, or unit-price basis.

Legal Problems in Contracts

Standard forms of contract, such as the AIA forms and the forms used by the Associated General Contractors of America, have been in use for many years and most contractors are familiar with their clauses. If a standard form is used, normally it is not necessary for an operating business person to send contracts about to be entered into to an attorney for review. The exception occurs when an unusual contract form is being considered or when the form of contract seems to include particularly onerous clauses.

Some of the most significant cost-additive items that may affect the financial or legal outcome of the agreement for the temporary-structure contractor are:

Hold-harmless clauses

Extraordinarily high insurance limits

Exculpatory clauses in which the contractor agrees not to make claims for changes, extras, or unexpected work

Waiver of lien

Fixed completion dates with liquidated damages for failure to complete

Requirements for furnishing performance and payment bonds

Many other clauses which impose special expenses or requirements appear in contracts and these have to be dealt with on a case-by-case basis.

Claims: The Unexpected in Underground Construction

The problem of who shall bear the risk of unexpected subsurface conditions in construction has plagued the engineering profession and construction industry for many years. If a contractor is engaged in construction for an owner in accordance with a previously prepared design, and the contractor encounters subsurface conditions materially different from what might normally be expected, resulting in increased costs, should the contractor absorb the increased cost or should the owner pay it?

Many owners or engineers attempt to pass on these unexpected costs to the contractor, under the theory that a "grandfather" clause making the contractor responsible for anything and everything that happens in the way of unexpected contingencies will amply protect them. Needless to say, when disputes arise, contractors will point to many unusual or extraordinary circumstances that will tend to contradict such "grandfather" clauses. On the other hand, by including such a clause in the agreement, the owner has requested the contractor to assume certain risks for which the contractor must provide contingency allowances in the estimate. These contingencies may or may not arise, and it is therefore an added cost burden which the owner is paying for something which may not come about.

The best view, and the one which is endorsed by most enlightened engineers, contractors, and owners, is that the contractor should provide in the estimate only for certain agreed-upon assumptions which all of the parties stipulate are the bases for the estimated cost and for the contract. Any situations or eventualities which are outside of these agreed-upon assumptions, which would have resulted in an additional cost to the owner in the original bid or estimate, will be considered as an unexpected which should be reimbursed.

A recommended contract clause for such a provision is as follows:

> The contract documents indicating the design of the portions of the work below the surface are based upon available data and the judgment of the engineer. The quantities, dimensions, and classes of work shown in the contract documents are agreed upon by the parties as embodying the assumptions from which the contract price was determined.
>
> As the various portions of the subsurface are penetrated during the work, the contractor shall promptly, and before such conditions are disturbed, notify the engineer and owner in writing, if the actual conditions differ substantially from those which were assumed. The engineer shall promptly submit to owner and contractor a plan or description of the modifications he proposes should be made in the contract documents. The resulting increase or decrease in the contract price, or the time allowed for the completion of the contract, shall be estimated by the contractor and submitted to the engineer in the form of a proposal. If approved by the engineer, he shall certify the

proposal and forward it to the owner with recommendation for approval. If no agreement can be reached between the contractor and the engineer, the question shall be submitted to arbitration as provided elsewhere herein. Upon the owner's approval of the engineer's recommendation, or receipt of the ruling of the arbitration board, the contract price and time of completion shall be adjusted by the issuance of a change order in accordance with the provisions of the sections entitled "Changes in the Work" and "Extensions of Time."

Extras

A contractor's claim for extras under a temporary construction contract is more difficult to justify and is more likely to be resisted than in a contract for the permanent structure. The reason is that the temporary structure is generally not fully detailed or designed on the contract drawings, and it is often on the basis of a performance type of specification that the contract is awarded for the temporary structure. Nevertheless, there are many instances in which an extra or change order in favor of the temporary-structure contractor is justified. Some instances are changed conditions, delays, interferences by the owner, interferences by separate prime contractors, changes or redesign to the permanent structure, or subsurface conditions materially different from those which were contemplated.

The legal aspects of the claims for extras are dealt with in Chap. 2 of this book.

After it has been established that a change order is justified, it will probably be in one of the following forms:

Unit Prices: Unit prices may have been established prior to entering into the agreement, for possible changes or revisions to the work. These unit prices are generally different for added work than for deducted work. The deducted work unit prices are usually somewhat lower than the prices for added work. Should the quantity of work applied under any one unit price be radically different from that which could conceivably have been contemplated, a renegotiation of the unit price is sometimes justified.

Negotiation by Lump Sum: This is the most common type of method of arriving at the amount of an extra or of a change order. Generally speaking, the negotiation will take place on the basis of an estimate which has been prepared by the contractor showing what is believed to be the probable cost of the work. Oftentimes the change order will be negotiated after the extra work is done. Here again, the negotiation may start with the contractor's estimate of the cost of the work as it was performed or with the contractor's actual time sheets and material sheets of the work as it progressed. A final price might be agreed upon after these estimates or actual time sheets and material bills are examined for correctness and proper allocation of costs.

Cost of Labor and Material Plus Markup: This type of change order can result when there is no agreement between the parties as to how best to resolve the value of a contemplated change. When work is done in this manner, costs are kept contemporaneously and daily as the work progresses, and the records are signed by representatives of both the contractor and the engineer or owner.

Settlement of Disputes

Disputes can be settled by preagreement through referral to arbitration, or they may be settled through the process of negotiation between the contractor for temporary structures and the owner. In the event of the failure of a negotiated settlement, resort to legal means and the courts may be required.

Claims for extras and for changes in the plans form the major area of disputes, as discussed in the previous paragraphs. Other areas of disputes which are sometimes the cause of litigation include: late completion, poor or unacceptable work, back-charges for failure to perform certain portions of the work, and strikes and labor disputes.

In order to avoid increasing the losses which are sustained when a dispute arises, it is often useful to have a clause in the contract to the effect that in the case of a dispute, the temporary-structures contractor will nevertheless continue working and complete contract obligations pending resolution of the dispute.

Arbitration

Arbitration as a method of settling disputes in contracts involving temporary structures has become more widespread in recent years. The American Arbitration Association, one of the organizations most frequently used for arbitration of disputes, is a nonprofit organization which was founded expressly for the purpose of providing rules for conducting arbitrations and for providing qualified arbitrators who can be selected by parties to hear the disputes. In an arbitration both parties must agree voluntarily to participate in the arbitration and to abide by the decision of the arbitrators. This agreement to arbitrate can be entered into either prior to a dispute as, for example, in the original contract or agreement between the parties, or it can be agreed to by mutual consent of the parties after the dispute has arisen.

Once an arbitration clause is agreed to and a dispute has arisen which one or the other party seeks to take to arbitration, a demand is served on the opposing party. All this paperwork, as well as the providing of arbitrators, hearing rooms, tribunal clerks, and written decisions, can be handled through the offices of the American Arbitration Association. Offices of the Association are located in principal cities of the United States.

When the arbitrators have made a decision, it is final and binding on all parties and it can be enforced in a court of law. No appeal is permitted under the terms of the arbitration laws, except for proof on the part of one of the parties that there has been dishonesty, bad faith, or fraud perpetrated by the arbitrators.

Insurance

Insurance requirements fall under three categories:

1. Insurance which is required by contract, such as liability insurance or Builder's Risk Insurance
2. Insurance which is mandated by law, such as Workmen's Compensation

3. Insurance which is neither required by contract nor mandated by law, but which the contractor may feel is prudent to have as additional protection for good business practice

Table 1.1 shows the most common types of insurance which might be required for a small- to medium-sized construction project by a general contractor. Special insurance which should be obtained by the designer is also mentioned. Table 1.1 also describes the purpose of each type of insurance, with suggested limits for a small-to medium-sized company, and describes the basis of calculating the premium for each type.

Insurance costs have risen spectacularly in the past few years because of the increasing awareness of juries of the high human costs of injury to persons and of the high material costs of damage to property. In addition, many jurors are also aware of the fact that in most instances recoveries for such damages are covered by insurance and there seems to be a tendency on the part of jurors to award higher recoveries when they are of the opinion that these will be paid for by an insurance company.

Because of these high present-day insurance costs, it is most important that contractors for temporary structures be particularly alert to constant reexaminations of their insurance programs. The broker or agent who handles the contractor's insurance is particularly charged in this instance with using intelligence in reappraising the policy each time it is being renewed. The contractor may attempt to be "self-insured" for some of the less likely eventualities that are capable of being covered by insurance. On the other hand, instances where there are frequent losses and where coverage has been inadequate should be boosted up and greater amounts of insurance must always be considered.

Examples of excess insurance are when the contractor is paying for an insurance limit in an amount that is not covered under the policy anyway. This situation can exist, for example, in insurance on buildings under construction under a Builder's Risk form. This type of policy excludes coverage for demolition work, portions of the foundation below the basement slab, and certain sitework and landscaping. Since these are not covered, it is unnecessary to insure such a building for the full contract price.

Bonding

Performance and Payment Bonds, when issued by a fidelity company, are a contract in which the company agrees to complete the project if the contractor fails to do so and agrees to pay all bills for material and labor if the contractor should not do so.

In order to obtain Performance and Payment Bonds, temporary-structure contractors must demonstrate to a bonding company that they have adequate finances and experience to perform and complete the jobs being bonded. Generally speaking, a bonding company will favor issuing bonds to those contractors with whom it has previously established a track record, and those who have been able to show continuing financial stability and profits over a period of years. Bonding company relationships are very much like banking relationships. They depend on confidence on the part of the surety in the ability of the contractor and they depend on experience which has been trouble-free and satisfactory in the past for future business to be conducted.

TABLE 1.1 Understanding Your Insurance Needs

Type of policy	Purpose	Suggested limits, small- to medium-size company	Basis for premium
Builder's Risk Fire Insurance with extended coverage, vandalism and malicious mischief	Protect building, fire and lightning, explosion, windstorm, etc. All risk coverage is the broadest and most expensive	Amount of contract less demolition, construction below lowest basement slab and landscaping	Dollars per hundred
Comprehensive Liability 1A. Premises, Operations of Contractor	Third-party bodily injury and property damage	$300,000 BI $100,000 PD	Payroll classification
1B. Independent Contractors Contingent Public Liability and Property Damage	Protect contractor against subcontractor's negligence	Same	Amount of subcontract
1C. Completed Operations Insurance	Claims by third parties on projects already completed	Same	Amount of coverage
1D. (Hold Harmless) Contractual Liability	Contractually assumed liability	Same or as per contract	Amount of coverage
2. Owner's Protective Liability Insurance	Third-party liability	As per contract	Amount of contract
3. Umbrella Liability	Third-party liability	$3,000,000	Various
Owned and Nonowned Motor Vehicles Insurance	Third-party liability and physical damage insurance	BI, PD, collision, fire and theft (no fault)	Number of cars and experience
Contractor's Equipment Insurance	Protects equipment, tools, etc.	$50,000	Coverage Amount
Material Floater	Losses to materials before installation in job (may be included in builder's risk)	$50,000	Coverage amount
Workmen's Compensation and Employers Liability	Employee protection	Set by state	Payroll
Employers Liability Insurance	Employee suits outside of Workmen's Compensation	Unlimited in some states or $100,000	Payroll

TABLE 1.1 Understanding Your Insurance Needs (*Continued*)

Type of policy	Purpose	Suggested limits, small- to medium-size company	Basis for premium
Disability Benefits Insurance	(Where required by state)	Set by state	Payroll
Fidelity Insurance	Embezzlement of funds	$100,000	Number of employees
Valuable Papers	Loss of valuable papers	$10,000	Amount of coverage
Key Man Insurance	Life Insurance Policy to cover corporate officer in case of death or disability	$100,000, or amount to be determined	Amount of coverage
Group Hospitalization	Employees protection (Blue Cross)	120-day extended coverage	Number of employees and coverage
Major Medical	Employees protection	Limits beyond Blue Cross	Same
Group Life Insurance	All eligible employees	6 months to 1 year salary	Age and number of employees
	Technically not insurance		
Pension and/or Profit-sharing Plan	Employees retirement	25% salary when combined with Social Security	Salary of employees
Unemployment Insurance	(Where required by state)	Set by state and federal government	Payroll
Social Security	Employees welfare	Set by federal government	Payroll
Street Obstruction Bond	Where required by municipality	As required	As required
	Architects & engineers insurance		
Architects and Engineers Liability Insurance	Damage to persons and property	$300,000 BI $100,000 PD	Amount of coverage
Errors and Omissions Insurance	Errors or omissions from the plans	Highest available	Amount of coverage
	Owner insurance		
Owner's Protective Liability Insurance	Public liability and property damage as a result of construction operations	$500,000/ $1,000,000	

TABLE 1.1 Understanding Your Insurance Needs (*Continued*)

Type of policy	Purpose	Suggested limits, small- to medium-size company	Basis for premium
Property Fire Insurance	When construction is being done to an existing building, fire, etc., general contractor and subcontractors	Full insurance value	
Loss of Use Insurance	Losses in property as result of fire or other loss	Approximate loss of income	
Steam boiler	Explosion, etc.	$500,000	
Nuclear Incident	Nuclear reactor loss (if required)	High	

Before issuing a bond, the bonding company will look at a number of factors:

The uncompleted work on hand by the contractor

The working capital of the contractor, particularly the contractor's net quick—which is the current assets less the contractor's current liabilities

Other bids which were received by the owner or by the one who is issuing the contract to the temporary-structure contractor for the same work

The size of the job and the location

The terms of payment

The amount that is to be subcontracted to others

The qualifications of each principal in the contracting organization, including personal history and integrity

The financial history of the company

The type of job

Other jobs recently contracted for

The relationship of the broker to the contractor and to the bonding company

Obviously, a contractor with strong financial resources, an unbroken record of experience in the type of work being bonded, and a program of work on hand that does not strain the contractor's resources will obtain a favorable decision for the issuance of a bond.

Cost Analysis

A contractor's cost-keeping system should provide the contractor with information for:

Estimating on future contracts

Checking the efficiency of operations to determine whether a completed contract is more or less costly than was estimated

Controlling the costs of jobs as they are in progress so as to pinpoint areas where costs are running over and so that corrective action can be taken

An adequate cost system is more than just allocating the total costs of the labor and material for each contract to a specific job number. This is, of course, the simplest type of cost system and is elementary in the contracting business. But an adequate cost system will attempt to provide information in all of the above three categories.

Such an adequate cost system would be based on the following two types of reports:

Daily work reports, which will show all types of work performed, the quantity of work performed, and the number of work hours and hourly pay rates that are being committed.

Unit cost reports, which will be prepared periodically to provide data on production, a check on costs at an early date, and records for estimating for the future

For an extended treatment of a contractor's cost keeping, refer to Sec. 25-24, *Building Design and Construction Handbook,* Frederick S. Merritt, ed., 4th ed., McGraw-Hill Book Company, New York, 1982.

Purchasing

In issuing a purchase order for materials to be delivered on a temporary construction contract, the purchase order must take into account the scope of the work, the quantities and unit prices or lump-sum price, and must list inclusions properly, note exceptions or exclusions, and, where practicable, state unit prices for added or deleted work. There must also be listed on the purchase order the time of performance and the dates of delivery.

In purchasing materials for temporary construction contracts, one must also take into account the availability of equipment to perform the work, the adequacy of the plant or supplier to deliver on time, and the reliability of the supplier's previous promises for materials which were purchased.

Purchase orders should also contain a provision for field measurements by the vendor, if this is required, and should indicate whether delivery and transportation charges and sales taxes are included in the prices.

Subcontracting

In many instances the temporary-structures contractor will be a subcontractor to a general contractor on the job. This will be the case, for example, in a temporary dewatering subcontract. The negotiation and award of a subcontract will proceed along the following lines:

Price

Time of performance

Availability of labor, material, and equipment

Exceptions, inclusions, and exclusions

Form and language of subcontract

Terms of payment

Other subcontract clauses which will be important and which will be dealt with during the course of the subcontract negotiation are:

Subcontractor to be bound by plans and specifications for the general contract

Submission of payment application and breakdown with each request for payment

Payment by subcontractor of all monies owed for materials and labor previously paid for by the general contractor on past requisitions

Making of claims promptly

Protection of other trades

Removal of debris

Safety requirements

Assignment of subcontract not made without consent of contractor

Guarantees, whether personal or corporate

Pursuit of work diligently

Use of the general contractor's equipment only at the general contractor's discretion

Reports on material in fabrication

Changes

Coordination

Shop drawings

Laws and permits

Work subject to approval

Hold-harmless clauses

Arbitration clause

A common form of standard subcontract is the AIA Form A401, "Contractor-Subcontractor Agreement."

Financial Controls and Banking

Proper financial controls and banking practices on the part of the contractor include safeguards for issuance of checks and payment of bills. These include, first, that a purchase order or subcontract be issued for each liability that is being incurred by the contractor. When invoices are received, it is important that they be accompanied by a delivery ticket or that a responsible individual who is familiar with the work being invoiced issues an approval to accompany the invoice before it is processed for payment. When the invoice is approved for payment, checks that are issued should be, if possible, on an account that requires two signatures. This will provide for a further safeguard and duplicate scrutiny of the invoice.

Often it is advantageous to open a separate bank account for each major project. This will aid in cost keeping and will also provide for a stricter control of the expenditure of funds versus the income received for a particular job.

It is also good practice to have banking relationships with more than one bank. These can then be drawn upon, for example, when borrowing or when requesting the issuance of such bank accommodations as letters of credit, with greater strength and facility than if only one bank is relied upon.

Technical and Business Advice

In the chapters that follow, experts in many of the fields of temporary structures in construction have described their specialties. If readers find themselves involved with temporary structures which have not been discussed in this book, perhaps because they are too novel, a few words of advice may be in order.

As a practicing general contractor with four decades of involvement in the civil engineering and construction industry, the writer still feels the inadequacy of his background and experience. Daily, in the conduct of his profession and business, he is assailed with problems that are new and have never been dealt with by him before. In each case he attempts to obtain a thorough understanding of the issues and problems involved before attempting a solution. But, most importantly, he is constantly seeking the advice and the help of others in finding these solutions. There are, in the broad practice of this profession, experts who are dealing with every conceivable problem. When they are sought out, they are usually more than anxious and willing to offer their advice. The writer is not ashamed to say that the advice of others is always given great weight in the conduct of his affairs. But the principles of careful investigation, adequacy of preparation and design, and careful checking of all work will help any practitioner in overcoming the most difficult problems.

2

The Legal Positions

by Robert A. Rubin

NOTE: The author acknowledges the assistance of Richard M. Konecky, Esq., and Todd Herbst, Esq., in the preparation of this chapter.

LEGAL RELATIONSHIPS

Temporary structures present as many legal pitfalls as they do construction challenges. This chapter deals with the myriad rights, responsibilities, and liabilities of the parties with temporary structures in construction.

There are three sources of these rights, responsibilities, and liabilities: common law, statute, and the contract, if any, between the parties. Common law is best understood as judge-made law, which is developed over the years on a case-by-case basis. Statutory law is the legislative enactments of the U. S. Congress, state legislatures, and municipal bodies, many of which are supplemented by the rules and regulations of administrative agencies charged with the enforcement of the legislative enactments. Examples are noise restrictions, building codes, and safety requirements. Contract law emerges from the agreements between the parties, setting forth their respective rights and obligations. So long as a contract provision does not contravene a statute or public policy, it will be enforced. Where a contract is vague or silent, common law often will provide the needed interpretation.

Construction people share concern about their liability to others, both on and off the jobsite. Liability arises when individuals fail to exercise the "standard of care" imposed on them. Temporary construction activities generally impose one or the other of two distinct standards of care upon the individual, depending on the circumstance of the work and the type of activity being performed. The negligence standard, usually imposed for work that is not deemed ultrahazardous, requires that the actor use *reasonable* care under the circumstances. The absolute standard, as the name suggests, imposes strict liability without fault should damage result from engaging in a particularly hazardous activity. The rationale for this severe standard is that one who subjects the public to highly dangerous conditions must be legally responsible for the full consequences, even though the individual exercised reasonable care. Blasting is the prime example of an ultrahazardous activity for which strict liability has been imposed by the courts. Strict liability for the safety of workers has been imposed by statute in some states.

Apart from negligence or strict liability standards, the rights and duties of the parties involved in temporary construction work are most often detailed in the contract documents, i.e., the general, supplementary, and special conditions. These generally start with one of several standard forms, such as those provided by the American Institute of Architects or the Engineers Joint Documents Committee, and then include modifications tailored to meet the specific needs of a particular project. In the event of a dispute, the court or arbitrator will look to the contract documents to determine what was required or to ascertain what was the actual intent of the parties. Consequently, all parties in a construction project should, indeed must, thoroughly familiarize themselves with the requirements of contract documents.

RESPONSIBILITIES

All Parties

All parties to a construction project have responsibility both to the other contracting parties and to a wide range of third persons, as well, such as workers, pedestrians, and adjoining property owners. As regards the contracting parties, their responsibilities to each other are governed by the contract. These include the duty not to interfere with the work of other trades and the duty to cooperate with all other parties to progress the work.

All contracting parties potentially can be held liable to third persons who are injured or damaged as a result of the contracting parties' failure to exercise the standard of care required by either common or statutory law. Sometimes, even third parties can be held to be the beneficiaries of a contract provision intended for their benefit, e.g., pedestrians protected by a sidewalk bridge. Similarly, an excavation contractor who fails to adequately underpin an adjacent structure will be liable to the owner of that structure for any damage caused by the contractor's neglect. In some states the violation of a statutory duty (e.g., to adequately underpin) is, in and of itself, conclusive proof of negligence. The practical effect of this is that the injured party has but to establish the amount of damages in order to recover; there can be no issue as to fault or liability. In other states, violation of a similar statute is considered merely some evidence of negligence to be considered with all other evidence, so that the injured party's prospect for recovery is far less certain.

The Owner

The owner is concerned with temporary structures only insofar as they are necessary to the construction of the finished product. Therefore, the construction contract documents will often simply provide what must be done rather than detailed instruction on how to do it; for example, "Earth banks will be adequately shored"; "Adjoining structures will be adequately protected." Under this performance-type specification, the involved contractors are responsible for meeting the required result, and they—not the owner—are responsible for the means and methods used to achieve it. This is in contrast to the technical or detailed specification wherein the owner or the designer specifies the method to be used and is deemed impliedly to warrant its adequacy. The difference in the owner's responsibilities can be seen as follows: If a proper performance specification is being used, then the owner bears no responsibility if the contractor's work fails unless the owner was negligent in hiring the contractor or, notwithstanding the written documents, the owner undertook direction or control of the contractor's work. If, however, a detailed specification is utilized, then the owner bears full responsibility for any deficiencies in the contractor's work, as long as the contractor performs that work as specified. An owner's responsibility to third persons is usually secondary, in that most lawsuits arising out of temporary construction work, while naming the owner as a party, are attributable to negligent performance by the contractors or design deficiencies by the responsible architect or engineer. Thus, even in the absence of a contractual indemnity provision, the owner can shift the loss to the party ultimately responsible, assuming that party has the financial resources to shoulder the responsibility. In other words, although the owner is primarily liable to the injured party, the owner has a right of indemnity or contribution from the party who was negligent.

The Contractor

The ambit of the contractor's responsibility is far-reaching. The contractor is responsible for performing all required work in accordance with the contract documents, generally accepted construction practices, and governing legislative enactments.

It is important to stress that it is not necessarily enough that the contractor has performed to the best of his or her own ability. When assessing liability, the courts do not simply accept at face value what the contractor did, but measure that against what an ordinary contractor of similar experience would have done under similar circumstances. Consequently, when determining whether a contractor exercised ordinary or reasonable care in doing the work, the court will look at whether that work favorably compares to that degree of skill then prevalent in the industry.

The Designer

The design professional is responsible for the accuracy and adequacy of his or her plans and specifications. By distributing these documents for use, the design professional impliedly warrants to both the owner and contractor that a cohesive structure will result, providing the plans and specifications are followed. But, the designer does not guarantee perfection; only that he or she exercised ordinary care. When a contractor who properly follows the designer's documents experiences a problem, liability will ultimately rest with the architect or engineer. The designer would be responsible for all additional costs of construction attributable to faulty design. For example, if an access ramp to a construction site proved unable to sustain the weight of heavy construction equipment, the engineer who designed the ramp would be required to redesign the ramp, or bear the cost of redesign, and would be held liable for all resultant damages to the heavy equipment. In a more drastic case, the design professional would be liable for all damages resulting from a retaining wall collapse attributable to deficient design. This includes liability to third parties for injuries, death, and property damage.

A design professional has no duty to supervise, inspect, or observe the performance of the work designed, absent a duty which the design professional has assumed by contract or by having filed plans with a local building department. Once some duty has been assumed, however, state courts are in wide disagreement as to the extent of responsibility and liability. At the very least, the architect or engineer will be responsible for making occasional visits to the site to verify technical compliance with the contract documents. In marked contrast, some courts have found that design professionals have a nondelegable duty to regularly supervise the means and methods of construction. Designers should be wary of so-called routine sign-offs of the contractor's work. The sign-off may include a certification by the designer that the contractor has complied with the plans and specifications and with applicable codes. Such a certification could impose on the designer liability equal to the contractor's in the event of a failure.

Designers should be particularly careful in requiring submission of and in "approving" the contractor's detailed plan for temporary construction. Exculpatory language is often placed on the designer's "approval" stamps to the effect that the designer has assumed no responsibility by having reviewed the submission. In the event the design was not adequate and there was injury to person or property, it is likely that the designer's stamp would not be sufficient to exculpate him or her from liability to injured third parties. If designers do not intend to carefully review each and every submission, they ought not require the submission at all.

The Construction Manager

The construction manager, although not involved in the performance of any actual construction work, may assist the owner in choosing the contractors who will perform work, coordinating the work of these trades, and in approving requisitions for payment. The manager ordinarily does not assume any design responsibility, which remains with the owner's architect or engineer. However, construction managers should also be wary of contractual assumption of responsibility for the contractor's performance or sign-offs or certifications related to the work.

INSURANCE

Liability insurance is often purchased to cover the owner and contractor for the risks assumed in temporary construction. The designer should also maintain professional liability (also known as "errors and omissions") insurance to cover these risks. However, it should be noted that liability arising from plans and designs is a common exclusion to most owners' and contractors' liability insurance. Hence, if the contractor is to make a design for temporary construction, it is important that the contractor procure special insurance coverage for the plans and designs. It should also be noted that liability insurance only covers the claims of third parties for personal injury or property damage and ordinarily does not cover damage to the owner's own property or the contractor's work.

Thus, if a scaffolding collapses, injuring a pedestrian and damaging the building being worked on, the contractor's liability insurance in most instances would only cover a claim by the pedestrian, and not the cost of repairing the building and replacing the scaffolding.

State workmen's compensation laws require employers to provide specific insurance coverage for injury to their employees. However, workmen's compensation benefits have not kept pace with inflation and in many instances are grossly inadequate. Injured construction workers are barred by law from suing their employers beyond workmen's compensation benefits, but they are not barred from suing others, such as the owner or the designer, for negligence or breach of a statutory or contractually imposed duty. For example, in New York, building owners have an absolute statutory duty to provide workers a safe place to work, even if they are the employees of an independent contractor hired by the building owner. For that reason, and also because of the owner's vicarious liability to third parties such as adjacent property owners, the owner will often insert a contract clause requiring the contractor to indemnify and hold the owner harmless from all claims and liability arising from the contractor's activities, even if the claim is based on the owner's breach of a statutory duty. Sometimes the indemnity is expanded to include the owner's own negligence. Likewise contractors can seek contractual indemnity from their subcontractors, and designers from contractors.

State laws vary on the enforceability of indemnity for one's own negligence and for defects in a design or plan. Nevertheless, an indemnity is not better than the indemnitor's ability to pay. For that reason, indemnitors are frequently required to provide specific "contractual liability" insurance to cover the added risks assumed by a hold-harmless agreement. This coverage is generally provided in a rider or supplement to the contractor's liability policy. Contractual indemnity and hold-harmless provisions, unless backed by adequate contractual liability insurance coverage, may be of little or no practical value.

Insurance coverage (and exclusions) for temporary construction are highly complex. The advice of a competent insurance professional should be sought.

PROTECTION OF ADJACENT PREMISES

Since construction can affect adjoining land and structures, the law has historically imposed certain duties of protection upon the constructing owner, differentiating, however, between protection of the adjoining land itself and protection of structures on that land. The constructing owner has a strict and absolute duty to provide lateral support to adjoining land in its natural condition, and the constructing owner's duty is limited to the use of reasonable care with respect to the buildings located on the land. Thus, the constructing owner will be held liable for any and all damage to adjacent land regardless of whether the owner, or the owner's contractor, was negligent. However, the owner would only be liable for any damage to buildings on that land if it is proved that the owner was negligent. If the contractor were negligent, the owner would be liable for damage to the adjoining building only if the owner retained supervision over the contractor or negligently prescribed the methods and procedures followed by the contractor.

An exception is made to this latter rule regarding buildings where "ultrahazardous" activities are conducted such as blasting and pile driving. In such circumstances, strict and absolute liability can be imposed regardless of whether negligence can be proved.

In many jurisdictions, such as New York City, statutes have supplanted the general rules of law referred to above. The New York City Administrative Code (§C26-71.0) contains what is commonly known as the "ten-foot" (3-meter) rule relating to the protection of adjacent premises during excavation. Interestingly, the historical distinction between the protection of the land itself and the structures thereon is retained in that statute. Thus, with respect to adjoining *land*, regardless of the depth of the excavation, there is an absolute requirement that the "person causing an excavation to be made" shall provide adequate sheet piling, bracing, and other support as is necessary to prevent the sides of the excavation from caving in before the new structure is constructed. However, that person must be afforded the right to enter the adjoining property to perform such work as may be required. If the adjoining owner refuses to give such right of entry to the person causing an excavation to be made, the adjoining owner must then protect his or her own property, and, in turn, is given a right to enter upon the property where the excavation is to be made to do so.

With respect to the protection of adjoinging *buildings,* the New York City ordinance differentiates between excavations that are shallower than 10 ft (3 m) below the legally established curb level and excavations that exceed 10 ft (3 m) below such level. For excavations less than 10 feet (3 m) in depth, the owners of adjoining buildings are required to protect their own buildings, provided they are given the right to enter the property where the excavation is to be made to perform such protective work. If excavating owners refuse such right of entry, then they must assume the duty to protect the adjoining buildings. They, in turn, are granted a right to enter upon adjoining property for such purpose.

Where the excavation is to be carried to a depth more than 10 ft (3 m) below the legally established curb level, the duty to protect an adjoining building falls upon the person causing such excavation to be made, provided he or she is given the right to enter on the adjoining property to do so. Otherwise, the duty falls upon the owner of the adjoining building, who is given the right to enter on the property where the excavation is to be made in order to do so.

Further requirements may be imposed such as the necessity for a permit to perform excavation work and fencing, filling, and other protection requirements, including notice to adjoining landowners prior to commencing operations.

TRESPASS

Some jurisdictions have enacted legislation providing that adjacent owners are permitted to enter upon adjoining property for the purpose of protecting their property and structures. This, however, is most unusual and contrary to the general common law rules concerning trespass. Normally, an unauthorized entry on to adjacent property subjects the invader to criminal liability as well as civil damages. Moreover, owners would be liable for such damage if their plans require the contractors to trespass on adjacent land.

Subsurface encroachments are generally treated similarly to surface encroachments regardless of whether the subsurface encroachment will impair any present use of the land or any foreseeable future use. Nevertheless, it has been held that an unauthorized subterranean trespass occasioned by construction of a municipal sewer 150 ft (45 m) below grade is not actionable unless some present or prospective use of the land is interfered with (*Boehringer* v. *Montalio,* 142 Misc. 560, 254 N.Y.S. Supp. 276). If the subsurface encroachment would interfere with the prospective use of the land, the adjoining owner would be entitled to recover the difference in the value of the land resulting from the encroachment.

A builder who seeks to place a foundation wall, fence, scaffold, or equipment shed directly on the property line must be aware that necessary entry upon adjoining premises will constitute a trespass. An easement from the adjoining property owner should be obtained. However, some jurisdictions have legislated certain relief with respect to one's right to enter on to the adjoining land in order to repair or improve one's property. For example, a New York statute permits an owner or lessee to apply to the court for temporary authority to enter on the adjoining land for the purposes of repair or improvement (N.Y. Real Prop. Actions Law §881).

In one case under that statute, a waterproofing contractor was permitted by court approval to erect his scaffolding temporarily on adjacent land in order to waterproof the wall located on the building line (*Chase Manhattan Bank* v. Broadway Whitney Co., 57 Misc. 2d 1091, 294 N.Y.S. 2d 416). The party going on to the adjacent land for such purpose is subject to liability for any resulting damages. Furthermore, the courts will not automatically grant authority to enter the adjoining property where objection is made by the adjoining property owner. The courts will weigh the relative inconvenience and expense to each landowner in determining whether authority should be granted. Thus, a request for permission to excavate on adjacent property to waterproof a basement may be denied in the absence of showing that the waterproofing was impractical or unduly costly without entry.

PROTECTION OF UTILITIES

Excavation near or in a public street invariably involves the protection of utility vaults, pipes, and ducts located under the street. Shoring and other means of support must be provided to prevent damage to these utilities. However, whether the owner of the utility company is required to pay for the protection often depends on whether the construction project is privately or publicly owned. Generally, where the purpose of the owner's construction is purely private, the burden of protection is on the owner. On the other hand, where the construction work is public or quasi-public, the utility company is required to bear the costs of shoring, supporting, or relocating its structures. Thus, for example, when a contractor laying city sewers was forced to provide support to an elevated railroad, the railroad company was required to reimburse the contractor. (*Necaro Co., Inc.* v. *Eighth Avenue R. Co.,* 220 App. Div. 144, 221 N.Y. Supp. 276). In addition, a public utility company which has been given a franchise to install its facilities in public streets has the correlative duty to relocate or protect its facilities when changes are required by public necessity; the company takes the risk of the location of its facilities if public inconvenience or security demand their removal or protection.

Statutes generally have not changed these rules, but rather have specified procedures for giving notice of impending excavation which will affect the utilities.

Many jurisdictions require that a contractor or person causing excavation to be done near a public street give specified written notice to the affected public utility companies. In the case of private construction, usually an engineer from the utility company will aid the excavator in locating the utility structures to be protected. If notice is given and the utility company fails to advise the excavator of the location of its subsurface structures, the utility company may not thereafter sue the contractor for negligent damage to the utility's lines while providing required support and protection. In the case of public construction, the utility company has the obligation of protecting, removing, or replacing the affected structures. If it fails to do so, the contractor will protect or remove the utility structures, as determined by the public entity, at the utility company's expense.

A good example of such a statute is the New York statute (N.Y.Gen. Bus. Law §761). Each town and city (and each borough of the City of New York) must now maintain a registry of the underground facilities located in the municipality. The actual excavator (not the owner or general contractor) must notify the utility listed in the registry in advance of proposed excavation. The utility companies must then advise the excavator of any underground facilities which might be affected by the proposed excavation or demolition. Thereafter, the excavator must support or protect the utility company's lines in the manner set forth in the Rules and Regulations of the State Board of Standards and Appeals. An excavator, in violating the statute, will be liable for any damages caused thereby. Violation of the statute is a crime, and will subject the excavator to civil penalties as well.

Thus, every excavator must contact the utilities listed in the registry. If any damages occur to the utility lines, the utility must be notified immediately. No repairs may be made in such case, except at the direction of the utility company. Similarly, no backfilling may be attempted without notification of the utility company. This statute fairly allocates the financial burden for location and protection of utility lines and structures during construction.

SAFETY CODES AND REGULATIONS

The *federal government* and many states and municipalities have established safety codes concerning construction activities. Failure to comply with the federal code can subject the contractor to substantial civil penalties by the Occupational Health and Safety Administration (OSHA). Moreover, willful disobedience can subject a contractor to criminal penalties. The federal regulations create a duty extending from the contractor to his own employees. This duty may not be delegated by the contractor to another. Violation of an OSHA safety provision does not, in and of itself, give an injured employee the right to bring a case in federal court against the employer or anyone else. Rather, this right will depend on the circumstances of each case.

OSHA requires, for example, that there be an engineer-designed system of support where the excavation will not have sloped earth banks. Further, OSHA requires an inspection of the shoring system on a daily basis and after every endangering storm. OSHA will permit any shoring system within accepted engineering standards.

The extent of safety requirements regarding construction activities varies greatly. Generally, *state requirements* are considerably less extensive than OSHA requirements. In some states, however, the duties owed are broader than under OSHA. In New York, for example, the duty extends from all contractors and owners connected with the construction or demolition of a building to employees of subcontractors. A violation has been held to entitle a contractor, sued by an employee of a subcontractor, to contribution from the subcontractor to the extent of the subcontractor's fault. Other typical provisions contained in the New York code are the requirement for daily inspection of shoring systems, and prohibitions against placing excavated material within 2 ft (60 cm) of the edge of the open excavation.

Municipal codes are generally somewhat between the state and federal codes in length, but deal with local problems not usually considered in the state or OSHA codes. For example, under the New York City Administrative Code, abandoned excavations must be filled; controlled inspection must be provided by a licensed architect or engineer for all construction relating to the support of adjacent properties or buildings; excavated material and other superimposed loads must be kept at a certain distance from the edge of the cut in proportion to the depth of the cut, unless the excavation is in rock or the sides of the cut have been sloped or sheet piled and shored to withstand such loads; and shoring or other protection must be provided for earth cuts below a certain depth unless the sides are of a specified gradual slope.

Because three sets of regulations have been issued at the federal, state, and local levels, contractors are best advised to follow the strictest requirement when the codes merely supplement each other. When there is a direct conflict between state, federal, or city regulation, the federal code should take precedence, followed by the state and lastly the municipal regulations.

3

Codes and Regulations

by Julian J. Karp

GENERAL OVERVIEW OF CODES AND REGULATIONS

Any element used in construction which will be removed or will not be used as a part of the finished structure is considered to be of temporary nature; for example, guys and braces in erection of structural steel frame, forms and shores in concrete, and well points in dewatering of construction sites. These elements are needed during the construction of a project and generally must comply with the same codes and regulations as the permanent members. The codes usually permit increased allowable stresses for temporary elements. The reason for this increase is that temporary members are generally used and stressed for a relatively short period of time as compared with permanent members of a structure. However, if a temporary member, for any reason, must remain in place for an extended period of time, then the stresses used in the design of such members shall be equal to those used for permanent members.

Safety is another subject which must be considered during construction of a project. First, one must comply with the applicable safety laws, and second, one must program and manage safety for the duration of a construction project. There are a number of texts on safety and accidents prevention. However, the Occupational Safety and Health Act (OSHA) of 1970[1] is a rigorous standard which must be complied with during construction. One must prepare to meet its requirements. Details of the OSHA act must be studied, and approaches and solutions to the problem developed and applied, so that the employees, supervisors, and management can live with OSHA standards. The military has its own manuals; for example, the Corps of Engineers, U.S. Army, developed a manual, *Safety, General Safety Requirements.*[2]

Generally, one should become familiar with the applicable laws and what they require. Periodic reading of the *Federal Register*[3] will provide additional information such as an update of the latest, revised occupational safety and health standards for construction and other Department of Labor construction regulations.

The codes most used in steel, concrete, and timber construction are the American Institute of Steel Construction, *AISC Specification for the Design, Fabrication and Erection of Structural Steel for Buildings,*[5] the American Concrete Institute *Building Code Requirements for Reinforced Concrete* (ACI 318),[6] *National Design Specifications for Wood Construction,*[10] with supplements,[11] by National Forest Product Association, and *Timber Construction Standards*[12] by the American Institute of Timber Construction.

As far as building codes are concerned, the most popular and widely accepted ones are the *Uniform Building Code*[7] by the International Conference of Building Officials, the *BOCA Basic Building Code*[8] by the Building Officials and Code Administrators, and the *National Building Code*[9] recommended by the National Board of Fire Underwriters. In addition, practically every state as well as most of the large cities in the nation have their own building codes.

The military works in accordance with regulations are listed in the *Index of Guide Specifications for Civil Works and Military Construction*[13] by the Department of the Army, Office of the Chief of Engineers. Federal housing projects are governed by *Minimum Property Standards,*[14] by the Department of Housing and Urban Development.

One must be very careful in deciding which codes and regulations apply for a particular project. An error in this decision may be very costly in time and money.

TECHNICAL MANUALS

The construction industry has a great number of technical manuals to choose from for both temporary and permanent types of construction. The most important and most widely used are: *Manual of Steel Construction,*[15] published by AISC (American Institute of Steel Construction); *Manual of Concrete Practice,*[16] published by ACI (American Concrete Institute); *Formwork for Concrete,*[17] also by the ACI; *CRSI Handbook,*[18] published by the Concrete Reinforcing Steel Institute; *Post-Tensioning Manual,*[19] by PTI (Post-Tensioning Institute); *Timber Construction Manual,*[20] by American Institute of Timber Construction; *Concrete Manual,*[21] by U.S. Department of the Interior; *OSHA Compliance Manual,*[22] by Dan Petersen; and many manuals published by the corps of engineers of the U.S. Army, Air Force, and Navy.

Most of the large cities in the United States have individual manuals of building laws. One of the most complete manuals of this type is the *New York Building Laws Manual*[23] published by the New York Society of Architects. This detailed manual covers most of the situations and problems encountered during construction of a project.

In addition, individual organizations, such as transit authorities, port authorities, federal and state departments, large cities, and a number of local authorities publish manuals, as do the Portland Cement Association,[24] the Asphalt Institute,[25] and the Prestressed Concrete Institute.[26] The U.S. Army Engineers has many manuals available that are listed in a multipage index and may be ordered from the Department of the Army, Office of the Chief of Engineers, Washington, DC 20314.

Depending on the individual specialty, a designer, engineer, or contractor will choose an appropriate manual to suit a particular assignment. However, the most common ones in the possession of professionals are the following five: *Manual of Steel Construction,*[15] *CRSI Handbook,*[18] *Manual of Concrete Practice,*[16] *Formwork for Concrete,*[17] and *Timber Construction Manual.*[20] These manuals are what one may call "reference" or "basic" manuals.

The AISC *Manual of Steel Construction* is one of the best known and most widely used manuals. It was produced under the guidance of the most experienced and knowledgeable engineers from the industry and fabricator companies for the most efficient and economical design and construction of structures. The manual consists of six basic parts. Part 1 covers dimensions and properties of structural steel sections and shapes. Part 2 covers beam and girder design, including selection tables for standard rolled sections and welded plate girders. Part 3 covers column and base plate design. Part 4 discusses connection design, including standard and moment connections and suggested detail, fasteners data, welded joints, and miscellaneous data such as fabricating practices. Part 5 covers specifications and codes for the design, fabrication, and erection of structural steel buildings, the AISC Code of Standard Practice,[5] specifications for structural joints, and a quality certification program. Part 6 gives miscellaneous data and mathematical tables.

The *CRSI Handbook*[18] was compiled with the objective to disseminate helpful reinforced-concrete design information for economical and efficient use of materials and to save time. All tables and charts are based on the latest applicable ACI code provisions and material specifications. The manual consists of 15 chapters: Chap. 1 gives the basis and the use of tables; Chaps. 2, 3, and 4 cover design of columns; Chaps. 5 and 6 cover the design of flexural members; and Chaps. 7–12 and 15 go into the design of different types of beams, slabs, and one-way and two-way joists (waffle) construction; Chap. 13 covers the design of square footings, pile caps, drilled piers, and caissons; and Chap. 14 covers cantilevered retaining walls.

The *ACI Manual of Concrete Practice*[16] is possibly the most complete manual available for concrete construction. Each standard contained in this manual bears a five-digit hyphenated number to identify it. The first three digits identify the ACI committee originating the standard, and the last two digits identify the year it was prepared. The ACI committee classifications are as follows:

100—Research and Administration

200—Materials and Properties of Concrete

300—Design and Construction Practices, including Specifications and Inspection

400—Structural Analysis

500—Special Products and Special Processes

A detailed description of this manual would be so wide in scope that it cannot be included here. This is the most complete, all-encompassing manual, and it is constantly being reviewed, revised, and expanded. Naturally, as with the previous manuals, it covers both temporary and permanent concrete construction.

Formwork for Concrete[17] is basically a manual exclusively for temporary construction. As a rule, forms are removed after concrete has hardened and has developed its desired strength. This manual was originally developed for improving the safety, efficiency, economy, and quality of formwork for concrete construction, with the basic goal of developing a specification for design and construction of formwork. In addition, this manual provides the how-to-do-it information useful to form designers, contractors, engineers, and architects, following guidelines established by ACI Committee 347.

The many photographs, drawings, diagrams, tables, and sketches make this manual most complete for planning, design, and construction of concrete forms. The manual not only covers lumber, which for many years was the predominant form material, but also includes instructions for use of plywood, metal, plastic, and other materials, together with the increasing use of specialized accessories.

Timber Construction Manual[20] was prepared by the American Institute of Timber Construction (AITC) for the use of architects, engineers, contractors, laminators, and fabricators concerned with engineered timber buildings and other types of timber structures. This manual covers both temporary and permanent types of construction. For many years this manual was accepted as the authoritative standard for all those in the construction industry having a need for accurate, up-to-date technical data on engineered timber construction. In addition, the manual provides up-to-date information on design stresses, latest lumber and timber sizes, and design procedures for wood structural elements and fastenings.

The manual is divided into two parts: Part I, which contains design data and construction information, has been divided into six major sections: (1) design consideration in the use of structural timber, detailing, and erection; (2) physical and mechanical properties of wood; (3) load and force specifications; (4) design features of wood structural systems; (5) mechanical fastenings and connections for timber joints; and (6) reference data. Part II contains the applicable official standards and specifications as adopted by AITC, which are based on the most recent information and the best engineering and commercial practices. Part II starts with the AITC 101 "Standard Definitions, Abbreviations and References" and ends (2d ed.) with AITC 120 "Standard Specifications for Structural Glued Laminated Timber using 'E' rated and visually graded lumber of Douglas fir, southern pine, Hem-fir and lodgepole pine."

Technical developments and the establishment of an engineered timber fabricating and laminating industry have had a profound effect on construction. Long spans of girders, trusses, arches, and decking are now commonplace. Engineered timber is widely used, and modern practices combine good engineering, quality control, and careful grading. Proper working stresses, dependable adhesives, and economical, efficient mechanical fastenings are required to provide reliable construction. The AITC has developed *Timber Construction Manual*[20] for convenient reference by architects, engineers, contractors, the fabricating industry, and all others having a need for reliable, up-to-date technical data and recommendations on engineered timber construction.

DESIGN AND BUILDING CODES

Familiarity with a specific code, standard, or specification is the prerequisite for all architects, engineers, and contractors. No intelligent decision can be made without a thorough knowledge of the governing code, standard, or specification.

In addition to the few national building codes mentioned earlier, there are other codes one must be aware of. Because of this country's size, diversity in climate, and other conditions, the design codes may vary from state to state and even city to city. However, all national and local codes usually refer to the few basic technical codes which were developed by national organizations such as the American Concrete Institute (ACI),[27] the American Institute of Steel Construction (AISC),[28] the National Forest Product Association,[29] and the American Society for Testing of Materials.[30] These three organizations, which are national in scope, have developed codes for concrete, steel, and wood. The important codes, which may also be called specifications or standards, are as follows:

1. ACI Standard 318 *Building Code Requirements for Reinforced Concrete.*[6]
2. AISC *Specification for Design, Fabrication and Erection of Structural Steel for Buildings.*[4]
3. AISC *Code of Standard Practice.*[5]

4. Specification for Structural Joints Using ASTM 325 or A 490 Bolts.[31]
5. AISC *Quality Certification Program.*[32]
6. *National Design Specifications for Wood Construction* (NFPA).[10]
7. *Design Values for Wood Construction,* supplement[11] to the 1977 edition of *National Design Specification for Wood Construction.*
8. American Society for Testing and Materials (ASTM)[30] *ASTM Standards in Building Codes.*[33] Building codes throughout the United States and Canada have adopted by reference a large number of ASTM standards as authentic sources of test procedure and as bases for acceptable quality for materials and constructions. This compilation includes ASTM[30] standards that are referenced in such nationally known codes as *BOCA Basic Building Code,*[8] Building Officials Conference of America; *Southern Standard Building Codes,*[34] Southern Building Code Congress; *Uniform Building Code,*[7] International Conference of Building Officials; *National Building Code,*[9] the American Insurance Association; and the *National Building Code of Canada.*[35]
9. ANSI A58.1 Standard,[36] *American National Standards Institute Building Code Requirements for Minimum Design Loads in Building and other Structures.*

How do these codes affect the design and construction of temporary structures? Let us review briefly.

The ACI Standard 318,[6] *Building Code Requirements for Reinforced Concrete,* makes a number of recommendations: Chap. 5 has some reference to forms and other temporary operations; Chap. 6 covers the design of formwork and the removal of forms and shores; Chap. 8 refers to analysis and design—general considerations which cover both temporary and permanent structures in concrete; Chap. 9 also covers temporary structures in concrete, as do Chaps. 10 through 19 and Appendices A and B.

Generally, the required strength of temporary structures to resist dead load, live load, wind, earthquake, and other horizontal loads, may be reduced by a factor given in the code, usually 25%. However, one must follow the code as far as the combinations of the different loads are concerned. The different load combinations must be investigated to determine the greatest required strength. If the "alternate design method" shown in Appendix B of the ACI Code[6] is used, all members may be proportioned for 75% of capacities required by other parts of this Appendix. What this means is that the allowable stresses specified for permanent members can be increased by 33⅓%.

The AISC *Specifications for Design, Fabrication and Erection of Structural Steel for Buildings*[4] apply to both temporary and permanent construction; the code does not differentiate between the two. Generally, the required strength of temporary structures to resist all applicable combination of loads may be reduced by a factor given in the code, or the allowable stresses indicated in the code may be increased by a factor (usually one-third) above the values provided in the code. No temporary members or connections in conventional buildings need to be designed for fatigue (Sec. 1.7). Generally, temporary members can be considered as "secondary" members with the appropriate stability and slenderness ratios (Sec. 1.8). The allowable deflection for temporary members shall be established by the requirements of the particular project (Sec. 1.13), and the same applies for vibration and pounding. Connections, stresses indicated in the code to be used in connection of temporary members, may be increased by one-third. Section 1.25 of the code covers erection and discusses temporary bracing. Such bracing shall be provided in accordance with the requirements of the *AISC Code of Standard Practice*[5] wherever necessary to take care of all loads to which the structure may be subjected, including equipment and the operation of same. Such bracing shall be left in place as long as may be required for safety.

Wherever piles of material, erection equipment, or other loads are supported during erection, proper provision shall be made to take care of stresses resulting from such loads. Adequacy of temporary connections in discussed in Sec. 1.25.

The practices defined in the AISC *Code of Standard Practice*[5] for steel buildings have been adopted by the AISC[28] as the commonly accepted standards of the structural steel fabricating industry. In the absence of other instructions in the contract documents, the trade practices defined in this code govern the fabrication and erection of structural steel.

The code goes into details of "Temporary Support of Structural Steel Frames" (Sec. 7.9). It talks about temporary guys, braces, falsework, cribbing, or other elements required for the erection operation. These temporary supports will secure the steel framing or any partly assembled steel framing against loads comparable in intensity to those for which the structure was designed, but not the loads resulting from the performance of work by others or the acts of others, nor such unpredictable loads as those due to tornado, explosion, or collision. In Sec. 7.9.3 the code discusses the non-self-supporting steel frames. Such frames shall be clearly identified in the contract documents. The contract documents shall specify the sequence and schedule of placement of such elements. The erector determines the need and furnishes and installs the temporary supports in accordance with this information. The owner is responsible for the installation and timely completion of all elements not classified as structural steel that are required for stability of the frame.

Section 7.9.4 discusses special erection conditions in areas where the design concept of a structure is dependent upon the use of shores, jacks, or loads which must be adjusted as erection progresses.

Section 7.9.5 discusses the removal of temporary supports. It is stated that all temporary elements which were furnished and installed by the erector are his property. It discusses the timing of the removal of the temporary supports for "self-supporting structures" and "non-self-supporting structures."

Section 7.9.6 goes into temporary supports for other work.

Section 7.10 discusses temporary floors and handrails for buildings. In this case, the erectors provide and remove floor coverings, handrails, and walkways as required by law and by applicable safety regulations for protection of their own personnel. The owner is usually responsible for all protection necessary for work of other trades. When permanent steel decking is used for protective flooring and is installed by the owner, all such work is performed so as not to delay or interfere with erection progress and is scheduled by the owner and installed in a sequence adequate to meet all safety regulations.

Section 7.17 discusses final cleanup. Upon completion of erection and before final acceptance, the erector removes all of the falsework, rubbish, and temporary building elements.

In Sec. 10.5.1 erection of architecturally exposed structural steel is discussed. If temporary braces or erection clips are used, care is to be taken to avoid unsightly surfaces upon removal. Temporary tack welds are to be ground smooth and holes filled with weld metal or body solder and smoothed by grinding or filing. As far as structural joints using bolts is concerned, unless required by contract documents or by stress, machine bolts may be used for all temporary connections.

Wood has the property of carrying substantially greater maximum loads for short duration than for long duration of loading. Recommended design values provided in *National Design Specifications for Wood Construction*[10] apply to normal duration of loading. Normal loading duration contemplates a load that fully stresses a member to its allowable design value by the application of the full design load for a duration of approximately 10 years, either continuously or cumulatively. For other than normal duration of loading, design values for wood members and design values for fastenings, when fastening load capacity is determined by the strength of the wood rather than the strength of the metal, shall be adjusted in accordance with Sec. 2.2.5.2 through Sec. 2.2.5.7 to take into account the change in the strength of wood with changes in duration of loading. (See also Appendix B of these specifications.)

All temporary members shall be designed for the combination of design loads and forces that may be expected to occur, and the most severe distribution or concentration of these loads and forces shall be taken into consideration. When both wind and earthquake loads may occur, only that one which produces the greater stresses need be considered, and both need not be assumed to act simultaneously. When the duration of the full maximum load is of a temporary nature, the normal design values for wood members and fasteners shall be multiplied by the following modification factors:

1.15 for 2 months duration of load (snow)

1.25 for 7 days duration of load

1.33 for wind or earthquake

2.00 for impact (except for "marine" exposure)

The above modification factors are not cumulative. The resulting sizes of structural members or load-carrying capacities of joints shall not be smaller than required for a lesser design load acting for a longer duration. The above provisions do not apply to modulus of elasticity. The above provisions apply to mechanical fasteners when fastening load capacity is determined by the strength of the wood rather than the strength of the metal, unless otherwise provided. Appendix B of *National Design Specifications for Wood Construction*[10] discusses the various durations of loading and related design values. The design values tabulated in this specification are for normal duration of loading (10-year loading). Since tests have shown that wood has the property of carrying substantially greater maximum loads for short duration than for long durations of loading, Appendix B gives, in Fig. B-1, the adjustment of working stresses for various durations of load.

Generally, all codes, standards, and specifications apply to temporary as well as to permanent structures, with the proper adjustments for design values to be used for temporary members because of their shorter loading duration.

FEDERAL AND STATE REGULATIONS

To encourage improvement in housing and residential land development standards and conditions, the U. S. Department of Housing and Urban Development (HUD)[37] has prepared and published *Minimum Property Standards.*[38] This standard has provided the department with a single unified set of technical and environmental standards. They define the minimum level of acceptability of design and construction for low-rent public housing as well as housing approved for mortgage insurance purposes.

As far as temporary members and structures are concerned, Chap. 6 of this standard (construction) goes into details of reference standards, general structural requirements, concrete, masonry, metals, and timber.

Different state or city agencies and authorities have published their own manuals, standards, state building codes, and regulations. For example, the New York City Transit Authority has issued *Field Design Standards,*[39] mostly concerned with temporary structures such as underpinning, temporary earth retaining structures, and allowable unit stresses for temporary structures.

The Port Authority of New York and New Jersey[40] has developed its standards and specifications for the design of temporary as well as permanent structures.

New York State has issued its *State Building Construction Code*[41] for (Part A) "One and Two-Family Dwellings," (Part B) "Multiple Dwellings," and (Part C) "General Building Construction," all applicable to permanent and temporary structures.

New York State has also published a *Code Manual* for the *State Building Construction Code*[42] to assist in the interpretation, application, and enforcement of the *State Building Construction Code.*[41] Where the code has the force and effect of law, the code manual is purely advisory. The manual presents construction details for all disciplines. It also specifies structural requirements for all typical construction procedures. It describes the safety measures during construction and demolition which are required by law to conform with rules relating to the protection of persons employed in the erection, repair, and demolition of buildings or structures.

U.S. ARMY GENERAL SAFETY REQUIREMENTS

The U. S. Army Corps of Engineers has published Manual EM 385, *Safety, General Safety Requirements,* divided into 32 sections and 13 appendices (available through the Superintendent of Documents, U.S. Government Printing Office, Washington, DC 20402). The purpose and scope of this manual are to establish the general safety requirements for all Corps of Engineers activities and operations. Application of this manual is mandatory to all missions under command of the Chief of Engineers. Pertinent provisions of this manual shall be applied to all work under jurisdiction of the Corps of Engineers, whether accomplished by military, civilian, or contractor forces. The term "pertinent provisions" is defined as those provisions which are applicable to the situation at hand. In situations where literal application of the requirements to a specific job has impractical aspects, division engineers, district engineers, and commanding officers of separate installations and activities are authorized to approve an adaptation which meets the obvious intent of the requirements.

Basically, Manual EM 385 concerns itself with the safety of each employee and all persons present at the site. The manual begins with initial instruction and training or indoctrination and such continuing instruction that will enable the employee to conduct his work in a safe manner. Accident reporting, sanitation, medical facilities, and physical qualification of employees are the subject of the next few sections.

Personal protective apparel, safety equipment, emergency plans, and poisonous and harmful substances are discussed next. Lighting, signals, and warning signs are subjects of the next two sections. The other sections are concerned with material handling, storage, fire prevention, fire protection, welding and cutting, electrical wiring and apparatus, hand tools and power tools, ropes, slings and chains, machinery and mechanized equipment, motor vehicles, aircraft, ramps, runways, platforms and scaffolds, excavations, tunnel excavations, blasting, floating plant and marine locations, work in confined or enclosed space, safe clearance procedure, formwork and falsework, access facilities, floor and wall opening, and noise control. The 13 appendices are concerned with special conditions encountered and safe resolution to the problems caused by these conditions.

OSHA CONSTRUCTION SAFETY AND HEALTH REGULATIONS

The regulations most profoundly affecting temporary structures in construction are those of the Occupational Safety and Health Administration of the U.S. Department of Labor.

The *Occupational Safety and Health Act*[1] of 1970 was an effort to provide all workers with healthful working conditions to the extent possible. The Occupational Safety and Health Administration[43] was established in 1971 to administer the Act. It is the most sweeping job safety and health legislation in the history of the United States. The OSHA Act applies to almost every employer in the country, but specifically to those engaged in business affecting construction and commerce. It requires every employer to provide a job environment that is free from hazards that cause or are likely to cause serious physical harm or death. These standards set forth in considerable detail requirements with respect to such industrial and occupational matters as hazardous materials, personal protective equipment, fire protection, and health and environmental control, including air contaminants, noise exposure, ventilation, and radiation. To provide such a job environment, each employer is required to comply with safety and health standards which are specifically spelled out by the law.

In addition, OSHA has teeth. Failure to meet standards or comply with provisions of the Act can subject an employer to heavy fines or imprisonment. The inspection and enforcement of OSHA regulations is carried out through regional and area offices located in all major cities. Unannounced inspections at industrial and construction work sites are conducted by compliance health and safety officers under the supervision of area directors, and the Secretary of Labor or his or her representatives are authorized to enter at reasonable times and without undue delay any construction site, factory, plant, or other work place where work is performed by employees and to inspect all pertinent conditions, structures, equipment, machines, and material therein, and to question any employer, employee, or operator. Generally, an OSHA inspector who is refused admission can obtain a search warrant. But in 1978, a Supreme Court decision suggested that the inspector must have a reasonable expectation of finding something amiss. Inspection can also be triggered by serious accidents or by employee complaints. Precompliance checks of construction sites and businesses are made by some local or state agencies, most insurance carriers, and special private consultants.

Standards are promulgated by the Secretary of Labor. The initial standards were published in the May 29, 1971, issue of the *Federal Register*[3] and additional provisions have been added since then. One of the most difficult requirements of OSHA is to keep abreast of current standards and methods. In each company, someone must be responsible for the standards which apply to the operation, because all standards are subject to change. The scope of standards is very broad, and regulations touch on such things as scaffolding, protection equipment, location and size of temporary and permanent facilities, and material handling. It is safe to say that the regulations are so broad that there is a chance that a company is not in complete compliance with all aspects of the law at all times.

Many companies have found their safety and their procedures in violation of law and the cost of corrections and improvements very high. The fines are minor to the outlays required. The cost of corrections is usually the most expensive part. The most frequent violations as far as construction sites are concerned are: missing ground plugs; frayed or improperly spliced cords; failure to guard belts or gears; missing guards on saws; no rails or guards at walkways, holes, pits, openings, and unprotected drops of 4 ft or more; failure to keep work areas orderly, clear, and sanitary; improper storage and handling of compressed gases used for welding and cutting; unmarked exits and means of egress; excessive storage or containers of flammable or combustible materials left uncovered; failure to provide proper sanitary facilities; failure to post or observe nonsmoking signs in paint spray areas; missing guards on portable power tools; blocked or inadequate exits; no load capacity posted on crane; noise exposure in excess of 90 dB on an 8-h time-weighted average; missing or inadequate first aid kit; failure to wear protective goggles where there is strong possibility of danger to the eyes from flying objects; missing hard hats where there is danger of dropped items; ladder rungs too widely spaced; and too little clearance to sides or rear of ladder.

OSHA also pays attention to so-called hidden health violations, such as concentration or use of carcinogens in work place atmospheres. Such violations are usually considered serious. The National Safety Council offers a self-evaluation form, which covers the law, to help companies trying to comply with the standards.

REFERENCES*

1. *Occupational Safety and Health Act* (OSHA), enacted in 1970, administered by the Occupational Safety and Health Administration under the Secretary of Labor, Department of Labor, Washington, DC.

*Use latest edition or printing.

2. *Safety, General Safety Requirements,* EM 385, Department of the Army, Corps of Engineers, U. S. Government Printing Office, Washington, DC 20402.
3. *Federal Register,* Office of the Federal Register, National Archives and Records Service, General Services Administration, Washington, DC, U.S. Government Printing Office, Washington, DC 20402.
4 *AISC Specification for the Design, Fabrication and Erection of Structural Steel for Buildings,* American Institute of Steel Construction, 400 North Michigan Ave., Chicago, IL 60611.
5. *AISC Code of Standard Practice,* American Institute of Steel Construction (AISC), 400 North Michigan Ave., Chicago, IL 60611.
6. *Building Code for Reinforced Concrete,* ACI Standard 318–77, American Concrete Institute, P.O. Box 19150, Redford Station, Detroit, MI 48219.
7. *Uniform Building Code,* International Conference of Building Officials, 5360 South Wollman Mill Road, Whittier, CA 90601.
8. *The BOCA Basic Building Code,* Building Officials and Code Administrators International, Inc., 1313 East 60th St., Chicago, IL 60637.
9. *National Building Code,* National Board of Fire Underwriters, 222 West Adams St., Chicago, IL 60606.
10. *National Design Specifications for Wood Construction,* National Forest Product Association, 1619 Massachusetts Ave., N.W., Washington, DC 20036.
11. *Supplements to National Design Specifications for Wood Construction,* National Forest Product Association, 1619 Massachusetts Ave., N.W. Washington, DC 20036.
12. *Timber Construction Standards,* American Institute of Timber Construction, 333 West Hampden Ave., Englewood, CO 80110.
13. *Index of Guide Specifications for Civil Works and Military Construction,* Department of the Army, Office of the Chief of Engineers, Washington, DC 20314.
14. *Minimum Property Standards,* Department of Housing and Urban Development, Washington, DC 20402.
15. *Manual of Steel Construction,* American Institute of Steel Construction (AISC), 400 North Michigan Ave., Chicago, IL 60611.
16. *ACI Manual of Concrete Practice,* parts 1, 2, 3, 4, and 5, American Concrete Institute, Box 19150, Redford Station, Detroit, MI 48219.
17. Hurd, M. K., *Formwork for Concrete,* Special Publication No. 4, 4th ed., American Concrete Institute, P. O. Box 19150, Detroit, MI 48219.
18. *CRSI Handbook,* Concrete Reinforcing Steel Institute, 180 North Lasalle St., Chicago, IL 60601.
19. *Post-Tensioning Manual,* Post-Tensioning Institute (PTI), 1701 Lake Ave., Suite 375, Glenview, IL 60025.
20. *Timber Construction Manual,* American Institute of Timber Construction, 333 West Hampden Ave., Englewood, CO 80110.
21. *Concrete Manual* by U. S. Department of the Interior, Bureau of Reclamation. U. S. Government Printing Office, Washington, DC 20402.
22. Petersen, Dan, *OSHA Compliance Manual,* McGraw-Hill, New York, 1979.
23. *New York Building Laws Manual,* New York Society of Architects, New York, N. Y.
24. Portland Cement Association, Old Orchard Road, Skokie, IL 60076.
25. The Asphalt Institute, Executive Office and Laboratories, University of Maryland, College Park, MD 20742.

26. Prestressed Concrete Institute, 20 North Wacker Drive, Chicago, IL 60606.
27. American Concrete Institute (ACI), Box 19150, Redford Station, Detroit, MI 48219.
28. American Institute of Steel Construction (AISC), 400 N. Michigan Ave., Chicago, IL 60611.
29. National Forest Product Association, 1619 Massachusetts Ave., N. W., Washington, DC 20036.
30. American Society for Testing of Materials (ASTM), 1916 Race St., Philadelphia, PA 19103.
31. Specification for Structural Joints Using ASTM A325 or A490 Bolts, *Manual of Steel Construction,* American Institute of Steel Construction, 400 North Michigan Ave., Chicago, IL 60611.
32. "AISC Quality Certification Program," *Manual of Steel Construction,* American Institute of Steel Construction, 400 North Michigan Ave., Chicago, IL 60611.
33. *ASTM Standards in Building Codes,* American Society for Testing and Materials, 1916 Race St., Philadelphia, PA 19103.
34. *Southern Standard Building Code,* Southern Building Code Congress, 1116 Brown University, Marx Building, Birmingham, AL 35203.
35. *National Building Code of Canada,* Associate Committee of the National Building Code, National Research Council, Ottawa, Canada.
36. *Minimum Design Loads for Buildings and Other Structures,* American National Standard A58.1-1982, U.S. Department of Commerce, National Bureau of Standards, Washington, DC 20234.
37. U. S. Department of Housing and Urban Development, Washington, DC.
38. *Minimum Property Standards,* Superintendent of Documents, U. S. Government Printing Office, Washington, DC 20402.
39. *Field Design Standards* by New York City Transit Authority, New York, N.Y.
40. The Port Authority of New York and New Jersey, One World Trade Center, New York, NY 10046.
41. *New York State Building Construction Code,* Building Codes Bureau, 393 Seventh Ave., New York, NY 10001.
42. *New York State Code Manual,* Building Codes Bureau, 393 Seventh Ave., New York, NY 10001.
43. The Occupational Safety and Health Administration, U. S. Department of Labor, Washington, DC.

4

Erection and Earthwork Equipment

by Stuart Wood, Jr.

With each construction project, a certain percentage of work must be done to prepare for the actual creative undertaking. This preliminary effort may well require some form of contractor's equipment. During the progress of the project, a wide variety of equipment may be necessary for the ongoing structural work as well as for satisfying requirements of supplemental support facilities. This equipment may be needed either for the life of the contract or for temporary use. It may be owned by the contractor or leased from an outside agency for the work to be done, or the work may be subcontracted to an organization that has the needed equipment. A substitute unit of equipment from the contractor's available inventory may accomplish the mission, even if the unit may not be the best or most applicable for that specific job.

For purposes of this presentation, the equipment is divided into the two major groups according to use: erection and earthwork. Separate discussions of equipment selection and cost and on equipment safety follow.

ERECTION EQUIPMENT

This class of mechanical devices temporarily used for the raising of any structure or facility, because of its primary purpose of serving to assist in the erection of structures, materials, or implements, is grouped here under the general heading of erection equipment.

Starting with the most simple individually operated apparatus, coverage runs through a broad spectrum of mechanical advantages. These advantages lift vertically, move horizontally, or combine to serve a specific need for the construction industry.

BASIC ERECTION EQUIPMENT

Handlines

Probably the simplest of erection equipment handlines are an age-old system of extending one's brute strength by using a rope or cable to pull or haul the item to be moved. This effort can be supplemented by the creation of an angle in the line of pull or by the use of "blocks" (Fig. 4.1). This provides an increased lifting or hauling capacity. Blocks, one or more rotating sheaves in a frame, may improve the pull capacity manyfold. With more than a single sheave in use, however, it becomes necessary to employ a cable-return device to complete the circuitry of the hookup.

Passing from the blocks to the source of power is the lead line. Those lines supporting the load itself are known as fall lines. If a line passes from the power source through a block to the load, it is a one-part line. If it passes through an upper block and a separate lower block (which is attached to the load) before being anchored or dogged to the upper block, the line then has two parts. This process may be repeated to obtain a many-part line. The use of a line or cable in conjunction with blocks in moving a load is known as "rigging." The pull required to lift or move the load is equal to the weight of the load divided by the

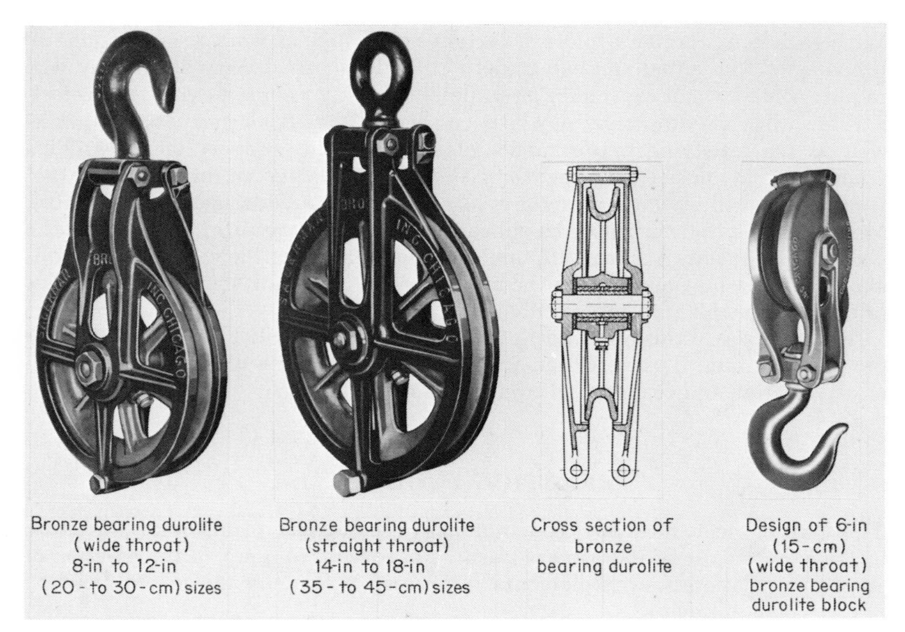

FIG. 4.1 Blocks. *(Sauerman Bros., Inc.)*

number of parts of the line plus the friction caused by passage of the line over the sheaves and elsewhere.

Hoists

To facilitate the use of the handline and "tackle" (composed of rope or cable and blocks), the contractor may employ the mechanical hoist or winch. This may be a simple hand crank with worm, screw, or spur gears, or a complex piece of machinery with multiple drums utilizing mechanical power. The most basic winch consists of a crank rotating an axle which turns a drum around which rope or cable is wound. A drum hoist, such as that shown in Fig. 4.2, may be located so as to provide a direct pull on the item to be moved, or may be worked through blocks to increase its efficiency. The unit ranges in size from the small hand hoist up to those with a capacity of over 3000 tons (2700 t). The type and size of hoist required depend on how many drums are needed, drum diameter and length, speed, and line pull. They may be adapted for a right-hand or left-hand "lay" cable, and provide power in both forward and reverse directions, as long as there is tension on the line.

The material hoist, or elevator, is of tremendous service in erection of structures. A braced and supported tower, through which passes a load-bearing platform controlled by a mechanical hoist, will facilitate the raising and lowering of construction loads. When properly controlled and with adequate safety precautions, the hoist may also serve as a lift for personnel. It may be self-supporting or affixed to the structure being erected.

FIG. 4.2 Hoist for steel erection. *(Clyde Iron.)*

Gin Pole

This pole is actually a centerpiece tipped up from the ground to a position slightly inclined from the vertical. It is held in place by four stays or guys reaching from the top of the pole to anchors on the ground (Fig. 4.3). These guys then

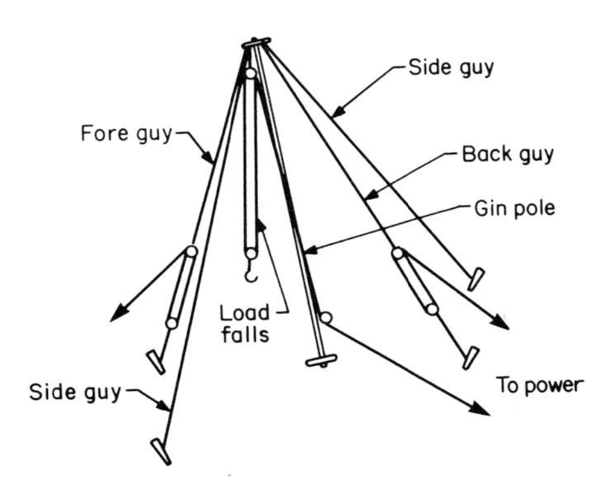

FIG. 4.3 Gin pole.

provide the stability for the pole itself. At the top of the pole may be affixed a block through which ropes or cables, extended from the base of the pole to the item to be lifted, are passed. This pole may be of wood or pipe or it may be built up of structural members. It is secured in position at the base, but is allowed to tilt as required through the loosening or tightening of the anchor stays. Capacity of the gin pole is based on the strength of the material comprising the pole, strength of the rope or cable used in the lifting, capacity of the block or blocks, and the capability of the motivating force. It is generally used for loading and unloading materials and erecting lightweight structural members. With its fixed-base location and limited inclination, the gin pole has restricted mobility and use. Because of its overall configuration, it also has limited height capability. A gin pole can handle loads, which must be located directly under the tip of the inclined structure, only in a vertical mode. Because of its simplicity, it can be readily rigged up on the job and relocated as needed.

A variation of the gin pole is the "shear" or A frame. Here two inclined poles, spread and anchored at the bottom, yet joined and secured at the top, are held in position by back stays. This provides a structure with more rigidity than the gin pole.

Booms

Whether these units are fixed in place or self-propelled, they are a necessity in the erection of structures (Fig. 4.4). The simplified boom is nothing more than an inclined pole, fixed at the base, used to lift a load. How it is rigged determines its capacity and uses. For small jobs, this boom may be truck-mounted and provide the necessary job coverage from the mobility of the truck itself and the length of the boom. A mounted boom may be a fixed "stick" or an articulated knuckle boom which gives added flexibility in reach. The truck mounting may be nothing more than a hinge at the boom base, allowing a raising or lowering of the boom, or it may include a rotating base plate allowing the boom to pivot 360°. The boom may be hydraulically powered to facilitate its telescopic extension or retraction. Booms have a relatively short length and restricted lifting capacity. However, for lifting purposes, they prove far more economical, when available,

FIG. 4.4 Hydraulic boom loading steel. *(Lorain Division, Koehring.)*

than tying up one of the larger tower cranes. They may be erected and torn down or relocated a number of times as the project increases both vertically and horizontally.

To assist with high lift requirements, this device may also be of the "Chicago Boom" configuration. Such a unit is affixed to the side of a structure, which serves as the mast for this boom, wherever the base or seat may be suitably anchored. The angle of the boom may be adjusted by allowing it to lean away from the vertical structure by extending the lines through the topping lift. The Chicago Boom is useful in raising, erecting, dismantling, and lowering the larger derricks used for heavier lift purposes.

DERRICKS

Within the broad classification of derricks, here are presented the boom, the stiff-leg, and the guy derricks. Power to all three types may be provided by an electric motor or, more frequently, a gasoline or diesel engine. With its fixed-base mast the derrick is, at best, only a semimobile piece of equipment. These derrick structures are temporary in nature in relation to construction of the facility itself. They are built or positioned before or during the construction process, relocated as needed during construction to serve the continuing lift requirements, and removed upon completion. Or, they may be employed as yard structures for long-term unloading, storing, or loading material. These derricks are not capable of over-the-ground movement without dismantling.

Boom Derrick

This device is actually an extension of the previously discussed gin pole. In this particular case, a boom is used to lift a load at some distance from the base of a vertical pole or mast. Cables or guys from the ground anchor the mast in a vertical position. The boom will be affixed to the mast at a point slightly up on the pole and may be held in place by lashing (truly a field expedient) or a trunnion-type mechanism. Lines run from the mast to the top of the boom and may be used to determine the boom angle. Other lines are run from the ground through sheaves up along the mast to the boom where they drop from the boom tip to secure the load on the ground. For heavy loads, this boom base may be lowered to rest on a support on the ground itself so as not to create undue bending forces in the mast. The boom itself should never be raised and anchored high enough on the mast so as to cause structural instability in the vertical member as the load is lifted.

Stiff-Leg Derrick

This device is also an extension of the gin pole. As shown in Fig. 4.5, the stiff-leg derrick consists of a center mast with two inclined, fixed, stiff-back legs supporting it from the rear. These supporting legs are separated by an angle of above 90° and may be themselves supported at midspan by added stiff members to provide added rigidity. The inclined legs may be attached at their base to horizontal sills. Extending from the base of the mast, and hinged to it, is a boom with

FIG. 4.5 Bridge construction with stiff-leg derrick. *(Clyde Iron.)*

an adjustable inclination. Flexible guys, approximately 3 times as long as the mast height, affix to and support the top of the mast. The boom may be pivoted horizontally, around from one back leg to the other back leg, on its heel at the bottom of the mast. The angle between the back legs should be adjusted, depending on load location and weight, to attain best overall stability. The boom is held in position by boom falls reaching its top from the peak of the mast. Because of the translation of the load into uplifting tensile forces on the supporting legs (as the derrick tends to pivot about its base), these legs must be securely anchored to prevent overturn of the derrick. Stiff-leg derricks are relatively immobile and are usually employed for heavy lifting requirements.

A variation of the stiff-leg derrick is the "Jinniwink." This device has its legs and mast attached to a T-shaped horizontal frame or sills. This frame can be fixed to the structure for operation, then relocated for further continued use. Another variation of the stiff-leg derrick is the "skid" derrick. This unit is usually smaller than the stiff-leg device and attached to a skid frame. It may be slid across the ground for mobility.

Guy Derrick

The guy derrick (Fig. 4.6) has more overall capability than the stiff-leg rig in that it may be rotated a full 360° horizontally around its base bearing. For this operation, it is necessary that the boom be short enough to clear the sloped supporting guys; hence it must be shorter than the mast itself. Tension in the anchored supporting guys may be adjusted by turnbuckles to resist the opposing load.

FIG. 4.6 Guy derrick assisted by stiff-leg derrick. *(Clyde Iron.)*

Assuming that the strength of the rigid structural members is sufficient, capacity of the guy derrick is determined by the guys and the lifting line. These wire rope guys connect to the "spider" at the top of the mast. Because of the derrick's rotational capability, the base, known as a bullwheel, is usually mounted with a roller bearing, race, or, in some cases, a bronze washer. Whatever is used must be able to reduce the friction caused by the weight of the equipment itself, the turning action, and the load being lifted. The "spider," secured at the mast top by the guys, is penetrated by a gudgeon pin which allows full mast rotation without effect on the spider. The bullwheel might be turned by the action of slewing cables attached to opposite points on the wheel.

The guy derrick has advantages other than its rotational capability over the stiff-leg and boom derricks. It is readily adaptable for use in conjunction with vertical structures, as the members may be closed together to pass through small apertures. It has only two main inflexible members as compared to the six members of the stiff-leg derrick. Because it may be set up easily, it may be raised vertically, or jumped, without serious problems, and consequently can be raised as a structure is raised. In such a case, it is anchored to the structure by both the base and the guys. Due to its vertical configuration, a guy derrick can be set up in close or relatively inaccessible areas.

In jumping to the next tier of a building (two floors), the boom is disconnected from the mast and set to act as a pole to lift the mast. This mast is raised the two floors to rest on a pair of horizontal members (known as jumping beams), where it is anchored and guyed. Once the mast is secured, it acts as a lifting pole to raise the detached boom. The base of the boom is then secured to the foot of the mast for full performance lifting. This process may be repeated every two floors as needed to peak out the building. Then the derrick is lowered and removed from the site. When more than one derrick is in use on a project, one may be used to assist another in the jumping or lowering operation. In this case, the boom and mast may fold together for ease in stepping up or lowering through constrained openings.

CRANES

Under this single heading, there are actually four types of cranes (Fig. 4.7) employed in erection of structures. A fifth type is discussed here, but is not truly classed as a crane: that unit is the sky crane or helicopter.

Cranes have increased mobility over the stationary derricks, but still require some plant erection before they can be used. The first type is the conventional

FIG. 4.7 Cranes placing girders. *(Northwest Engineering Company.)*

mobile crane with its several boom or tower variations. The second type is the rail-mounted unit. The third type of crane is the self-climbing tower crane. And last, there is the gantry. The gantry may serve as the base for a crane or derrick, in which case it becomes known as a "whirley."

Because of the need for high reach, these four cranes have the long boom (with jib and supporting mast) or vertical mast configuration. However, their vertical reach is limited. When the height required for operational support surpasses their reach, the cranes must be raised, replaced, or supplemented by the aforementioned derricks. While the various types of cranes have the same basic capability in lifting materials to the elevated work site, there is a definite difference as far as their mobility is concerned. The mobile crane, as its name implies, has full mobility to travel where its size and weight may allow. The rail-mounted crane's mobility is governed strictly by the location of the rails on which it travels. These rails are generally restricted to the construction site and must be laid out so as to give the crane maximum capability, as far as its reach is concerned, within the actual construction project itself. The tower crane is the most limited of the cranes in that once it is in place, its reach is constrained to the length of the job boom and the arc covered by boom rotation. The gantry may be free-traveling, but in the largest sizes, it is usually rail-mounted. Supply deliveries should be arranged so that unloading and storage areas facilitate future employment of the crane without double handling of steel or other construction material.

Mobile Crane

Mobile cranes may be either crawler- or track-mounted. Where the structure is not overly high or where there is sufficient space surrounding the construction site, the crane may utilize its normal lifting boom with a jib attached for added overreach. This requires some offset, from the base of the structure, of the crane body as it lifts material over the ever-heightening roof edge. The crane has to back away from the structure so as to be able to obtain reach over the uppermost edge of the framework. Therefore, the lifting capacity becomes less as the boom is lowered from the near vertical. The horizontal, vertical, and weight limitations in this situation are apparent.

Boom support and the hoisting lines may be run from the operating drums over an A frame to the boom. This A frame provides added stability. Use of outriggers and the crane's internal counterweight may be insufficient to assure balance of the crane when raising heavy loads with the boom at an angle. Supplemental counterweight may be needed, such as attached, surface-mounted water-filled tanks connected to the boom by lines passing over the mast.

An adaptation of this conventional lifting boom is the tower, which may be mounted directly on the crawler- or truck-mounted frame (see Fig. 4.8). These towers allow the unit to work in close proximity to the actual construction site and still provide a vertical lift capacity. Atop this tower will be a horizontal jib. Rotation is achieved by turning the revolving superstructure of the crane itself.

Because of the necessity for extremely large counterweights and outriggers to provide additional stability to the unit, it is not feasible to fully utilize the mobile aspects of the crane when lifting vertically. Short moves may be possible, while longer moves require removal of the tower or boom and probably the counter-

FIG. 4.8 Mobile cranes with jib. *(Lorain Division, Loehring.)*

weights. However, this crane's versatility is largely due to its capacity for a relatively rapid setup and takedown. Because of this convenience and its adaptability, the mobile unit is generally more expensive than two other basic crane types (rail and tower).

Rail-Mounted Tower Crane

This unit is functionally similar to the mobile crane already discussed. However, it is confined to the traveling limits of the in-place trackage. Such a rail-mounted lifting device is shown in Fig. 4.9. For a long-term operation, this unit may be less expensive than the mobile crane. The rail-mounted tower crane does not have the capacity for conversion to other uses (clamshell, dragline, pile driver, etc.) which are available to the mobile crane. A rail-mounted regular crane used with railroad construction does have this capability, however, and may serve as a primary excavator or driver of piling. The mobilization and demobilization of this unit is far more extensive than that of the more mobile unit. Terrain must be prepared and rail must be in place before the crane can be brought in and used. Upon completion of the job and after the crane is removed, this trackage must, likewise, be removed or otherwise covered so as not to obstruct the area surrounding the construction or impair the final grade.

Thorough planning is essential to assure that the trackage is in the optimum location for maximum use of the tower. Stockage of steel and other supplies must be located within reach of the track-bound lifting device. Storage must be in

FIG. 4.9 Rail-mounted tower crane. *(Courtesy of Robert T. Ratay.)*

close proximity to the tower crane, or double handling of materials will result as it is brought in from a distant stockpile or shakeout area.

For use with large dams, long structures, or the like, it is possible to run the trackage along the longitudinal axis of the structure for maximum utilization. If the ground is level, then track placement is facilitated. However, on uneven terrain, it may be necessary to lay the tracks at differing elevations. This may cause the rail-mounted crane to have legs of differing lengths to compensate for this gradient differential.

Climbing Tower Crane

This crane consists of a fixed tower, a climbing tower, a swing ring, and a rotating jib boom (see Fig. 4.10). The horizontal jib boom protruding in one direction is offset and balanced by a heavy counterweight protruding from the opposite end.

These units are limited in their capability to the span reached by the rotating boom. The radius to which the load may be delivered can be controlled by a trolley traveling along the boom. The tower crane may be erected adjacent to or on the structure under construction. When placed separate from but in proximity to the structure, its capacity for work is limited by its maximum vertical

FIG. 4.10 Tower cranes at work on power plant. *(Liebherr Crane Corp.)*

expansion capability. When erected on the structure, it continues to offer service throughout the life of the project. This tower crane continues to expand upward as construction elevates the structure. Or, rather than be raised to full extension, the tower may be relocated upward on the structure itself. For faster lifting, the height to which the load must be raised should be minimized. Upon job completion, the tower is disassembled and lowered to the ground for reuse elsewhere.

While this type of tower crane is initially the most inexpensive, it is limited in the area which it is capable of servicing due to its fixed base and boom length. With large structures, it may be practical to use several of these tower cranes on a single project. A cost trade-off should be investigated to determine the most economical system of crane support for a job.

Gantry Crane

The gantry is a unit which straddles the load to be moved (Fig. 4.11). While it is self-propelled, it may be wheel-, rail-, or track-mounted. It will be propelled by an internal power source to the load to be lifted (structural steel, for example), straddle the load, secure and lift, then move the load to the location where it might be needed. In the case of the rail-mounted gantry, as with rail-mounted cranes, it is restricted in its travel by the location and run of the rails. In the more mobile case of the crawler- or wheel-mounted gantry, its limitations are governed strictly by the accessibility of the load and unload sites and the capacity for lifting. Vertical lift is provided by a hoist-and-cable system from the main beam or beams of the gantry. Longitudinal movement is provided by the mobil-

FIG. 4.11 Gantry with precast bridge segments.

ity of the gantry itself. Transverse movement is provided by the ability of the hoist to travel along the main beam or beams of the rig.

When the gantry supports a rotating crane, it is known as a "whirley" (Fig. 4.12). The gantry may be high for clearance, reach, and visibility by operator or have a low configuration for added stability. The rotation of the body of the crane, raising or lowering of the boom, lifting or lowering of the hook, and longitudinal travel of the gantry provide the desired mobility of the unit. The whirley affords the long boom length and high lift capacity of a derrick, yet supplies the mobility of a crane. Whirleys may have a lift capacity exceeding 3000 tons (2700 t).

In its simple load-straddling configuration, the gantry is probably the least expensive of the crane systems. Yet, when supporting a rotating crane, or whirley, the gantry may exceed all other crane types in cost.

Helicopters

Where it may be impractical or uneconomical to utilize a crane of any sort as a lifting device, it may prove necessary to take advantage of the versatility of a helicopter, or flying crane. They can reach points far higher than those attainable with the boom of a conventional crane or tower. Helicopters are also ideally suited for free egress into remote reaches. Otherwise an extensive road network or access system may be required to bring in a crane, other equipment, or supplies. The high costs of these units, coupled with their restrictions imposed by weather, limited lift capacity, and area clearances, require that helicopter employment be closely examined before utilization is attempted. This high cost of use may be most economical, however, when considering short-duration lift requirements in unreachable situations.

FIG. 4.12 "Whirley" in dam construction. *(Clyde Iron.)*

SUPPLEMENTAL ERECTION EQUIPMENT

While the basic units of equipment already discussed are an absolute requirement for structure erection, they have certain distinct limitations. Costs are high, their sizes create accessibility problems, they may not be suited for the specific job to be done, terrain conditions may preclude their use, or they may actually be unable to accomplish the task.

With these constraints in mind, the manager must look to other systems or items of equipment to supplement those already discussed: to the *conveyor* for long-distance haul within the construction site; to the *cableway* for access to the bottom of gorges or ravines; to the *pump* for moving concrete; or to the *skid, roller*, or *jack* for localized movement of heavy loads. One of these might well be the answer. Overall, the manager must seek the most cost-effective means of expeditiously accomplishing any construction requirement.

Conveyors

In construction, materials used or to be removed may be transported by many methods. When high capacities are required for job accomplishment, the belt conveyor comes into serious consideration. Of course, the materials to be moved must be of a type or configuration enabling them to be carried on a moving belt. Typical of the variety of items which might be moved are earth, ore, aggregate, sand, brick, block, cement, and concrete. Depicted in Fig. 4.13 is a conveyor system employing a hopper, 15 field conveyors, a stationary conveyor, and a radial stacker to excavate material from an area. Distances covered may be short, for local stockpiling or in-plant movement, or long for excavation, backfill, or aggregate movement in mass earth construction. The configuration of the belt conveyor depends on the materials to be moved and distances to be covered.

Basic to the conveyor is the rubber-covered, fiber-cored belt, or for added strength, a belt with a steel cable core. The rubber must be able to resist both the impact of the dropped load and the abrasive wear of materials carried, while the core provides the strength to withstand the tensile pull on the belt. An electric motor is very likely used in driving this continuous belt. At the conveyor section terminals are located the pulleys about which the conveyor belt travels. Intermediate support is provided by a series of idlers, troughed to contain the load. Spacing of these idlers depends on the loads to be carried and the width of the belt. Weights which can be carried may exceed 100 lb/ft^3 (1600 kg/m^3). Belt widths may reach up to 120 in (300 cm). Speeds may vary from 100 to 1000 ft/min (30 to 300 m/min). Supporting the entire system is a structural frame which is placed to reach from point of load to point of discharge. For very long hauls, there may be a series of these frames or flights. For short close-in use, there may be a gasoline-driven self-powered conveyor system. The power required for a smooth conveyor operation must overcome the friction of idlers, pulleys, etc., as well as carry or lift the load.

When the construction project is analyzed, the advantages and disadvantages of the belt conveyor must be considered. For long-term use, it offers a low operating cost and relatively lower investment cost, due to the high tonnages to be moved. It can be used across a wide variety of terrain features with reliability even though the belt itself is subject to damage. The operational staff required

FIG. 4.13 Conveyor system used to remove material being excavated. *(Kolman.)*

to manage the belt conveyor is less than that required with other transportation systems. The conveyor's most serious limitation is its inability to transport supplies, personnel, and equipment or to be diverted to other uses. Vertical lift with the belt conveyor is limited to a maximum of approximately 22°, or material slippage on the belt may occur. For temporary or semipermanent employment, it is ideal for transporting materials rapidly. The belt conveyor may be moved into position (self-propelled or hauled) to cover miles of unimproved terrain, connected, and placed in operation in far less time than that required by other haul systems. The frame may be run continuously for long distances, or portable frames may be aligned for long haul or used singly. Due to the nature of this framing, the conveyor system may be disassembled upon job completion for storage or reuse elsewhere.

Although the belt conveyor is limited in its vertical lift capability, it may be replaced or supplemented by a bucket or other lifting conveyor. Such a self-loading, self-dumping system is an integral part of certain pieces of construction equipment (loaders, trenchers). But it may also be employed for jobsite lifting of granular materials. It might be well suited for elevating brick or block for masonry construction.

Cableways

This system is basically a method of hoisting, conveying, and lowering materials from a carriage supported by a cable which runs between two end towers, one tower and a fixed support, or two fixed supports. These towers may be secured in place, or either or both may be movable. Where one tower moves radially about the other, adequate backstays on the immobile tower must be used to resist increased cable tension. When both towers move in a direction perpendicular to the direction of the conveying cable, they must move conjunctively. Otherwise, there will be undue tension of the connecting lines. The tower near which is located the operating machinery is known as the head tower, while the opposing end is the tail.

Mobility of the towers is desirable to allow the entire system (or one end) to be moved so as to increase the lateral area available for pickup and delivery of the load. The run of the cable carrying the suspended carriage or conveyance provides the longitudinal area coverage. Vertical movement from the movable carriage is provided through a fall-rope system with block and hook. Figure 4.14 identifies the parts here discussed. The three major increments of a cableway system may be identified as the track cable with supports, the carriage and

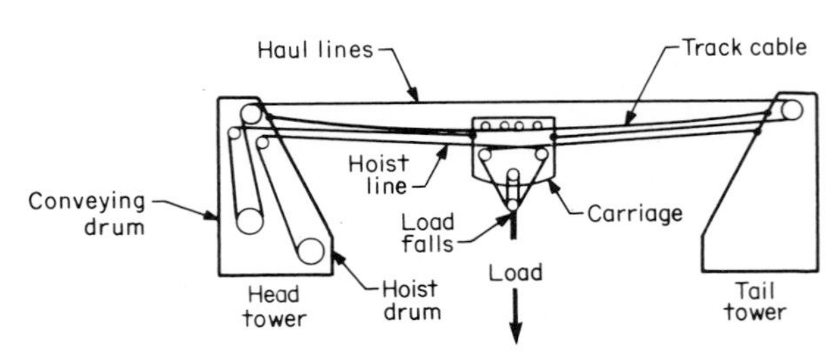

FIG. 4.14 Cableway.

appurtenances for handling the load, and the machinery for providing the necessary power.

The most common employment for the cableway is to assist in areas where the construction effort is located in deep valleys or canyons inaccessible by road or split by rivers. Another use may provide access to mountain peaks. While the cableway may reach over 3000 ft (900 m), the normal span will approximate 2000 ft (600 m). One type of work typical for such a locale is a large concrete dam. While the cableway can also be used for earth movement (removal or delivery), it is far more practical for delivery of concrete to the bottom of the gorge. It may also assist in delivery of supplies and construction forces. Once the construction project is completed, the cableway may remain as a part of the permanent equipment for access to unreachable areas.

The cableway will usually remain in place for the life of a long-duration concrete dam project. Even so, it is intended as a temporary structure needed and used only during the active life of construction. This situation is more frequent than the permanent utilization of the cableway for prolonged site access. Even if roads can be created to reach the bottom of a narrow canyon, this cableway system will be advantageous to supply concrete to the ever-heightening structure. It is also far faster in delivery than trucking concrete to the bottom for rehandling by crane for upward placement. Concrete bucket sizes range up to 8 yd^3 (6 m^3) normally, or up to 12 yd^3 (9 m^3) if the system is so designed.

Smooth-surfaced lock-coil track cable is commonly used to avoid excess cable wear. The greater the load, the higher the number of track wheels on the dolly to better distribute the carriage load to the main cable. The operating cables, as they perform the most work, must be observed closely for wear and maintained regularly.

Cost must be carefully investigated, as the initial expense for a cableway is high. Yet for large volumes of concrete or materials to be moved over long distances of inaccessible terrain, the cableway may be highly cost-effective.

Concrete Pumps

With high-rise structures, and even those inaccessible or with hard-to-reach areas, some system must be available to facilitate the placement of structural concrete. Time may preclude the use of workers and buggies. The concrete pump offers a ready solution. Concrete may be delivered by transit-mix trucks to a base pump. This unit receives in its hopper the green concrete and forces it, through a pumping action, along the delivery pipeline or large-diameter hose for placement at the desired location.

While such pumps are of a mobile configuration (truck, trailer, skid, or tower), a delivery pipeline requires pre-use installation. This pipeline will far exceed the normal radius of operation and reach of the integral boom. With a pump and pipeline, the material may be forced approximately 2000 ft (600 m) horizontally or lifted over 500 ft (150 m). Where it may be necessary to move the concrete over longer distances, it may be possible to station and employ an intermediate pump. Appropriate arrangements must be made in the mix of the concrete to prevent it from setting up in transit, however.

For short-duration use, the more portable setup is advisable, as the truck boom itself and flexible hose will provide a vertical reach of 120 ft (36 m) and a horizontal reach of over 100 ft (30 m). With long-term use (approximately that time required to construct a major structure), the tower-mounted boom may be recommended (Fig. 4.15). This unit not only has its boom for reach, but may

FIG. 4.15 Concrete pump, tower-mounted boom. *(Allentown Pneumatic Gun Company.)*

employ the preplaced pipeline for added distance between feed and discharge. A skid- or trailer-mounted hopper and pump provide the feed from transit-mix haul trucks.

Air Skids

The use of air in construction runs the broad gamut from the pneumatic jacks and tires to airbags for support and protection; from tool operations to transporting materials; from cooling and heating to air skids. While these other subjects are more properly considered elsewhere, the air skid requires some clarification here.

The basic principle behind this skid or caster (Fig. 4.16) is the elevation of the skid or skids with the load so that the overall weight rests on a thin film of air. Once elevated, the load is movable in any direction with minimal effort, against negligible friction, which facilitates precise positioning. For use of these

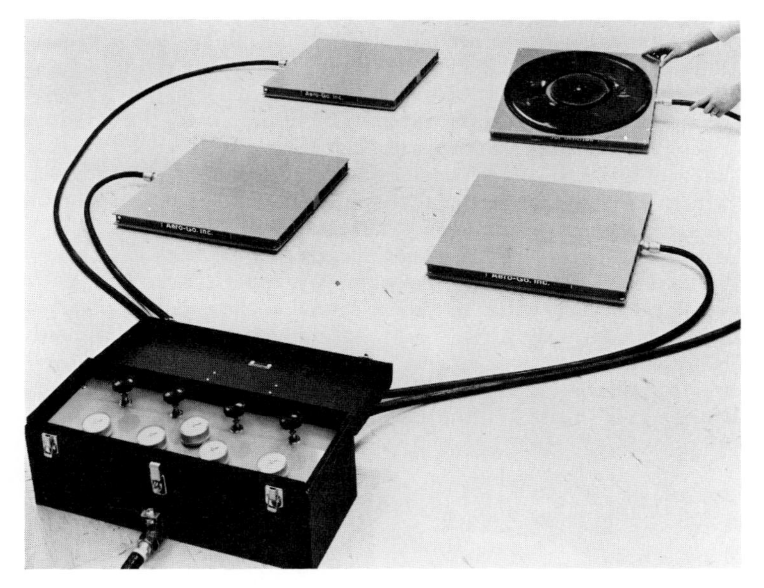

FIG. 4.16 Caster principle, air movement. *(Aero-Go, Inc., Seattle, Wash.)*

air casters, the load is elevated by a system of leveling screws, jacks, or lifting eyes, or it may be pre-positioned to allow insertion beneath it of the casters. The skids are then inflated using sufficient pressure depending on the load to be lifted and the number of casters used. When the air pressure exceeds the weight of the load, the air will seep beneath the air bag to elevate the system. The load is then moved to position where some type of lifting device is used to suspend it, permitting removal of the casters for further use.

The requirements for such a system might include an overhead compressed air drop, or other access to compressed air, and a nonporous floor. A floor covering which assures even air escape is satisfactory. Where heavy loads are to be carried, several casters may be combined and formed to support a pallet which provides a firm load-spreading base for the object to be transported.

These air skids may be employed in a temporary vein for movement of heavy equipment (transformers), semipermanent installation (ship movement in dry docks or shipyards), or may be permanently fixed (support and move stands at Denver's Mile High Stadium).

Rollers

Rollers moved mammoth rock for construction of the Pyramids and Stonehenge; current construction practices still use them. Rollers offer advantages over other transporting systems due to their low profile and a reduced requirement for motivation. While they may be employed individually, they generally are used collectively. In some cases, they may satisfy an end in themselves; in other cases, they may be a component in a major system (troughing rollers in a conveyor setup). They can be readily added to an existing system to increase load-carrying capacity. Because of the load's relatively unaltered center of gravity when raised on rollers, control is easily maintained and the tipping potential is reduced. The roller's low physical profile minimizes the height to which the load must be raised for insertion of the carrying system. Because of the load-spreading aspects of rollers, which can come in almost any length, they are able to traverse varying terrain configurations with minimal site preparation.

Rollers come in two broad classifications: those with axles and those without axles. The axle may be fixed, around which the roller rotates (on bearings), or the axle itself may rotate with the roller (secured at the ends in a race-type bearing). When wide rollers are used to carry a rigid load, mobility is somewhat limited. It is difficult to turn or redirect the load. If the rollers are employed with a dolly configuration for added flexibility, the height-saving advantage is reduced.

Without an axle (Fig. 4.17), the roller might be of the endless-chain configuration. This track-laying principle eliminates or reduces the torque encountered with use of the axle. Such an endless-chain system drastically lessens rolling resistance and maintenance of the bearings or other wearing parts common to the axle roller.

FIG. 4.17 Rollers used to move heavy equipment. *(Hilman Equipment Co., Inc.)*

Roller systems may be employed in the temporary needs of construction (such as movement into position of steel members or large generators) or incorporated into the permanent structure itself (facilitating movement of the stands at the Superdome sports complex).

Jacks

While one realizes the importance of the jack in construction use, seldom is its full scope envisioned. These handy, compact devices can be used for raising or lowering heavy loads over short distances. They can be used to exert outward pressure against opposing restraints, such as with trench jacks. Jacks are instrumental in lifting slabs into place in "lift-slab" construction. The movement of the vertical slipform may well be dependent on the hydraulic jack. Being small and easily transported, they can be employed in very restricted areas. Their overall use may be relegated to the needs of construction, or they may become a permanent part of a structure or facility.

The jack may have a single plunger or piston to move the load (Fig. 4.18) or a series of sequentially exercised nested rams, each with a limited stroke. It may have a screw-thread plunger activated by bevel gears which turn the plunger through a stationary collar. With the rack-and-lever jack, pawls engage the rack on the plunger to secure the position of the bearing plate. (See Fig. 4.19 which reflects both the ratchet—or rack and lever—and screw jack.) In the case of the hydraulic jack, the activating lever, through a piston, forces the fluid from a small reservoir into one with greater capacity which, in turn, activates a larger piston. With lift-slab or slipform jacking, the activating pump may be separate from the jack itself. Flexible lines will connect to the jacks, raising the load upward on lift rods. Lifting collars provide anchors to secure the load and prevent slippage. Care must be taken to lift evenly to preclude binding.

With long-term use under continued pressure, the hydraulic jack may leak slightly, allowing a lowering of the load. Screw-type or rack-and-lever mechanical

FIG. 4.18 Hydraulic jack to raise heavy loads. *(Hilman Equipment Co., Inc.)*

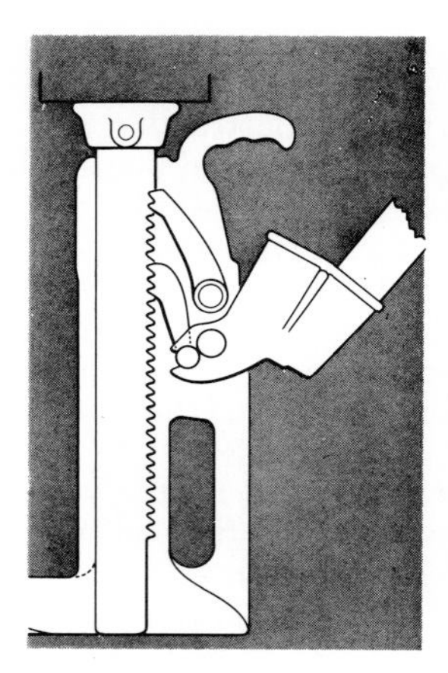

1. Ratchet

Simplest construction. Employs the basic lever and fulcrum principle. Downward stroke of the lever raises rack bar one notch at a time and the pawls spring into position, automatically holding the load and releasing the lever for the next lifting stroke. Use should be limited to lighter loads, usually to 20 tons, because of physical effort required. Advantages are less expensive construction, fast operation.

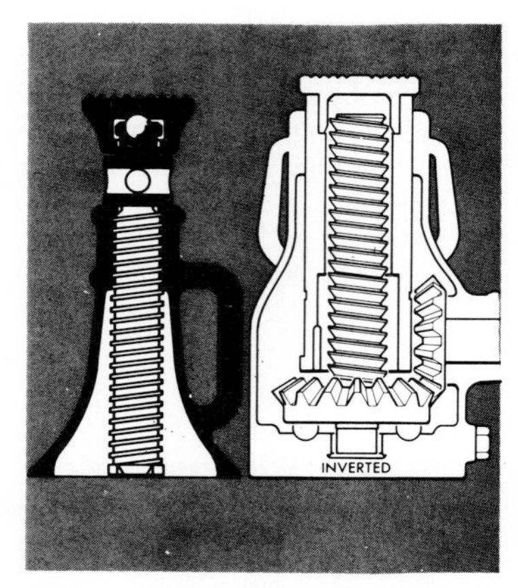

2. Screw

For lifting heavier loads, the screw and nut principle is used. Two general types—regular and inverted. For lighter loads a simple lever bar will apply sufficient power to turn screw but as loads increase, gear reductions and ratchet devices are used to multiply the operator's strength. In the heaviest jacks the screw is operated by an air motor for faster lifting and lowering without effort.

FIG. 4.19 Ratchet and screw jacks. *(Duff-Norton Co.)*

devices better serve the long requirements. Hydraulic jacks, however, are usually simpler to operate than the others and require less effort.

Jacks may be individually employed or joined to operate collectively. They may also be rigged onto wheels to increase their mobility. This mobility requires a smooth, hard surface on which to operate.

Overall individual lifting capacity may well range over 400 tons (360 t). Stroke movement, usually a function of the length (or height) of the jack frame, may be anywhere from ¼ in (6 mm) up to over 3 ft (1 m). As cited earlier, the nested ram or multiple screw will increase the stroke in relationship to jack height.

ERECTION ACCESSORIES

With all of the equipment types already discussed, it is necessary to have some means of securing, handling, connecting, and rigging the parts to be lifted to the piece of engineering equipment. Items to be included within this segment are: wire rope, slings, connectors, and the tools utilized with these erection materials.

Wire Rope

In common use with almost all of the equipment discussed in this chapter is wire rope, or cable. This wire rope fills the need of load line, falls, guys, topping lift, haul line, and even with the load securing ties or slings. With light loads, manila hemp may be employed similarly. Although cables may be manufactured by wires wound into strands which are, in turn, wound around a core, the most effective wire rope is that pre-formed or set into strands which are set in the rope. This restricts wickering of the wires themselves and improves serviceability.

Where the wire rope encounters rough usage due to wear and abrasion, the 6 × 7 class is most frequently recommended (6 strands of wire with 7 wires per strand). For improved flexibility, the 6 × 19 class is used. Although its resistance to abrasion is not equal to the two classes already mentioned, the 6 × 37 class wire rope provides more flexibility. Strength varies according to the constitution of the rope. In the 6 × 19 class, a 1-in (25-mm) wire rope might have a breaking strength of 46 tons (40 t) with a fiber core, yet may reach 52 tons (46 t) with a wire or metallic core. A similar size rope of the 6 × 31 class might have a breaking strength of 44 to 50 tons (39 to 45 t), depending on the respective core composition.

When using the wire rope, the equipment operator must be careful not to allow or cause kinks in the cable. This is particularly true when removing the rope from a spool, when it should be unrolled rather than pulled from the spool.

Load spreaders or softeners help provide protection by avoiding extremely sharp bends in the cable. Eyes may be spliced into the cable to provide a passage for rope or other connector. Cable ends may be secured by looping and splicing into the cable itself. The cable ends should be seized or bound to prevent unraveling. Two pieces of wire rope may be joined by the use of cable clamps or wire rope clips. Where the rope is looped, it should be protected with the insertion of a thimble to prevent kinking or undue wear. Where extra strength is needed, the fittings may be secured by swagging them in place.

Type (right or left hand, regular or long lay, locked coil or strand), class (6 × 19), and size [¼ in (6 mm) up to 4 in (10 cm)] depend on the job to be performed. The rope is generally purchased in reels and cut to the length desired. Care should be taken to replace as wear, wickering, or fatigue become apparent.

Material-Handling Tools

With the wire rope rigged for lifting, there must be some system for attaching the line to the load. Frequently this attachment is predicated on a hook affixed to the end of the line. This hook is usually provided with sufficient weight (by blocks or ball) to establish and maintain tension of the line. The hook may be paired at the ends of rods or lines (pointed hook sling) so that when attached to the load at opposite ends, the resultant horizontal compressive forces prevent load slippage. Or, there may be a specially configured J hook with one long horizontal tine for lifting pipe sections. Two hooks may also be formed with a pivot in the middle, as with ice tongs, to grip the load. But the single hook, with safety shackle, is the most prevalent single load-carrying device.

To this lifting hook may be attached a sling or slings. The single sling, where there is no chance for load shift, is usually choked (or wrapped around the load), while several slings may be spread and choked to form a bridle hitch arrangement. This multiple sling is used to resist load shift and maintain a balanced load. Capacity of the bridle hitch is dependent on the separation of the hitches or the angle formed at the hook. Ideally the sling arrangement or hookup should approximate an equilateral triangle for optimum lifting capacity.

Other tools relating to material handling are those particularly associated with steel erection. For obtaining relatively permanent position of the members, it may be necessary to employ a coffing hoist or ratchet jack with draw chain and hook and a connecting bar.

EARTHWORK EQUIPMENT

Employment of construction equipment is generally envisioned as a part of the overall effort to build the primary structure. Frequently this equipment is utilized in the creation or erection of the temporary structures necessary to the overall project. These temporary needs may vary extensively in type and scope. Among these typical nonpermanent requirements, or phases of work, might be an earthen fill to surcharge a foundation area; drainage trenches or pipes and pumps to dewater an excavation; caissons to hold back surrounding water; pilot bores to investigate underground conditions; bypass structures for roads and railroads to keep the traffic moving; plant erection to supply aggregates, concrete, or asphalt; or the simple field office or shop from which to originate the operation and support the constructive effort.

Because of these considerations, it is appropriate to broadly address certain types of equipment commonly used in the construction industry. These units will not be presented as regards any particular use or structure, but in a wide sense of employment. Thus, an understanding of the equipment itself will help open the door to the great variety of potential employment opportunities.

TRACTORS

The single item most frequently used in earth movement is the off-highway tractor with its many attachments and allied supporting equipment. A mechanical piece of equipment, the tractor is designed to convert engine power into drawbar or rim pull (traction) to move itself, other equipment, or an acquired load.

Both crawler and wheeled tractors are available with either the direct-drive or power-shift transmission. The direct-drive unit will have greater drawbar pull than the power-shift unit; however, the speed range for the power-shift transmission will be greater than that of the direct-drive.

Each tractor may be equipped with cable, hydraulic, or electric controls. Cables were employed initially for control, but subsequently, with hydraulic developments, hydraulic controls have predominated. Currently, cables are used only where the towed equipment so requires. Hydraulic and electrical controls have a distinct advantage in that, for example, they may exert a positive downward pressure and a more precise blade setting, whereas cable controls allow a floating blade. Hydraulic and electrical controls also afford added versatility in that they can perform several operations simultaneously, such as providing a down pressure along with simultaneous blade tilt.

Crawler Tractors

The tracked or crawler tractor (Fig. 4.20) is a basic item of equipment in itself, but it may be adapted to satisfy innumerable other requirements by the attachment of assorted blades, hoes, shovels, booms, hoists, and rippers, or by supple-

FIG. 4.20 Crawler tractor, D-10, with bulldozer. *(Caterpillar Tractor Co.)*

menting it with towed equipment. This towed equipment might include such items as scrapers, loaders, trailers, tillers, scarifiers, and compacting devices. The crawler tractor develops pull, in pounds, at the drawbar. Pull, for this type of machine, is great with respect to the machine's weight and size; however, almost irrespective of the pull available, the crawler's maximum forward speed will be limited to around 7 mi/h (11 km/h). Speed is sacrificed with this equipment in order to obtain its large drawbar pull and excellent traction. With its weight transmitted to the driving tracks, the compact crawler tractor is able to go almost anywhere, across any terrain, regardless of the weather or soil condition. Even with all its versatility, the crawler has a major drawback in its slow speed. Another disadvantage is in the requirement for a transporting unit to move the tracked equipment, due to the likelihood of the crawler's grousers tearing up finished road surfaces; hence, tracked units are limited to off-highway use.

Wheeled Tractors

Whether the two- or four-wheeled model, the wheeled tractor provides the speed not truly available to the crawler (see Fig. 4.21). Because of the desire for speed and mobility, most wheeled tractors have power-shift transmissions. It is this speed and mobility that makes the wheeled tractor a perfect complement to the

FIG. 4.21 Two-wheel tractor with elevating scraper. *(Deere & Company.)*

crawler's power and traction. The wheeled tractor is also used as a mount for certain attachments and, like the crawler, it is used as a prime mover for towed equipment. The speed of these wheeled units may result in loss of tractive power. The tractor may end up with slipping wheels before reaching its full pull capacity.

The two-wheeled unit is limited as a puller of scrapers or other haul units because of its inability to act independently. It must be connected to other wheeled units in order to maintain its balance. With two wheels serving both drive and steering requirements, all the weight of the prime mover is transmitted to these drive wheels. This produces better traction, mobility, and decreased rolling resistance than with the four-wheel unit.

Versatility of the two-wheeled rig is less than that of the four-wheel tractor, and the contractor who is looking for a minimum equipment requirement to handle a multitude of jobs usually looks at other than the two-wheel prime mover.

One of the most common uses of the two-wheel tractor is in the role of motivator for the self-propelled scraper. This self-propelled unit's tractive effort is usually assisted by a crawler pusher; however, there are numerous models that have power trains providing drive to the rear or the bowl-support wheels. This rear-wheel drive assistance also may be similarly utilized with the four-wheel tractor.

Functional capability is broader in the four-wheel tractor than in the two-wheel model. The four-wheel unit can operate independently with improved stability under its own motive power. Due to lesser bounce and, consequently, increased driver confidence, it is a better traveler. Its four wheels give it better speed potential than other tractors. But, due to its weight distribution over four rather than two wheels or tracks, this vehicle has less traction. Costs are generally greater than that of the comparable two-wheel tractor.

Speeds of two- or four-wheeled tractors are as much as 4 times that of crawler tractors. Therefore, for a long haul, the wheeled units will reduce costs due to time saved at travel speeds of up to 30 mi/h (48 km/h) versus the 7 mi/h (11 km/h) of the crawler. As stated above, the loading traction lost to provide the overall speed may be supplemented by use of pusher units. Thus, crawler and wheeled tractors may complement each other.

When considering which type of machine to get, the two main factors to be weighed are speed versus pull. To have rim pull equal to drawbar pull, the wheeled tractor must be considerably heavier than the crawler because its coefficient of traction is less. Also, due to this needed heavier weight, the wheeled tractor must have added horsepower to ensure that the horsepower-to-weight ratio is acceptable. All of this is summed up by the fact that for equal pull, the wheeled unit is more expensive. Its speed and production must therefore offset the higher owning and operating costs. With sufficient funds, a contractor can purchase or rent both, or either, wheeled and tracked machines to handle any job. However, many contractors are not that fortunate. They are looking for versatility or a particular type of equipment to perform satisfactorily on a wide variety of jobs. This versatility then supplants in importance the speed-versus-pull considerations.

Employment

The blade is probably the tractor's most important accessory. This becomes most apparent when power requirements are great and the distances are short. A blade ranges in approximate size from 6 ft (1.8 m) to 24 ft (7.3 m) in length and from 2 ft (60 cm) to 6 ft (180 cm) in height. The term "bulldozer" is often used in a broad sense to include many attachment modifications such as the angle blade, brush rake, rock rake, and stump rooter. Each of these blades may be used on either wheeled or crawler tractors rigged with either cable or hydraulic controls. Under positive control and adjustment, these blades may be raised, lowered, angled, tilted, or canted. One corner of an adjustable blade may be lowered as much as 12 in (30 cm) or more and provides an excellent tool for rock excavation and removal. Such a blade inclination is recommended for penetration into hardpack or frozen earth. It can also be used for the initiation of a ditching action. The blade may be tipped forward or backward as much as 10°. This is advantageous under certain conditions in controlling the material boiling up before the blade as the dozer moves the material forward. Other basic blade types are the angling (A) and the cushion (C) blades. These are used, respectively, for sidecasting operations and as push cups for loading assistance. A typical operation for the bull blade is general excavation. This should be limited to pushes under 300 ft (90 m). For anything over that distance, it would be more economical to change to a self-propelled scraper or some other type of excavation and haul unit.

A first step in most construction projects is *clearing* the site. Here the dozer blade is employed to remove the surface vegetation (including brush and trees). In *grubbing,* the root system required for plant growth is removed. *Stripping* is the next step in that the top soil, or vegetation-sustaining earth, is removed. Not only do grubbing and stripping preclude further organic growth, but the unpredictable engineering properties of this organic top soil are removed. The existing subsurface soil may then be consolidated, or additional fill may be brought in as necessary for the job.

When scrapers have difficulty loading under their own power, they will require the assistance of a pushing unit. Through the use of the pusher providing additional tractive effort, production is improved in obtaining both larger loads and a shorter loading time. For best operation the *pusher,* equipped with a cushion blade or a push cup, must make smooth contact with the scraper's push block while matching the scraper's speed. Then the pusher provides added momentum to send the loaded equipment on its way. A ratio of one push dozer to four scrapers is not uncommon. An exact selection depends on variables of the job itself.

Ripping involves adjustable teeth, used to rend the earth's surface, mounted on the equipment (dozer). The tractor, or other equipment weight, assists in the penetration of the single or multiple ripper teeth. These teeth consist of a shank covered in the front by a protective shield and at the end by a replaceable tip.

The tractor has, in addition to the aforementioned mounted tools, a further capacity for a wide range of rear- and side-mounted equipment. To the rear may be mounted the hoe for excavating; the tiller, screen, or harrow for earthwork finishing; the cherry picker or boom for lifting. To the side may be mounted trenching, pipe-laying, or cable-burying appendages. Employment opportunities are as broad as one's imagination.

Besides acting as a base for these mounted tools, a tractor has the desirable capability of serving as a *prime mover* for towed equipment. This equipment is limited only by the tractor's capacity for pull and control. These towed items may range from earth haul, through compaction and water transportation, to trailer movement. This added capacity for service as a prime mover greatly enhances the versatility of the tractor.

FRONT LOADERS

Basically, the loader has either a crawler- or wheel-type tractor frame. Stability must be increased to prevent tipping or rollover due to increased lifting load and raised center of gravity. The center of balance may be changed through the use of longer equipment frames. Wheels or tracks may be wider spread than usual to provide the equivalent of outrigger support. With the wheeled front loader, the engine is usually rear-mounted to provide it protection and have it serve as needed counterweight.

As with the tractor, the front loader may be either crawler- or wheel-mounted, as shown in Fig. 4.22. Front end, rear, and side attachments are almost limitless. The variety of uses depends on the appendages such as the basic bucket, push and drag blades, grapples or clamps, forklift, backhoe, boom (or lifting device), mower, rake, and the pipe layer or side boom. With working implements which drastically raise the loader's center of gravity (boom, hoe, bucket, etc.), the tractor should be equipped with outriggers for improved stability.

Bucket sizes range up to 5 yd^3 (3.8 m^3) for the crawler and over 20 yd^3 (15 m^3) for the wheeled loader. As with tractors, wheeled loaders have considerably more speed and mobility but less available traction than crawler units. They may be articulated for a shorter turning radius and improved mobility. Employment of the loader bucket itself is normally as a front-dump unit. However, in an effort to preclude time lost due to turnings, the rear- and side-dump capabilities were developed. These buckets are generally limited in use to specific requirements such as mine or tunnel work. Here, the rear dump is ideally suited as it can dig, move the load overhead, and dump to the rear into muck buckets or haul units, all within the constraints of the narrow tunnel width.

Loading operations are dependent on the position of the equipment (truck,

FIG. 4.22 Articulated front loader loading trucks. *(Caterpillar Tractor Co.)*

grizzly, hopper, etc.) being loaded. The shortest possible travel time from loading operation to dump and return is desirable to reduce cycle time and improve overall production. Improved results can be obtained by using a spotter to ensure that trucks, wagons, or trailers are properly positioned so as not to delay the loader.

In normal operations, the lowered bucket (with or without attached teeth) is forced into the bank or stockpile. The bucket is rotated upward, breaking loose the material being worked. As it is filled, the bucket is leveled and lifted; meanwhile the loader is backed and turned away from the charging area. It then moves forward to the spotted vehicle for dumping. The bucket is rotated forward, permitting the material to spill forward, or in some cases, the clam-type bucket is opened to release the load. The entire bed of the transporter should be loaded through alternately dumping from the side, in the front, then the rear of the bed. The load should be spread and heaped as necessary to provide full equipment utilization.

Four broad, basic categories of work handled by the bucket loaders are: excavation, loading of materials, lifting and transporting, and general earthwork. Employment of any of the many attachments will usually fit within one of those categories.

MOBILE CRANE SHOVELS

Shovels are composed of three basic components: (1) the revolving superstructure, (2) the propelling mechanisms, and (3) the assortment of attachments. The revolving superstructure, typical with most crane shovels, consists of an engine, controls, and cab—all mounted on a platform. The platform is ring-mounted, permitting 360° rotation without interfering with the chassis or operation of the attachments. The engine is mounted to the rear from the attachments so as to balance the combined weight of the attachments and the load. Fixed to the platform opposite the attachment is a counterweight which helps to keep the machine in balance. Counterweights may be added or removed, but are usually joined and left in place. They may weigh in excess of 100,000 lb (45,000 kg) and may be ballastable. The propelling mechanism separates the crane into its three classes of truck, self-propelled, and crawler equipment.

The crane shovel as a lifting device has been addressed already in this chapter. Although the crawler, because of its better stability, probably serves more successfully as a tower base for high lift requirements than does the wheel unit, both propelling mechanisms are employed with the long boom or tower configuration. Further, the long boom, or stick, is needed for a number of additional purposes beyond the high lift for erection of temporary structures.

Wheeled Crane Shovel

The wheeled crane shovel, self-propelled or with truck frame, is rubber-tire-mounted and provides better speed and overall mobility than the crawler. The self-propelled unit is driven by the same engine that operates the boom, shovel, or other attachments. On the other hand, the truck-mounted machine has the same rotating superstructure as does the self-propelled unit, but is, in addition, mounted and propelled on a truck frame (see Fig. 4.23). As the unit is road-mobile, it is limited in its wheel width. To keep this limited width and resultant potential stability problem from constraining the lift capability, wheel cranes are equipped with load-spreading outriggers. These beams may be extended from

FIG. 4.23 Wheel crane erecting precast concrete panels. *(Lorain Division, Koehring.)*

the body and supported by a float and leveling jack to better distribute the load and assist in maintaining stability.

Crawler Crane Shovel

The crawler (Fig. 4.24) is driven by the same engine that powers the attachments. Here speed is not required, nor is it desirable to have the excavator propel itself long distances. The long, widespread base of the heavy crawler tracks, with their

FIG. 4.24 Crawler crane placing bridge girders. *(Crane and Excavator Group, Koehring.)*

excellent load distribution, provides improved stability over the other mobile cranes. These tracks also afford rough terrain maneuverability through their low ground pressure. Traction and flotation are better with the crawler than with the wheel unit, but a transporting rig (low boy) is needed for movement between job locations.

Attachments

Thus far, the crane shovel has been addressed completely in the long boom or tower mode. These are the crane configurations common to the erection of permanent and temporary structures. With the boom, the vertical conformation satisfies the maximum lift. Once there is a lowering of the boom, the lift capacity of the unit is reduced. The shorter the boom, the less the reduction of capacity with the increase in angle. Therefore, for heavy loads, a short hammerhead boom at near vertical inclination will best do the job. Although there are a number of attachments usable with the crane, the overall most efficient employment is in the lifting configuration. With this setup, the rig can devote its capabilities to the load itself and not have to overcome the dead weight of the attachment. Boom height and length may be adjusted by adding or deleting sections. For added height or overreach, a jib may be added to the boom top. Wire rope, or cable, must be adjusted to satisfy boom employment.

A crane serves as base for a greater variety of uses than just the lifting capability of the boom. While retaining the boom, although not in the lengths needed for the high lift of building erection, the dragline and clamshell may be employed for excavating or earth-moving operations.

The dragline (Fig. 4.25) operates as an open-ended bucket which is cast out and drawn back while its teeth dig into the material to be excavated. Once filled, the bucket is raised and moved to the dump zone for truck loading or stockpiling.

Employment of the clamshell depends on the location of the boom tip. The clamshell is operated vertically in that it is spotted over the area to be worked. It is then lowered with the clam jaws open at the bottom. Equipment weight will cause a penetration of the bucket jaws or teeth. Closure of the jaws (with or without digging teeth) will entrap and encase the material to be moved. The shell is then raised and the crane rotated to the dump zone.

FIG. 4.25 Dragline. *(Northwest Engineering Company.)*

Numerous other attachments depend on the boom also. Included among these are the pile driver; the various grabbing devices such as the grab, clamp, tongs, and grapples; the bulk material tools such as the concrete bucket, skip, and platform; and digging tools such as the orange peel bucket. All these devices are suspended from the boom tip by a rigging of wire rope (already discussed).

Two other attachments common to crane shovel uses are the shovel front and the backhoe. Both of these require the removal of the boom and the attachment of a special front piece. A shovel (Fig. 4.26) main boom is attached at the front

FIG. 4.26 5-yd^3 shovel loading truck. *(Northwest Engineering Company.)*

of the rotating superstructure; through it, and pivoting near the boom center, passes a secondary boom known as the dipper stick. At the forward end of this stick is the dipper bucket. In shovel front operation, the bucket with teeth (dipper) is hoisted into position and crowded or thrust forward into the mass being worked. It is loaded as it is raised while digging outward and upward. When the bucket is filled, it is retracted and the superstructure is rotated to the point of load discharge. Here the bottom of the bucket is triggered open to release the load.

The backhoe attachment (Fig. 4.27) is also secured to the front of the revolving superstructure. In operation, the bucket is extended forward with its loading opening pointing downward. The bucket is lowered below track level and pulled in toward the main body while the teeth dig into the face. Thus it loads itself through this upward and inward pull on the bucket. Then the bucket is rotated upward to retain the load while the unit is turned toward the discharge point. Here the bucket is extended to the dump point and pivoted downward to drop the load.

FIG. 4.27 Backhoes at work in mass excavation. *(Northwest Engineering Company.)*

As the hoe is digging toward itself, it must be constantly on the move, backing away from the point of work. On the other hand, the shovel front moves forward as the work continues.

Hydraulics

While here listed as a separate entity from the crane shovels, hydraulic units serve the same general purposes. In lieu of gears, chain drives, and wire rope, these units employ hydraulics (cylinders, valves, and pumps). Problems of leaking or ruptured hydraulic lines replace the wickering or breaking of the wire rope. Pump deterioration replaces clutch and brake wear. Attachments for the basic hydraulic unit include the telescoping lifting boom with its many subattachments, the shovel front, the backhoe, and the telescoping grading hoe. Advantages in using the hydraulically actuated attachment lie in the increased maneuverability of the bucket itself due to its available wrist action and the ability to rapidly expand boom length. Hydraulic systems also may be used to provide an independent hydraulic drive for each track of a crawler and to power the placement of the outriggers of the wheel unit. An important advantage in hydraulic units is the capability of asserting positive control over and pressure on the attachment at all times.

EARTHMOVERS

There is hardly a construction project which does not require the movement of certain quantities of earth. This movement might be as insignificant as the leveling of an area for an athletic field or as impressive as the placement of hundreds of thousands of cubic yards (cubic meters) for an earthen fill dam.

A number of systems are available for this earth movement, starting with the short-distance dozing capability of the tractor with blade. For slightly longer distances and larger quantities, the self-loading scraper may be employed. As distances get longer and quantities even larger, the wagon or trailer may be utilized. At the far end of the spectrum, for most rapid roadability, is the truck. Selection of the system to be used depends on a number of variables. These include quantities of material to be moved, distances they are to be transported, uses to which the material will be put, and availability of equipment to do the work.

Scrapers

A scraper as a single item of equipment is capable of self-loading and hauling operations (see Fig. 4.28). It also can deposit its load in uniform thickness or layers. When returning for a subsequent load, scrapers can use their cutting

FIG. 4.28 Tractor scraper with push dozer. *(Terex Division, General Motors Corp.)*

blade to remove high spots in both the deposit area and the haul roads. But they should not jeopardize their turnaround times. Capacities of these machines may run over 50 yd^3 (38 m^3). This capacity may be specified as *struck,* leveled with the top of the bowl, or *heaped,* above the bowl level. When loading to a heaped capacity, one must take care not to exceed the bowl load limit, resulting in damage to the equipment.

The scooplike bowl hangs from a frame supported by rubber tires at the rear and by the prime mover at the front. This bowl can be rotated vertically approximately 20° to lower the forward cutting edge to penetrate the soil or to raise it for travel. The load may be discharged in one pile or spread through the forward motion of the scraper. The thickness of the spread depends on how high the containing apron is raised, how fast the material is expelled by the ejector, how rapidly the scraper is traveling, and how high the bowl is from the earth.

Either the crawler or rubber-tired tractor may power the scraper. In the case of the rubber-tired scraper, units are available with rear-mounted engines to drive the rear wheels. This two-engine scraper has improved traction and speed over the single-engine configuration. Necessary traction, travel speed, and distance to be covered are factors to be considered in tractor selection. If the decision is made to go with the rubber-tired unit, then determination is needed as to whether to use the two-wheel or the four-wheel tractor, the single or twin engine, and the multiple pusher or a push-pull configuration (two units mutually assisting).

Although the crawler-towed scraper is the more capable of self-loading, even it may occasionally need a pusher. With wheeled prime moving units, the pusher becomes almost mandatory.

The *elevating scraper,* a variation of the self-loader, loads well in most materials, but is not usable in hard earth or with large rock present. The material being loaded is cut by the cutting edge of the bowl, is then lifted upward by the elevating flights, and is dumped back into the bowl. There is no need for a pusher; the unit can work alone. As the elevating mechanism is dead, unproductive weight when the elevator is idle during haul or on return, the unit is best suited for short hauls.

Wagons

The rubber-tired hauling unit, or wagon, is the heavy-duty unit to satisfy the intermediate-to-long range, large-quantity haul requirement (see Fig. 4.29). It consists of a tractor prime mover and a semitrailer hauler unit. Depending on the type of trailing wagon, the tractor can be of either the two- or four-wheeled configuration. As these prime movers are capable of motivating either wagon or scraper, they improve the versatility of any equipment spread. Capacity of the wagon may range to over 100 tons (90 t). The wagon is able to carry a relatively greater payload than the scraper in that none of the wagon's capacity is taken up by the weight of loading and ejecting equipment. However, the wagon is not a self-loading unit; it must be filled by an outside source.

The type of wagon can be identified and characterized by the method used to dump the load. The three wagon types are rear-dump, bottom-dump, and side-dump. While the bottom- and side-dump haul units are moved by the four-wheel tractor, the rear-dump is propelled by the overhung-type of engine, the two-wheel tractor. Rear-dump wagons are best suited for rock hauling. Their beds are reinforced. The rear dump never has to pass over or through any of its dumped load. It is very maneuverable; the tractor can turn 90° without even moving forward. Bottom-dump wagons are best with free-flowing loads which they dump in long, narrow windrows. As dumping may be done on the run, these

FIG. 4.29 Tractor with bottom-dump wagon. *(Challenge-Cook Bros.)*

wagons have a better cycle time than the rear-dump ones. When dumping into drive-over hoppers, the machine must stop to discharge. Side-dump wagons, however, are better able than the other types to dump at high speeds. They are particularly suited to dumping over a hillside. With wider beds than the other style wagons, these units are easier to load. The two types of side dumps are the *bathtub* unit, which has fixed, outlaying sides over which it dumps, and the *opening-gate* type which has a downfolding side to chute the material away from the bed.

HAUL UNITS

Long-distance hauling employs both trucks and trailers used on the local road network, in the field, or in both situations. This haulage is accomplished at relatively high speeds (compared to scrapers or wagons) with large capacities, to result in low hauling costs. Cross-country movement is restrained for the highway unit, although it has unlimited mobility on the road network. Those units of excess size and weight (when loaded) are prohibited from highway travel and are employed for off-highway use.

Four main components of a truck are the chassis, power train, cab, and body. The frame and other chassis members, generally similar for all vehicle classes, are stronger and more durable for cross-country or off-highway types of equipment. Diesel power serves the heavier, larger vehicles, while the lighter trucks employ the gasoline engine. Cabs serve the same overall purpose in all vehicles. They provide a protected consolidated control and operational area for the driver. The body of the truck serves the haul capacity (see Fig. 4.30). Protection of the cab and occupants from the dangers of overhead loading may be provided

FIG. 4.30 Off-highway, rock bed, haul unit. *(WABCO Construction and Mining Equipment Group.)*

by an upper bed extension known as a cab guard. The bed is formed of reinforced sheet metal. Corner posts and rigid side bracing resist the outward thrust of the encompassed load. Tailgates may be provided to secure the cargo, yet may open for dumping and spreading the load. To withstand impact loading of large material, the heavy-duty rock bed was developed. These vehicles, because of their design to carry heavy, bulky cargoes, may not be equipped with tailgates but rather have an upward sloping chute.

Haul Bed Configurations

The varied types of haul beds include rear-dump, bottom-dump, side-dump, and lift-off box. These beds may be valid both for trucks proper and for the towed trailer.

Rear-dump units are boxes which, when dumping, pivot vertically near the rear of the frame. Extension devices, such as hydraulic rams mounted on the frame beneath the truck's bed, raise the forward end of the bed. The tailgate can be adjusted to control the discharge of the dumped material. These units may haul almost everything, including large rock, ores, earth, sticky materials, or free-flowing loads.

In areas where a broadside discharge is desired, the *side-dump units* may be most appropriate. These units operate similarly to the side-dump wagons already discussed. Rams fixed to the frame below the bed will lift the box from the side, tilting it to dump from the opposite side.

Bottom-dump trucks and the *lift-off* container are not as prevalent in material haul as are the rear-dump units. Although not specialty items, they have a more limited use. The bottom-dump unit, used with free-flowing material, discharges through bottom gates into a long windrow. Lift-off containers are removed for further treatment while the bare truck continues to serve a need through the pickup and movement of additional containers.

Construction Employment Configurations

Equipment in this category is not utilized in the earth-haul function. Included are the highway tractor, the truck configured with other than dump bed, and the assorted trailer configurations used for transporting equipment or construction material.

The short-frame tractor, or prime mover, is equipped with a trailer hitch, or a connecting fastener (fifth wheel). Tractors serving as prime movers may pull wagons, trailers, semitrailers, or even large items of mobile equipment used in construction work. If time is available, a single tractor may shuttle the numerous increments of a large mobile plant (asphalt, concrete, aggregate production, etc.).

Other vehicle types used within the construction industry may have the same basic characteristics as the rear-dump unit. However, the dump bed may be supplanted by items such as the platform, stake bed, concrete mixer drum, water tank, pump, elevating frame, or some other utilitarian replacement. These types are as varied as the needs of the construction industry itself and run in all sizes from the half-ton pickup truck on up to the limitations of the highway.

Trailers, usually self-supporting, are towed through a drawbar by the motivating equipment. Their bed, as with truck variations, ranges from the platform, through the enclosed cargo area, to the frame carrying special equipment. Semitrailers differ from trailers in that part of their load is transmitted through their connection to the prime mover. These semitrailers, which have as wide a use as trailers, need support assistance when detached from the tractor.

COMPACTION

Compaction of a soil mass results in void reduction with expulsion of the air and excessive moisture. Results of this compaction are soil consolidation, the reduction of future settlement, and the improved strength of the soil structure itself. Several methods, such as impact forces, static weight, vibration, or kneading action are available to provide this compactive effort.

Lift thickness should be limited to assure adequate compaction due to the restricted load distribution through the soil itself. The number of passes to be made by the compacting equipment depends on specifications, materials being compacted, and moisture content of the material.

There are four major types of either towed or self-propelled compactors for mass construction use: the smooth-wheel roller, pneumatic roller, segmented-pad roller, and grid roller.

Smooth-wheel rollers are used for finishing granular or granular-plastic base and subbase material. The two-axle rollers may be either two-roll tandems or three-wheel units. There is also a three-axle roller which is known as the two-wheel tandem unit with three drum rolls.

Pneumatic rollers provide a high degree of compaction good for granular and clayey materials. The tires may provide a kneading action to the soil as well as a direct compaction effort.

Segmented or *tamping rollers* (sheepsfoot) are used for compacting cohesive materials (see Fig. 4.31). Pressures exerted by the feet exceed that exerted by

FIG. 4.31 Compaction with a segmented pad roller. *(Caterpillar Tractor Co.)*

rubber tires. The feet, initially penetrating the soil, will proceed to walk out of the compacting material.

Grid rollers appear like heavy open-wire mesh drums. Working best with granular material, the heavy mesh produces a high-pressure concentration that may shatter the larger particles to improve surface smoothness.

Vibratory rollers may include any of the aforementioned classes. This item of equipment combines the normal static weight of the compacting equipment with cyclic dynamic force. Vibration may be induced by weights mounted eccentrically on the horizontal shaft or by separately mounted motors attached to the equipment. These combined forces provide deeper penetration and better compaction. The greater the number of vibrations in a period of time, the greater the production. Greater weights of vibrating unit will also produce greater depths of compaction.

PRODUCTION

Getting material, hauling it to the use area, depositing the load, and returning for another load all take time. The time required to do this is known as *cycle time*. It is a combination of fixed and travel time.

Fixed time is that time spent when equipment is not actively engaged in either load hauling or returning. Fixed time factors, fairly constant, include spotting of the equipment; loading operation; turning at each end of the haul and turns between as needed; accelerating or decelerating to or from travel speed; dumping or disposing of the load; and reversing the direction of travel.

Travel time is the time spent hauling the load or returning empty. With optimum speed relatively constant, the variables affecting travel time are altitude and ambient temperature; rolling resistance of the material traveled; resistance due to grade; coefficient of traction; distance covered in hauling the load and dead heading; and the actual travel speed itself.

Tables are available providing the fixed time values, or they may be clocked, while the travel times may be calculated. Allowance must be made for interferences in production (breaks, tire problems, etc.) so a working or efficiency hour of 40 to 50 effective minutes is established. Production can then be evaluated by trips per hour times load per trip. In calculating the load, the loose cubic yards must be used in lieu of the in-bank quantity. Depending on the material, a swell factor multiplier will convert from bank to loose volume.

In many cases, the actual effective production is somewhat less than described, due to standby of the haul equipment while waiting for the loading device to complete its requirements. Constant charging from a continuous conveyor or an overhead bin may well be faster than the repeat loading of a shovel or loader.

Further, loads (of earth, at least) may be placed only as fast as the distributing and compacting equipment may receive and service it. All of the various factors must be considered and analyzed for optimum production. Standby items of equipment may well be an economical necessity to keep the more expensive units operating at full capacity.

EQUIPMENT SELECTION AND COST

With any construction effort, there is the continual need to evaluate equipment requirements, including selection, costs, and economic life, and to assure safe operations of the equipment and the job. Although these several areas of importance are not necessarily interrelated, they transcend the broad image of equipment both in erection of temporary structures and in general construction needs. These areas are discussed here to guarantee their close association with the temporary employment of equipment on the construction project. The proper selection and use of both erection and other equipment includes the choice, for whatever reasons, of a particular item or unit; its acquisition by a variety of means; its utilization on this and subsequent projects; and its future. Meanwhile, safety is essential across the broad spectrum of construction activities.

EQUIPMENT SELECTION

Obtaining proper equipment to be used on a project depends on numerous variables. Among these variables are: the types of equipment under consideration; impact of the forthcoming project itself; analysis of the contractor's current assets; project location; relationship of the equipment with future requirements of the contractor's organization; and versatility of selected machinery. While these various factors are part of all initial selection criteria, they must constantly be reevaluated to assure that performance supports that initial selection. Where or when it is found that the selection was in error, for whatever reason, the process must be repeated. Such a reanalysis should hopefully reveal the misconception resulting in the initial erroneous choice and should provide guidance for a possible substitution or replacement of equipment. Remember, however, that additional cost and use factors will require consideration beyond that of the original plan.

Types

The actual types of equipment available on the market are as wide in capability as construction requirements themselves are wide in scope. However, there are two broad types relating to frequency of use which warrant discussion at this time. These two types are the "standard" and the "special" equipment.

Depending upon the contractor and the field of effort, a unit of equipment may fill either categorical type. An earthwork contractor would probably consider a tower crane a special piece of equipment needed only infrequently, if at all. For a large building contractor, the tower crane would likely be a standard item needed on every high-rise structure undertaken. Frequency of use is one criterion which might be applied to differentiate between the two types. The standard unit generally is more frequently used than the special equipment. It is adaptable to serve a number of requirements throughout the construction industry. Another criterion is the manufacturer's approach. If the unit is in general production for sale on the open market, the equipment is usually classed as standard.

If the item is produced or tailored for a specific performance, it is probably classed as special. This does not mean that it will not be extensively used. A massive "whirley" ordered and manufactured specifically for use in construction of a large concrete dam would be classed as a special item of equipment. It would be employed daily for the multiyear extent of the construction effort and would serve a variety of needs relating to its lift, reach, and place capability. This whirley would not be economical or even feasible for use on another project of lesser scope or magnitude.

Even within the standard and special types of equipment, it is advantageous to standardize as to make, model, or component. This standardization provides a commonality as far as training, operation, maintenance, repair parts, and interchangeability are concerned. In the long run, such a standardization provides a more economical operation.

Operators and maintenance crews spend less time becoming familiar with the equipment. The training needed to educate the unfamiliar potential operator is simplified greatly. Mass, or quantity, procurement reduces both the initial and upkeep costs and the need for extensive stockage of spare parts. A common fuel filter, usable with most or all of the engines available to the contractor, would certainly be easier to handle than a large number and variety of filters.

Parts and components for standard items of equipment are also more generally available than those for the special unit. This becomes of great importance when the project is in a remote or inaccessible location. The standard part, as well as the equipment itself, being more readily available than the special counterpart, can generally be delivered more expeditiously, thus reducing downtime or wait time for the contractor.

Where special items of equipment are needed or selected, lead time for acquisition must be evaluated. If there is a need beyond the contractor's resources, the contractor may desire to obtain guidance of experts in the manufacture of that specialized, extra large, or expensive item. Short mobilization time and impending contract initiation may preclude the wait for special fabrication, however. Shipping delays for overseas projects may seriously hinder equipment acquisition also.

The actual selection of equipment type is based on job needs. With the wide scope of equipment from which to choose, it is difficult for even the most experienced to know all the makes and models. When questions arise, the best source of information is the equipment representative or distributor. The contractor should turn to some of the many publications that provide the needed specification information. While each manufacturer has specialized publications, those of general interest might include the 18 volumes of the Equipment Guide-Book Company's *Comparative Data* or Caterpillar's *Performance Handbook*. Other similar publications are readily available on the market.

Project Factors

While the factors considered by a contractor in preparing the cost evaluation are numerous, there are several which more specifically relate to the job itself rather than the manner of accomplishment. As all use factors must be examined, there are no firsts or seconds in importance.

The manager looks at the project for its overall duration. For short-duration projects, the contractor will probably look at the available resources to handle the work even if it might mean a short period of misuse or overkill. Equipment rental and time of actual physical employment on a project will also be examined. If the project extends over a long enough period, acquisition of new equipment to be amortized over the life of the job might be seriously considered. Even on a long-term contract, if an item cannot be gainfully worked for a major part of the contract period, it may be advantageous to make do with whatever is readily available, to lease the item, or to subcontract out that phase of work.

Project location is also a necessary consideration. In large population centers where equipment and parts are readily available, the contractor's concern is greatly reduced from what it might be in a remote site locale. With overseas projects, where equipment acquisition is geographically difficult, the contractor should carefully weigh the probability of single-source new plant procurement (to facilitate training, operation, and maintenance).

Analysis

The project must be analyzed in its bid state to derive a proper equipment selection. It must be reanalyzed upon receipt of the contract to adjust or validate the initial bid and for further investigation of equipment employment. Throughout the life of the job, equipment usage must constantly be reevaluated to assure availability to that which is most advantageous, whether on hand or to be acquired. This analysis is difficult enough when considering equipment for prolonged use on the job itself. When that analysis relates to the use of equipment for temporary employment or of structures needed to support construction, the evaluation becomes more difficult. Justification of selection is often tenuously based on a learned "best-judgment" factor.

This evaluation depends largely on initial costs (to the organization) of the equipment itself, costs incurred during the time of employment, costs of the deprivation of that existing equipment from serving as an available resource for other jobs, depreciation, and a forecast of future use for the equipment. An examination of the contractor's available plant should readily reveal whether or not the proper equipment, or equitable substitute, is within the contractor's current inventory.

One of the major principles of selection is to have the best and best-suited machine for the planned work. Each piece of equipment is designed and built to satisfy a specific purpose. While using the best-adapted equipment for the forecast project is ideal, in many instances it is not realistic. The economic point of view—the monetary situation—usually guides the contractor to purchase equipment that satisfies multiple requirements to get maximum utilization of resources.

The contractor must also consider costs per unit of production. When a front loader is working soft material in a bank, the life of the bucket, teeth, hydraulic system, and other parts which might be affected by the wear are prolonged. If that loader is shut down for mechanical or other reasons, its production would be reduced.

When production decreases, the cost per unit of production increases. If the loader is simply stockpiling or performing a solo task, the cost per unit of production can be readily determined based on overall costs relating to that quantity produced or effort for a period of time. If the loader has been servicing a fleet of trucks, production of the truck fleet stops also. Thus, the costs of the truck fleet are included with the loader for a unit cost determination. A work stoppage impacting the dependent equipment will make a drastic change in the unit cost of production. It strongly behooves the contractor to schedule equipment use so that the newer or more reliable units service the older units, while the least reliable units operate alone.

Disposition of the equipment upon termination of the project also requires analysis. Equipment may be written off at the end of a long tough job or even sold upon job completion. The contractor may be able to obtain additional work in that area of the world to retain and prolong equipment utilization. If economically justified, the equipment may be shipped from the completed project back to home base or even to another overseas job.

COSTS

A careful, current record of the costs incurred relative to an item of equipment on a project is vital in determining the well-being of that project. The record of previous projects employing construction equipment helps establish the bid and budget of the current effort. An up-to-date equipment monetary status also reflects the success, or lack thereof, of the equipment operation. If the equipment costs are within or on budget, there may well be success on the project. If the budgeted costs are exceeded by the equipment operations, then immediate remedial action should be taken to rescue the project.

Equipment costs are frequently divided into ownership and operating costs. Those which occur whether or not the equipment is operated are known as ownership costs. Generally they are fixed fees including the base cost, interest on the investment, taxes, insurance, protection, and replacement costs. All of these costs are one time or annual in charge against the equipment. Within this category is depreciation, which combines the costs involved in downgrading of the equipment due to use with degrading of the equipment due to the simple passage of time.

Operating costs are those resulting from utilization of the equipment. These costs include wages of the operator, use of petroleum products (fuels and oils), maintenance, and repairs. These costs are variables which depend on usage of the equipment.

Owning

If the equipment will pay for itself accomplishing the work for which it was acquired, then it should be owned. Further use on other projects, or even its disposition through sale at a good price, will enhance the profit margin. But with equipment costs rising, potential owners are looking closely at the merits and demerits of equipment ownership. Ownership cost, any which is incurred or raised due to owning the equipment (less the expected resale value), must be closely evaluated. If the contractor can favorably reflect on a prolonged need with a continuous requirement and subsequent reuse, then ownership should certainly be considered. If these areas do not measure up to expectations, then the prospective buyer might look to leasing or renting the item of equipment.

Factors other than monetary ones enter into the picture. Ownership is recommended if equipment control and availability are essential to the contract operations. Upkeep, with its resultant on-site availability, would tend to be improved when the equipment is owned. Each prospective buyer's special requirements deserve mention; what is good for one contractor, might not satisfy another.

Leasing

Short-term usage of equipment should be examined with the thoughts of renting or leasing a unit. Equipment rental also precludes the need for a substantial monetary outlay undertaken with the purchase of the machine. This investment is also saved in negating the need for spare parts and special tool acquisition. Each new owned item of equipment requires trained maintenance personnel. This requirement can be avoided through renting to satisfy equipment needs.

Leasing provides a chance to try out a wide variety of equipment without a large financial commitment.

The contractor can generally have equipment available when needed without worrying about obsolescence, which is a major concern in ownership.

A lease-purchase option plan in which the contractor can lease equipment with an option for purchase provides an opportunity to try the machine without a large initial investment, while getting a better reading on what the future holds for the company. Then, if ownership becomes practical, the rent paid applies to the purchase price. Because only an agreed portion of the rental fee is applied to the purchase price, lease-purchase options result in total costs somewhat higher than those for outright purchase.

ECONOMIC LIFE

Each equipment owner is intent on keeping the cost per unit of production (discussed earlier) at an absolute minimum—a requirement for a productive, profitable enterprise. The owner must keep complete and accurate cost records to reflect the true costs of a unit of production. Required items to evaluate include maintenance and repair, replacement, depreciation, and downtime.

Maintenance and Repair

This aspect of the economic life of equipment is difficult to forecast accurately. Certain minimum standards should be adopted to assure an ongoing program. Equipment life degrades rapidly if maintenance and repairs are strictly on an as-needed basis. The squeaking wheel getting the grease does not serve to prolong longevity. Manufacturers usually recommend maintenance programs for their machines. Depending on the equipment type, the maintenance program may be based on passage of time (change filters every 30 days), on hours of operation (grease after every 60 h of use), on distance traveled [replace part every 6000 mi

(10,000 km)], or on combinations of these and other criteria. While such guidelines are necessary and helpful, they must be tempered with on-site judgment. Extremely adverse working conditions, such as continuously blowing dust, may require a more frequent changing of oil, cleaning of a filter, and shorter greasing periods. Parts may wear more rapidly than programmed. Cold weather operations require attention to antifreeze, while hot weather operations will require added attention to cooling and hydraulic systems.

The important point is that for improved economic life of the equipment, a maintenance program must be followed. Maintenance can be scheduled far enough in advance so as to minimize impact on other aspects of the work effort. Repairs, of course, should take place as soon as possible after the requirement. Establishment of a preventative-maintenance program will assist in reducing needed repairs and result in an improved economic life of construction equipment.

While attention is necessary to keep the equipment operating in a satisfactory manner, the daily servicing can be scheduled to allow for maximum production. Service vehicles, maintenance, oil, and fuel tankers can supply the equipment after or between shifts so that the unit is always ready for use.

Where there are major equipment concentrations and the project duration warrants, it is frequently advisable to provide a nearby maintenance workshop for equipment servicing. This shop will enable the maintenance and repair crew to better perform, which will result in reduced equipment downtime. Importance of this shop increases with the remoteness of the project. Overseas projects generally require almost complete self-sufficiency of operations; hence, maintenance and repair procedures must likewise be self-sufficient.

Replacement

There are a number of reasons leading to equipment procurement beyond job-related requirements for new machine acquisition. Among these might be the need to supplant the equipment made unserviceable due to accident or damage. Another is the need to replace outworn or obsolete machinery. A third reason might be the requirement to relocate existing equipment to other phases of the work where they may be more efficiently utilized.

With equipment acquisition, for upgrading or replacing worn-out items, such factors as need, maintenance, and expense should be considered. When satisfied with the equipment type of the existing inventory, a contractor who needs to upgrade or replace will likely remain with a similar, yet newer, make and model. Unsatisfactory performance or upkeep of available equipment or a new requirement requires comparisons of equipment available on the market. The primary consideration in routine acquisition of equipment that meets production needs will likely be of a monetary nature.

The economic life of on-hand equipment must be closely tracked. Continued improvements in production capability of newly developed equipment result in lower costs of production. At some point in the life of existing equipment, before its efficiency begins to noticeably decrease and its production costs rise substantially, the contractor must examine the needs for a possible replacement by evaluating equipment costs relating to the investment, maintenance and repairs, replacement, downtime, and obsolescence.

Depreciation

A loss in value of equipment as a result of its age or use is known as depreciation. Depreciation is determined from the total cost, the economic life, and the salvage value. It is, in reality, composed of two parts: *physical* deterioration due to the wear and tear and *economic* downgrading due to obsolescence. Depreciation values vary with both hours of operation and the passage of time itself.

There are a number of methods used for determining the cost of depreciation. Any reasonable system may be used, depending on the contractor's accounting and bookkeeping procedures. Several types currently in practice are straight-line, declining-balance, sum-of-the-digits, production hours or units, or length of job.

Straight-line methods

This system assumes that the value of a piece of equipment decreases at a uniform rate. The original cost is reduced by the estimated salvage value at final disposition and is divided by the useful life of the equipment. This gives a constant equal yearly depreciation for the life of the equipment.

Declining-balance method

This system is based on a depreciation rate of 200% over the estimated life for new equipment and 150% for used equipment. For a new equipment with a 5-y life, the yearly rate of depreciation would be 40%; for used equipment it would be 30%. Total cost of the equipment would be reduced by 40 or 30% the first year and a new book value (original less the 40 or 30%) would be established for the next year. Each year the book value would be newly established. Initial year depreciation will be higher than with other systems.

Sum-of-the-digits method

For calculations, the estimated life in years is determined. These digits (1 + 2 + 3, etc.) are then totaled. The first year's depreciation would be the original cost, less reasonable salvage value, times the expected life in years divided by the total sum of the digit values. This provides a reasonably high initial depreciation with lesser depreciation in subsequent years. The procedure would be repeated each year, using a multiplier of the remaining years of life.

Production hour or unit method

In this system, using production hours, the cost less salvage value would be spread over the expected lifetime in hours of operation. Then, each hour of use through the year would be used to determine the yearly depreciation value. Production *units,* e.g., cubic yards (cubic meters), can be similarly used for determining yearly depreciation.

Length of job method

In this system, the equipment is forecast to depreciate completely over the length of the project. Upon job termination, the value of the equipment is determined. It is then credited to that project. If the equipment is sold, that value is credited to the job.

Downtime

All times when the equipment is not producing work—while undergoing repairs or adjustments or is inoperable due to bad operating and maintenance practices—fall in this category. Downtime tends to increase with equipment age and usage but can be reduced through efficient management. It should not be countered at the expense of the equipment. An overloaded haul unit might provide a short-term increase in production but the resultant increase in maintenance will more than offset that overload increase.

With proper overall control of the project, downtime can be minimized. An adequate stockage of reinforcing steel adjacent to the point of need might allow sufficient repair time needed for the lifting rig. The lift device might undergo downtime, but there would be negligible impact on the other dependent functions.

In project-operation evaluation, consideration should be given to determining an acceptable level of downtime. This enables the manager to review the options available for the most efficient and economical correction of the downtime when it might occur.

SAFETY

Safety issues relating to the use of erection and construction equipment and practices are far too numerous for complete itemization here. Instead, certain pertinent aspects of general safety selected for discussion appear here followed by more specific points relating to equipment types in the general order of the equipment types presented in the preceding discussions.

General Requirements

With equipment employed for any purpose, controls must be established to govern the use of that equipment. These controls may be made available and implemented through published documents, operator indoctrination prior to use of the machinery, use of signs to impart the desired information, or employment of signals to immediately control the action.

Warning Signs, Tags, and Signals

All warning signs, tags, and signals should conform to an existing uniform set of standards. Printed materials should be placed only where necessary to warn both the workers and the public, and then should be removed when the potential problem areas no longer exist. Lighting may be needed to illuminate the sign. Lighting may also be needed to illuminate the person giving the signs or signals.

Signals during construction operation are distinctly needed for guidance and control. These needs are due to visual obstructions of the equipment, noise confusion, and difficulties the operator might have in seeing an intended target when working in congested or remote vertical or horizontal situations. Illustrations of the appropriate signals should be made available and posted for the signaler and operator. One person only should be responsible for the signals so as to preclude confusing or conflicting guidance. Where distances are over 100 ft (30 m) between signaler and operator, radios, telephones, or other communication systems should be employed. This signaler should control only a single item of equipment at any one time.

Safety Belts, Lifelines, and Nets

When working with equipment in general, and more specifically erection equipment where height is a factor, safety belts, lifelines, and nets may be required. Safety belts or lifelines have an established breaking strength of 5400 lb (24,000 N), according to many states, and are a minimum of ¾ in (19 mm) manila or equivalent. Where these belts or lines have been subject to impact loadings, they shall be removed from service. All equipment of this nature shall be reinspected daily before use. Where scaffolds, platforms, floors, ladders, etc. are impractical, safety nets shall be installed at a distance of no more than 25 ft (7.6 m) below the working surface. Impact-load testing must be employed to assure clearance from contact with objects below the net. When workers are exposed, this net must extend 8 ft (2.4 m) beyond the edge of the work. Net mesh have to be less than 6 × 6 in (15 × 15 cm), and the nets shall meet 17,500-ft·lb (23,700-N·m) minimum impact resistance.

Erection Equipment Requirements

Hoists, Derricks, and Cranes

As with other equipment, hoists should have covers or guards over gears, screws, cables, sprockets, manifolds, exhausts, or other hazardous parts. The operator must be able to reach easily all controls. Disconnects and overspeed preventers should be integral parts of the hoist. Tag and handlines are recommended for load controls.

One key point with all hoisting equipment is the requirement for performance and load testing before use. Results of these tests (capacities, speeds, warnings, and instructions) must be posted to assure that the limitations are not exceeded. No capacity-altering modifications are authorized without the manufacturer's approval.

An operator must (1) remain with lifting equipment at all times until the load is released, (2) assure that no one rides the loads, hooks, buckets, or materials being hoisted, and (3) control access and clearances beneath loads to avoid injuries due to spilled contents.

Stiff-leg derricks must have anchors capable of withstanding 1.5 times the maximum load. Derrick masts and guys must be separated by 30° or greater, or the rated load must be reduced. The top of the guyed derrick mast should be secured by at least six equally spaced guys. Foundations for derricks should be designed to resist 1.5 times the manufacturer's rated load.

Where applicable, boom stops, boom angle indicators, and jib stops should be employed. Cabs are to be equipped with safety glass. Surrounding clearances should be checked to assure that the rotating structures do not contact fixed objects and to allow passage of employees around the equipment. Hoisting equipment and booms are not to operate within 10 ft (3 m) of an energized power line or the lines should be deenergized. The rotating superstructure should be grounded. Solidly set outriggers are to be employed for all side-lifting operations. Power traveling mechanisms (overhead and gantry cranes) must be provided with audible warning signals.

Proper security of slung loads is of vital importance with all lifting operations. Do not use the hoist line to lift loads and secure loose materials. These steps are of particular importance when employing helicopter cranes. Each day's operation involving helicopters should commence with a thorough briefing of pilot, crew, and ground personnel. These assisting ground forces are to be afforded eye protection and must assure that all loose items within 100 ft (30 m) of the load lift or deposit areas are secured. No one should work under the hovering craft except to engage or disengage the cargo slings, and only authorized persons are to be allowed within 50 ft (15 m) of a craft with turning blades.

Supplemental Erection Equipment

Of particular importance with conveyor operations is the need for devices to preclude runaway or backup of the belt. Conveyors are for material haul and not for personnel.

With cableways, only authorized personnel can ride in the secured carriages, and no one is to ride the load blocks or loads. Communications between the signalers and the cableway operator are a requisite.

Discharge pipes for concrete pumps should be on properly designed supports. Pressure hoses must have fail-safe connectors. Concrete buckets are to have safety latches to avoid accidental discharge.

The use of jacks must be controlled by the rated capacity. Safety devices are necessary to prevent overtravel and to allow continuing load support in the event of malfunction of the jack. A jack is only as good as its foundation. The base should be blocked or cribbed.

Rigging equipment should be inspected before each use. Hooks require a safety closure. Slings should be protected from sharp edges and, depending on their make, will have a minimum length of clearance from the load. Wire rope with excessive broken wires; evidence of corrosion, kinking, bird caging; reductions in nominal diameter; evidence of heat damage; or certain other problem conditions must be removed from service. This wire rope should be one continuous piece when under load, except for the eye splices. Load limitations of these various items should not be exceeded.

Earthwork Equipment

With regular construction equipment, probably the single most universally required safety feature is the reverse signal alarm. This automatically initiated device on all self-propelled construction equipment sounds distinctly when the unit commences backward motion. Dozers, tractors, loaders, earthmovers, trucks, and the like must have protection for the operator through canopies or shields from flying objects, rollover protection, and seat belts. The rollover protection structure (ROPS) must support a load of at least twice the equipment weight. Blades, buckets, dump bodies, and suspended loads shall be lowered, blocked, or cribbed when not in use so as not to fall or close on the unwary. Equipment controls are to be unengaged (in neutral) with the brake set and the motor off when the unit is not being utilized.

Haulage vehicles, where the load is placed by overhead equipment (shovels, loaders, etc.) shall have overhead protection such as a cab shield. Dump equipment must be equipped with a positive device to preclude accidental lowering of the bed while maintenance or inspection work is under way. Operating levers which control the dump or hoist action shall be secured to prevent accidental engagement.

Safety, as concerns all erection and construction equipment as well as the safety aspects required of all construction efforts, remains a vital part of any project. Each person on the project must be indoctrinated in requirements of a safe operation. A job can not be too safe.

BIBLIOGRAPHY

The Aero-Go System, AGS 374, Aero-Go, Inc., Seattle, n.d.

Bethlehem Wire Rope, Catalog 2305-A, Bethlehem Steel Corporation, Bethlehem, Pa., n.d.

Havers, John A., and Frank W. Stubbs: *Handbook of Heavy Construction,* 2d ed., McGraw-Hill, New York, 1971.

Hilman Rollers Move the Heavies, Hilman Equipment Co., Inc., Wall, N.J., n.d.

Huntington, Whitney C.: *Building Construction,* 3rd ed., Wiley, New York, 1966.

Kellogg, F. H.: *Construction Methods and Machinery,* Prentice-Hall, New York, 1954.

Merritt, Frederick S.: *Building Construction Handbook,* 4th ed., McGraw-Hill, New York, 1982.

Mobile Power Crane and Excavator Standards, PCSA Standard No. 1, Power Crane and Shovel Association, a Bureau of the Construction Industry Manufacturer's Association (CIMA), Milwaukee, 1973.

Peurifoy, R. L.: *Construction Planning Equipment and Methods,* McGraw-Hill, New York, 1970.

Rossnagel, W. E.: *Handbook of Rigging,* 3rd ed., McGraw-Hill, New York, 1974.

Wood, Stuart, Jr.: *Heavy Construction Equipment and Methods,* Prentice-Hall, Englewood Cliffs, N.J., 1977.

5

Loads Created by Construction Equipment

by Enno Koehn

Construction equipment can exert substantial loads on both temporary and permanent structures. The design of some members such as those supporting the first floor of buildings and those located in specific areas on certain bridges may be controlled by construction loads. Work bridges must be built to perform under a multitude of conditions caused by the movement of construction equipment. Concrete formwork and shoring must also be able to resist moving loads due to construction activity. The failure of either temporary or permanent structures may result in injuries and death in addition to substantial expense due to lost time and repairs. In order to reduce the probability of failure, realistic and comprehensive design loads must be utilized. The equipment and corresponding specifications tabulated in this chapter may be used to compute actual live loads for use in stress and deflection calculations.

IMPACT LOADS

Live-load impact may be produced by moving equipment or by falling material, such as concrete, being emptied from a bucket. In addition, precast members have been known to exert large impact forces on the falsework upon which they are supported. This can occur not only by dynamic loading during placement, but by sudden relaxation of the tension in the cables used to transport the members. Impact effects should be taken under consideration when designing temporary facilities such as formwork, falsework, work bridges, and other structures. The magnitude of these forces may have been underestimated in some instances in the past, since a number of failures attributable to impact loading have occurred.

The standard specifications adopted by the American Association of State Highway and Transportation Officials (AASHTO)[4] contain a provision for impact. For highway bridges the amount of impact allowance or increment is expressed as a fraction of the live-load stress, and may be determined by the expression:

$$I = \frac{50}{L + 125}$$

where I = impact fraction (maximum 30%)
L = length in ft of the portion of the span which is loaded to produce the maximum stress

Both the AASHTO Specifications and the *Timber Construction Manual*[5] indicate that unless there is a steel connection through which the load must pass, impact effects need not be applied to timber structures. Temporary structures, however, compared with more permanent type construction, tend to have little dead load, and impact loading should therefore be relatively more significant. It may be reasonable therefore to consider impact effects in the design of temporary facilities built of wood.

FORMWORK LOADS

Concrete formwork must be designed to support all lateral and vertical loads that are applied during construction.

The American Concrete Institute (ACI) recommends a live load of 50 lb/ft^2

($2400\ N/m^2$) of horizontal projection to provide for the weight of workers, equipment, and impact.[1] When powered concrete buggies are utilized, a live load of 75 lb/ft^2 ($3600\ N/m^2$) may be justified. Some authorities, however, recommend that the combined design dead and live load must never be less than 100 lb/ft^2 (4800 N/m^2).

The impact force due to concrete falling on a deck can be of sufficient magnitude to cause overload on certain portions of the structure. This force may be calculated from the following expression:[3]

$$F = \frac{(0.249)\ W\ (h)^{1/2}}{T}$$

where F = impact force in lb
W = weight of concrete in bucket in lb
h = height of fall of concrete in ft
T = time required to empty bucket in s

Since the force may be applied to a relatively small area, high pressures may cause localized failures which in some instances could lead to total collapse.

Table 5.1 lists the horizontal forces for three weight classes of concrete buggies. The vertical components of these values should probably be increased in order to take the effect of impact under consideration.

ACI recommends that 100 lb/lin ft (1460 N/m) of slab edge or 2% of the total slab dead load distributed as a uniform load along the slab edge, whichever is greater, should be used as minimum lateral design load for a slab. However, the localized horizontal forces due to the starting and stopping of motorized concrete buggies, as illustrated in Table 5.1, may be significant and should therefore also be taken under consideration since they may control the design.

EQUIPMENT LOADING

Tables 5.3 through 5.13 contain listings of construction equipment along with the corresponding loads and other specifications required for design. For some equipment the braking, starting, and centrifugal forces are not tabulated. An approximate value for these horizontal forces may be obtained by utilizing the coefficient of traction between the tires or crawler tracks and the haul surface. The maximum force that may be developed before slipping occurs is the product of the total equipment weight and the respective coefficient of traction shown in Table 5.2.

The total weight at tipping is given for some equipment. The effect of this force may be considered to be concentrated on one axle for tire vehicles and approximated as a triangular distribution for track vehicles.

The bending moment for the loads listed in Tables 5.3 through 5.13 may be readily computed. For the simply supported case, the maximum moment beneath any wheel occurs when the distance between that wheel and the resultant of all the wheels on the structure is bisected by the centerline or midspan of the structure.[2] For example, the maximum bending moment applied to the simply supported bridge shown in Fig. 5.2 when loaded by a full 769C Caterpillar truck, illustrated in Table 5.5 and Fig. 5.1, may be computed as follows:

1. Determine the location and magnitude of the resultant of the axle loads (Fig. 5.1).

TABLE 5.1 Horizontal Forces Due to Concrete Buggies

Weight, lb (kg)	Speed, m/h (km/h)	Horizontal force for 2-s stop, lb (N)	Horizontal force for 4-s stop, lb (N)
2000	4	182	91
(907)	(6.5)	(810)	(405)
	6	273	137
	(10)	(1214)	(610)
	8	364	182
	(13)	(1619)	(810)
	10	455	228
	(16)	(2024)	(1014)
	12	547	273
	(19)	(2433)	(1214)
3000	4	273	137
(1360)	(6.5)	(1214)	(610)
	6	410	205
	(10)	(1824)	(912)
	8	547	273
	(13)	(2433)	(1214)
	10	683	342
	(16)	(3038)	(1521)
	12	820	410
	(19)	(3648)	(1824)
4000	4	364	182
(1815)	(6.5)	(1619)	(810)
	6	547	273
	(10)	(2433)	(1214)
	8	729	364
	(13)	(3243)	(1619)
	10	911	455
	(16)	(4052)	(2024)
	12	1093	547
	(19)	(4862)	(2433)

TABLE 5.2 Friction Coefficients for Various Road Surfaces

Surface	Rubber tires	Crawler tracks
Dry, rough concrete	0.80–1.00	0.45
Dry clay loam	0.50–0.70	0.90
Wet clay loam	0.40–0.50	0.70
Wet sand and gravel	0.30–0.40	0.35
Loose, dry sand	0.20–0.30	0.30
Dry snow	0.20	0.15–0.35
Ice	0.10	0.10–0.25

SOURCE: From R. L. Peurifoy, *Construction Planning, Equipment, and Methods,* 2d ed. McGraw-Hill, New York, 1970.

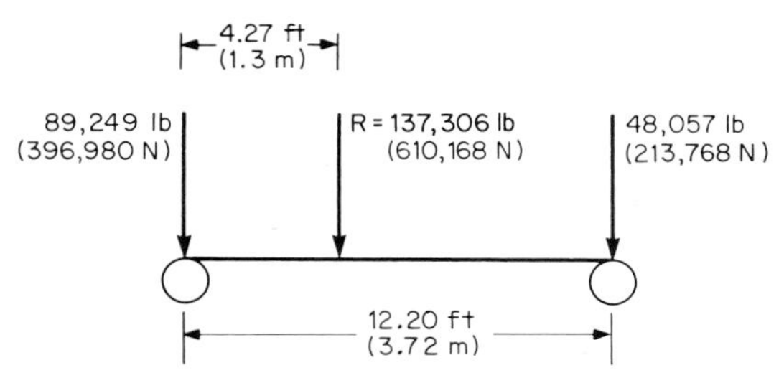

FIG. 5.1 Wheel loads.

2. Calculate the bending moment at a wheel or axle located such that the distance between the wheel and the resultant of all the wheels on the structure is bisected by the midspan or centerline (Fig. 5.2).

Bending moments at different sections or for other than the simply supported case may be calculated using the concept of influence lines and effect of moving concentrated loads.

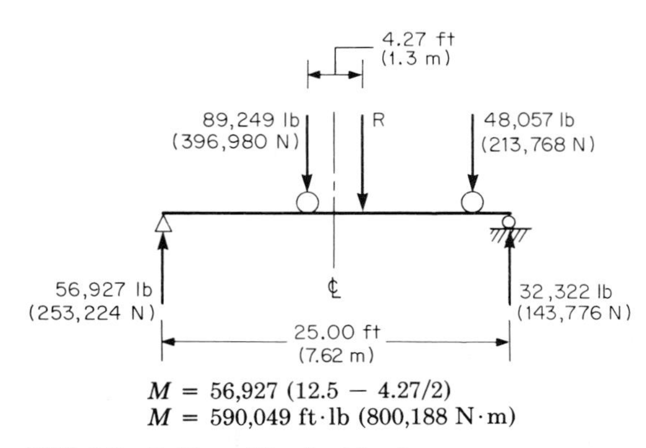

$$M = 56{,}927\ (12.5 - 4.27/2)$$
$$M = 590{,}049 \text{ ft}\cdot\text{lb } (800{,}188 \text{ N}\cdot\text{m})$$

FIG. 5.2 Bridge with wheel loads.

The effect of both impact and vibration should be taken under consideration in the design of temporary structures. The AASHTO specifications may be used to calculate impact forces. The unbalance associated with centrifugal and reciprocating machines can cause vibrations which will apply dynamic loads to supporting members. The magnitudes of the forces involved are dependent upon the operational characteristics of the equipment. Since the age and maintenance records of machines vary widely, the operational characteristics and the resulting forces cannot be specified. Care should be taken, however, such that vibrational effects are considered in areas where they may cause significant loading problems.

TABLE 5.3 Crawler Dozers

Model no.	Operating weight,* lb (kg)	Ground contact, ft^2 (m^2)	Ground pressure, lb/ft^2 (kg/m^2)	Length of track, ft (m)	Load on each track, lb/ft (kg/m)
J. Deere					
JD 350C	10,150	13.47	763	5.78	878
	(4,604)	(1.25)	(3,730)	(1.76)	(1,307)
JD 450C	14,230	16.17	878	6.06	1,174
	(6,455)	(1.50)	(4,290)	(1.85)	(1,745)
JD 550	15,480	16.17	950	6.06	1,277
	(7,022)	(1.50)	(4,640)	(1.85)	(1,898)
JD 350C	12,050	31.76	389	5.78	1,042
wide track	(5,466)	(2.95)	1,898)	(1.76)	(1,552)
JD 750	29,335	22.50	1,303	7.50	1,956
	(13,307)	(2.09)	(6,367)	(2.29)	(2,905)
JD 850	37,120	26.39	1,407	7.92	2,343
	(16,837)	(2.45)	(6,900)	(2.41)	(3,493)
JD 750	31,375	42.50	734	7.50	2,092
wide track	(14,232)	(3.95)	(3,580)	(2.29)	(3,107)
Fiat-Allis					
8	21,300	19.28	1,105	6.53	1,631
	(9,660)	(1.79)	(5,397)	(1.99)	2,427
14-C	35,430	26.06	1,360	7.91	2,240
	(16,070)	(2.41)	(6,668)	(2.41)	(3,334)
16-B	48,385	32.87	1,472	8.96	2,700
	(21,944)	(3.06)	(7,171)	(2.73)	(4,020)
21-C	71,645	41.34	1,733	10.34	3,465
	(32,498)	(3.84)	(8,463)	(3.15)	(5,158)
31	110,840	51.56	2,149	11.50	4,819
	(50,277)	(4.79)	(10,496)	(3.51)	(7,162)
41-B	131,675	63.78	2,064	11.96	5,505
	(59,727)	(5.93)	10,072)	(3.65)	(8,182)
Komatsu					
D10A	4,170	5.69	732	3.84	542
	(1,890)	(0.53)	(3,566)	(1.17)	(808)
D20A	7,340	10.80	675	5.49	668
	(3,330)	(1.00)	(3,330)	(1.67)	(944)
D21A	7,560	10.80	696	5.49	689
	(3,430)	(1.00)	(3,400)	(1.67)	(1,023)
D31A	14,000	13.36	1,044	6.17	1,135
	(6,350)	(1.24)	(5,100)	(1.88)	(1,689)
D40A	20,550	17.68	1,168	6.76	1,520
	(9,320)	(1.64)	(5,700)	(2.06)	(2,262)
D45A	21,050	17.68	1,188	6.76	1,557
	(9,550)	(1.64)	(5,800)	(2.06)	(2,318)
D50A	25,130	18.88	1,331	7.22	1,740
	(11,400)	(1.76)	(6,500)	(2.20)	(2,591)
D53A	26,230	18.88	1,392	7.22	1,816
	(11,900)	(1.76)	(6,800)	(2.20)	(2,705)
D60A	34,240	26.72	1,290	7.98	2,145
	(15,530)	(2.48)	(6,300)	(2.43)	(3,195)
D60E	35,740	28.95	1,228	8.64	2,068
	(16,210)	(2.69)	(6,000)	(2.64)	(3,070)

TABLE 5.3 Crawler Dozers (*Continued*)

Model no.	Operating weight,* lb (kg)	Ground contact, ft² (m²)	Ground pressure, lb/ft² (kg/m²)	Length of track, ft (m)	Load on each track, lb/ft (kg/m)
D65A	34,680	26.72	1,290	8.64	2,006
	(15,730)	(2.48)	(6,300)	(2.64)	(2,979)
D65E	36,180	28.95	1,250	8.64	2,094
	(16,410)	(2.69)	(6,100)	(2.64)	(3,114)
D80A	48,040	32.85	1,454	8.96	2,681
	(21,790)	(3.05)	(7,100)	(2.73)	(3,991)
D80E	50,040	36.70	1,351	10.00	2,502
	(22,700)	(3.41)	(6,600)	(3.05)	(3,721)
D85A	48,700	32.85	1,475	8.96	2,718
	(22,090)	(3.05)	(7,200)	(2.73)	(4,046)
D85E	50,710	36.70	1,371	10.00	2,536
	(23,000)	(3.41)	(6,700)	(3.05)	(3,771)
D150A	73,190	38.01	1,925	10.37	3,529
	(33,200)	(3.54)	(9,400)	(3.16)	(5,253)
D155A	73,190	38.01	1,925	10.37	3,529
	(33,200)	(3.54)	(9,400)	(3.16)	(5,253)
D355A	98,550	44.17	2,232	11.04	4,461
	(44,700)	(4.11)	(10,900)	(3.37)	(6,642)
D455A	150,840	63.91	2,355	12.83	5,878
	(68,420)	(5.94)	(11,500)	(3.91)	(8,749)
D20P	7,940	18.42	432	5.50	722
	(3,600)	(1.71)	(2,100)	(1.68)	(1,071)
D20PL	8,270	25.26	329	5.50	752
	(3,750)	(2.35)	(1,600)	(1.68)	(1,119)
D21P	8,160	18.43	452	5.50	742
	(3,700)	(1.71)	(2,200)	(1.68)	(1,104)
D21PL	8,490	25.26	329	5.50	772
	(3,850)	(2.35)	(1,600)	(1.68)	(1,149)
D31P	14,880	28.18	533	7.17	1,038
	(6,750)	(2.62)	(2,600)	(2.19)	(1,545)
D40P	23,570	39.67	593	8.50	1,386
	(10,690)	(3.68)	(2,900)	2.59)	(2,064)
D45P	23,900	39.68	593	8.50	1,406
	(10,840)	(3.68)	(2,900)	(2.59)	(2,093)
D50P	28,110	48.22	573	9.18	1,531
	(12,750)	(4.48)	(2,800)	(2.80)	(2,277)
D50PL	27,560	57.86	471	9.18	1,501
	(12,500)	(5.38)	(2,300)	(2.80)	(2,232)
D60P	37,480	64.20	573	10.30	1,819
	(17,000)	(5.97)	(2,800)	(3.14)	(2,707)
D60PL	35,710	75.78	471	9.63	1,854
	(16,200)	(7.04)	(2,300)	(2.94)	(2,760)
D65P	39,220	64.20	615	10.30	1,904
	(17,790)	(5.97)	(3,000)	(3.14)	(2,832)
Caterpillar					
D3B	13,980	11.97	1,168	5.98	1,169
	(6,340)	(1.11)	(5,712)	(1.82)	(1,738)

TABLE 5.3 Crawler Dozers (*Continued*)

Model no.	Operating weight,* lb (kg)	Ground contact, ft² (m²)	Ground pressure, lb/ft² (kg/m²)	Length of track, ft (m)	Load on each track, lb/ft (kg/m)
D4E	19,450 (8,820)	13.00 (1.20)	1,496 (7,300)	6.00 (1.83)	1,621 (2,411)
D5B	25,800 (11,700)	19.51 (1.81)	1,322 (6,464)	7.25 (2.21)	1,779 (2,647)
D6D	31,450 (14,270)	23.34 (2.17)	1,347 (6,576)	7.75 (2.36)	2,029 (3,023)
D7G	44,600 (20,230)	29.66 (2.76)	1,504 (7,330)	8.92 (2.70)	2,500 (3,746)
D8K	70,500 (31,980)	37.76 (3.51)	1,867 (9,111)	10.33 (3.15)	3,412 (5,076)
D9H	94,300 (42,780)	43.25 (4.09)	2,155 (10,459)	11.00 (3.35)	4,286 (6,385)
D10	171,680 (77,870)	59.89 (5.56)	2,867 (14,005)	12.83 (3.91)	6,691 (9,955)
D3LGP	14,750 (6,660)	28.26 (2.63)	516 (2,532)	6.78 (2.07)	1,088 (1,610)
D4ELGP	22,110 (10,030)	36.28 (3.37)	609 (2,976)	7.25 (2.21)	1,525 (2,269)
D5LGP	32,400 (14,700)	52.44 (4.87)	618 (3,018)	9.25 (2.82)	1,751 (2,605)
D6DLGP	38,300 (17,370)	56.68 (5.27)	675 (3,296)	9.45 (2.88)	2,026 (3,016)
D7GLGP	51,300 (23,270)	56.52 (5.26)	908 (4,424)	10.00 (3.05)	2,565 (3,814)
DD9H	185,420 (84,110)	88.03 (8.18)	2,106 (10,282)	11.00 (3.35)	8,428 (12,542)

*Operating weight is the weight of standard unit with general purpose blade, ROPS, and operator. Actual weight may vary depending upon use of other equipment.

EQUIPMENT MANUFACTURERS

The following equipment manufacturers supplied data in the form of company reports and brochures for the specifications compiled in Tables 5.3 through 5.13.

1. Case, Construction Equipment Division, J. I. Case Company, Racine, Wisc.
2. Caterpillar Tractor Company, Peoria, Ill.
3. Chevrolet Motor Division, General Motors Corporation, Detroit, Mich.
4. Euclid Incorporated, Cleveland, Ohio.
5. Fiat-Allis Construction Machinery Incorporated, Springfield, Ill.
6. Ford Tractor Operations, Ford Motor Company, Troy, Mich.
7. Gehl Company, West Bend, Wisc.
8. International Harvester, Melrose Park, Ill.
9. John Deere Company, Moline, Ill.
10. Komatsu Limited, Tokyo, Japan.
11. Scoot-Crete, Getman Brothers Manufacturing Company, Marion, Ohio.
12. Stow Manufacturing Company, Binghamton, N.Y.
13. Terex Division, General Motors Corporation, Hudson, Ohio.
14. Wabco Construction and Mining Equipment, Incorporated, Peoria, Ill.

REFERENCES

1. Hurd, M. K., *Formwork for Concrete,* special publication no. 4, 4th ed., American Concrete Institute, Detroit, Mich., 1979.
2. Norris, C. H., and J. B. Wilbur, *Elementary Structural Analysis,* 2d ed., McGraw Hill, New York, 1960.
3. Peurifoy, R. H., *Formwork for Concrete Structures,* 2d ed., McGraw Hill, New York, 1976.
4. *Standard Specifications for Highway Bridges,* American Association of State Highway and Transportation Officials, Washington, D.C.
5. *Timber Construction Manual,* American Institute of Timber Construction, Wiley, New York, 1974.

TABLE 5.4 Crawler Loaders

Model no.	Operating weight,* lb (kg)	Total weight at tipping,† lb (kg)	Ground contact, ft² (m²)	Ground pressure,‡ lb/ft² (kg/m²)	Length of track, ft (m)	Load on each track,§ lb/ft kg/m
J. Deere						
JD 350C	12,400 (5,625)	19,550 (8,866)	11.54 (1.07)	1,037 (5,060)	5.78 (1.76)	1,073 (1,598)
JD 450C	16,700 (7,582)	25,900 (11,755)	14.78 (1.37)	1,123 (5,480)	6.34 (1.93)	1,317 (1,964)
JD 555C	18,255 (8,281)	28,855 (13,089)	14.78 (1.37)	1,180 (5,760)	6.34 (1.93)	1,440 (2,145)
JD 755	32,005 (14,517)	51,505 (23,362)	18.75 (1.74)	1,706 (8,330)	7.50 (2.29)	2,133 (3,170)
JD 855	48,400 (21,954)	83,600 (37,920)	27.75 (2.58)	1,786 (8,700)	9.25 (2.82)	2,616 (3,893)
International						
250 C	45,165 (20,487)	76,835 (34,853)	26.75 (2.49)	1,688 (8,228)	8.91 (2.72)	2,535 (3,766)
Fiat Allis						
FL 9	24,005 (10,889)	39,310 (17,831)	16.58 (1.54)	1,447 (7,071)	7.08 (2.16)	1,695 (2,521)
FL 10C	29,210 (13,250)	47,710 (21,642)	18.75 (1.73)	1,557 (7,659)	7.50 (2.29)	1,947 (2,893)
FL 14-C	38,065 (17,265)	63,105 (28,625)	23.79 (2.21)	1,600 (7,848)	8.52 (2.60)	2,234 (3,320)
Komatsu						
D 10S	3,900 (1,770)	7,030 (3,190)	5.67 (0.53)	697 (3,400)	3.84 (1.17)	508 (756)
D 20S	8,380 (3,800)	13,780 (6,250)	10.81 (1.00)	778 (3,800)	5.50 (1.68)	762 (1,134)
D 21S	8,600 (3,900)	14,000 (6,350)	10.81 (1.00)	800 (3,900)	5.50 (1.68)	782 (1,164)
D 31S	14,880 (6,750)	24,030 (10,900)	13.36 (1.24)	1,106 (5,400)	6.17 (1.88)	1206 (1,795)
D 45S	24,230 (10,990)	39,930 (18,110)	17.74 (1.65)	1,372 (6,700)	6.75 (2.06)	1,795 (2,667)
D 50S	29,760§ (13,500)	46,560 (21,120)	18.95 (1.76)	1,577 (7,700)	7.21 (2.20)	2,064 (3,068)

D 53S	31,200§	49,390	18.95	1,639	7.21	2,164
	(14,150)	(22,400)	(1.76)	(8,000)	(2.20)	(3,216)
D 57S	32,260	53,030	20.88	1,556	7.95	2,029
	(14,630)	(24,050)	(1.94)	(7,700)	(2.42)	(3,022)
D 60S	38,910	62,720	26.05	1,495	8.64	2,252
	(17,650)	(28,450)	(2.42)	(7,300)	(2.63)	(3,349)
D 65S	39,570	64,480	26.05	1,516	8.65	2,287
	(17,950)	(29,250)	(2.42)	(7,400)	(2.63)	(3,406)
D 75S	45,830	76,690	27.08	1,679	8.98	2,552
	(20,790)	(34,790)	(2.52)	(8,200)	(2.74)	(3,801)
D 95S	62,130	106,660	33.49	1,863	10.00	3,107
	(28,180)	(48,380)	(3.11)	(9,100)	(3.05)	(4,619)
D 155S	92,150	154,320	43.64	2,110	11.88	3,878
	(41,800)	(70,000)	(4.05)	(10,300)	(3.62)	(5,773)
D 20Q	8,930	14,020	18.40	491	5.49	813
	(4,050)	(6,360)	(1.71)	(2,400)	(1.67)	(1,209)
D 21Q	9,150	14,400	18.40	491	5.49	833
	(4,150)	(6,530)	(1.71)	(2,400)	(1.67)	(1,239)
D 31Q	15,650	25,550	24.28	635	6.16	1,270
	(7,100)	(11,590)	(2.26)	(3,100)	(1.88)	(1,888)
Caterpillar						
931	15,300	24,200	12.08	1,266	5.98	1,280
	(6,940)	(10,975)	(1.12)	(6,196)	(1.82)	(1,902)
941B	24,740	41,050	14.65	1,689	6.75	1,833
	(11,225)	(18,623)	(1.37)	(8,193)	(2.06)	(2,728)
951C	27,390	44,670	16.41	1,669	7.03	1,948
	(12,425)	(20,263)	(1.53)	(8,121)	(2.15)	(2,896)
955L	34,950	57,410	21.89	1,597	7.72	2,264
	(15,885)	(26,040)	(2.01)	(7,903)	(2.36)	(3,365)
977L	47,030	78,734	27.78	1,693	9.25	2,542
	(21,330)	(35,710)	(2.58)	(8,267)	(2.81)	(3,795)
983	75,980	130,730	40.51	1,876	11.05	3,438
	(34,460)	(59,290)	(3.76)	(9,165)	(3.37)	(5,113)

*Operating weight is the weight of empty standard unit with general purpose bucket, standard tracks, ROPS, and operator. Actual weight may vary depending upon use of other equipment.

†Total weight at tipping equals operating weight plus SAE straight tipping load. Actual tipping weight may vary depending upon use of other equipment.

‡At operating weight and with standard size tracks.

§With long lift arms.

TABLE 5.5 Trucks/Haulers

	Operating weight		Loaded weight				
Model no.	Front, lb (kg)	Rear, lb (kg)	Front, lb (kg)	Rear, lb (kg)	Wheel base, ft (m)	Maximum rimpull, lb (kg)	Maximum centrifugal force, lb (kg)
Komatsu							
HD180	36,930		76,610		13.13	22,000	16,618
	(16,750)		(34,750)		(4.00)	(9,977)	(7,538)
HD200	40,780		84,870		12.30	70,000	18,351
	(18,500)		(38,500)		(3.75)	(31,746)	(8,325)
HD320	59,970		130,520		12.30	45,000	26,987
	(27,200)		(59,200)		(3.75)	(20,408)	(12,240)
HD680	102,420		252,420		15.58	45,000	46,089
	(46,500)		(114,500)		(4.75)	(20,408)	(20,925)
Caterpillar							
769C	32,979	34,326	48,057	89,249	12.16	62,000	30,287
	(14,960)	(15,570)	(21,799)	(40,483)	(3.71)	(28,118)	(13,739)
773B	39,313	44,332	61,705	121,940	13.75	95,000	37,640
	(17,832)	(20,109)	(27,989)	(55,312)	(4.19)	(43,084)	(17,073)
777	57,224	67,176	97,152	197,248	15.00	125,000	55,980
	(25,957)	(30,471)	(44,068)	(89,472)	(4.57)	(56,689)	(25,393)
WABCO							
35C	28,898	28,212	41,184	85,926	10.83	46,000	25,700
	(13,108)	(12,797)	(11,681)	(38,976)	(3.30)	(20,861)	(11,657)
Euclid							
R-25	17,060	19,790	22,700	64,150	12.92	38,000	16,583
	(7,740)	(8,975)	(10,295)	(29,100)	(3.94)	(17,234)	(7,522)
R-50	37,625	39,450	59,975	117,100	13.75	78,000	34,683
	(17,070)	(17,890)	(27,200)	(53,120)	(4.19)	(35,374)	(15,732)
R-75	48,000	52,000	83,000	167,000	14.50	100,000	45,000
	(21,800)	(23,600)	(37,600)	(75,800)	(4.42)	(45,351)	(20,430)
Terex TS-18 Tractor w/ T2233R Trailer							
	37,300	16,500	56,568	63,232	20.16	45,000	24,210
	(16,919)	(7,484)	(25,659)	(28,682)	(6.15)	(20,408)	(10,981)

TABLE 5.6 Small Trucks

Model no.	Gross vehicle weight rating,* lb (kg)	Wheel base,† ft (m)	Maximum centrifugal force,‡ lb (kg)
Chevrolet, 1980			
LUV 2WD	3,550	8.53	1,598
	(1,610)	(2.60)	(725)
LUV 2WD (long bed)	4,150	9.83	1,868
	(1,882)	(3.00)	(847)
LUV 4WD	3,750	9.83	1,688
	(1,700)	(3.00)	(765)
L-10 Blazer 2WD	6,050	8.88	2,723
	(2,744)	(2.71)	(1,235)
K-10 Blazer 4WD	6,200	8.88	2,790
	(2,812)	(2.71)	(1,265)
L-10 Pickup 2WD	5,600	10.95	2,520
	(2,539)	(3.33)	(1,143)

TABLE 5.6 Small Trucks (*Continued*)

Model no.	Gross vehicle weight rating,* lb (kg)	Wheel base,† ft (m)	Maximum centrifugal force,‡ lb (kg)
C-10/Big C-10 2WD	6,200 (2,812)	10.95 (3.33)	2,790 (1,265)
C-10 Pickup 2WD (diesel)	6,200 (2,812)	10.95 (3.33)	2,790 (1,265)
C-20 Pickup 2WD	7,500 (3,401)	13.70 (4.18)	3,375 (1,531)
C-20 HD Pickup 2WD	8,600 (3,900)	13.70 (4.18)	3,870 (1,756)
C-20 Bonus pickup 2WD	8,600 (3,900)	13.70 (4.18)	3,870 (1,756)
C-30 Pickup 2WD	10,000 (4,535)	13.70 (4.18)	4,500 (2,041)
C-30 Bonus pickup 2WD	10,000 (4,535)	13.70 (4.18)	4,500 (2,041)
K-10 Pickup 4WD	6,200 (2,812)	10.95 (3.33)	2,790 (1,265)
K-20 Pickup 4WD	6,800 (3,084)	10.95 (3.33)	3,060 (1,388)
K-20 HD Pickup 4WD	8,600 (3,900)	10.95 (3.33)	3,870 (1,756)
K-30 Bonus 4WD pickup	10,000 (4,535)	13.70 (4.18)	4,500 (2,041)
G-10 Van	6,000 (2,721)	10.42 (3.18)	2,700 (1,224)
G-20 Van	6,600 (2,993)	10.42 (3.18)	2,970 (1,347)
G-30 Van	8,600 (3,900)	10.42 (3.18)	3,870 (1,755)
G-30 RV Van	10,500 (4,762)	12.16 (3.71)	4,725 (2,143)
G-30 Commercial van	10,000 (4,535)	12.16 (3.71)	4,500 (2,041)
G-30 Hi-cube van	10,000 (4,535)	12.16 (3.71)	4,500 (2,041)
C-10 Suburban 2WD	6,800 (3,084)		3,060 (1,388)
C-20 Suburban 2WD	7,500 (3,401)		3,375 (1,531)
C-20 HD Suburban 2WD	8,600 (3,900)		3,870 (1,755)
K-10 Suburban 4WD	7,300 (3,311)		3,285 (1,490)
K-20 Suburban 4WD	8,600 (3,900)		3,870 (1,755)
Ford Courier (Pickup)	4,100 (1,859)	9.40 (2.87)	1,845 (837)
F-100 Pickup 2WD	5,150 (2,336)	11.08 (3.38)	2,318 (1,051)

TABLE 5.6 Small Trucks (*Continued*)

Model no.	Gross vehicle weight rating,* lb (kg)	Wheel base,† ft (m)	Maximum centrifugal force,‡ lb (kg)
F-150 Pickup 2WD	6,300 (2,857)	12.92 (3.94)	2,835 (1,285)
F-250 Pickup 2WD	8,000 (3,628)	12.92 (3.94)	3,600 (1,633)
F-350 Pickup 2WD	10,000 (4,535)	12.92 (3.94)	4,500 (2,041)
F-150 Pickup 4WD	6,350 (2,880)		2,858 (1,296)
F-250 Pickup 4WD	8,000 (3,628)		3,600 (1,633)
F-350 Pickup 4WD	9,100 (4,127)		4,095 (1,857)
E-100 Van	5,550 (2,517)	10.33 (3.15)	2,498 (1,133)
E-100 Van	5,500 (2,495)	11.50 (3.50)	2,475 (1,123)
E-150 Van	6,300 (2,857)	10.33 (3.14)	2,835 (1,286)
E-150 Van	6,300 (2,857)	11.50 (3.50)	2,835 (1,286)
E-250 Van	8,250 (3,742)	11.50 (3.50)	3,713 (1,684)
E-350 Van	9,750 (4,423)	11.50 (3.50)	4,388 (1,990)

*Maximum gross vehicle weight rating for each type.
†Maximum wheel base for series.
‡Centrifugal force = 0.45 × GVWR.

TABLE 5.7 Medium/Large Trucks

Model no.	Gross combined weight rating,* lb (kg)	Maximum wheel base,† ft (m)	Maximum centrifugal force,‡ lb (kg)
Chevrolet Trucks, 1980			
C5D042	18,500 (8,390)	13.92 (4.24)	8,325 (3,776)
C6D042	45,000 (20,408)	18.16 (5.53)	20,250 (9,184)
C7D042	60,000 (27,211)	21.16 (6.45)	27,211 (12,341)
C6D062 Tandem	41,000 (18,594)	17.42 (5.31)	18,450 (8,367)
C7D064 Tandem	60,000 (27,211)	17.42 (5.31)	27,211 (12,341)
N9E062	80,000 (36,281)	13.67 (4.17)	36,000 (16,326)

TABLE 5.7 Medium/Large Trucks (*Continued*)

Model no.	Gross combined weight rating,* lb (kg)	Maximum wheel base,† ft (m)	Maximum centrifugal force,‡ lb (kg)
N9F042	80,000	15.83	36,000
	(36,281)	(4.83)	(16,326)
N9E064	80,000	19.17	36,000
Tandem	(36,281)	(5.84)	(16,326)
N9F064	80,000	21.33	36,000
Tandem	(36,281)	(6.50)	(16,326)
Ford			
LN 600	42,000	20.83	18,900
	(19,051)	(6.35)	(8,573)
LN 700	50,000	20.83	22,500
	(22,680)	(6,35)	(10,206)
LN 7000	50,000	18.33	22,500
Diesel	(22,680)	(5.59)	(10,206)

*Gross combined weight rating is truck chassis and optional body and payload.
†Maximum wheel base for series.
‡Centrifugal force = 0.45 × GCWR.

TABLE 5.8 Wheel Loaders

Model no.	Operating (empty) weight,* lb (kg)	Maximum loaded weight (at tipping),† lb (kg)	Wheel base, ft (m)	Maximum centrifugal force, 1 lb (kg)
Caterpillar				
910	14,400	24,285	7.67	6,480
	(6,530)	(11,010)	(2.34)	(2,939)
920	18,600	31,990	8.33	8,370
	(8,440)	(14,510)	(2.54)	(3,798)
930	21,200	37,000	9.00	9,540
	(9,620)	(16,790)	(2.75)	(4,329)
950	28,500	49,970	9.58	12,825
	(12,930)	(22,660)	(2.90)	(5,819)
966C	36,900	65,070	10.16	16,605
	(16,740)	(29,510)	(3.10)	(7,533)
980C	58,000	98,760	11.58	26,100
	(26,310)	(44,800)	(3.53)	(11,839)
988B	93,650	148,300	12.50	42,143
	(42,480)	(67,200)	(3.81)	(19,116)
992C§	187,800	294,280	15.83	84,510
	(85,180)	(133,480)	(4.82)	(38,331)
Komatsu				
W90	27,670	49,390	9.00	12,452
	(12,550)	(22,400)	(2.74)	(5,648)
W170	41,840	73,300	10.00	18,828
	(18,980)	(33,250)	(3.05)	(8,541)
John Deere				
JD 444	19,280	31,375	7.87	8,676
	(8,745)	(14,231)	(2.40)	(3,935)

TABLE 5.8 Wheel Loaders (*Continued*)

Model no.	Operating (empty) weight,* lb (kg)	Maximum loaded weight (at tipping),† lb (kg)	Wheel base, ft (m)	Maximum centrifugal force, lb (kg)
JD 544B	21,775 (9,877)	38,170 (17,314)	7.87 (2.40)	9,798 (4,445)
JD 644B	28,100 (12,746)	51,005 (23,136)	8.67 (2.64)	12,645 (5,736)
JD 24A	5,483 (2,487)	7,983 (3,621)	2.91 (0.89)	2,467 (1,119)
Fiat Allis				
345 B	16,284 (7,386)	26,732 (12,123)	8.00 (2.44)	7,328 (3,323)
545 B	20,410 (9,255)	34,390 (15,595)	9.16 (2.80)	9,184 (4,164)
605 B	22,295 (10,111)	39,369 (17,855)	9.16 (2.80)	10,032 (4,549)
745 C	40,360 (18,300)	69,550 (31,540)	10.83 (3.30)	18,162 (8,235)
945 B	65,300 (29,619)	111,130 (50,407)	12.33 (3.76)	29,385 (13,328)
International				
H400 C	124,980 (56,691)	214,740 (97,406)	15.00 (4.57)	56,241 (25,511)
Gehl				
2600	3,000 (1,360)	4,600 (2,085)	2.67 (0.81)	1,350 (612)
2800	3,250 (1,474)	5,050 (2,992)	2.67 (0.81)	1,370 (621)
4300	3,975 (1,805)	5,975 (2,713)	3.00 (0.91)	1,789 (812)
4500	4,200 (1,907)	6,500 (2,951)	3.00 (0.91)	1,890 (858)
4600	4,385 (1,991)	6,785 (3,079)	3.00 (0.91)	1,973 (896)
4700	4,250 (1,930)	6,550 (2,974)	3.00 (0.91)	1,913 (869)

*Operating weight is empty weight of standard unit with ROPS, operator, and standard bucket. Actual weight may vary depending upon use of other equipment.

†Maximum loaded weight (at tipping) equals operating weight plus SAE straight tipping load. Actual tipping weight may vary depending upon use of other equipment.

‡Maximum centrifugal force = (0.45) × (operating weight).

§Fully equipped for rock work.

TABLE 5.9 Wheel Scrapers

Model no.	Operating (empty weight*)		Maximum loaded weight†		Wheel base,† ft (m)	Maximum rimpull,§ lb (kg)	Maximum centrifugal force,¶ lb (kg)
	Front, lb (kg)	Rear, lb (kg)	Front, lb (kg)	Rear, lb (kg)			
Komatsu							
WS16	74,070			122,570	24.28	NA	33,332
	(33,600)			(55,600)	(7.40)	NA	(15,120)
WS235	76,720			149,470	27.52	NA	34,524
	(34,800)			(67,800)	(8.39)	NA	(15,660)
John Deere							
JD 762	22,310	12,140	31,110	30,840	20.15	48,000	15,503
	(10,120)	(5,507)	(14,111)	(13,989)	(6.14)	(21,769)	(7,031)
JD 860B	32,120	15,720	44,110	41,230	22.77	108,000	21,528
	(14,570)	(7,130)	(20,010)	(18,700)	(6.94)	(48,979)	(9,763)
Caterpillar							
621 B	44,940	19,260	61,710	50,490	25.33	82,000	28,890
	(20,384)	(8,736)	(27,992)	(22,903)	(7.72)	(37,189)	(13,102)
631 D	61,341	27,559	88,506	75,394	28.67	100,000	40,005
	(27,824)	(12,500)	(40,146)	(34,199)	(8.74)	(45,351)	(18,143)
641 B	82,110	36,890	115,020	97,980	31.00	135,000	53,550
	(37,246)	(16,734)	(52,175)	(44,445)	(9.45)	(61,225)	(24,286)
				110,832			
651 B	85,023	41,873	120,068		31.92	130,000	57,103
	(38,565)	(18,995)	(54,462)	(50,272)	(9.73)	(58,957)	(25,897)
627 B	43,247	30,053	60,650	60,650	25.33	87,000	32,985
	(19,619)	(13,633)	(27,513)	(27,513)	(7.72)	(39,455)	(14,959)
637 D	63,378	37,222	89,556	86,044	28.66	140,000	45,270
	(28,750)	(16,885)	(40,624)	(39,031)	(8.74)	(63,492)	(20,531)
657 B	88,260	58,840	123,039	12,806	32.92	210,000	66,195
	(40,035)	(26,690)	(55,811)	(58,089)	(10.02)	(95,238)	(30,020)

TABLE 5.9 Wheel Scrapers (*Continued*)

	Operating (empty weight*)		Maximum loaded weight†				
Model no.	Front, lb (kg)	Rear, lb (kg)	Front, lb (kg)	Rear, lb (kg)	Wheel base,† ft (m)	Maximum rimpull,§ lb (kg)	Maximum centrifugal force,¶ lb (kg)
613 B	19,239 (8,726)	12,301 (5,579)	27,619 (12,528)	29,921 (13,572)	20.83 (6.35)	35,000 (15,873)	14,193 (6,436)
623 B	49,100 (22,273)	22,060 (10,007)	65,426 (29,678)	55,734 (25,282)	26.16 (7.98)	85,000 (38,549)	32,022 (14,523)
633 D	67,541 (30,639)	33,269 (15,091)	91,421 (41,470)	84,389 (38,280)	29.16 (8.89)	100,000 (45,351)	45,365 (20,574)
Fiat Allis							
161	29,400 (13,336)	15,410 (6,990)	41,450 (18,801)	41,450 (18,801)	22.58 (6.88)	57,000 (25,850)	20,165 (9,145)
260 B	38,950 (17,665)	19,450 (8,820)	56,750 (25,738)	52,050 (23,607)	23.50 (7.16)	70,000 (31,746)	26,280 (11,918)
261 B	41,400 (18,775)	21,500 (9,759)	58,400 (26,486)	57,600 (26,124)	24.42 (7.44)	70,000 (31,746)	28,305 (12,836)
262 B	40,500 (18,370)	27,400 (12,430)	59,900 (27,120)	58,400 (26,480)	25.00 (7.62)	110,000 (49,886)	30,555 (13,857)
263 B	41,700 (18,910)	32,500 (14,740)	60,700 (27,530)	66,500 (30,160)	25.83 (7.87)	110,000 (49,886)	33,390 (15,143)
WABCO							
101 G	20,633 (9,359)	10,868 (4,929)	29,900 (13,590)	27,600 (12,519)	20.13 (6.13)	33,500 (15,192)	14,176 (6,429)

*Operating weight equals weight of machine with ROPS and operator. Actual weight may vary slightly.
†Maximum loaded weight equals operating weight plus load-rated capacity.
‡Wheel bases are measured with the bowl in up (travel) position.
§Maximum rim pull should be derated for actual field conditions.
¶Centrifugal force = (0.45) × operating weight.

TABLE 5.10 Motor Graders

Model no.	Operating weights*: On front wheels, lb (kg)	Operating weights*: On rear wheels, lb (kg)	Wheel base front wheels/rear wheels, ft (m)	Maximum centrifugal force,† lb (kg)
John Deere				
JD570 A	6,755 (3,063)	14,148 (6,418)	17.33 (5.28)	9,406 (4,266)
JD670 A	8,828 (4,004)	18,252 (8,279)	19.58 (5.97)	12,186 (5,527)
JD672 A	9,668 (4,385)	18,507 (8,395)	19.58 (5.97)	12,678 (5,751)
JD770 A	9,564 (4,338)	21,755 (9,868)	19.58 (5.97)	14,093 (6,393)
JD772 A	10,404 (4,719)	22,010 (9,984)	19.58 (5.97)	14,586 (6,616)
Caterpillar				
120 B	26,460 (12,002)		19.16 (5.84)	11,907 (5,401)
120 G	25,295 (11,485)		18.67 (5.69)	11,383 (5,168)
130 G	27,220 (12,350)		19.42 (5.92)	12,248 (5,558)
12 G	29,525 (13,390)		19.42 (5.92)	13,286 (6,025)
140 G	30,030 (13,620)		19.42 (5.92)	13,513 (6,129)
14 G	40,650 (18,440)		21.16 (6.45)	18,292 (8,298)
16 G	54,060 (24,520)		22.83 (6.96)	24,327 (11,034)
Komatsu				
GD22-H	11,240 (5,100)		11.80 (3.60)	5,058 (2,295)
GD30-5M	17,350 (7,870)		15.42 (4.70)	7,808 (3,542)
GD31RC-3A	20,940 (9,500)		16.08 (4.90)	9,423 (4,275)
GD310RC-3A	22,580 (10,250)		18.70 (5.70)	10,161 (4,613)
GD37-5H	24,010 (10,890)		19.19 (5.85)	10,804 (4,900)
GD37-6H	24,910 (11,300)		16.75 (5.11)	11,210 (5,085)
GD40-HT	33,160 (15,040)		20.18 (6.15)	14,922 (6,768)
Fiat Allis				
65	3,380 (1,532)	8,220 (3,728)	14.33 (4.37)	5,220 (2,367)
100 C	8,170 (3,706)	19,990 (9,068)	19.57 (5.97)	12,672 (5,748)
150 C	8,380 (3,800)	20,780 (9,425)	19.57 (5.97)	13,122 (5,951)
200 C	8,610 (3,905)	21,140 (9,590)	19.57 (5.97)	13,387 (6,072)

*Operating weight equals weight of machine with ROPS and operator. Additional equipment such as rippers will increase weight substantially.

†Centrifugal force is based on $F = MV^2/R$ where M = operating; V = maximum unloaded velocity; and R = minimum turning radius. Maximum centrifugal force is taken as μN where μ = coefficient of friction (0.45) and N is normal (empty) operating weight.

TABLE 5.11 Vibratory Rollers

Model no.	Operating weight, lb (kg)	Vibratory force, lb (kg)	Cycle of vibration, cpm	Wheel base, ft (m)
Komatsu				
JV 16	2730 (1240)	F 3530 (1600)	3300	2.95 (0.90)
JV 25	5310 (2410)	R 5510 (2500)	3000	5.25 (1.60)
JV 32W	6830 (3100)	F 4410 (2000)	3000	4.93 (1.50)
		R 3090 (1400)	2500	

Model	Empty weight,* lb (kg)	Loaded weight, lb (kg)	Length weight, ft (m)
Econoroll	1140 (517)	2275 (1032)	6.25 (1.90)

*Includes 175-lb (80-kg) operator.

TABLE 5.12 Backhoe Excavators

Model no.	Operating weight, lb (kg)	Ground contact area, ft^2 (m^2)	Ground pressure, lb/ft^2 (kg/m^2)	Length of track, ft (m)	Load on each track, lb/ft (kg/m)
Caterpillar					
215	38,100 (17,282)	39.77 (3.70)	958 (4671)	11.93 (3.64)	1597 (2373)
225	51,600 (23,405)	46.37 (4.30)	1113 (5443)	12.84 (3.92)	2,009 (2985)
235	84,430 (38,297)	71.46 (6.63)	1182 (5803)	16.50 (5.03)	2558 (3807)
245	130,800 (59,330)	83.94 (7.80)	1558 (7606)	18.25 (5.56)	3584 (5335)

TABLE 5.13 Concrete Power Buggies

Model no.	Empty weight, lb (kg)	Maximum weight,* lb (kg)	Wheel base, ft (m)	Maximum centrifugal force,† lb (kg)
Scoot-Crete				
WG-12 (walking type—no operator included)	660 (299)	2660 (1206)	3.33 (1.09)	1197 (2639)
N-12	1000 (454)	3175 (1440)	3.67 (1.20)	1430 (649)
T-52	1300 (590)	4475 (2029)	4.92 (1.61)	2014 (913)
T-70	1450 (658)	5825 (2642)	4.92 (1.61)	2621 (1189)

*Maximum weight = empty weight + 175-lb (80-kg) operator + weight of maximum load.

†Centrifugal force = 0.45 × (normal loaded weight).

6

Construction Dewatering

by Robert G. Lenz

Construction dewatering has as its purpose the control of the surface and subsurface hydrologic environment in such a way as to permit the structure to be constructed "in the dry." Although construction dewatering is a key element of the construction program, it is important to remember that the overall purpose is to build a structure—not to dewater the excavation. The entire dewatering program and all of its components should therefore be oriented toward allowing the construction operations to be conducted in an environment where the groundwater and surface-water problems are under control.

Dewatering means "the separation of water from the soil," or perhaps "taking the water out of the particular construction problem completely." This leads into concepts like predrainage of soil, control of groundwater and surface water, and even the improvement of physical properties of soils and the directing of loads on coffer-dam structures, all of which are included in the general subject of dewatering. These go far beyond the idea of pumping water out of a hole, which can be termed "unwatering." Pumping is the easy part of dewatering. The important part is the collection of groundwater and of surface water so as to achieve the purposes of the dewatering program.

History

Modern dewatering probably goes back to the soil-stability problems in connection with constructing the Kilsby Railroad Tunnel in England in the 1830s. In that project a pocket of "quicksand" 1200 ft (366 m) long was encountered and the material was stabilized by pumping from a series of shafts and boreholes. That project, one of the first recorded engineering efforts to understand the movement of water in the soils and to analyze methods of accomplishing the dewatering results, was described by the engineer in charge, one Robert Stephenson, in a report dated 1841:

> As the pumping progressed, the most careful measurements were taken of the level at which the water stood in the various shafts and boreholes; and I was much surprised to find how slightly the depression of the water level in the one shaft influenced that of the other, notwithstanding a free communication existed between them through the medium of the sand, which was coarse and open. It then occurred to me that the resistance which the water encountered in its passage through the sand to the pumps would be accurately measured by the inclination which the surface of the water assumed toward the pumps and that it would be unnecessary to draw the whole of the water off from the quicksand, but to preserve in pumping only in the precise level of the tunnel allowing the surface of the water flowing through the sand to assume that inclination which was due to its resistance.
>
> The simple result, therefore, of all the pumping was merely to establish and maintain a channel of comparatively dry sand in the immediate line of the intended tunnel, leaving the water heaped up on each side by the resistance which the sand offered to its descent to that line on which the pumps and shafts were situated.

These careful observations and deductions eventually led to the discovery of the parabolically curved drawdown gradient and to the concept that a construction excavation may be predrained by maintaining a balanced relationship between pumping rates and gradients.

With beginnings like this, the use of shafts and boreholes (today called sumps and wells) spread throughout the construction industry. Also developing at this time was the art of handling unstable soils with fantastic results with only the crudest of tools. Gravel, salt hay, French drains, sheeting, sandbags, and steam-powered pumps were the tools of the trade until the wellpoint concept came along in the mid 1920s.

During the 1930s, dewatering was largely an art, and its practitioners did not really try to incorporate the physical sciences into the procedures. In the late 1930s and certainly in the 1940s and 1950s, the various disciplines of science including soil mechanics and hydrology were integrated into the art of dewatering. Today, the dewatering practitioner can and does draw regularly from all the engineering disciplines so as to produce the best possible result. Fortunately or unfortunately, as the case may be, dewatering operations still require people with the practical sense and ability to visualize the movement of water in the ground; the art has not been replaced by science entirely. We still need people who know and understand water and soils and who have the practical experience to work with the materials at hand to accomplish the end result.

Scope of This Presentation

It is assumed that the readers have a background in construction or engineering or both and that they are familiar with some of the more elementary concepts of soil mechanics and hydrology. This chapter is intended to give readers an overall picture of the elements that are involved in construction dewatering. It starts with analysis, touches on the techniques of design, and discusses dewatering methods, their principal features, and advantages and disadvantages. The fac-

FIG. 6.1 An excavation for a very large drydock with a peripheral deep-well system augmented by an interior wellpoint system to dewater close to an impervious bottom. Excavation, pile driving, and concrete placement are proceeding unencumbered by dewatering operations. *(Moretrench American Corporation.)*

tors influencing costs are discussed, as well as the procedures and techniques commonly employed in the execution of dewatering work.

This presentation is not intended to enable readers to become dewatering experts. Rather, it is intended to provide an overview of a construction specialty which invariably is on the critical path and is frequently one of the more important determining factors in the success or failure of the project as a whole. References at the end of the chapter will be helpful to readers who wish to go beyond the scope of this presentation. The purposes of this presentation will be satisfied if construction and engineering people who encounter dewatering problems are aware of the methods of predicting necessary efforts, costs, and results, thereby improving their judgment with respect to construction dewatering.

FACTORS INVOLVED IN DEWATERING ANALYSIS

Dewatering Objective

The objective of a dewatering program is not always defined. To fully understand the problems that might be associated with building a structure below the groundwater table, all relevant facts and conditions must be gathered and understood. These include the physical characteristics of the structures to be built; their length, width, and depth; the types of cofferdams that may be considered (earthen dikes, steel sheet piling, H beams and lagging); the location of deeper parts of the structure, including sumps, foundation piles, potential overexcavation, and backfilling (to remove unsuitable materials); and whether the work will be conducted as a single excavation or the particular structure is one of several that may be done in sequence or as a group.

With these facts in hand, one can begin to appraise the nature and scope of the dewatering problem. For example, is the dewatering problem simply a question of lowering the water level in a free-draining granular aquifer of substantial depth with no particular concerns for pressure relief in underlying aquifers or stability of questionable soils? Or, to the contrary, are there underlying aquifers which will require independent or supplemental pressure relief, and are there soils in the subsurface picture which, in their natural state, will present construction stability problems, and should the dewatering program address itself to their improvement?

Figures 6.2 and 6.3 are intended to depict the evolution of a dewatering analysis for a hypothetical but somewhat typical structure, introducing perhaps an excessive number of factors and complications which bear on dewatering analyses and procedures. In plan and in section, the physical characteristics of the structure to be built are indicated along with the adjacent structures and the excavation's proximity to an open body of water. Superimposed on this takeoff initially are the soils typical of this site as well as construction features of the excavation dictated by its environment. This includes support systems and slopes. Typically, a takeoff such as this can be expanded after the design phase to include most, if not all, of the salient features of the dewatering program. Figures 6.2 and 6.3 will be referred to in subsequent parts of this chapter.

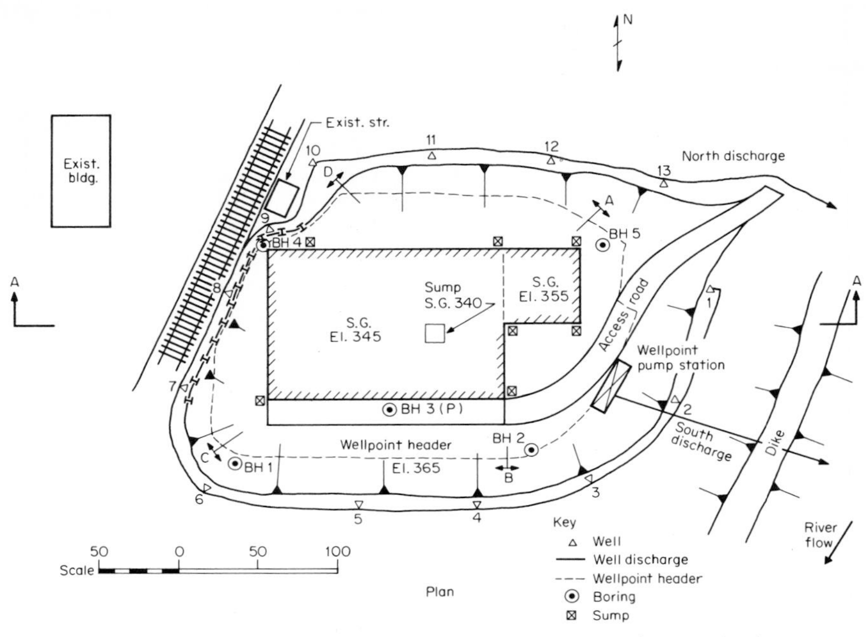

FIG. 6.2 Hypothetical dewatering job. *(Moretrench American Corporation.)*

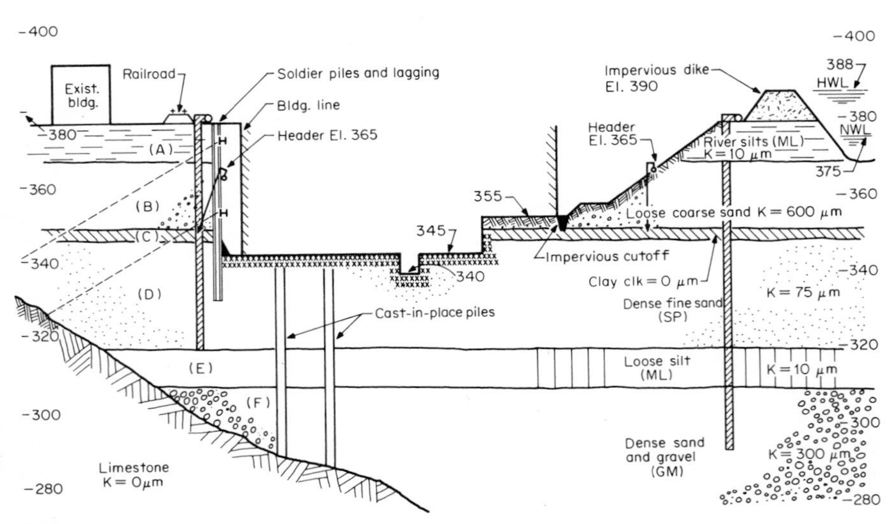

FIG. 6.3 Section AA hypothetical dewatering job. *(Moretrench American Corporation.)*

Soils and Their Characteristics

The nature and extent or continuity of the soils involved in an excavation and below it is the single most important determining factor with respect to how much water must be pumped and what techniques will be used in collecting the water. To this end, one must assemble the available soils information and determine whether the scope of the project requires additional information.

With the availability of a series of borings with soil classifications that are understood, one can construct a soil profile, paying particular attention to the stratification in the underlying soils. The continuity of all layers is important, because that has a profound effect on groundwater movement. The densities of various soil strata are also important. Information concerning density can then be obtained from descriptive adjectives, as well as from blow counts or other information contained in the borings. Occasionally, soils can be described with colloquial terms, which can convey significant meaning to the experienced analyst. Bulls liver, sugar sand, and gumbo are terms that one encounters in local areas, and each of them has a specific meaning and a characteristic which can be very important. Other descriptive information can be significant, for example, cohesive, loose, dense, soft, hard, cemented, and varved.

Is there permeability information available that can be used reliably in the dewatering analysis? In Fig. 6.3 such information has been included in the soil profiles and the reader will appreciate the significant variations in permeability that are characteristic of different soils. Numerical values used are typical, and one can readily see that in most soils, the horizontal movement of water is frequently determined by the most pervious layer, whereas the vertical movement of water is determined by the least pervious layer. Actual samples of the soils are often an aid to the evaluation of the continuity of layers. The color and other characteristics of the material can give the analyst a good idea of continuity between borings.

Rock and Its Characteristics

In addition to the soils, the underlying rock and its location should be known and understood. Is the rock an aquiclude or perhaps an aquifer, such as some of the limestone formations existing in various parts of the country? If, as in Fig. 6.3, the piles for the structure will extend to the rock, one must be sure that penetration of otherwise not dewatered or pressure-relieved soil strata will not cause a vertical flow along the pile, leading to a boil or a blow in the excavation. Quite frequently the specific location of rock is not known in dewatering investigations, but again the analyst must have a concept as to where the rock is, because the thickness of the overlying aquifers will have a sizable effect on the magnitude of the dewatering problem.

Geology

As indicated previously, the soils information from borings and other sources should be developed into a geological profile which will enable the analyst to understand the movement of groundwater in the ground. The analyst should be aware of the glacial history of the site, should that be a pertinent factor, and should be aware of the deposition mechanisms with soils of alluvial or sedimentary origins. Other characteristics of the soils should be understood: Has there been a compaction by natural means? Has there been preloading through glacial action or previous structures at the site? If the site is in a river valley, what are the deposition characteristics of the river? What has been its flooding history? How has scour and redeposition affected the dewatering problem?

These are all questions that can never be fully or positively answered, but an understanding of the site and its geological history can frequently help in avoiding problems.

Hydrology

Having assimilated the information concerning the nature of the excavation to be made and the soils in which it is made, one should understand where the water levels are and where the source of water will be when the dewatering system is functioning.

The water pumped by the dewatering system generally has two components: first, the storage that is required to be evacuated from the site and nearby areas to lower the groundwater table or tables; second, the steady-state recharge to the dewatering system while it is maintaining the lowered water level. Commonly in construction dewatering, the steady-state pumping volume is only one-third to one-half of the initial pumping volume. This is particularly the case in pervious soils with substantial depth of aquifer where the gradients will be very flat and the dewatering effect widespread. Time available for lowering the water level, of course, also bears on the relationship between initial pumping volume and steady-state recharge. The hydrological characteristics of the site can be evaluated from simple soils and geological analyses, assuming certain characteristics of the soils involved, or they may be measured directly by pumping tests, both pump-out tests and recharge tests. Occasionally it is important to measure the permeability and transmissibility of a particular stratum rather than the whole profile. Should this be the case, pumping tests must be designed accordingly. Frequently, a pumping test at one location can be augmented by test pumping of piezometers at other locations. Sometimes this procedure can be used to sustain the geographical inferences that are frequently drawn from pumping tests. It is important to understand the sources of water that might be encountered in pumping tests, and if necessary, make provisions to grout holes caused by drilling which are connections between aquifers so as to avoid contributory problems during the actual dewatering program. As a case in point, it is common practice in deeper cofferdammed structures in Florida limerocks to be sure that all borings are taken outboard of the cofferdam so that vertical flows from untapped aquifers will not present problems during dewatering programs.

Geography and Meteorology

To fully understand the dewatering problems and to make a good design for their solution, one must be aware of all of the environmental factors involved in geography and meterology. The climate can be rather important and a major factor in selecting dewatering methods. The Arctic has its problems and the tropics have theirs.

Chemistry

Corrosion and encrustation are always of concern in major construction dewatering projects, particularly those of extended duration. Occasionally, groundwaters and some surface waters are extremely corrosive to the degree that even short-term dewatering programs must entail the use of specially designed equipment and materials.

The analyst should understand the corrosive effects of the water to be pumped and should understand the corrosive effect of the soils that the dewatering system may be installed into. If surface waters or wastewater from industrial processes leach into the ground or are to be pumped by segments of the dewatering system, their corrosive characteristics should be understood. Water temperature is a very important factor in the corrosion process. The temperature variations that may occur during the life of the dewatering program should be estimated and its effect on the corrosion process evaluated.

Encrustation of well screens, pumps, and piping systems can also contribute to the difficulties in achieving a successful dewatering program. Encrustation is influenced by pressure drops and by temperature, and occasionally dewatering systems must be designed to minimize pressure changes so as to minimize encrustation.

The gas and mineral content of the water to be pumped should be evaluated and its effect on the program known. Hydrogen sulfide, free CO_2, iron, and alkaline hardness are some of the more common characteristics of water which can have a great bearing on corrosion and encrustation. If the water to be pumped from the ground is to be used for recirculating purposes or groundwater-recharging purposes, it becomes critical to understand the corrosion and encrustation problems so that suitable chemical control or alternative procedures may be adopted.

Adjacent Structures

Adjacent structures can dictate dewatering techniques and methods and may even require artificial recharging of groundwater levels under them. The effect of the current job on the adjacent structure is always a topic of first priority, but the effect of the adjacent structure on the current job is sometimes very significant too. Was that structure built with dewatering methods that involved underdrains and sumps which can bring local dewatering problems that must be recognized and faced? Some of the most difficult dewatering problems are occasioned when a new structure goes deeper than an older existing structure and is directly adjacent to it. Is there previous dewatering experience in the area? This can frequently be a helpful piece of information with respect to pinpointing problems to be encountered.

Local Labor and Materials

What is the availability and the quality of local labor, materials, and equipment for a dewatering program? Where are the sources of sand and gravel, of cranes, drill rigs, loaders, other construction equipment, and, of course, people? All these factors should be understood so that the dewatering analyst's designs can produce a practical program that can be executed.

Construction Methods

The construction methods which will be used to excavate and to construct the structure must be understood by the dewatering analyst. Will the excavation be made in open cut or with steel sheet piling, a cellular steel cofferdam, or perhaps with H beams and lagging? In the case of a conduit, will it be cut and cover, or will the work be done by tunneling? Is a tunnel job to be done in free air, partial air pressure, or full air pressure? Perhaps a combination of methods will be used. What kind of excavating equipment is available and will be used—scrapers, draglines, trucks, bottom dumps, clamshells, dredging? Each of the various techniques and equipment employed in excavation jobs may have its own particular impact on the dewatering program and the need for a particular dewatering result. Keep in mind that the purpose of the program is to build a structure, not to design and build a beautiful dewatering system. The dewatering system will serve its purpose if the remaining portions of the construction program can proceed unimpeded by water problems.

Sequence and Duration of the Work

Can the dewatering system be installed external to the excavation? Must portions of it be installed internal to the excavation? Must portions of the dewatering system be relocated as construction proceeds? What are the requirements for pressure relief as the structure is being built? Is uplift a consideration even after the foundation and the backfill are completed? Must the superstructure be in place prior to discontinuing pumping? These are some of the questions that have to be answered, and to do so, some knowledge of excavation, fine grading, pile driving, mud mats, base pores, wall pours, backfilling, machinery installation, and superstructure programs should be available.

Occasionally, the removal of a dewatering system becomes a critical feature of the program. Sometimes it is not feasible to remove dewatering systems which are buried under structures, and consideration should be given to grouting up of piping so as to prevent voids in the future and possible detrimental effects to subsequent construction work. The technique of abandoning a dewatering system must consider the effect of the interconnection of several aquifers and the avoidance of problems as a result of it.

Summary

In this section scores of questions were raised and very few answers given. The reader should understand that any of these items may directly influence the success or failure of a dewatering program. In the writer's experiences, conditions which cause dewatering difficulties are invariably unrecognized at the outset of a program, and the difficulties are not corrected until an accurate understanding of the problem is achieved. If one knows the problem, the solution is generally easy. If one does not really understand the problem, the success of corrective procedures can only be hit-and-miss.

ANALYTICAL TECHNIQUES—DEWATERING DESIGN

Having assembled the information and facts discussed so far and having decided upon the dewatering objective or objectives, we are now ready to use analytical techniques to determine (1) pumping quantities, (2) the manner in which the aquifer or aquifers will perform, and (3) the methods of collecting the water.

A number of parameters will determine pumping quantities and aquifer behavior. They may be determined by full-scale pump tests, by results from previous relevant experience, by careful analysis of soils and hydrology and extrapolation of their characteristics into quantitative assumptions, or perhaps by simple assumptions made by experienced designers. The writer hesitates to treat this subject in great detail because a proper understanding of the limitations of various mathematical approaches is necessary to avoid expensive mistakes. A number of the references listed at the end of this chapter treat the subject in detail, and the following presentation is intended only to give the reader an overview of the various factors which determine pumping volumes and aquifer characteristics.

The basic mathematical relationships concerning flow in soils assume that the aquifer is an ideal aquifer. This means that it extends horizontally in all directions beyond the area of interest without encountering recharge or barrier boundaries. Its thickness is uniform throughout. It is isotropic, i.e., its permeability in the horizontal and vertical directions is the same. Water is released from storage, the head is reduced instantaneously, and the pumping well is frictionless, very small in diameter, and fully penetrates the aquifer. In the ideal water-table aquifer, the additional assumption is that the phreatic surface will rise and fall with pumping operations. The basic flow relationships for ideal confined aquifers (Fig. 6.4*a*) and water-table aquifers (Fig. 6.4*b*), respectively, are:

$$Q = \frac{2\pi K f(H - h)}{\ln R_o/r_w}$$

and

$$Q = \frac{\pi K(H^2 - h^2)}{\ln R_o/r_w}$$

where Q = pumping quantity
K = permeability
f, H, h, R_o, and r_w are as shown

Construction dewatering jobs seldom qualify as ideal aquifers; therefore, a number of modifications of the mathematical relationships have been developed for use in calculating flows in actual situations. Again, this subject is extremely involved and the reader is urged to consult other reference works, listed at the end of this chapter, to pursue the matter in depth. A very good presentation of this subject is given in Powers, *Construction Dewatering.*

As may be seen from the above formulas, pumping quantities are determined by dimensional factors and the characteristics of the soil. Obviously, the further one wishes to lower the water level, the more water one must pump. The larger the dewatered area is, the higher the pumping volume will be. Pumping volumes will also vary directly with the permeability of the soil. Permeability, in turn, is a function of the gradation, density, and grain shape of soils. Empirical relation-

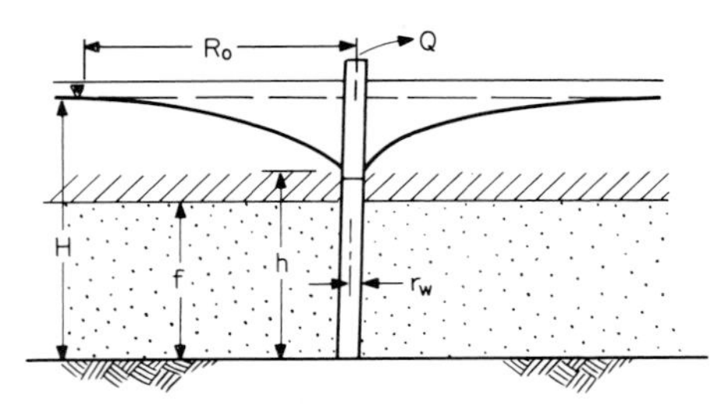

FIG. 6.4a Confined aquifer. *(Moretrench American Corporation.)*

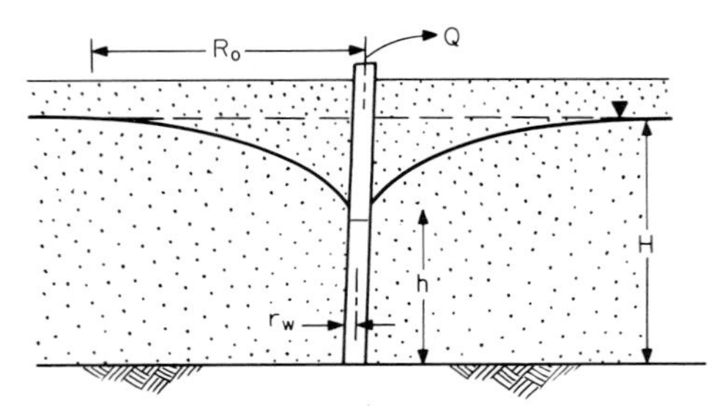

FIG. 6.4b Water-table aquifer. *(Moretrench American Corporation.)*

ships have been developed which take these factors into account, and accurate estimates of permeability may be made from grain-size curves of most soils, making allowances for particle shape and density. Seldom is an aquifer isotropic. Invariably, the horizontal permeability is many times that of the vertical permeability. Suitable recognition of this must be made when establishing a value for permeability.

Transmissivity is defined as the product of permeability and aquifer thickness, and is a useful term which can be used to describe the potential of an aquifer to produce water. In a confined aquifer, transmissivity is a constant. In a water-table aquifer, it decreases as the water level is drawn down, and the saturated zone through which the water flows decreases.

The radius of influence of a pumping system is frequently very extensive, particularly if recharge boundaries such as rivers or other bodies of water are not present. Barrier boundaries created by discontinuities in aquifer characteristics will affect calculations concerning pumping volumes. Often in construction dewatering, the radius of influence will be different in different directions. This too must be considered when calculating pumping volumes. The pumping quantity will be very much affected by the radius of influence, which of course determines the gradient which is produced when the water flows from its replenishment source to the dewatering system. In a given aquifer, the steeper the gradient, the higher the flow. It is not unusual to conduct dewatering operations in alluvial plains where the radius of influence can sometimes be many thousands of feet and the gradience very flat—1:200 to 1:400 not being unusual. By way of contrast, dewatering for, say, a subaqueous tunnel approach on an island in a bay surrounded by water may have a radius of influence not much higher than the depth of cut. Successful dewatering has been performed under these circumstances with gradients as steep as 1 to 1.

Sometimes in the course of a dewatering program, an aquifer will act as a confined aquifer until the water is drawn down below the top of the layer, at which point it will react as a water-table aquifer. Mathematical analyses must be adapted to this situation when applicable.

The basic mathematical relationships have been developed for fully penetrating pumping systems. The well fully penetrates the aquifer and draws from it evenly over its entire depth. Frequently in practice this is not the case, and a suitable mathematical adjustment must be made in the calculations.

A dewatering program involving only pressure relief theoretically involves a linear relationship between the pumping quantity and the depth to which the pressure is relieved. In theory, the pumping quantity remains constant with time, perhaps being influenced by factors such as water temperature and seasonal changes in the normal static head caused by variations in rainfall, river stages, etc. Occasionally a construction dewatering job will have these characteristics. If that is the case, the job will be sensitive to interruptions of pumping. Piezometers can go up and down drastically on an almost instantaneous basis as pumping is started and stopped.

In the more common water-table situation, the pumping quantity is actually composed of two primary components: The first is the steady-state flow to the pumping system that will continue after drawdown has been achieved. The second is the quantity of water that must be pumped to effect storage depletion. Even in the case of aquifers with high transmissivities, storage depletion can be the dominant factor influencing system design, particularly when the time available for dewatering is limited. This, incidentally, sometimes determines the dewatering method. For example, a peripheral deep-well system can be activated and pumped prior to the beginning of excavation, and storage depletion can be caused over a long period of time. In the typical wellpoint application, it is important that the drawdown be achieved quickly so as not to delay the excavation. Under these circumstances, sufficient capacity to cause storage depletion must be provided. It is not unusual for construction dewatering systems to pump only one-third to one-half of the initial pumping quantity after drawdown and after the bulk of the storage has been depleted. Again, mathematical relationships can be used to calculate storage volumes and depletion quantities.

Pumping tests are sometimes conducted when the success of dewatering programs is critical. The test generally involves the installation of a well or a series of wells with piezometers radiating in directions significant to the project. The test may be conducted by steady-state pumping or by transient-state pumping, and mathematical relationships are used to determine aquifer characteristics as a result of the pump test. Sometimes the extrapolation of pump-test results on a particular project into areas not properly instrumented during the pumping test can lead to costly errors. Concepts such as the specific capacity of a well or even a piezometer which may be test-pumped can be valuable in terms of sustaining assumptions with respect to the continuity of aquifers and their characteristics. Frequently, the recovery characteristics of a well when pumping is stopped can be indicative not only of the well's efficiency and performance, but also of aquifer characteristics. Plots of drawdown and recovery can be made and used to classify aquifers as confined or water table, to determine recharge and barrier boundaries, and in general, to convey to the designer a good understanding of the behavior of the aquifer. Intelligent instrumentation can identify well losses and eliminate this as a major source of error in dewatering computations. In some instances, pump-in tests are valuable. Generally, this is the case when the soils or rock in question contain extremely large void systems. Under these conditions the technique of preparing the hole for injection testing has only a small bearing on the results.

Several different methods may be used to determine the pumping volume for any particular project. Those which are based on mathematical relationships must be carefully adapted to the problem or the job at hand. The experienced designer has the added ability to correlate mathematical results with a practical experienced concept of pumping quantities. It enables the designer to at least know that the decimal point is in the right place. Occasionally, dewatering designs made by people without such experience are indeed wide of the mark by as much as a factor of 10. The experienced designer uses other techniques in evaluating dewatering flows. For example, the designer could calculate the flow to a dewatering system that might have a circular configuration by simply assuming that the radius of the pumping well is the radius of the dewatering periphery. The designer can calculate a dewatering flow by utilizing a cumulative drawdown method, wherein the drawdown at any particular point on the project is the accumulative drawdown from each of the pumping wells. In this case, appropriate assumptions are made with respect to the pumping quantity for each well, and a sufficient number of wells is included in the design to give the required drawdown. This latter procedure is particularly valuable when the well array does not conform to a circle or to a rectangle, which can be equated to an equivalent circle.

The cumulative drawdown concept is also helpful on line jobs, where the dewatered subgrade constantly moves. In this situation, the most economical design incorporates spacing of dewatering devices such that the well remote from the particular point of dewatering contributes only in a minor way at that time, and in a major way at the time when the dewatering point is adjacent to that well. While this entails pumping large quantities of water, installation costs are minimized.

In many dewatering situations the pumping quantity is not particularly critical. The key factor may be the spacing of dewatering devices and the ability to effect a drawdown close to an impervious layer. Here, designs are generally matters of judgment and based on knowledge and experience with different dewatering devices and the practical realization of what can be accomplished. In still other cases the paramount factor might be the degree to which soils can be stabilized by removing minute quantities of water. Again, design experience and judgment play very important roles, particularly if the designer intends to make use of more pervious layers to assume that the dewatering effect can be propagated away from the dewatering device.

Of course, aquifer performance is a very important ingredient in determining pumping quantities, but there are other reasons for investigating aquifer performance. One may be interested in the gradients away from the pumping system relevant to the effects of the operation on adjacent structures, other pumping operations, soil consolidation, or any number of factors which could be influenced by construction dewatering. One may be concerned with failure of a portion of a pumping system and its effect on the overall program. If one well in a well array fails, what happens to the water table in that area? If a segment of a wellpoint system becomes inoperative because of a break in a header system, what happens? The designer must consider these possibilities, and the design should recognize reasonable probabilities of difficulty and incorporate features appropriate to the problem.

When a design has established the pumping quantities and the aquifer-performance characteristics, the next logical step is to determine the method of collecting the water. Here, of course, one must be aware of the options and their characteristics; the advantages and limitations of each dewatering approach. The method of collecting the water can influence the pumping volume. For example, fully penetrating wells will pump more water per well than partially penetrating wells but they will also be more expensive. However, a well whose capacity and efficiency deteriorates significantly when the water levels are lowered can be a very expensive well. The designer must have a feel for costs when drawing conclusions as to how to design the dewatering system. Subsequent sections will cover these considerations in detail.

Readers are again urged not to plunge into dewatering design by working out a formula which may appear in this or other texts. To utilize the correct mathematical approach requires a thorough understanding of all of the factors involved in the project and of the basis on which the mathematical relationships have been developed. It is hoped that from the foregoing, readers have obtained sufficient insight into the variables and the relevant factors so that if it is necessary for them to prepare a dewatering design, they will seek proper counsel.

DEWATERING METHODS, FEATURES, ADVANTAGES AND DISADVANTAGES

It may be useful to consider different dewatering methods in the same manner as one considers different tools. A given method or tool may be most appropriate to a particular problem, but sometimes one may be forced to use other approaches that may do the job at some lower efficiency or to some lesser degree. Awareness of the different tools or methods of dewatering can help one decide on the most appropriate procedure or, if circumstances warrant, combination of procedures.

Sumps, Ditches, Trenches, Surface Water

Sumps are important to any dewatering job for use in collecting surface water, storm water, and perhaps seepage which does not readily find its way to other predrainage facilities. In almost every major construction job, sumps are necessary because of the limited ability of soils to accept major quantities of storm water and because surface water falling within the excavation limits will accumulate at the deep places and should be removed quickly to facilitate the work.

Occasionally, sumping procedures can be the major dewatering method, and sumps are established and pumped in advance of the excavation so as to depress the general water table. In various parts of the country, techniques have been developed to cope with local soil conditions, and sometimes the art of sumping is practiced to a very sophisticated degree. For example, in the New York area years ago, a common method of handling bulls liver soils was referred to as a "peel-the-onion" technique. Under this procedure, very shallow slices would be taken out of an excavation, and after each slice, additional dewatering would be accomplished with the sumps by lowering them within the excavated area. In the South, "rim ditch and sump" is an expression that describes a way, under certain conditions, to conduct a satisfactory dewatering operation.

Occasionally groundwater can be controlled at the toe of the slope or at and below subgrade by means of ditches, in which either gravel or a pipe is placed for the purpose of collecting the water and conveying it to a sump. When this procedure is used to intercept seepage and stabilize the subgrade, it is frequently referred to as a French drain.

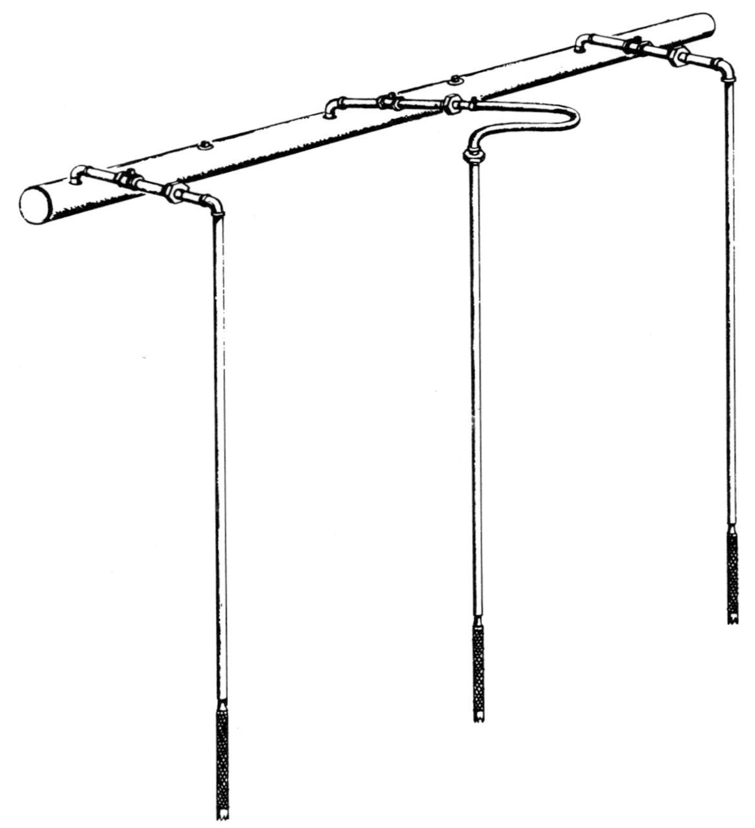

FIG. 6.5 Wellpoint systems. The basic, most versatile dewatering tools. Five styles of wellpoints to suit varying needs. System capacities from 500 to 25,000 gal/min (2 to 95 m^3/min) or more. Adequate for 15- to 20-ft (4.6- to 6.1-m) suction lift per stage. *(Moretrench American Corporation.)*

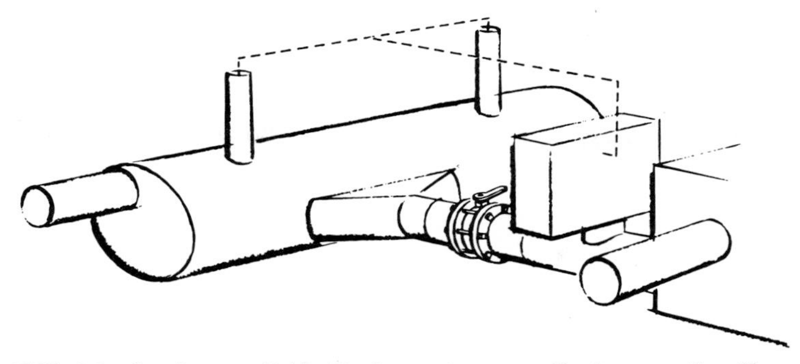

FIG. 6.6 Suction manifolds. Devices to increase effective capacity of horizontal and vertical wellpoint pumps. Provide air separation and smooth water flow to the pumps. *(Moretrench American Corporation.)*

Toe drains are used at the bottom of a slope so as to collect the water which may flow over impervious layers such as clay or rock. This water must be collected and the slope contained so as to preserve its integrity. In the circumstance where an impervious layer exists near or above subgrade, toe drains are frequently necessary even when the major dewatering work is accomplished with deep-well systems or even wellpoints. Alternatively, an impervious dike of clay may be built up at the toe of the slope so as to build up the water behind the dike and cause the predrainage system to function more effectively.

Wellpoints

Small pipes, up to 2½ in (6.3 cm) in diameter, connected to screens at the bottom and to a vacuum header pipe at the surface constitute a wellpoint system. A few to several thousand individual wellpoints may be connected to the manifolds, along which is placed a pump (combination centrifugal and vacuum) that has as its function the separation and disposal of air and water so that the system can be constantly and effectively primed and the water removed. The effective suction lift of a wellpoint system is governed by the available vacuum and the atmospheric pressure, and by friction losses in the system. Effective lifts of 15 ft (4.5 m) are quite common at sea level, and under certain circumstances, lifts can be increased to as much as 25 ft (7.6 m). Wellpoints typically have capacities ranging from a fraction of a gallon (0.0038 m^3) a minute to 100 gal (0.38 m^3)/min, and they may be used in single stages or in multiple stages to accomplish deep dewatering. Each wellpoint is equipped with a cock for regulating flow and limiting air intrusion. Systems are made of steel, aluminum, or PVC, depending on the need for strength, durability, light weight, and corrosion resistance. Typically, horizontal pumps are used with shallow wellpoint systems and vertical turbine pumps used with multiple-stage or high-capacity wellpoint systems. Occasionally in a cofferdam, several stages of header pipe may be used with single wellpoints, which in turn can be connected to the headers at different elevations (see Figs. 6.5, 6.6 and 6.7).

FIG. 6.7 A typical two-stage wellpoint system dewatering a sewage-pumping station in very fine silts. Pumping small quantities of water stabilized a treacherous soil and permitted excavation in the dry. *(Moretrench American Corporation.)*

Deep Wells and Shallow Wells

Wells may be employed for dewatering singly or in multiples. They may be as shallow as 25 ft (7.6 m) and as deep as several hundred feet. They may be powered with line-shaft turbine pumps either electrically or engine driven, or by electric submersible pumps (Figs. 6.9 and 6.10). Well casings may be operated under vacuum, or they may be vented to the atmosphere. The characteristics of the pumps used in the well should be considered from the point of view of proper submergence and its net positive suction head characteristics. Wells commonly pump as little as 10 gal (40 L)/min and as much as several thousand gallons (liters) a minute. They may be placed with or without a filter pack around the screen. They may employ various types of screens from simple slots to sophisticated wire and mesh configurations. Wells can be as small as 4 in (10 cm) in diameter and as large as 36 in (90 cm) in diameter. Developing procedures are generally employed to bring wells up to their maximum potential, particularly in well-graded soils. Each well may have a single discharge, or multiple wells may have a common discharge system.

Electrical distribution can be a major component increment of the dewatering system if a large number of units are to be electrically powered. This may also entail consideration of standby power sources, generally provided by generating sets.

FIG. 6.8 A peripheral deep-well system with 100 wells on a 1-mi (1.6-km) long perimeter dewaters a large excavation 60 ft (9 m) below mean sea level in variable sand, silt, and limerock soils; pumping quantity about 12,000 gal/min (45 m^3/min) *(Moretrench American Corporation.)*

Ejector Wellpoints

The ejector (sometimes called eductor) wellpoint overcomes the ordinary limitations of suction lift by employing a nozzle and venturi located within the ejector body, which in turn is positioned just above a wellpoint screen in the ground (Fig. 6.11). The apparatus requires separate supply and return risers, which may be concentric or parallel. The risers are connected to separate supply and return headers, and water under substantial pressure flows from the supply header down the supply riser and through a nozzle and venturi, creating a vacuum within the ejector body. Passages within the body permit water flow from the wellpoint screen generally through a foot valve. This water then joins the supply water and flows up the return riser and in the return header to a tank. Generally the supply water is drawn off the bottom of the tank, and the overflow of the tank becomes the net yield of the system. The ejector mechanism can create very high vacuums in the wellpoint screen, which makes ejectors a rather effective tool in fine-grain soils.

Ejectors are commonly used to lift water as much as 100 ft (30 m). They are not generally used for higher lifts because the efficiency of ejector systems is rather poor—generally in the order of 5 to 15% on a water horsepower basis. Other dewatering systems may approach 75% efficiency. Ejector systems are frequently operated when the volumes of water to be pumped are within the lower ranges, say up to 2000 gal/min at 100 ft (7.5 m^3/min at 30 m) Total Dynamic Head (TDH). In most ejector systems it is particularly important that the piping, nozzles, and venturi be sized to suit the circumstances because of the inherent low efficiency of the method.

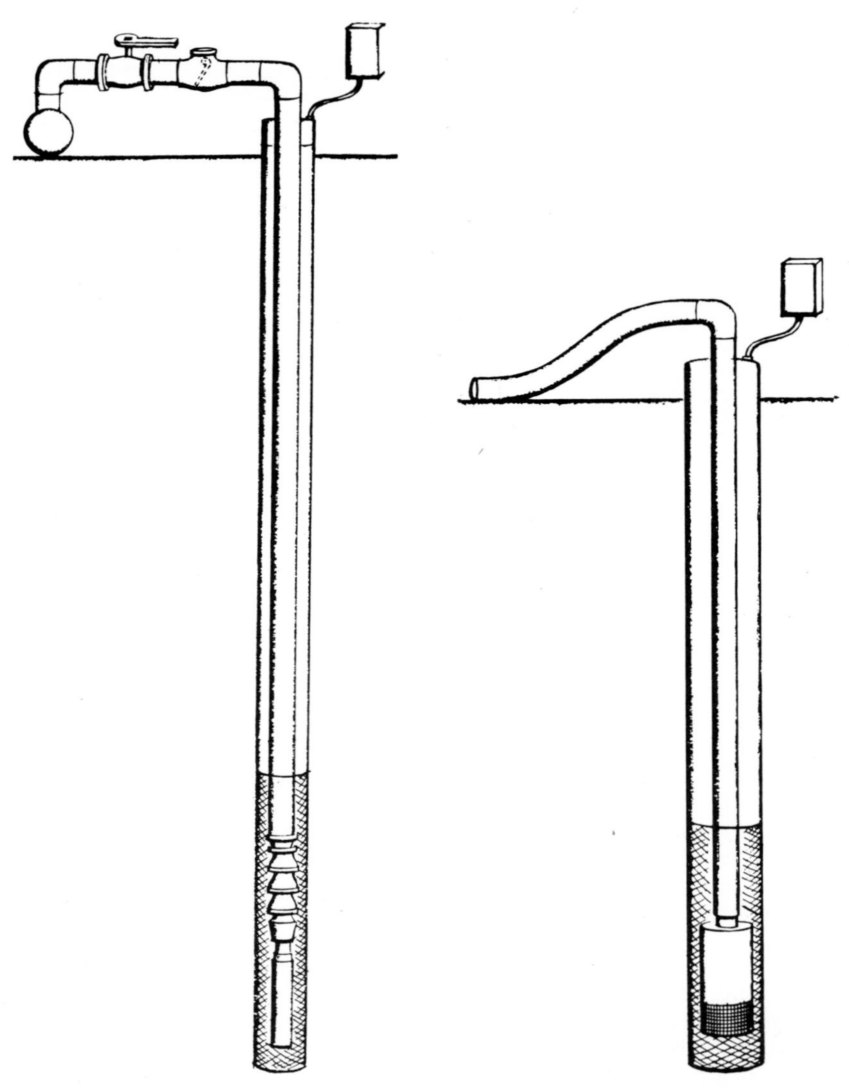

FIG. 6.9 Deep wells. Under favorable soil conditions, effective in deep dewatering and pressure relief. Equipped with turbine-type submersible pumps, or lineshaft pumps in capacities to several thousand gallons per minute. Can be drilled or jetted. *(Moretrench American Corporation.)*

FIG. 6.10 Shallow wells. Equipped with centrifugal-type submersible pumps, useful under favorable soil conditions for shallow structure excavations and trenches. Can be drilled, self-jetted, or jetted with a hole puncher. *(Moretrench American Corporation.)*

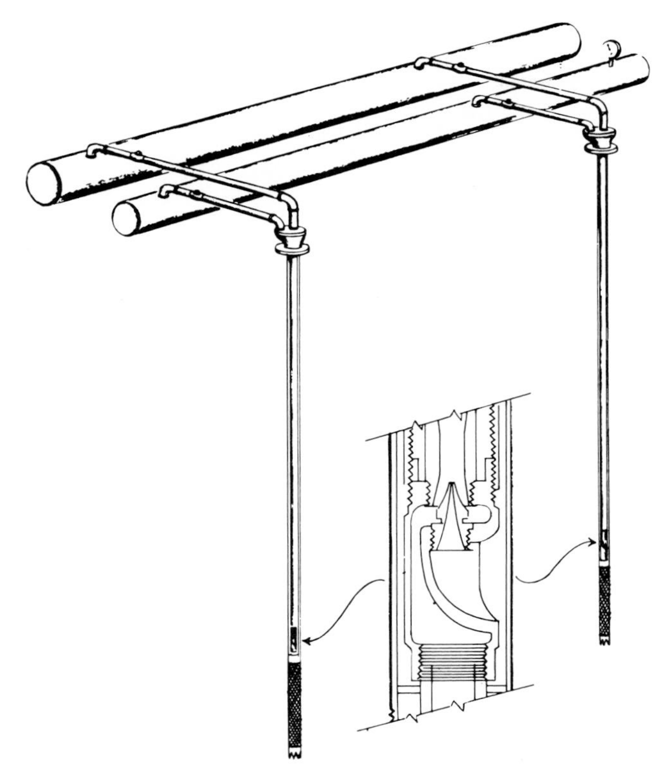

FIG. 6.11 Ejector system. Recommended in variable or stratified soils where depth exceeds 20 ft (6.1 m). Low unit cost per collection point offsets higher power consumption under many conditions.

Suction Wells

When a series of shallow or deep wells are connected to a vacuum manifold, they are referred to as suction wells (Fig. 6.13). Under this procedure, large quantities of water can be pumped very efficiently, and the limitations of suction lifts improved by reducing friction in the piping systems. Frequently, vertical turbine pumps are used with suction wells so as to pump large quantities of water efficiently. Suction wells have been used in multiple stages, and occasionally with multiple stages of header pipe connected to a single suction well in a cofferdam (Fig. 6.14).

Horizontal Drains

These are occasionally used to drain unstable soils such as highway cuts, spoil piles, etc. While modern drilling techniques readily permit the insertion of horizontal pipes into predrilled holes, the installation of proper filters and screens

FIG. 6.12 Five stages of suction wells pumping as much as 30,000 gal/min (113 m^3/min) lower the groundwater more than 80 ft (24 m) below high-river stage for a lock-and-dam excavation adjacent to the Mississippi River. *(Moretrench American Corporation.)*

and the techniques for developing the hole are very limited, and therefore horizontal drains are not used frequently for construction dewatering.

Vertical Sand Drains

Vertical sand drains are frequently utilized to improve the drainage of layered soils. The technique evolved from the common line wellpoint job or sewer job where wellpoints might have to be installed on both sides of a trench to adequately cut off lateral seepage under certain layered soil conditions. It was found that in some instances it would not actually be necessary to install the wellpoints on the opposite side of the trench, but simply to jet a hole and sand it, removing the jet pipe at the conclusion of the sanding process. This vertical column of sand extending through layers sometimes provided a path by which lateral flows could migrate to a deeper layer and thus to the dewatering system on the opposite side of the trench. Under some conditions, hundreds of sand drains can be placed in a single day very cheaply. When this method is employed properly, it can be extremely effective.

Vertical sand drains, of course, are the key element of many soil-consolidation projects. In this procedure, compressible soils may be consolidated during the construction program by improving the vertical drainage and causing a pressure differential to exist between the pores of the compressible soil and the sand drain. This differential pressure may be caused by common surcharge or sometimes by supplemental pumping operations with or without vacuum effects. Sand drain projects require an integrated design involving diameter, spacing, sand characteristics, surcharge time, and installation technique. The earliest sand drains used for consolidation purposes were both jetted and driven with pile-driving equipment. Hollow-stem augers were also devised and introduced to install sand drains, although on a very limited basis

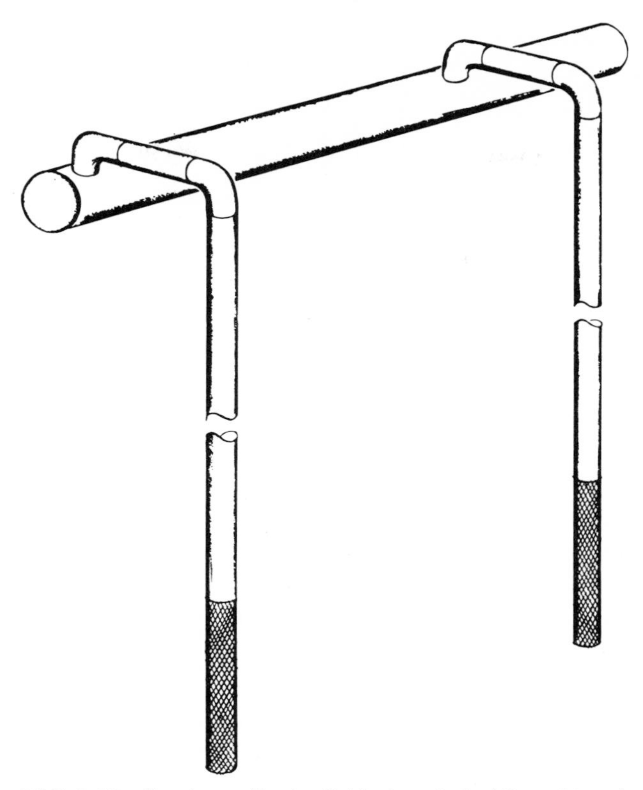

FIG. 6.13 Suction wells. Available 4- to 8- in (10- to 20-cm) diameter. Effective in handling large volumes of water. Can be staged for deep excavations. *(Moretrench American Corporation.)*

In recent years, improvements in jetting equipment and procedures have permitted jetted sand drains to be placed as economically as other methods, and occasionally at less cost in competitive-bidding situations. Jetted sand drains are inherently more effective than sand drains installed by other methods, and are installed using one of three basic methods:

1. Jetting an open-ended casing with either an internal or an external boil, or a combination of the two. When the casing is in place, it is filled with sand and the casing extracted.
2. Utilizing a bog cutter or a reamer on a jet pipe or hole puncher (which may or may not have rotating capabilities for clay and cemented soil penetration). This method relies on a positive head of water to support the hole during sanding.
3. Jetting an open-ended jet pipe to which is attached a pressurized sand vessel

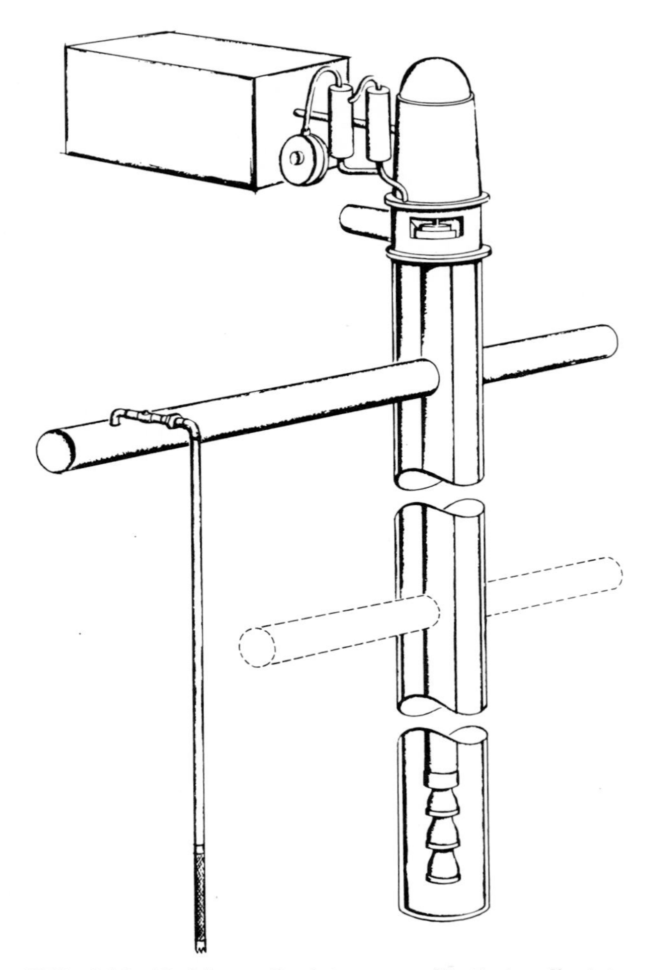

FIG. 6.14 Turbine wellpoint pumps. Vertical wellpoint pumps, used for handling large volumes of water with minimum operating units, or to conserve space in cramped cofferdams. Convenient for multistaging. Available in diesel or electric line-shaft configuration to 10,000 gal/min (38 m^3/min) and as submersibles to 500 gal/min (1.9 m^3/min). *(Moretrench American Corporation.)*

and valving apparatus to permit the direct application of the sand into the jetted cavity during withdrawal of the mandril.

Each of these three methods results in a nondisplacement sand drain, generally free of smear, with a washed periphery establishing good contact between the sand drain and the native materials. It is not unusual for sand drain projects

FIG. 6.15 Sand drain apparatus with 150-ft (46-m) leads capable of installing 18-in (45-cm)-diameter and 122-ft (37-m)-long sand drains. Sand pressure vessel has a capacity of 13 yd^3 (10 m^3). *(Moretrench American Corporation.)*

to involve many thousands of individual drains with diameters ranging from 6 to 24 in (15 to 60 cm), and depths to 125 ft (38 m). The sand drain technique has proven to be an extremely effective way of coping with very difficult soils in certain stabilization problems.

Electroosmosis

Electroosmosis is a process wherein a low voltage dc current is applied to the soil, utilizing anodes and adjacent cathodes which are generally part of a water-collection system, which may be one of the previously discussed systems. The application of electric potential to the molecules of water can overcome capillary attraction and provide an additional head, causing water to migrate through the soils to the cathodic collection system. Electroosmosis generally involves most of the cost of the dewatering system with the additional cost of a rather elaborate electrical distribution system and the power to operate it. However, it can be a way of stabilizing soils which are not otherwise drainable.

Steel Sheet Piling

Interlocked steel sheet piling is frequently used in an effort to cut off or limit the flow of water. When installed in open water, as in a river cofferdam, it is frequently necessary to seal the interlocks with cinders or sawdust so that the flow through interlocks can be reduced to the point that available pumping capacity can draw the water down. When installed in soils, continuous interlocked steel sheet piling can sometimes eliminate dewatering problems. For this to be so, the sheeting should make a positive cutoff with an impervious layer of soil or rock capable of resisting the differential pressures, and the interlock leakage should be small. In fine- or well-graded soils, interlock leakage is generally quite small. In clean permeable soils, it could be quite large unless interlock tension seals water passages. Quite frequently, the use of steel sheet piling does not eliminate dewatering problems, but simply minimizes the scope of a dewatering program. For example, it may be necessary to cause pressure relief below subgrade and below an impervious cutoff even though the walls of an excavation may be constructed of interlock steel sheet piling.

Slurry Trenches

In recent years, the use of a slurry trench backfilled with an impervious material to control groundwater seepage has gained prominence. Typically, a narrow trench is excavated to cutoff, using a backhoe or sometimes a clamshell. During excavation, the trench walls are supported by a bentonite water slurry which provides a positive head and prevents groundwater from entering the excavated trench and causing sloughing. As the trench is excavated, a selected backfill is placed. Generally, the backfill is mixed with the slurry prior to placement, and the backfill-placement techniques can be extremely critical in assuring that voids will not exist in the wall. The procedure results in a trench backfilled with an impervious material whose characteristics are generally such that the wall is flexible and even self-healing. The specific gravity of the slurry must be controlled so as to avoid having a liquid in which the backfill cannot be properly placed. Particular concern should be paid to the ability of the impregnated backfill to resist differential head without the migration of slurry through the backfill granular material, thus causing a small breach and eventual failure of the cutoff wall.

Diaphragm Walls

The procedures for digging the slot in which a diaphragm wall is placed are similar to those described for the slurry trench. The wall generally is a reinforced concrete structure which is placed in the panels by the tremie method. The panel joints can be formed by structural members placed vertically on 8- to 20-ft (2.4- to 6-m) centers, or alternatively, by excavating the trench for the wall in alternate panel sections. The diaphragm wall can serve the same purpose as a slurry trench with respect to the control of groundwater, although usually its primary purpose is that of a structural component of the work serving as the cofferdam during the construction period.

Ground Freezing

This is a process that has been used widely in Europe for over 100 years and has recently been introduced into the United States. The procedure entails the installation of freeze pipes along the line of the wall to be frozen. Pipes are generally installed on about 3-ft (0.9-m) centers, and each one consists of an external casing and an internal pipe. The two portions of the freeze pipe are connected to separate headers, one a supply header and one a return header. A refrigerated brine solution is circulated through the header system and through the individual freeze pipes. The brine is then returned to the freeze plant, and the heat which it has absorbed in its travels through the ground is removed. Ground freezing is a rather positive method of providing an impervious cutoff, provided the conditions relevant to the project are completely understood. Among the most important conditions are natural groundwater velocities, which have been known to prevent the effective formation of a freeze wall; the possibilities of subsequent movement of water around the toe of a freeze wall; the possible presence of extraordinary heat sources, such as pipelines adjacent to or penetrating a freeze wall; and the characteristics of the groundwater to be frozen. On rare occasions, a very severe problem can result if the water is very highly mineralized. The formation of the freeze wall requires substantial energy and substantial time. However, once formed, the wall is not subject to rapid deterioration unless a significant heat source is concentrated at one point. Most frequently, the maintenance of the freeze wall can be done on something less than a continuous basis, and the use of freeze walls in construction sometimes has a decided advantage in terms of the sensitivity to interruption of pumping that sometimes causes great dewatering difficulties. A typical freeze layout is shown schematically in Fig. 6.16. Ground-freezing applications are most common in connection with small, deep structures.

Grouting

Grouting procedures seldom serve as the primary means of controlling water. However, grouting can be an effective way of dealing with local dewatering inadequacies. Grouting programs generally take either of two forms: the filling of voids which would otherwise provide conduits for large volumes of water and subsequent piping problems or the penetration of soils so as to render them stable and not subject to movement under modest flow conditions.

In the first case, a number of different kinds of grouts have been developed for the purpose of filling voids economically. These can run from bentonite cement solutions which have little structural value and simply have the characteristics of a soft clay, to cement grouts which may be quite effective in imparting structural characteristics to the void system. With respect to grouts used for penetration of soils, again in soils with large voids, bentonite cement can be used cheaply and effectively to seal off water flows and to provide cohesion to gravel-sized particles. Cement grouts can do the same thing, but at a substantially higher cost for materials. Fine-grained soils are not effectively penetrated by bentonite cement or cement grout. A number of chemicals have been developed for the purpose of penetrating fine soils and some are quite effective under controlled conditions.

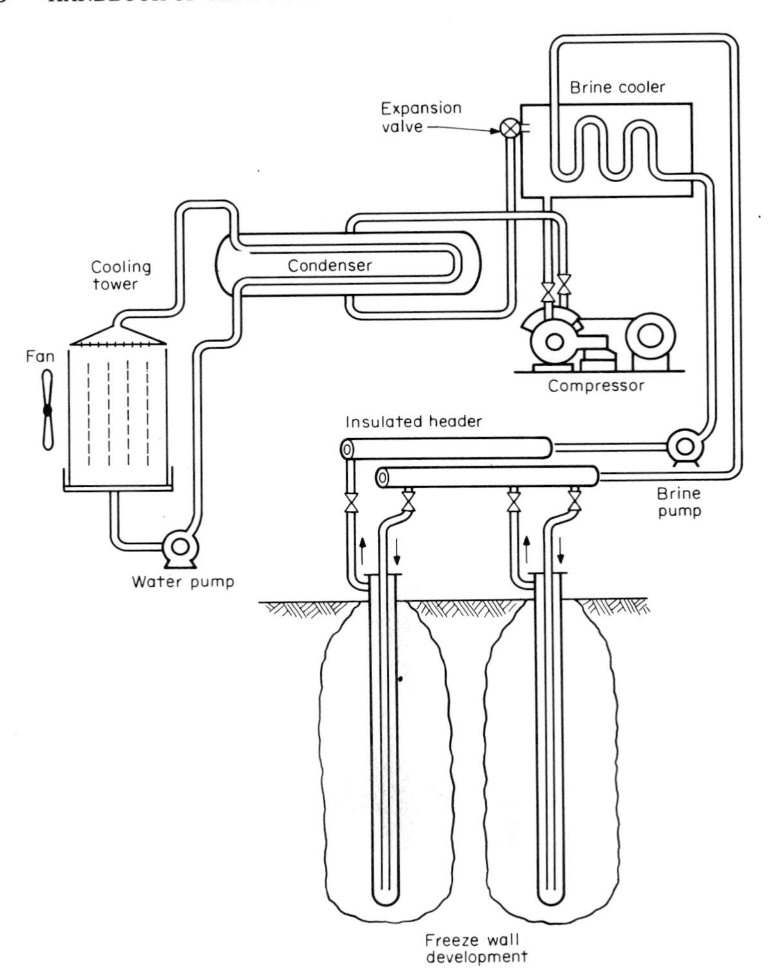

FIG. 6.16 Schematic freeze plant and system. *(FreezeWALL Corp.)*

Additives can be added to most grouts so as to cause gelling or setup of the liquid in a controlled time frame going down to seconds. Some chemical grouts are mixed with additives as catalysts to cause the set time to occur in a predictable fashion. Others are mixed in the grout pipe itself, again to control gelling.

The success of a grouting effort is difficult to predict, particularly under dynamic conditions. When groundwater is moving, there are techniques that have been developed to stop the movement. However, procedures are sometimes less than 100% successful. It is necessary to approach a grouting problem with the understanding that the injected grout will migrate in the directions of least resistance. This may or may not be the direction which will effectively deal with the water problem. If conditions were uniform and if soils were completely isotropic, it would be easier to predict the results of grouting and to utilize the tool with more confidence. This, unfortunately, is not the case, and grouting procedures should generally be looked upon as a remedial expedient which can be helpful in an overall dewatering problem.

Recharging

Groundwater recharging may be done for a number of reasons, including (1) the control of saltwater or other objectionable intrusion into a groundwater aquifer; (2) the return of good potable water to the groundwater "bank" during temporary dewatering procedures; (3) control of the characteristics of groundwater which may be migrating through a soil or rock capable of having adverse solutionizing effects; (4) prevention or avoidance of settlement and consolidation of compressible soils. The latter reason is the most common in connection with temporary dewatering work, although there have been major recharging programs established to control saltwater intrusion in a number of different localities.

Water can be returned to the ground through a number of different procedures. Recharging basins can provide a means of adding storm water to the groundwater table over a long period of time simply by preventing runoff into adjacent bodies of water. Occasionally, ditches can be effective in returning the water to the ground along the periphery of a temporary dewatering project. More frequently, though, groundwater is injected into the ground by means of wellpoints or wells, and the procedures for designing and constructing these recharging wells are similar to those of dewatering devices. Water can be injected into a recharging system under pressure, although the pressure is generally limited by the tendency of water to bypass seals which may be installed in the injection holes, or even to heave the ground in local areas because of excessive pressures. Reasonable pressures are ensured by utilizing standpipes so that it is impossible to overpressurize a recharge system. It is also important to avoid air entrainment.

Perhaps the single most important factor in a recharging program (other than the permeability of the soils through which the water is being injected) is the character of the water itself. Frequently, it is necessary to chemically treat even potable municipal water supplies for recharging purposes. One should understand that most waters contain dissolved and suspended solids. In the dewatering mode, the water migrates from the voids in the soil through the voids in the filter material, through a well screen, and then through progressively larger pipes until it is discharged. In the recharging process, the procedure is reversed and the water must flow through ever-decreasing voids to accomplish the purpose. Whenever the temperature or pressure of the water is changed, as happens in pumping operations, the characteristics of the water with respect to materials in solution can change. The control of sediments and precipitation of solutionized minerals is the key to successful recharge. It is not unusual to see a recharging system's efficiency deteriorate in several weeks because of problems with the water quality. The deterioration can be almost instantaneous if, for example, the recharging water is supplied from a municipal water system that might be subject to occasional heavy pumping, as during a fire, with its accompanying surging of the mains and agitation of encrusted materials. A shot of rusty, dirty water can totally foul up a recharging system. Acidizing may restore or improve the effectiveness of a system under such conditions.

A discussion of the need to recharge is in order. The most common cause of settlements associated with construction dewatering is not due to the lowering of groundwater tables under an adjacent structure and the subsequent increase in load that that causes. Rather, most frequently the water being pumped from an excavation contains suspended solids and fines because of inadequate sumping procedures or dewatering systems which have not been designed and installed properly. If a pumping operation is moving fines, and if it is to continue for a substantial length of time, it is not hard to compute how much material will be removed from the ground by the pumping operation. This can create voids in the ground and cause settlements. On the other hand, lowering the groundwater table does remove the buoyancy from soil particles, and the effective weights of those soils are increased with respect to the loads on lower layers. This can be roughly translated into an increase of load of about 40 lb/ft^3 (6280 N/m^3) of soil per foot (0.3 m) of water-table lowering. For example, lowering the water table 20 ft (6 m) can result in an increase in load of 800 lb/ft^2 (550 N/m^2) on an underlying soil layer. If that layer is compressible under that increment of loading, difficulties could ensue. Groundwater recharging can be a means of minimizing these problems.

The foregoing brief description of various dewatering methods only touches on the many facets of their use in construction dewatering. There are always exceptions to general rules, and sometimes various dewatering applications warrant unusual utilization of methods and procedures. The foregoing cannot be a complete set of criteria for the employment of dewatering procedures. We can only stimulate the reader's thinking and provide a basis for deciding whether or not professional consultation is warranted.

CHOOSING DEWATERING METHODS

Preceding sections have discussed in a general way the various factors, techniques, and tools involved in defining, analyzing, and selecting various dewatering programs. This section will deal with the actual decisions which are made in formulating the dewatering program. In order to illustrate the various concepts, reference will be made again to Figs. 6.2 and 6.3. The hypothetical dewatering job illustrated presents a complex (perhaps abnormally so) series of conditions and situations which may be used to illustrate the material presented thus far.

Dewatering Objectives—Logical Methods

The job in question is a major structure having a perimeter at the top of the excavation of approximately 1100 lin ft (335 m). The general subgrade is about 30 ft (9 m) below normal water level and the maximum anticipated river stage would involve a head of about 43 ft (13 m) over general subgrade and an additional 5 ft (1.5 m) over the sump, located in the interior of the structure. The excavation is to be made largely in open cut. A line of H beams and lagging on the westerly side will be used to support an existing railroad and structure. The dewatering objectives may be enumerated as follows:

1. Layer B should have its water level lowered to permit the excavation to be done in open cut and to minimize the residual flow over the top of layer C around the entire periphery. This latter point could be extremely significant in the northwesterly corner where lagging must be placed to support soils under the existing structures.
2. Layer D should be dewatered below elevation 340 so as to permit the work at subgrade to be done in the dry.
3. Layer F should be pressure-relieved to elevation 340 or thereabout to avoid excessive uplift pressures or conceivably a boil during the drilling of the cast-in-place piles.

Concerning the first objective, one should keep in mind that the total dewatering of layer B is virtually impossible, and the dewatering system chosen can only achieve the results of making the toe of the slope workably dry. This might entail simply lowering the water level in the layer close enough to the underlying clay so as to install an impervious cutoff as indicated. In the northwest corner where this procedure is impossible, one must consider other means of coping with what could be a very difficult situation. These might include a concentrated cluster of wellpoints socketed into the clay and chemical injection to make the sands immediately over the clay stable enough to place the lagging.

In dewatering layer D, initially this would be a confined aquifer, but as the water level was drawn down below the bottom of layer C, dewatering and storage depletion would occur. The general water table should be lowered at least 3 ft (0.9 m) below the prevailing subgrade so as to provide some time cushion should there be an interruption of pumping. The pressure relief of layer F would entirely consist of pumping from a confined aquifer.

The dewatering methods which have been chosen have been illustrated in the sketch. They fit the conditions of the site and the objectives of the program. A peripheral deep-well system with wells installed and screened through layers B, D, and F will accomplish objectives 2 and 3, and will permit sufficient lowering in layer B to install a single-stage wellpoint system around the periphery so as to accomplish objective 1. In all probability, the wells on the west side of the job will be shorter than on the east side because of the rock. Logically, pumping quantities on the east side will be highest, on the north and south sides quite significant, and on the west side minimal because of both the location of the river and the elevation of the rock.

Pumping Quantities, Storage Depletion, Steady-State Flows

Using the analytical techniques discussed earlier, modified to suit the conditions inherent in this project, pumping quantities can be calculated and, in part, estimated for each of the three aquifers in question. A typical procedure would be to tabulate the design conclusions as shown in Table 6.1.

In our hypothetical job, it is rather obvious that layer B will communicate

TABLE 6.1 Summary of Pumping Quantities

Strata	*K* per second per second	Expected *Q* at normal water level, gal/min	Increased *Q* due to high water level, gal/min	Remarks
Dike	0	0	0	Impervious material scour protected
A	10	negligible	—	Consider sand drains between wells 1 and 3
B	600	4000	5000	Impervious cutoff to be provided where possible
C	0	0	0	—
D	75	1200	1500	—
E	10	negligible	—	Concern with moving fines through well filter
F	300	2500	up to 3000	—
Rock	0	0	0	—
Totals		7700	9500	

NOTE: 1 gal/min = 0.00006 m^3/s

directly with the river and that the radius of influence for a dewatering system will be very small. This situation is likely even if the river is silty river, because of the scour that might occur during flood stages. On the other hand, layer D would seem insulated from the river, and the radius of influence there might be substantially larger even though the soils are not very pervious. Layer F again should be well-insulated from the effects of the river, and one might expect a rather large radius of influence. The well system depicted would fully penetrate layer D, but only partially penetrate layer F. Storage depletion would be a factor in layer B, particularly if the floodplain of the river extended far to the west of the excavation. This would also be a factor in layer D, but probably not too significant considering the shallow penetration of the excavation into layer D. Storage depletion is not a factor in layer F because it is a confined aquifer. The steady-state flows are computed in each direction for each layer. The design of the pumping system should include sufficient excess capacity to overcome the problems of storage depletion and to provide a good engineered factor of safety and system flexibility for the whole program. This will be further discussed.

Impact of the Dewatering Method on the Environment

In our hypothetical job, there might be a number of things to consider concerning the impact of the dewatering on the environment. Consolidation which could occur in layer A under the existing structures are probably minimal, past river fluctuations having provided the same loading cycles that would be caused by the dewatering system. This is probably not true of the loose silt (layer E). Here, the lowering of the water level could be expected to provide a substantial surcharge on this layer, causing potential consolidation. However, its depth and the location of the rock under the existing structure would probably minimize this as a concern. One probably would not be concerned about water-supply wells in layer B unless they were very close to the project. The connection of layer B to the river should confine the radius of influence. Water-supply wells are not likely in layer D because of its low permeability, but are possible in layer F. We would expect, too, a rather large radius of influence from a long-term pumping project, and this could cause problems. Incidentally, problems with adjacent water-supply wells rarely involve a drying up of the well, but rather a lowering of the water levels so that well pumps are no longer capable of providing the required supply. In conducting this dewatering program, the 13 wells would be installed prior to beginning the excavation, and pumping could begin immediately. The H beams along the railroad should be installed in the early phases of the work, and it will be necessary to conduct the excavation in such a way as to prepare a wellpoint berm at approximatly elevation 365, or about 15 ft (4.5 m) above the top of the clay (layer C). As the excavation is brought down, utilizing the ramp shown in the plan drawing, lagging should be placed between H beams, and as portions of the excavation hit 365, the wellpoints should be installed. A delay in the whole program because of the wellpoint installation is not inevitable, particularly if the perimeter of the excavation is brought down first and the wellpoint installation is carried on while the dirt in the middle is excavated.

As the excavation proceeds down toward the top of the clay (layer C), its location should be verified around the entire perimeter, and an impervious cutoff placed as shown. This will minimize the number of wellpoints in the wellpoint system and will make it practical to maintain a good vacuum on that system. If such a cutoff is not placed, it will be necessary to deal with a very difficult tuning situation with the wellpoints, possibly necessitating sheet-pile cutoffs in the high section of the structure in the northeast corner. Conducting the lagging operation in the northwest corner, again, could be very difficult because there is no opportunity to place a cutoff. It would probably be prudent to concentrate many wellpoints in that corner to minimize the problem, but it may be necessary to go through special treatment of the loose, coarse sand immediately above the clay in that corner. As the excavation progresses below the clay and into the fine sand layer (layer D), the dewatering program's adequacy will be proven. As drilling for piles commences, piezometers in layer F should be carefully monitored so as to avoid boils during the drilling operation.

Spacing and Depth of Dewatering Devices

The deep-well system contemplated in our hypothetical dewatering job must satisfy the three purposes enumerated previously. To repeat, it should lower the water level sufficiently in layer B to establish the wellpoint system at elevation 365; it should dewater layer D deep enough to conduct the excavation operation to subgrade in the dry; and it should pressure-relieve layer F so as to avoid boils.

The tabulations with respect to pumping quantities made in Table 6.1 constitute the design criteria for the well system. The number of wells required to achieve these objectives is determined by the ability of each well to pump water from the ground and by a number of other considerations. One must anticipate pump failure in a well, and if the dewatering system does not have the ability to maintain the dewatered result with a reasonable number of pumps out of service, the design would have to be considered deficient.

In well systems, a larger number of wells provides more redundancy in the system as a whole. Keep in mind that the pump in each well must lower the water level sufficiently in the well to achieve the design purposes. With factors such as storage depletion entering into the process, the steady-state pumping quantities are invariably only a fraction of the pump's capacity. In the illustration, we have chosen a well system with wells spaced as far apart as 100 ft (30 m) on centers, and as close as 50 ft (15 m) on centers. Obviously, the wells on the east, or river, side of the job will pump far more water than the ones on the west side. However, the conditions particularly in the northwest corner are perhaps the most severe with respect to dewatering the project.

Well-system design conclusions have been tabulated in Table 6.2 and the position of the screens in the various layers indicated. Anticipated pumping quantity from each stratum is indicated. Thus the design is verified, or if necessary, the number of wells is modified. Each of the aquifers must have sufficient screen and available pump capacity to achieve the necessary results.

A number of secondary considerations enter into the detailed design. Wells 6 through 10 have to be fairly short, and there is no real point in penetrating below elevation 320 because the amount of screen that could be placed in layer F would be minimal. These wells might well be installed with a bucket rig. The fact that the pumping quantities would be very low in these wells would permit the use of inexpensive shutter screens, and because of its low cost, we have chosen a fairly large diameter screen so as to provide more open area for water passage. Wells 1 through 5 and 11 through 13 must go quite deep, and the outboard wells must pump substantial quantities from layer F to achieve our design purposes. This probably entails using a rotary or reverse rotary procedure for installing the well, and would indicate that maximum screen efficiency is desirable. Using a 14-in (35-cm) screen, one can put sufficiently large pumps in the well to satisfy the requirements. A tabulation of expected operating levels which may be used to calculate the head on the pumps completes the table.

Wellpoint-system design conclusions are summarized in Table 6.3. We have made a judgment concerning the contribution to total flow in layer B from both the well and wellpoint systems, varying the contribution on each side of the job according to the prevailing conditions. River-side wellpoint spacing is determined by capacity limitations and interference. On the land side, spacing might average 5 ft (1.5 m) on centers, but in the northwest corner spacing down to 2½ ft (0.75 m) on centers could be required, and it might even be advantageous to use a short wellpoint screen so as to get the maximum amount of drawdown close to clay layer C in that segment of the project where the cutoff cannot be installed. Along the north and south sides, wellpoint spacing can be increased somewhat and the maximum capacity of the wellpoint utilized to advantage. The tabulation indicates a follow-through using this reasoning, and a conclusion then that 124 wellpoints would be about right. However, the factor of safety in utilizing this number of wellpoints on the river side would not be sufficient. Further, it should be kept in mind that the rather uniform and level profiles are probably not representative of the local variations which might be expected. Hence, it would be a good idea to increase the number of wellpoints by 25 and have a system flexible enough to cope with local variations.

TABLE 6.2 Well-System Design Conclusions

Well	Expected maximum Q, gal/min				Depth of well, ft	Screen diameter, in	Screen length, ft			Screen type	Operating level	Pump setting
	B	D	F	Total			B	D	F			
1	200	150	500	850	110	14	15	25	50	W	320	300
2	200	150	500	850	110	14	15	25	50	W	320	300
3	200	150	500	850	110	14	15	25	50	W	320	300
4	150	150	300	600	90	14	15	25	30	W	320	300
5	150	150	200	500	80	14	15	25	20	W	320	300
6	150	150	—	300	60	18	all	all	—	S	325	320
7	100	100	—	200	60	18	all	all	—	S	325	320
8	50	100	—	150	60	18	all	all	—	S	325	320
9	50	100	—	150	60	18	all	all	—	S	325	320
10	150	100	—	250	60	18	all	all	—	S	325	320
11	150	150	200	500	80	14	15	25	20	W	320	300
12	200	150	300	650	90	14	15	25	30	W	320	300
13	200	150	500	850	110	14	15	25	50	W	320	300
Totals	1950	1750	3000	6700								

*B = B layer, D = D layer, F = F layer, S = shutter screen use, W = wire wound screen use.
NOTE: 1 gal/min = 0.00006 m^3/s; 1 in = 2.54 cm; 1 ft = 0.305 m.

TABLE 6.3 Well-Point-System Design Conclusions*

Header segment	Length, ft	Probable Q required, gal/min	Well-point spacing and number of well points		Probable well-point capacity, gal/min	Header size,† in	Pump and capacity†
							2 to 12 in at
A-B	250	900	7½ ft	37	1100	12	3000 gal/min
B-C	250	600	10 ft	25	1000	6	
C-D	210	300	5 ft	42	800	6	
D-A	200	500	10 ft	20	800	8	
	910	2300		124	3700		
				+ 25	+ 750		
				149	4450		

*Total Q layer B, 4000 gal/min, total Q wells B, 1950 gal/min; required Q wellpoints, 2150 gal/min.

†At relatively low cost, all headers could be made 12 in and another pump added (2 operating 1 space) giving a system with a capacity of 5500 gal/min and pumping continuity in the event of header disruption.

NOTE: 1 gal/min = 0.00006 m^3/s; 1 in = 2.54 cm; 1 ft = 0.305 m.

Interference, Gradients, Crowns

Dewatering is generally accomplished by one dewatering device interfering with another. Thus, the available supply of water is lower than the capacity of the dewatering device and system as a whole. In our well system, interference is a necessary ingredient, if only to arrive at a system which will not fail in its purpose should one of the pumps fail. Our design has contemplated this. In a wellpoint system, when the wellpoints are placed very close together, as is the case on the northwest corner, the water each wellpoint pumps is limited by the interference effect of the adjacent wellpoints. On the other hand, on the north and south sides with wellpoints on 10-ft (3-m) centers, the interference effects will be minimal. The gradients which might be expected in layer B are such that they would be rather flat and extensive to the west and fairly steep to the east, or river, side where the radius of influence is determined by a proximate source of recharge, namely, the river. On the west side the wells in layer B will have minimal wetted screen because of their proximity to the wellpoints. On the east side, the wetted screen will be increased because of the steeper gradient and the distance from the wellpoint system. This has been reflected in the various tables. The crown effect one might expect in layer D and in layer F must be investigated. In the illustration at hand, it should be minimal because of the loose silt layer and because of the rock, each minimizing the recharge from below in layers B and F, respectively. Occasionally, it is advisable to install a dewatering device interior to the system to depress the crown. This sometimes causes interference with other construction work, but in very large and extensive excavations, this may be a necessary procedure, particularly in an excavation with great depth of aquifer below and one in which the peripheral system does not fully penetrate the aquifer.

Cost Considerations

All through the decision-making procedures previously described, the designer must have an awareness of costs. This is a tremendously involved subject and can only be glossed over in a presentation such as this. There are some very basic factors that must be considered: Does the use of two different dewatering sytems present duplicate operating labor costs because of local union contract agreements? Should this be the case, a more economical solution might be to deal with one system or the other, sacrificing some of the results of the program, or substantially decreasing the efficiency and increasing the installation cost. The designer should be aware when the size and complexity of a particular well is such that two more simple wells could achieve better results at a lower cost. Similarly with a wellpoint system, cost factors involved in very close wellpoint spacing are not necessarily linear, because of the reduced installation cost in a highly concentrated production area. With a wellpoint system, one must consider costs relative to concentrating pumps and having large-size header pipe, or alternatively, spacing out the pumps along the periphery of a system with a smaller header. Where will the water be discharged? Should a storm-water-handling component of a system have a separate piping system, or should it be combined with the primary dewatering system discharge? There are no simple answers to these questions. The factors involved in any particular job determine the answer, but the designer must be aware of the costs when considering the alternatives.

Design of the System

Capacity — Sizes

The designer determines individual wellpoint capacity in each particular application based on experience, the size and type of wellpoint, the spacing, the heads, and the soils in question. Riser, swing-header, and discharge piping must be sized in accordance with their flow rates and their lengths, making the appropriate calculations for friction losses. Wellpoint header-piping calculations must allow for air space as well as water. The tabulations indicated in Table 6.3 are typical pumping sizes with the appropriate flow rates. Similar procedures are followed in the detail design of the well piping.

Spacing

The spacing can be determined either by the required flow rates or by the effect that is desired from the dewatering device. Specific considerations relative to spacing have been discussed.

Filters and Screens

Both the screen and its filter can be key elements in the success of any dewatering program. Screens can be utilized which minimize the need for filtering-in cleaner, coarser soils. They can be chosen for their low cost, or they might be chosen for their capacity to pass high quantities of water. A filter can be chosen for its ability to retain the aquifer formation while moving water through it, or it may be used to provide the vertical connections between several layers so as to improve the vertical drainage. There are many different kinds of screens available for dewatering devices and an infinite number of filter possibilities. Filter design is treated in *Ground Water and Wells* and in Powers, *Construction Dewatering,* both listed at the end of this chapter. Different screen configurations are available and commonly used. Each has its own advantages and disadvantages relevant to efficiency, cost, and its need for filters.

Piping Design

Piping design must be made in accordance with standard practices for evaluating flow rate, frictional considerations, nature of the pipe, and the precision with which it might be laid. Commonly, temporary dewatering systems are not laid carefully and frequently fittings are placed in such a way as to cause adverse hydraulic characteristics in the flow; sometimes air pockets can develop in the pipe seriously diminishing its flow capacity.

Pump Selection

Pump selection is based on the design requirements for the pump, of course. However, frequently pumps are selected because of their cost, availability, or perhaps because of their favorable characteristics under difficult NPSH (net positive suction head) conditions. The total dynamic head under which the pump must operate should be evaluated and an allowance made for the fact that the pump, even if it is new when first utilized, will wear and its ability to pump the water "over the hill" during the pumping period must be evaluated. The total dynamic head under which the pump must function is composed of the static head, the friction head, the velocity head, and the net positive suction head.

Material Selection

Standard screens, piping, and pumps are not effective under very corrosive conditions. Commonly, wellpoint systems are constructed of PVC, aluminum, or steel. Steel is the most rugged, but subject to corrosion problems; aluminum is lightweight and resistant to damage; PVC is lightweight and resistant to corrosion, but susceptible to damage. Wellpoint pumps commonly have cast-iron, bronze, and stainless-steel components. Well screens can be made of PVC, galvanized steel, or plain steel. Well pumps can be cast iron or bronze, or in the smaller size, PVC with stainless-steel fittings. Special corrosion-resistant materials have been developed, such as zincless bronze and stainless steel, for pump construction. Pumps composed of materials such as this may have a cost factor several times that of ordinary pumps.

Power

The mode of power for a dewatering system must be considered throughout the design process. If electric power is not available, perhaps it can be economically generated at the site; if not, the use of submersible pumps or electric-powered wellpoint pumps cannot be considered. It is not unusual to generate power for construction dewatering where public supplies are limited or extremely expensive. Where public power is used, it is generally advisable to provide standby generating capacity in the event of power failure. Years ago gasoline, liquid natural gas, and diesel were commonly used in combustion engines for powering dewatering systems. Today engines are generally diesel-powered. In wellpoint systems, if diesel power is used, 50% standby is generally considered adequate; if electric power is used, 100% standby should be provided. In deep-well systems 100% standby power should be provided if the system is powered electrically. If a well system has individual diesel power on each well, it may not be necessary to provide any standby power, but simply to be in a position to repair power outages.

Factor of Safety — Redundancy

Every dewatering design should include a factor of safety for a number of different reasons. Some of them have been enumerated, but perhaps the most compelling reason is that one is dealing with materials and conditions which were determined by nature and are interpreted through investigations and experience. Designs cannot be completely accurate. Any dewatering system design must have flexibility so that while the system is being installed, the variations which occur in nature can be appraised and the installation modified to make an efficient system. For example, one might change wellpoint spacing to more nearly suit soil conditions or to be sure that some screens are located in the low spots of an aquifer. One should be in a position to modify screen location and length, should the drilling activities of a well indicate substantial changes from the design program. Well design should always contemplate the need for a larger pump in the well. Sometimes wells can be installed so that multiple pumps can be put in one well. This is a particularly useful device where the installation method requires using small-diameter casing or where the operating characteristics of the well might be subject to great fluctuation.

Redundancy elements should be built into every dewatering system. Sometimes it can be done at very little cost. For example, the wellpoint system illustrated could substitute 12-in (30-cm) pipe for the various smaller sizes at relatively small cost. This would result in a system which could pump much more water and be capable of easily adding pumping capacity; also, in the event of pumping interruption at a point along the perimeter, the system would be hydraulically capable of pumping the other way and maintaining a result almost as good as if the system had not had the local difficulty. Valving on any pumping system should be included so as to provide alternate flow routes in the event that pipes become damaged. The experienced designer will give redundancy a great deal of thought, keeping in mind the duration of the dewatering job, the susceptibility of the structure or the excavation to damage in the event of dewatering failure, and any number of other factors which determine the value of adding redundant features to the dewatering system.

Instrumentation

Every dewatering system should have separate consideration given to the instrumentation of the operating characteristics of that system and of the results expected to be obtained. Piezometers should be installed in all aquifers and, where necessary, sealed off from extraneous sources of water. Pumps should be equipped so that pressure gauges may be utilized to monitor operating conditions and performance. Vacuum gauges should be spaced out along the wellpoint header to monitor characteristics of the system. Flow meters or other measurement devices should be included in discharge lines so that an evaluation can be made of the performance of a system. The instrumentation not only has the value of appraising the characteristics of the system as an installation proceeds, but after the job has been dewatered, it can be an excellent means of monitoring the continuing adequacy of the dewatering system.

ESTIMATING DEWATERING COSTS

The economics of a particular dewatering situation play a large part in the choice of dewatering methods. Frequently more than one type of dewatering approach is possible, and it is necessary to make an appraisal of the costs and relative merits of more than one system. The characteristics and limitations of each dewatering method must be interrelated with the other construction procedures to be employed on the project and the decisions made on the basis of many different factors. The following items have very direct effects on the cost of a dewatering system:

1. Type of dewatering system
2. Size of dewatering system
3. Pumping volumes and heads
4. Duration of pumping
5. Depth of installations
6. Weather
7. Availability, quality, and cost of local labor
8. Availability, quality, and cost of local materials
9. Union work rules
10. Surface conditions
11. Sequential and schedule problems

In any given situation, there may be other factors pertinent to the cost of dewatering.

In Table 6.4 portions of a dewatering estimate form are reproduced. In order to illustrate some of the concepts which have been previously discussed, the estimate form is completed for the dewatering program contemplated in the hypothetical project. The form has been adapted to indicate the costs associated with both the deep-well and wellpoint systems. Included also are surface-water-handling costs. The forms are generally self-explanatory, and the reader desiring to go through the details should find correlations between costs, procedures, and techniques discussed previously.

At this point, our dewatering system is designed. A list has been produced of the equipment involved and a bill of material for the supplies which are necessary. The equipment has had a lease or purchase value assigned to it, and such items as freight costs, possible job fabrication work, supplies, construction equipment costs, etc. have been researched and accumulated. In addition, labor rates and union jurisdictional considerations have been evaluated, as have any site characteristics pertinent to the costs. It is then desirable to indicate any pertinent conditions or qualifications on which the cost estimate will be predicated. Also, a decision must be made on the time frame for the dewatering program. Inasmuch as most dewatering costs are directly dependent on time, a basic time period is selected for the estimate, and then these costs are evaluated for time periods more or less than the base period. The mobilization costs can then be summarized using the checklist indicated on the estimate sheet.

The next step is to list the major items of dewatering equipment that will be installed and to consider the makeup and cost of the crew which will make this installation. The estimator then must decide—on the basis of all the information available as to difficulty, production rates, etc.—how many crew days will be required to complete this installation.

In the space provided on the estimate sheet, additional crew-day allowances are made for setup, site preparation, deliveries, cleanup, removal, weather, and other delays. The result is the number of days which will be required to complete the installation and removal work. This figure is multiplied by the applicable cost per crew day (perhaps using different size crews for different items of work) to arrive at the total labor cost. The other components of installation and removal costs are estimated using the applicable time frame in the space provided. The items shown on the estimate sheet are those which are common to many dewatering installations. On any particular job, of course, there may be additional cost items to be considered.

Operating costs are considered in the same manner and suitable allowances are made for the total operating cost in the base dewatering period and in the monthly variation from that base period. Operating costs may or may not be concurrent with installation and removal activities. This will influence the time applicable to job overhead items such as supervision.

At this point, we have established costs relative to mobilization, installation and removal, and operation, all in a specific time frame. It would be well to stipulate the assumptions and qualifications on which the estimate was made. For example, does the estimate presume that all excavating and backfill necessary for the dewatering system's installation will be done by others? The various cost elements are then summarized. Insurance, taxes, overhead, and profit are calculated and applied to the costs to provide a total price. The nature of the job may or may not demand a contingency. It is not good practice to build significant or extreme contingencies into the various phases of the estimate. The dewatering design, of course, should be able to stand on its own and should include a factor of safety relative to pumping volumes, deterioration due to wear, and perhaps other unknown situations. The installation costs should be based on a reasonable expectation of degrees of difficulty. Operation costs which are based on knowledge and experience should preclude too many surprises. In any event, the estimator must evaluate the need for contingencies, keeping in mind the purposes of the estimate which may have a bearing on the application of contingency funds.

The forms obviously do not lend themselves to all dewatering programs. Slurry-wall and freezing estimates, for example, must be made considering all components of the specific program contemplated for a job.

Like most forms of estimating, dewatering estimates should be made by people experienced and knowledgeable. The variables are many and the range is great.

TABLE 6.4 Cost Estimate

MORETRENCH AMERICAN CORPORATION ESTIMATE PACKAGE	Job:		
	Location:		
	By:	Date :	Bid Date:

Estimate Summary

	Months		Per	Thereafter
	Labor	Material	Labor	Material
Total mobilize				
Total install & remove				
Total operate				
Subtotal				
P. R. tax & ins. ___ % S&U tax __ %				
Subtotal				
	→		→	
Subtotal				
Overhead __%				
__%				
Profit __%				
Subtotal				
Contingency				
Contract price				

Estimate analysis		
Based on ____ Months:	Optimistic	Conservative
Direct cost		
Contingency		
Overhead		
Total cost		
Profit		
Bid range		

Final bid price ________________ _____ Months

________________ per month

Conditions on which estimate is based	
	1. All std. cond. except;
	2.
	3.
	4.

Bidders	Comments/results
1.	
2.	
3.	
4.	
5.	

Attachments: ☐ Dew. equip. Other ________________

Quote No. ________________

Page 1

Job:	By:	Date:

Crew makeup			Crew A				Crew B				Crew C			
	Wage		No.	Hrs/	Cost/day		No.	Hr/	Cost/day		No.	Hr/	Cost/day	
Trade	L	M	men	day	L	M	men	day	L	M	men	day	L	M
Engineer	9.50	1.50	3	8	228.	36.	2	8	152	24	4	8	304	48
Laborer	7.00	1.00	1	8	56.	8.					2	8	112	116
Equpt. Oper.	10.50	1.50												
Pump Oper.	9.00	1.50									1	8	72	12
Welder	10.00	1.50	1	8	80.	12.								
Notes: Escalate 7%			Cost/day		364.	56.	Cost/day		152.	24.	Cost/day		584.	88.
After 6/30/79			No. days		x 35	x 35	No. days		x 19	x 19	No. days		x 10	x 10
			Total		12740.	1960.	Total		2888.	456.	Total		5840.	880.

Item				Crew Days			
Wells	Wellpoints	Qty	Qty	Drill rig	Crew A	Crew B	Crew C
Drill & set well	Wellpoints	13	150	8+3		6	6
Develop	Header	13	910		3	7	
Set pump	Pumps	13	2		4 1		
Discharge line	Discharge	1300'	100		3 1		
Substation	Pump house	2	1		2 1		
Elec. distrib.	Eject. hookup	1050'					
Elec. hookups	Elec. hookup	13	1				
Sumping	Mops/sumps	5	2			4 2	
	Sand bragging		150 lf		2		
Job setup & site prep				1	2		1
Unload & fabricate					2 1		
Holidays Delays Weather					4 3		1
Cleanup					1 1		
Removal					3 1		2
Total days				9+3	21 14	11 8	10

Install & Remove Expenses

Acct	Item	Labor	Material
7020	Labor	21468	3296
7325	Crane 10 days @ 140. 1270 + 24 + 30	1400	3240
7325	Hydrocrane 35 days @ 84. 11.25 + 12 + 20	2940	5495
7339/40	Drill rig 9 @ 1200. / day oper. 3 @ 900. / day		13500
7345	Loader 10 days @ 84. 1125 + 12 + 30	840	1670
7327	Compressor 10 days @ - 1125 + 30		1550
7435	Misc. fuel 1000 gal. @ 46¢		460
7307	Filter mat'l. 150 cy @ 15. 200 cy @ 8.		3850
7346	Forklift		
7237	Welding costs 1500 + 1000		2500
7433	Revert 5 wells @ 100		500
7026	Supt. sal./bene./exp. 5 wks. @ 400/125	2000	625
7328	Pickup truck 5 wks. @ 90		450
7344	Backhoe		
7321	Piezometers	2000	1600
7434	Bentonite		
7431	Grout	400	1000
	Total install & remove	34548	44136

Total I & R 78,684	Cost/well /WPT	Cost/ft

Quote No. _____________________________

Page 2

TABLE 6.4 Cost Estimate (*Continued*)

Job:	By:	Date:

Mobilization Expense

Acct	Item	6 Months Material	Mo. after Material
7321/2	Dew. equip.	41000	3900
7321/2	Jet. & dev. equip.	3000	
7321/2	Sump equip.	5000	700
7321/2	Standby gen. 100kw + 75kw	17500	1750
7321/2	B/A paytime rent	3000	
7321/2	Shortages	4000	
7361	Freight 7 Loads @400 each way	5600	
	Dew. Equip. Wpts.	24500	3400
	Off. & tool trailers	2000	300
7301	Tools	2000	
7301	Spec. rigting	1000	
7010	Clerical services	800	
7415	Utilities	900	150
7012	Security		
7302	Winterization/pump shanty	500	
7304	Elec. substation	4000	
7304	Elec. distr. system	18500	
7341	Subcontractor		
7342	Subcontractor		
7168	Eng. expenses	500	
7325	Mob. exp. – crane 800 Hydrocrane 400	1200	
7345	Mob. exp. – loader	300	
7339/40	Mob. exp. – drill rig Bucket R. 400 - Rev. Rot. 100	1400	
7327	Mob. exp. – compressor	200	
7361	Special freight ______ Loads @ ________	500	
7167	Proj. mgr. exp.	500	
	Total mobilize.	137900	10200

Operation & Maintenance

Acct	Item	6 Months Labor	6 Months Material	Per Month Thereafter Labor	Per Month Thereafter Material
7022	Oper. labor 3-8-9-(26+1)- 9.00/1.50	52488	8748	8748	1458
7024	Maint. labor 1-8-7 (26+1)- 7.00/1.00	10584	1512	1764	252
7308	Maint. Mat'l. 7-26-10		1820		308
7028	Supt. sal./bene./exp. (26+1) 400/125	10800	3375	1800	568
7330	Pickup truck 26-90		2340		390
7436	Fuel ____ G/D @ ____ = ____ x ____ days		1000		150
7436	Oil ____ G/D @ ____ = ____ x ____ days		100		15
7436	Elec. 300 kwr/D @.0375 = 270 x 183 days X24		49410		8235
7310	Chem./water anal.		1000		
7308	Repairs elec.		1800		300
7308	Repairs mech.		1200		200
7308	Overhaul				
	Escalation 7% 12,312			1231	
	Total operate	73872	72305	13548	11866

Quote no. ________________

Job:	By:	Date:

Basic Cost Data

Wage rates eff. dates					Notes:
	L	M	L	M	
Oper. engineer					
Oiler					
Laborer					
Electrician					
Welder					
Teamster					

		Labor	Material
Crane rental	Per ____		
size ____	Labor		
	FOGM		
	Total		
Hydrocrane rental	Per ____		
size ____	Labor		
	FOGM		
	Total		
Loader rental	Per ____		
size ____	Labor		
	FOGM		
	Total		
Compressor rental	Per ____		
size ____	Labor		
	FOGM		
	Total		
Drilling rig rental	Per ____		
size ____	Labor		
	FOGM		
	Total		

Basic design assumptions & conclusions

1.

2.

3.

4.

Notes:

Quote No. ____________________

Page 4

MOBILIZATION COORDINATION AND RESPONSIBILITIES

Mobilizing a dewatering job involves the planning and execution of logistics, designs, and procedures necessary for the work. Long in advance of the actual work, the various logistical problems must be handled. Equipment pertinent to the system should be secured and scheduled for delivery. The designs upon which the programs have been based should be updated in the light of new information and should be clearly understood by those responsible for the installation and operation of the system so that if field conditions differ materially from design assumptions, suitable changes may be made. The construction equipment and labor necessary for the installation must be arranged for and programmed so as to be in the right place at the right time. Perhaps additional borings or even a pumping test should be done prior to making actual commitments to installation procedures or to pumping equipment.

Another ingredient of the program concerns itself with the standards which are to be followed in conducting the dewatering program. Most frequently job specifications will contain only performance requirements with respect to the dewatering result. This is actually a prudent procedure on the part of most design engineers, because the actual techniques of conducting the dewatering program should properly be within the province of the general contractor or a dewatering subcontractor. This will allow maximum flexibility with respect to available equipment, methods, techniques, etc. The owner's engineer should be properly concerned with the result, but not necessarily with the means of creating the result. There are exceptions to this, of course, generally relating to situations where temporary-construction dewatering systems are incorporated into the permanent function of a structure, such as might be the case in a drydock. Occasionally, to expedite major construction work, contracts covering different phases of the work are let in sequence, in which case the dewatering involved in the initial contract might extend into later contracts. Under these and perhaps other circumstances, the owner has a direct interest in the type and quality of the dewatering system.

However, in the usual case of a simple performance specification, it would be well to draw up performance and operational standards for one's own reference with respect to the conduct of the work. Having in mind the objectives of the dewatering system and the design assumptions and criteria, one should know what constitutes a satisfactory yield from a dewatering device and what constitutes satisfactory results from each component of a dewatering system. Should initial results not sustain these requirements, installation procedures could be reexamined and perhaps even repeated or fundamental design assumptions could be challenged. It is important that each component of the system be installed properly and function in the manner in which it was intended to function. Mobilization of a job might also include the preparation for record keeping and cost accounting. Estimates can conform to cost-accounting procedures, and accurate checks can be kept of dewatering progress versus cost consistent with budgetary planning.

Liaison with other activities at the site is extremely important. Coordination of dewatering and excavating can be such as to be helpful to both the dewatering contractor and the excavating contractor. If they do not cooperate, each can be hindered by the other. In the early phases of the work, it is frequently important to verify assumptions with respect to water quality. Water samples should be taken and checked out so that anticipated corrosion conditions can be verified.

Frequently, the owner will require a dewatering plan, even if the method of accomplishing the dewatering is a contractor option. It is entirely logical that submittals of this nature be done in two parts—the first part dealing with the designs and anticipated procedures, and the second part being the verification of the design and the proof of adequacy as determined by testing during the installation and after its completion. The dewatering plan should always follow the dewatering objectives and it should provide for mechanisms to achieve flexibility. Soil conditions rarely are as uniform as they are depicted in soil profiles. The system must cope with reasonable variations in soils. However, if changed conditions develop, they should be recognized as soon as possible and suitable changes made in the dewatering program to contend with these conditions. It is during the mobilization phase and perhaps the installation phase that contingency considerations are reevaluated and investigated so as to assure the success of the dewatering program.

The various considerations discussed in this section are exemplified in the dewatering program conducted at Lock and Dam no. 26 replacement near Alton, Illinois on the Mississippi River (Fig. 6.17). In this project, a major suction well system which pumped in excess of 100,000 gal/min during the floods shown in the photograph to keep the water level 84 ft below the then current river stage was installed in such a manner as to provide unobstructed working space in a very large cofferdam extending well out into the Mississippi River from the Missouri shore. The suction well dewatering system was augmented by a slurry trench cutoff to rock along the Missouri shore and by bentonite cement grouting in open-work gravel strata along other portions of the cellular cofferdam periphery.

This dewatering program conformed to a performance specification. Submittals to the owner were made in phases verifying previous assumptions. Several design revisions were necessary to cope with the very severe effects of scour which occurred in the river due to the diversion scheme, but attention to detail provided an end result permitting construction work within the cofferdam to proceed unimpeded by inadequacies in the dewatering program.

INSTALLATION METHODS

Dewatering devices are installed into the ground utilizing many different methods. At one time or another, every method of putting a hole in the ground has been employed in the installation of dewatering systems, sometimes with very poor results. Methods that cause compression of fine-grain soils, remolding of cohesive soils, smear, and other negative characteristics that inhibit the flow of water from the soil into the hole in which the dewatering device is placed should be avoided. Some methods will entail a mandatory developing procedure to remove drilling additives. In others, this need is minimized.

Wellpoints, ejector wellpoints, and even deep wells are frequently jetted into place through the use of a self-jetting device built into the tip of a screen section. Valves regulate flow direction for jetting or for pumping. A chain or other kind

FIG. 6.17 Dewatering program at Lock and Dam no. 26 near Alton, Illinois, on the Mississippi River. *(Moretrench American Corporation.)*

of reamer may be temporarily affixed to the tip of the wellpoint to produce a 6- to 10-in (15- to 25-cm)-diameter hole. Continuity of sanding is important in order to minimize segregation of the filter sand, particularly if it is well graded.

There are other methods of installing wellpoints in wells, including jetting pipes and hole punchers. A hole puncher is a 6- to 24-in (15- to 60-cm) heavy-duty pipe equipped with air and water connections. It may be equipped with a removable head, so that the dewatering device may be inserted and then the hole puncher withdrawn. A hole puncher may be used with an external sanding casing so as to facilitate the placement of the device and its filter. Casings 24 in (60 cm) in diameter and 100 ft (30 m) deep have been used for the installation of wells. Sometimes a single hole-puncher head is used with multiple barrels for more rapid installation. Methods which employ the jetting processes generally result in a better installation, particularly in fine-grain soils, than methods involving drilling and driving, simply because smear, remolding, and compression of the soil is avoided.

It is frequently more advantageous to install wells with well-drilling equipment. Commonly, shallow wells are installed with rotary bucket rigs and deeper wells installed with reverse-rotary equipment. Wells which are to be installed in soils or rock that have very difficult penetrating characteristics may be installed with direct-rotary-drilling equipment. Other types of drilling equipment including augers and percussion drills are occasionally used, but present potential difficulties in terms of the quality of the hole and the soil's ability to communicate with the dewatering device. The choice of the type of screen used on the dewatering device is sometimes affected by the installation method. The need to develop the hole will influence the choice of screen. The use of drilling additives such as bentonite and revert can make drilling procedures economical and even feasible, but present problems in connection with the condition of the hole. It is important that the choice of screen and filter, installation method, and developing technique all be coordinated so as to produce an effective dewatering device.

Dewatering piping systems frequently utilize aluminum, plastic, or steel piping, coupled with both flexible and rigid couplings. There is an endless variety of connectors that has been developed for special purposes in connection with dewatering-system applications. In connection with pipe laying, it is important to remember that the line frequently must handle air as well as water, and the lines should be sloped so as to minimize air pockets and other undesirable hydraulic characteristics. In the case of wellpoint headers, devices such as line flow chambers and suction manifolds are frequently used to improve vacuum and hydraulic conditions. A typical well-piping system should include a check valve so that when the pump is shut off, water cannot flow back into the well. It would include a throttling valve so that the flow may be regulated without causing severe oscillations of water level within the well, and it should be equipped with connections for pressure gauges and flow instrumentation so that the performance and condition of the well might be monitored throughout the life of the job.

A typical wellpoint pumping station will have large headers entering the area from the system. Air-water separation manifolds, fuel tanks, and pumps should be properly supported. On major jobs, standby pumps are essential. Generally in dewatering systems, pumps run 24 h/day, 7 days/week. This creates the need for standby facilities and preventive maintenance. The pump station should be located in a convenient place for access and fueling. Discharge lines which go through earth cofferdams should be equipped with suitable water stops around the pipe, and if there is a possibility of river rise, the discharge lines should be equipped with check valves to prevent siphoning and to permit pump relocation at high-river stages.

Turbine pumping units are employed in situations where large quantities of water must be handled. They may also be employed where multiple stages of header can be handled with one pumping station, as is the case in some cofferdams. Characteristically, a turbine pumping unit will have superior air-water-separation capability, and its hydraulic operating characteristics are excellent.

Pumping units are generally set up in a battery. It is prudent to valve each unit so that it may function as an operating or a standby pump. In these circumstances, 50% standby capacity is generally adequate when employing diesel-powered pumps.

Electrical power sources for pumping operations should be investigated from a cost-and-reliability standpoint. Invariably, major jobs will require 100% power standby. This may consist of motor generating sets or sometimes diesel-operating pumping equipment. Large deep wells can be adapted to both electric and diesel operation through a well head which will permit the mounting of the motor in line with the pump shaft on the top of the device and a diesel engine on a right-angle drive.

The electrical distribution of power to a large number of operating electric wells around the perimeter of a large excavation is an important and costly component of a dewatering system. Not only must the power be distributed with regard to amperage draw and voltage drops, but the job should be divided into multiple circuits permitting disconnection of individual circuits should repair or relocations be required. In this connection, redundancy concepts should be explored. For example, if an accident knocks out the power substation and adjacent generating facilities, what would this do to the dewatering program? Should the dewatering system have power brought in at two locations, and should the standby facilities be subdivided into two components?

It is important to remember the relationship of the dewatering system to the project as a whole. The dewatering program is intended to improve the environmental conditions under which the structure is to be built. The installation procedures should not monopolize job facilities. Ramps should be kept open. In the case of sequential installations, dewatering work should be coordinated with other operations at the site so as to minimize interference and waiting. Jetting water should be controlled so as not to flood out other operations. If dewatering devices are to be installed into deep pressure strata which might be under positive artesian heads, safeguards should be taken, such as sandbagging and diking around the early installations. Frequently, a procedure is worked out so that in this type of situation the dewatering device can be pumped immediately upon penetration into the pressure aquifer.

OPERATIONAL CONSIDERATIONS

Dewatering-system operations are important too. Some jobs are sensitive in that a malfunction of the pumping system can produce trouble in minutes; other jobs may take hours or even days for trouble to show. Standby pumps should be provided and valved so as to permit them to operate effectively and quickly. Engine and pump maintenance procedures should be set up so as to minimize downtime. The instrumentation of the operating characteristics of a system can be the basis for detecting trouble before it evidences itself as a problem. Piezometer readings, vacuum and pressure readings, fuel consumption, and flow measurements should be taken periodically and recorded. The variations or trends in the various measurements can suggest deterioration of pumps due to wear, clogging of well screens due to chemical encrustation, leaks in the system due to corrosion, or a multitude of other conditions which, if detected early on, may simply constitute maintenance and replacement chores. If, however, the eventual progression of a deteriorating condition is such as to cause system failure, the results can be catastrophic. On important jobs, qualified dewatering superintendents can earn their keep by preventing trouble.

Occasionally on a dewatering job the quantity of water to be pumped will vary, for example, with tide levels or with river stages. Depending on the type of dewatering system, this can be a very difficult problem to cope with. It could manifest itself, for example, in inadequate capacity at high tide and vacuum losses at low tide in the case of a wellpoint system whose wellpoints were tuned for a particular condition. It might also manifest itself, in the case of wells, in operating levels within the well which are too high at high-flow periods, and so low at the low-flow period as to possibly cause excessive cavitation or cycling of the pump. Occasionally, the severity of such a condition can be influential in choosing the dewatering method or particular components of equipment.

The number of pumping units is sometimes determined by local union regulations, as is the distance between units. A pump operator around the clock, 7 days/week, represents a substantial element of the total dewatering cost. If working rules impose limitations on the number or location of pumping units, this can influence the design of a dewatering system and the choice of methods.

Housing of a pump station is dependent on weather and the duration of a job. If internal combustion engines are used, they will generally provide the required heat in the pump house during the winter. If electric pumps are used in very cold weather, heat must be provided separately, particularly for vacuum piping systems and inactive water lines. Sometimes insulation of piping systems is necessary. Dewatering programs have operated under weather conditions in all parts of the world. Systems can be operated dependably at −40°F (−40°C), although this obviously entails special precautions and procedures.

Generating facilities which are used for standby purposes should have a source of heat in extremely cold weather. Damp conditions should be avoided. All standby units should be checked out and tested at regular intervals. This testing should be done under load, so as to ensure the adequacy of the equipment in an emergency. When diesel engines play a role in the dewatering system, the dependability of fuel supplies and its storage on the job is a key consideration. A bad lot of fuel that immobilizes all of the engines on the job could indeed lead to a catastrophe. Fuel storage and piping components can be organized so as to minimize difficulties attendant to bad fuel or interruptions of delivery due to weather, strikes, etc.

Most dewatering systems which are adequately designed and carefully installed may be operated on a rather routine basis. However, the dewatering superintendent should always be on the alert for changes which could affect the performance of the system. Commonly, storm water and other forms of surface water present the most problems. This is particularly so when an excavation is being dug. The site drainage is often not complete and the conditions in the hole are geared to excavating efficiencies, not to drainage. A good superintendent will anticipate each evening what would happen if it rained heavily overnight and, within reason, do the small amount of preparatory work that might be necessary to channel rainwater into areas not critical to the excavation or into ditches and sumps provided for such purposes.

In conducting an excavation program in a large area, it is generally wise to take the perimeter down deeper than the middle so that water drains off to be collected and pumped. Pumping of the water in this connection is not the difficult part; collecting it is. Frequently, excavations can become bogged down for days after a rain if the water is allowed to accumulate and soak into previously dried fine-grain soils. It might take a long time to extract that water again, and it is generally advisable to try to collect the water so it does not seep into the ground.

Overall site drainage should be considered as well. Water which might tend to flow into the excavation should be kept outside by a dike around the top of slope. Where ramps enter into an excavation, it is a good idea to put in an artificial hump at the top of the ground so that traffic on the ramp cannot wear grooves that would let the ramp drain the adjacent site area. This is sometimes accomplished by laying a log across the ramp at ground level and filling above it with a stabilized material. Thought should be given to slope erosion. Frequently on long-term jobs, it is entirely practical to seed slopes so that the vegetation can provide resistance to erosion. Occasionally, storm water is handled on slopes by means of collecting ditches, culverts under roadways and berms, and drop pipes to pumping facilities.

On some projects, particularly cofferdams adjacent to large bodies of water, it is necessary to provide controlled flooding facilities so as to avoid the overtopping of the cofferdam, or to permit flooding in the event of dewatering-system failure. Occasionally, jobs on rivers will be designed so as to flood the cofferdam at high water and interrupt construction operations until the river recedes. In this event, the dewatering system must incorporate features that permit its submergence and the rapid flooding and reentry of the cofferdam. More often than not, it is very difficult to convey large quantities of water into an excavation quickly. Typical procedures involve the use of large-diameter pipe siphons which can be activated with vacuum pumps. Occasionally, pipes are installed through the cofferdam, with suitable valving to prevent accidental flooding. The difficulty with procedures of this nature concerns itself with the inability to test-operate the facility, and sometimes after extended periods of time valves corrode or shut or other malfunctions develop. Occasionally, an earth dike can be breached so as to cause controlled flooding simply by digging it out. A good way to control the erosion processes that develop in circumstances such as this is to install pipes through the primary dike and have a secondary dike, outboard of the primary dike, which can be breached by digging as necessary. Unless the character of the excavation lends itself to the flow of large quantities of water to the bottom, suitable erosion protection should be installed along the probable flow path.

Sometimes the importance of the operational aspects of dewatering systems is well realized at the start of the job, but months or even years afterward people forget that the dewatering system has unbalanced natural forces. Nature is permissive, but not forgetful. That conditions have been good for months at a time does not remove the need for vigilance.

REMOVAL OF DEWATERING SYSTEMS

Dewatering-system removal is seldom given proper attention. Frequently, perhaps most of the time, there is nothing complex about removing dewatering systems, the primary considerations being the disassembly and cleaning of expensive equipment so that it may be used again another time. Occasionally, however, dewatering-system removal can be very complex. Sometimes the installation sequence is such that a withdrawal sequence must be prepared as well. This is particularly the case with multistage wellpoint systems. One must then decide whether backfilling and hydrostatic uplift-pressure considerations will permit the removal of the lower portions of the system and the shifting of operations to the upper portions, or whether the lower portions must be continually operated and ultimately abandoned in the backfill or within the completed structure. Should this be the case, suitable thought should be given to the grouting of piping and screens and perhaps even filters.

Sometimes it is possible to relocate components of a dewatering system simply to facilitate the backing-out procedures. When it is necessary to bury equipment and grout the piping, the program should be prepared carefully and executed properly. Sometimes control valves are extended up through the backfill so that proper operating control of the system may be retained. When backfill is placed over operating segments of a dewatering system, care should be taken to avoid breakage. The system itself should be designed and installed so as to function under this circumstance, and the backfilling should be done carefully with proper materials.

Deep-well systems which are installed around the periphery of an excavation present the fewest removal difficulties. The pumps and piping can readily come out, although very deep wells which have been installed for a long time are almost impossible to remove. Frequently they are left in place, in which case the well itself should be backfilled with sand, concrete, or other material that will prevent the occurrence of a void in the soils after the piping rusts out. Sometimes it is necessary to grout the filter pack as well as the well, so as to avoid a permanent connection between different aquifers. Occasionally regulatory authorities will be concerned about contamination of one aquifer by water from another, particularly if the one aquifer is a water-supply aquifer and the other is subject to contamination from surface sources.

In the case of dewatering programs associated with structures which have continuing additions planned, for example, where subsequent units of a generating station will be built either directly adjacent to or close by the existing units, the removal of a dewatering system should be given very serious thought. Occasionally the dewatering system can be designed and constructed so as to be useful for dewatering subsequent units. At other times this is not the case, and unless properly removed, sealed, or otherwise contained, old buried piping and gravel trenches can be the source of tremendous problems in dewatering a new adjacent excavation.

As mentioned previously, removal problems are not common, but they are sometimes very serious and should be given proper priority.

BIBLIOGRAPHY

Bell, F. G.: *Methods of Treatment of Unstable Ground,* Newnes-Butterworth, London, 1975.

Bouwer, Herman: *Groundwater Hydrology,* McGraw Hill, New York, 1978.

Carson, A. Brinton: *Foundation Construction,* McGraw Hill, New York, 1965.

———: *General Excavation Methods,* McGraw Hill, New York, 1961.

Day, David A.: *Construction Equipment Guide,* John Wiley & Sons, New York, 1973.

Ground Water and Wells, Edward E. Johnson Co., St. Paul, Minn.

Legget, Robert F.: *Cities and Geology,* McGraw Hill, New York, 1973.

Lenz, Robert G.: "Suction Wells Dewater Large Construction Site," *Civil Engineering,* ASCE, February 1970.

Powers, J. Patrick: *Construction Dewatering,* Wiley, New York, 1981.

———: "Groundwater Control in Tunnel Construction," Rapid Excavation in Tunnels Conference, Chicago, 1972.

Tschebotarioft, Gregory P.: *Foundations, Retaining and Earth Structures,* McGraw Hill, New York, 1973.

7

Cofferdams

by Ben C. Gerwick, Jr.

Cofferdams are temporary enclosures to keep out water and soil so as to permit dewatering and the construction of the permanent facility (structure) in the dry.

This chapter confines itself to the structural cofferdams and cribs used to permit subsurface construction of bridge piers, intake structures, pump houses, and locks and dams in the aqueous environment.

A cofferdam involves the interaction of the structure, soil, and water. The loads imposed include those of the water, both hydrostatic and dynamic, due to current and waves. The soil imparts loads against the walls of the cofferdam and provides passive (internal) support during construction as well as after completion of the cofferdam. The structure itself interacts with water and soil, restricting the soil and water and exerting forces on the soil. Finally, the cofferdam must be dewatered, which involves the flow of water through the soil.

The failure of a cofferdam, should it occur, not only is catastrophic from the point of view of the work going on inside the cofferdam, but may also involve disruption of the surrounding area with damage to adjoining structures. It may seriously affect the practicability of reconstruction of a replacement cofferdam. Therefore, more than usual precaution has to be taken to prevent failure or collapse.

With a cofferdam in water, and also in weak soils, the load follows the deformation; hence, yielding does not relieve the forces. Therefore redundancy needs to be provided within the structural system, so as to ensure that partial failure does not lead to progressive collapse. Large and deep cofferdams should be subjected to a "failure mode effect analysis" to verify that progressive collapse will not occur.

Since cofferdams are constructed at the site of the work, operations must be carried out stage by stage under adverse conditions. The loads acting on the cofferdam are usually more severe during intermediate stages of construction than after the completion of the cofferdam. Each stage must be analyzed and evaluated to ensure its stability and safety.

As the cofferdam is constructed, stresses and deformations are built in, which usually are not released during subsequent stages. Thus residual stresses also need to be considered.

Because the cofferdam may be constructed under adverse conditions and because significant deformations may occur at various stages of construction, it is difficult to maintain close tolerances. Ample provision must be made for deviations in dimensions so that the finished structure may be constructed according to plan.

The loads imposed on the cofferdam by construction equipment and operations, both during installation of the cofferdam and during construction of the structure within, must be considered. These are generally located high up, hence induce critical loads at adverse locations, and often involve a dynamic component.

Removal of the cofferdam must be planned and executed with the same degree of care as its installation, on a stage-by-stage basis. The effect of the removal on the permanent structure must be considered. For this reason, sheet piles extending below the permanent structure are often cut off and left in place, since their removal may damage the foundation soils adjacent to the structure.

Safety is a paramount concern, since workers will be exposed to the hazard of flooding and collapse. Safety requires good design, proper construction, verifi-

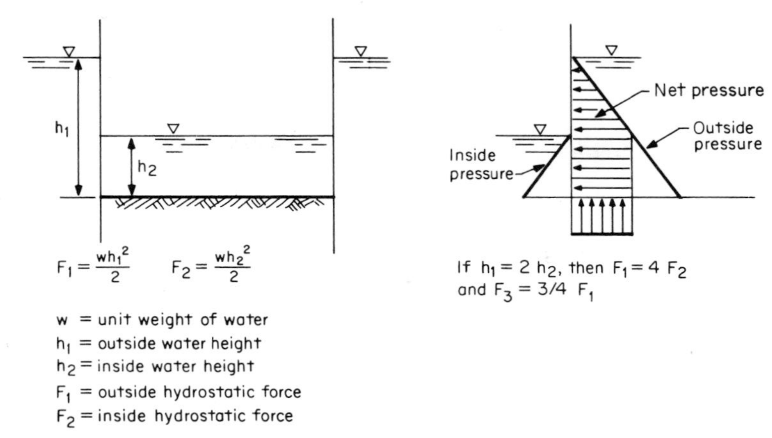

FIG. 7.1 Hydrostatic forces on partially dewatered cofferdam.

cation that the structure is being constructed as planned, monitoring of the behavior of the cofferdam and surrounding area, provision of adequate access, light, and ventilation, and attention to safe practices on the part of all workers and supervisors.

DESIGN CONSIDERATIONS

Hydrostatic Pressure

The maximum probable water height outside the cofferdam during construction and the water heights inside the cofferdam during various stages of construction need to be considered. These result in the net design pressures as shown in Fig. 7.1.

Current Force on Structure

The drag force D is given by:

$$D = A \cdot C_d \cdot \rho \cdot \frac{V^2}{2g}$$

where g = acceleration of gravity
ρ = density of water
A = projected area normal to the current
V = current velocity
C_d = drag coefficient

(Since in English units ρ is approximately numerically equal to $2g$, we can write $D = A \cdot C_d V^2$ where A is in ft^2, V is in ft/s, and D is in lb force.) The value of C_d depends on the overall shape of the structure and the roughness of the surface.

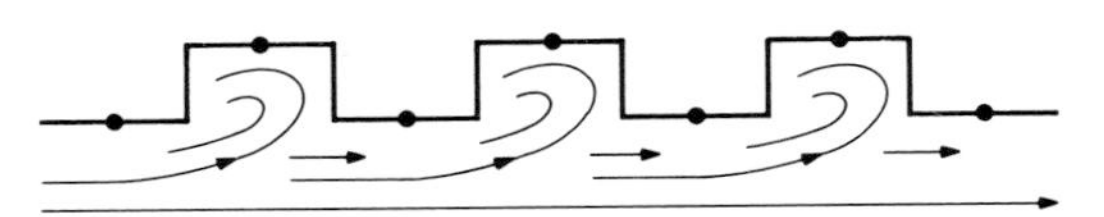

FIG. 7.2 Current flow along sheet piles.

With a typical cofferdam, the current force consists not only of the force acting on the normal projection of the cofferdam, but also on the drag force acting along the sides. With flat sheet piles, this latter may be relatively small, whereas with Z piles it may be substantial, since the current will be forming eddies behind each indentation of profile, as shown in Fig. 7.2.

As a practical approach, therefore, the use of a drag coefficient $C_d = 2.0$ will include the effect of the sheet piles; hence, the drag force $D = 2\ AV^2$ (English units).

During construction the current force also acts on bracing and support piles and on the sheets which are partially installed.

If the cofferdam will be subject to accumulating debris or ice floes during a flood, consideration must be given to current forces acting on the debris.

Wave Forces

Waves acting on a cofferdam are usually the result of local winds acting over a restricted fetch and, hence, are of short wave length, not subject to breaking. Waves can also be produced by passing boats and ships.

In most cases the waves acting against the face of a cofferdam are nonbreaking, that is, they reflect from the face of the sheet piles, forming a clapotis or standing wave. The effect is essentially hydrostatic in character. Attention is directed to the fact that since the water level inside the cofferdam is generally constant, the trough of the clapotis may lead to a net outward hydrostatic force on the sheet piles; hence, they must be well tied at the top.

Cyclic wave loads, particularly if allowed to cause play in the connections, may lead to fatigue.

Methods of calculating these wave forces are given in the *Shore Protection Manual* of the U.S. Army Engineers Waterway Experiment Station. The manual also gives a method of calculating the impact of breaking waves, which may develop a dynamic component due to the compression of entrapped air. Figure 7.3, which is from that manual, shows the typical profile and pressure distribution of waves acting on a cofferdam. Especially in the surf zone, with breaking waves, the dynamic forces can be much larger, due to the high velocity of the breaking or plunging wave.

In certain exposed and open sea areas, long-period swells may develop significant inertial forces, acting on the structure as a whole. In this case, the added mass (a mass of water equivalent to that displaced) must be added to the displaced mass in calculating the inertial component. Such forces can be very large. Evaluation is best made by diffraction analysis, in a manner similar to that used on offshore caissons.

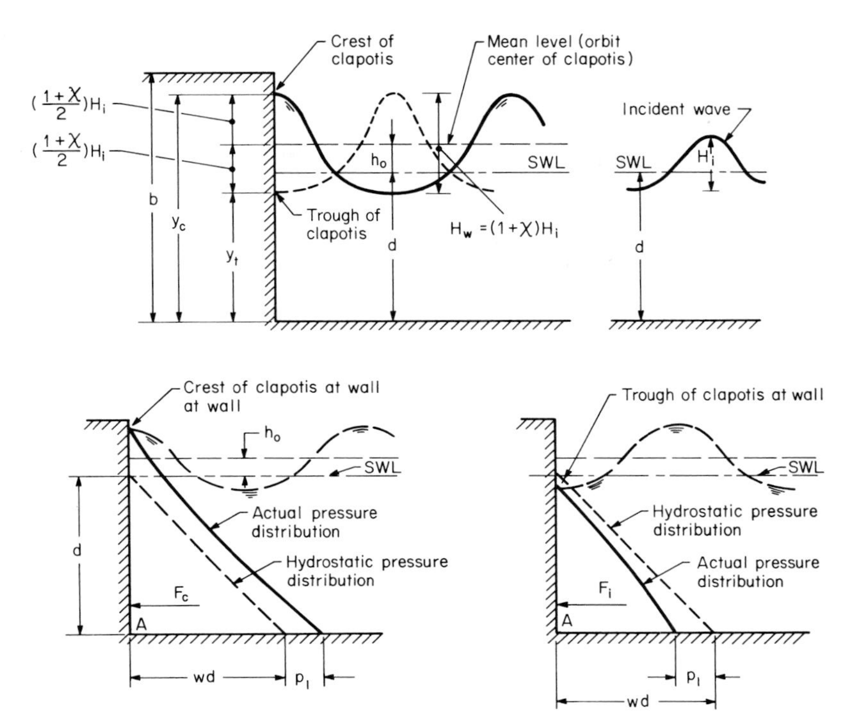

FIG. 7.3 Pressure distribution: non-breaking waves.

Ice Forces

These are of two types: the force exerted by the expansion of a closed-in solidly frozen-over area of water surface, and the forces exerted by the moving ice on breakup.

Static ice force

This depends on the coefficient of thermal expansion of ice and the elastic modulus of ice, as well as the degree of confinement and temperature differentials. As an example, a value of 4000 lb/ft^2 (190 kN/m^2) has been used on cofferdams and structures on the Great Lakes.

Dynamic ice forces

The values depend on the mode of failure of the ice against the structure, the strength of the ice, planes of fracture, etc. Average values acting on a cofferdam-type structure are often taken at about 12,000 to 14,000 lb/ft^2 (570 to 670 kN/m^2) of contact area, although extreme values may reach 3 times these values.

While this accounts for the normal force acting on the cofferdam, there is additional drag force along the long sides. On the favorable side, a shear or cut-water nose may serve to break the ice in tension or shear at about one-third of the value due to crushing.

Forces Due to Soil Loads

The soils impose forces, both locally on the walls of the cofferdam and globally upon the structure as a whole. These forces are additive to the hydrostatic forces.

Addressing first the global loading, in the most common case where the bottom is relatively level, the forces balance. When, however, a cofferdam is constructed on a slope, such as the bank of a river, there will be a net force trying to slide and overturn the cofferdam. A cofferdam is a relatively poor structure to resist such unbalanced loads because it has little weight and often is not founded on competent soils. Further, the necessary structural rigidity can be obtained only with additional bracing and connections. At one stage, just prior to placing the tremie seal, there is no bottom to the structure and no support from the bearing piles.

However, satisfactory solutions can often be obtained by providing one or more of the following:

- Increased penetration of sheet piles so as to develop bearing, uplift (skin friction), and lateral resistance (passive resistance of soil).
- Increased stiffness (section modulus of sheet piles).
- Diagonal bracing (in vertical plane) to ensure rigid truss action of bracing frame.
- Adequate connections of bracing frame to sheet piles so as to make sheet-pile walls act like flanges of a truss.
- Installation of batter piles and vertical piles to support the bracing frame. (These are sometimes driven through pipe sleeves in the bracing frame.)
- Installation of batter piles, connected to sheet piles at the top.
- Installation of ground anchors or anchors to deadmen on shore.
- Temporary placement of sand and gravel surcharge on low side of cofferdam.

Local soil forces are a major component of the lateral force on sheet-pile walls, causing bending in the sheets, bending in the wales, and axial compression in the struts. These forces are proportional to the effective unit weight of the soil, so that submerged soils contribute only about half the force that the same soil would contribute if located above water.

A critical case may occur when soil is exposed at low tide.

A detailed analysis of soil forces on sheet-pile walls requires the use of advanced principles of soil mechanics applied to carefully determined values of the various properties. However, preliminary values may be computed on a semiempirical basis.

For sands, a unit weight of 100 to 115 lb/ft^3 (1600 to 1840 kg/m^3) can be assumed. The angle of internal friction ϕ can be taken at 30 to 35°, giving an active pressure coefficient, K_a, of 0.33 and a passive pressure coefficient, K_p, of 1.5.

For cohesive soils, such as clays and silts, typical of marine seafloors, ϕ should be taken at 0 to 10°, the shear coefficient c varies from 300 to 3000 lb/ft^2 (14 to 140 kN/m^2), and the unit weight is 115 to 125 lb/ft^3 (1840 to 2000 kg/m^3).

Because the softer of these soils will tend to creep as the sheet piles deflect, K_a may be as high as 0.8 and K_p as low as 1.2. With firmer clays, the values of K_a and K_p approach 0.4 to 0.5 and 1.4, respectively.

For submerged soils the unit weight should, of course, be reduced by the buoyant effect of the water, 62.4 lb/ft^3 (1000 kg/m^3).

Satisfactory approximate computations of soil forces acting on the structure may usually be carried out by the method of equivalent fluid pressure. This, of course, gives a triangular distribution of pressure, which is approximately valid for determining total loads on a cofferdam and for the distribution of loads on the sheet piles. However, it is inadequate for proportioning the loads on the bracing, particularly at the top level. Therefore, for the bracing it is usually best to compute the total load by the method of equivalent fluid pressure, then redistribute it on a trapezoidal or rectangular pattern, from which the loads acting on the upper levels of bracing may be found.

For fully submerged soils, the following values of equivalent fluid pressure may be assumed for preliminary calculations of the loads acting on the wall.

Medium dense sands	74 to 80 lb/ft^3 (1180 to 1280 kg/m^3)
Firm clays	70 lb/ft^3 (1120 kg/m^3)
Soft clays and silts	80 to 85 lb/ft^3 (1280 to 1360 kg/m^3)

Firm clays may temporarily exert much lower forces due to their sealing off the water head. However, progressive deflection of the sheet piles may allow the full water head to act.

Soft clays and silts are very sensitive to surcharge loads as, for example, sloping bottom or an adjacent river bank, etc. For these cases a more in-depth geotechnical analysis is essential.

Seismic Loads

These have not been normally considered in design of temporary structures in the past. For very large, important, and deep cofferdams in highly active areas, seismic evaluation should be performed. The goal is to assure that collapse will not occur, although local damage may be acceptable.

For the portion embedded in the soil, an imposed total value equal to the passive pressure sets an upper limit. This is approximately valid for soft sediments such as those normally encountered in harbors.

For dense sands and for the firmer clays, an allowance of a 50% increase in active soil pressure is often assumed for an earthquake.

Since earthquake forces are of a temporary nature, allowable stress levels can be increased 33%. This is safe for wales and sheet piles, but struts may have to be treated more conservatively because of their buckling mode of failure.

For very extensive structures in plan, greater than 300 ft (90 m) across, the seismic waves may arrive on the two sides at different times, i.e., out of phase, causing extreme cyclic stress changes in the bracing.

For large cofferdam structures in deep water in a highly seismic area, the stability of the structure as a whole can be computed by applying an added mass coefficient to reflect the acceleration of adjacent water by the structure. For the typical cofferdam configuration, an added mass of 0.5 times the mass of displaced water will be found approximately valid.

Earthquake responses and effects are minimized for highly flexible structures such as most sheet-pile cofferdams. For the more rigid cofferdams now being considered for some deep-water bridge piers, there will be greater responses.

In loose sands, there is the possibility of liquefaction. This can be minimized in a practical way by a blanket of crushed rock placed around the cofferdam site.

Accidental Loads

These are the loads usually caused by construction equipment working alongside the cofferdam and impacting on it under the action of waves. One approach is to assume a local velocity, say 2.0 ft/s (60 cm/s), a distortion of the impacting vessel of 12 in (30 cm), and an added mass coefficient of 1.5 to be applied to the mass of the vessel.

The impact can be significantly reduced by fendering.

Logs can impact at the ambient current velocity. They can hit end-on; hence, require consideration of local punching shear.

Scour

Scour of the river bottom or seafloor around the cofferdam may take place due to river currents, tidal currents, or wave-induced currents. Some of the most serious and disastrous cases have occurred when two or three of these currents happened to combine.

Although the bottom of a river or channel may be stable before the cofferdam is constructed, the very act of constructing cofferdams, along with moored barges, etc., may cause blockage of a significant portion of the cross-sectional area of the stream, resulting in a dramatic increase in bottom current. The sediments on the bottom were previously in a condition of dynamic stability, so any increase in the bottom current leads to erosion and scour.

Scour can cause a rapid lowering of the seafloor over the entire area adjacent to the cofferdam or local pockets at the corners. These latter are due to eddy currents and can be minimized by streamlining, that is, rounding of the upstream corners and providing a fin or tail to prevent vortices at the downstream corners.

Scour is, of course, a greater potential when the bottom is fine sand or silt.

A very practical method of preventing scour, suitable in many cases, is to deposit a blanket of crushed rock or heavy gravel around the cofferdam, either before or immediately after the cofferdam sheet piles are set. Since scour can take place very rapidly, a delay in this latter case may not be acceptable when the bottom is very loose sand or silt.

A more sophisticated method is to lay a mattress of interwoven willow or plastic filter cloth, covering it with rock to hold it in place. This is laid on the bottom before starting construction. The support piles and sheet piles are driven through it.

Protection in Flood and Storm

When a flood or storm occurs which exceeds the preset limits, certain radical steps should be taken promptly in order to prevent loss of the cofferdam. These limits should be established by the designer of the cofferdam and be shown on the plans in writing, so that there can be no confusion or delay.

Practical steps have consisted of the following: flooding the cofferdam; preinstallation of a log boom to prevent debris from lodging against the cofferdam; provisions for clearing debris during the flood or storm; installation of extra ties diagonally across the bracing, so as to ensure that it will act as a space frame (wire rope plus turnbuckles can be quickly installed); tying sheet piles to top wale, if not previously done as part of the design; removing all barges from the cofferdam.

Ice floes can exert a very heavy and potentially damaging impact on cofferdams. In the past it has been common to attempt to deflect them by a boom or barge moored independently of the cofferdam. However, the possibility of the barge or boom breaking loose under the ice floe or storm, etc., and itself impacting the cofferdam must be considered. Moorings for such a barge or boom must be carefully designed to ensure against such a failure.

Similarly, when attempting to construct a cofferdam in a swift river current or where wind-driven waves are coming from one direction, the mooring of a large barge can effectively still the surface water. The moorings must, of course, be adequate for the design storm or flood, or else the floating breakwater must be removed at the onset of extreme conditions.

COFFERDAM CONCEPTS

The typical cofferdam for a compact structure, such as a bridge pier or pump house, consists of sheet piles set around a bracing frame and driven into the soil sufficiently far to develop vertical and lateral support and to cut off the flow of soil and, in some cases, the flow of water (see Fig. 7.4).

The structure inside may be founded directly on rock or firm soil or may require pile foundations. In the latter case, these generally extend well below the cofferdam.

An underwater concrete seal course may be placed prior to dewatering in

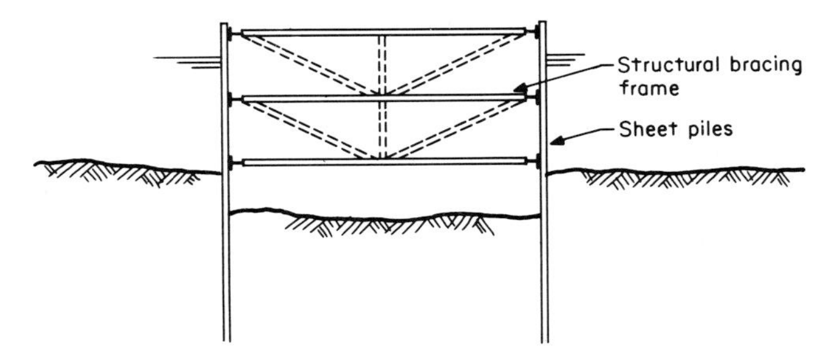

FIG. 7.4 Typical cofferdam (without seal).

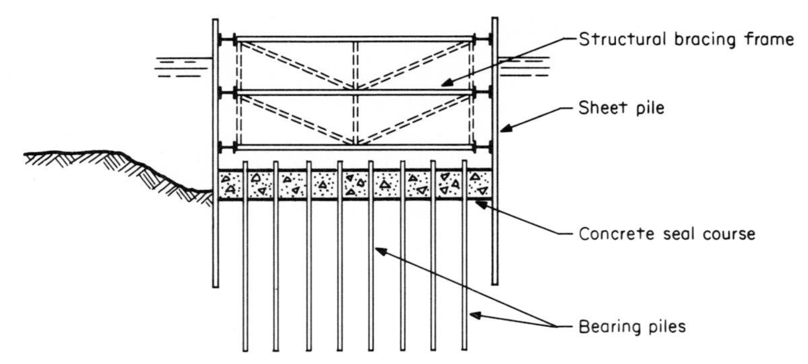

FIG. 7.5 Typical cofferdam (with seal).

order to seal off the water, resist its pressure, and also to act as a slab to brace against the inward movement of the sheet piles. The concrete seal may be locked to the piles in order to mobilize their resistance to uplift under the hydrostatic pressure (see Fig. 7.5).

For larger areas, such as locks and dams, cellular cofferdams or double-walled cofferdams are often used. These develop their resisting force by acting as a gravity retaining wall (see Fig. 7.6). In this concept parallel walls of sheet piles are tied together, either with sheet-pile diaphragms or tie-rods, and the space between is filled with granular material. In such cases, the dewatering is usually carried out by pumping.

Cellular cofferdams may be built on bare rock, since penetration is not necessarily required for stability, although it is required in sands and clays. The underlying soils must be capable of providing the required bearing support and lateral shear resistance.

COFFERDAM CONSTRUCTION STEPS

Typical Cofferdam Construction Sequence

For a typical cofferdam for a bridge pier or intake structure, the construction procedure follows the listed pattern:

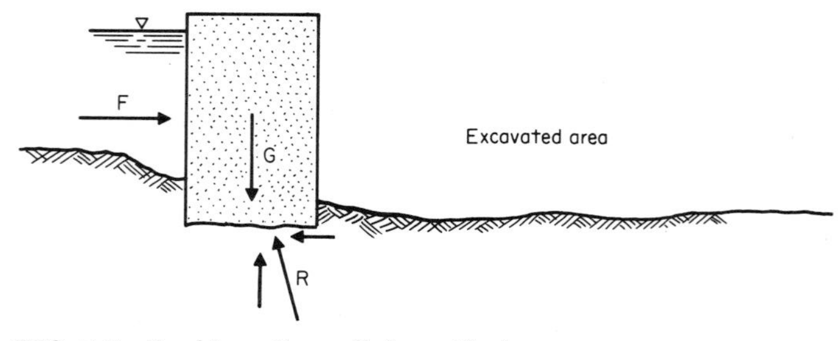

FIG. 7.6 Double-wall or cellular cofferdam.

1. Predredge to remove soft sediments and to level the area of the cofferdam (Fig. 7.7*a*).
2. Install temporary support piles (Fig. 7.7*b*).
3. Set a prefabricated bracing frame and hang on the support piles (Fig. 7.7*b*).
4. Set steel sheet piles, starting at all four corners and meeting at the center of each side (Fig. 7.7*c*).
5. Drive sheet piles to grade (Fig. 7.7*c*).
6. Block between bracing frame and sheets; provide ties for sheet piles at the top (Fig. 7.7*c*).

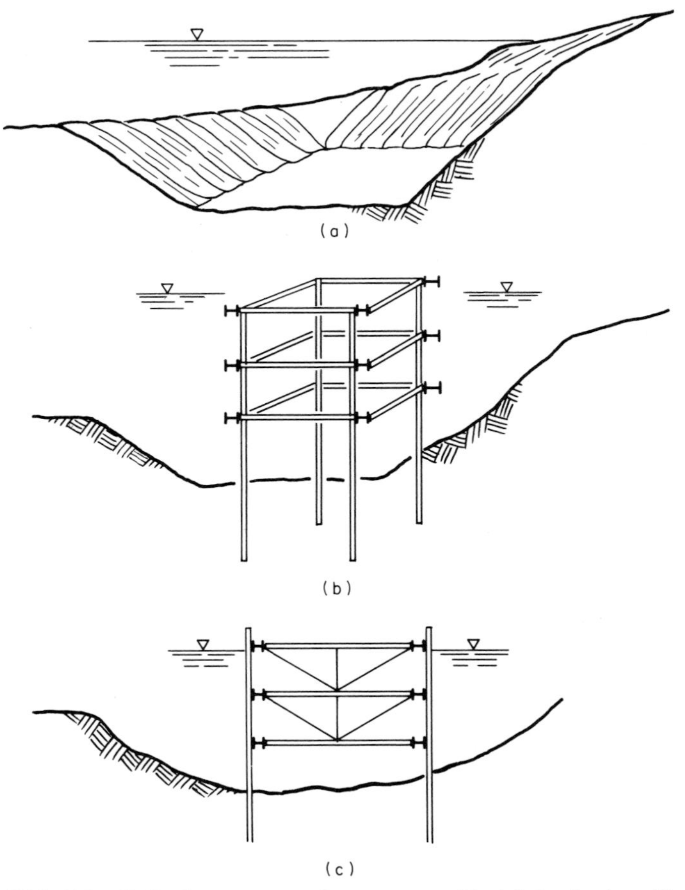

FIG. 7.7 Cofferdam construction sequence (I). (*a*) Predredge. (*b*) Drive support piles; set prefabricated bracing frame and hang from support piles. (*c*) Set sheet piles; drive sheet piles; block and tie sheet piles to top wale.

7. Excavate inside the grade or slightly below grade, while leaving the cofferdam full of water (Fig. 7.8*a*).
8. Drive piling (Fig. 7.8*b*).
9. Place rock fill as a leveling and support course (Fig. 7.8*b*).
10. Pour a tremie concrete seal (Fig. 7.8*c*).
11. Check blocking between bracing and sheets (Fig. 7.9*a*).
12. Dewater (Fig. 7.9*a*).
13. Construct new structure (bridge pier or intake structure) (Figs. 7.9*a* and 7.9*b*).
14. Flood cofferdam (Fig. 7.9*c*).

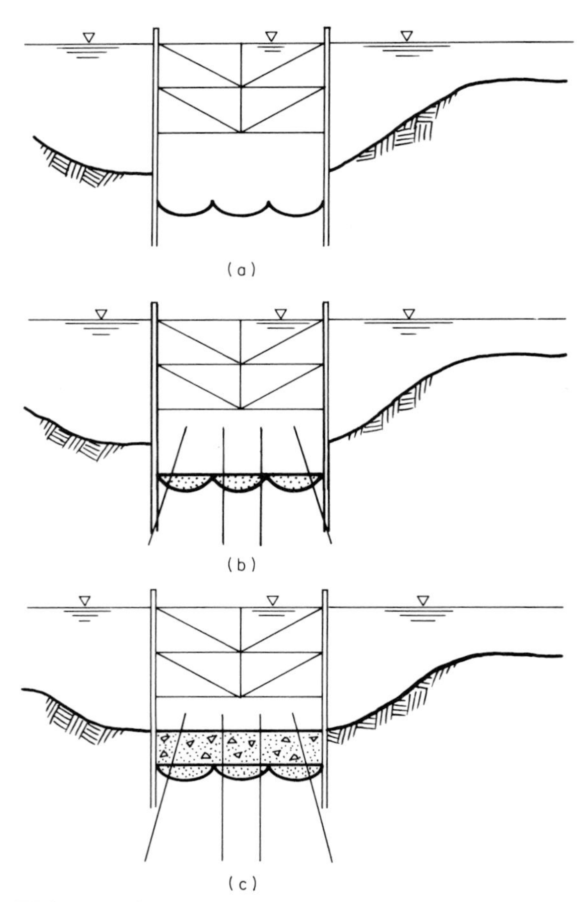

FIG. 7.8 Cofferdam construction sequence (II). (*a*) Excavate inside to final grade. (*b*) Drive bearing piles in place. (*c*) Pour tremie concrete.

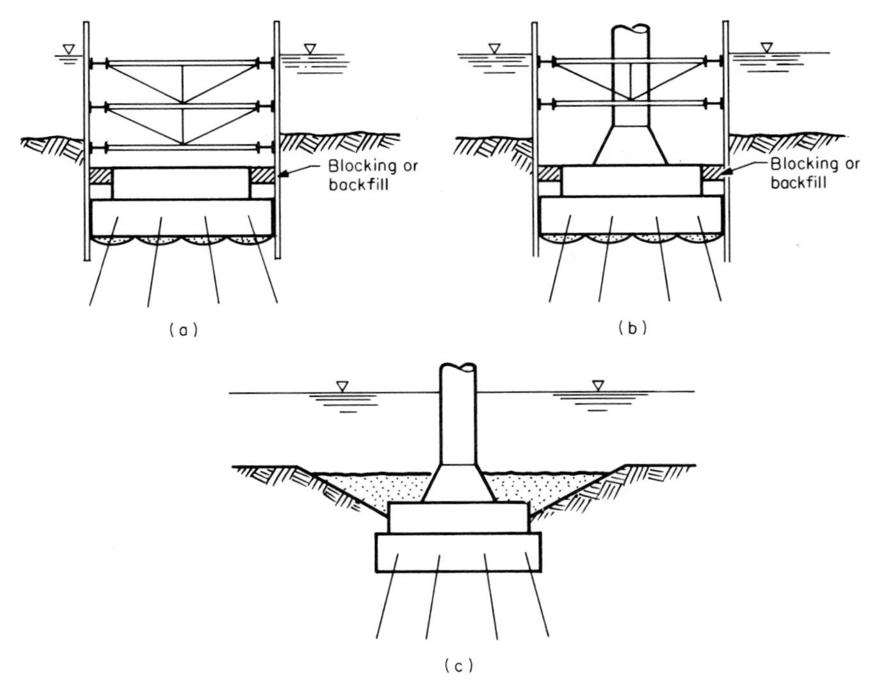

FIG. 7.9 Cofferdam construction sequence (III). (*a*) Check blocking; dewater; construct footing block; block between footing and sheet piles. (*b*) Remove lower bracing; construct pier pedestal; construct pier shaft. (*c*) Flood cofferdam; pull sheets; remove bracing; backfill.

15. Remove sheet piles (Fig. 7.9*c*).

16. Remove bracing (Fig. 7.9*c*).

17. Backfill (Fig. 7.9*c*).

Obviously, there are many variants depending on the circumstances.

Predredging (Step 1)

Prior removal of soil to form a deep basin means that the cofferdam will not have to resist temporary soil pressures in addition to water pressure.

When the seafloor is sloping, prior dredging removes unbalanced soil loads acting on the structure as a whole. This predredging, to be effective, must extend well back from the cofferdam wall, so as to leave a berm at the toe of the slope.

Predredging also permits the setting of a completely prefabricated bracing frame. Such a frame comprises two or more sets of horizontal bracing tied together with vertical and diagonal trussing and, hence, acting as a space frame (see Fig. 7.10).

Finally, it is generally significantly less costly to predredge than to dig inside, even though the quantity is obviously greater.

FIG. 7.10 Large bracing frame from Anchorage cofferdam, second Delaware Memorial Bridge.

When predredging all the way to grade, the subsequent outward pressure of the tremie concrete seal course must be considered. Often it will be desirable to backfill outside the cofferdam with granular material to at least the height of the seal prior to placing the tremie concrete.

Predredging also facilitates pile driving inside the cofferdam. If it is possible to predredge to grade, then it may be possible to drive the piles before the cofferdam bracing frame and sheet piles are installed, moving step 8 up to 2.

On the negative side, predredging obviously increases both dredging and backfill quantities. It may extend the cut too close to existing structures. Dredging in the open water may stir up bottom sediments into the water, although this normally will be of little ecological significance.

Predredging can also facilitate the removal of surface or subsurface layers of hard material, such as boulders or limestone layers. In fact, if heavy excavation is required, such as blasting or use of very heavy buckets, or if the removal of obstructions or old piles is required, these can best be done by preexcavating before the cofferdam is started. Performance of such heavy excavation inside a cofferdam structure is rendered very difficult due to restrictions of space between bracing and the need to exercise care to prevent damage to the structure.

When the site conditions and requirements of the project do not permit predredging to grade, then, of course, excavation will have to be carried out inside the cofferdam, and the sheet piles and bracing will have to be designed to resist the forces from the soil during all stages of construction.

In competent soils, such as sands and firm clays, this can be accomplished in a practical manner by selecting sheet piles of adequate length and strength to resist the soil pressures during excavation, transferring the load to the bracing above and to the passive pressure zone of the soil below. Then, when the excavation is completed, the piles can be driven and the underwater concrete seal can be constructed.

If the depth of inside excavation is too great for practical spanning of the sheet piles between the bracing and the soil, a third level of bracing may be placed after partial excavation. Such a lower level of bracing may be particularly useful if the driving of piles requires jetting, or the vibration may otherwise temporarily destroy the passive resistance of the "passive-zone" soils below the excavation limits.

When constructing cofferdams in very weak silts and muds, there may be inadequate passive resistance. In such a case the sheet piles must be very long and very strong, often exceeding the limits of practicability and economy. This becomes particularly critical when relatively shallow excavations are to be carried out in very deep muds. Many failures have occurred when there was inadequate recognition of this problem. The failure may be characterized either by excessive rotation or excessive deflection, as shown in Fig. 7.11.

Two economical and practical solutions have been devised. In one, the site is predredged several feet below grade, and an artificial "stratum" of sand and gravel is dumped to act as a support for the sheet piles and to develop lateral support for the sheet piles (see Fig. 7.12). Then the sheet piles need only be long enough to penetrate the sand-gravel stratum. Such a system was successfully employed for 24 shallow cofferdams for the west side of the San Mateo-Hayward Bridge near San Francisco, reducing the required length of sheet piles from 80 ft (24 m) to 35 ft (11 m).

However, the above solution assumes that predredging can be carried out. If predredging is not permissible (the initial premise), then the sand-gravel stratum can be created by injection, that is, by using sand-drain equipment and an excess of air pressure to force the sand bulbs out at the desired elevation (see Fig. 7.13).

The second solution is to cantilever the sheet piles, using a two-level bracing frame. The sheet piles are tied back at the top level, and cantilevered below the lower level. The efficacy of this system depends on using high-strength sheet piles (high section modulus) and on maximizing the distance between top and lower bracing levels. In some cases, it may be desirable to raise the top level

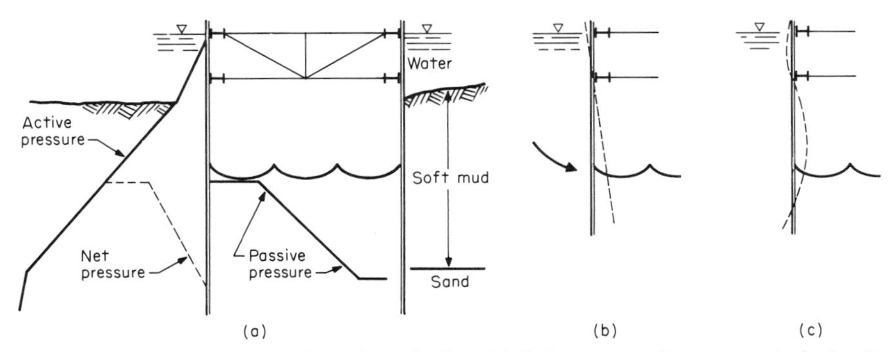

FIG. 7.11 Soil forces in soft muds and silts. (*a*) Soil pressure diagram typical of soft soils. (*b*) Rotational mode of failure. (*c*) Bending mode of failure.

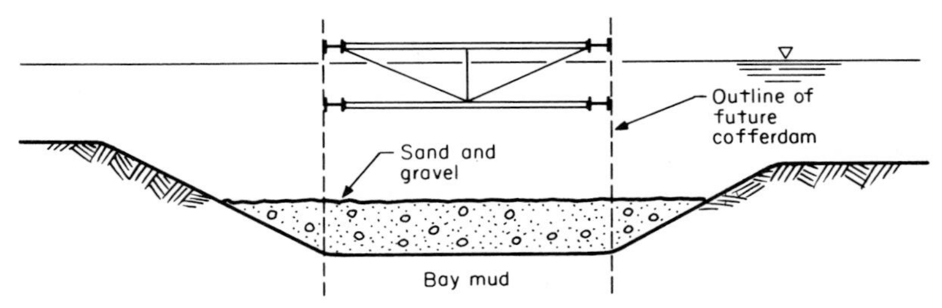

FIG. 7.12 Use of sand and gravel mat to provide temporary lateral support.

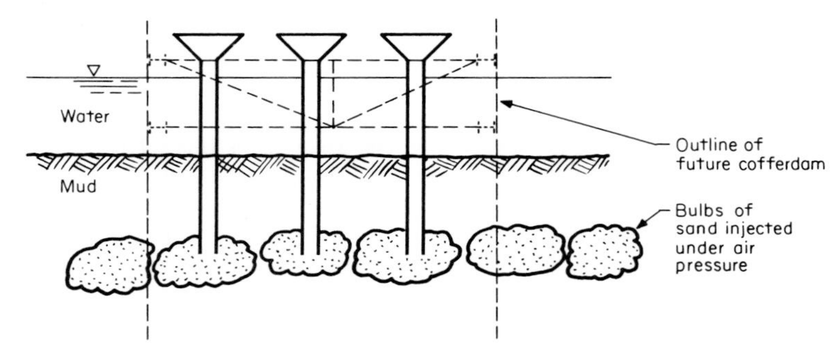

FIG. 7.13 Creation of firm stratum to provide temporary support for cofferdam.

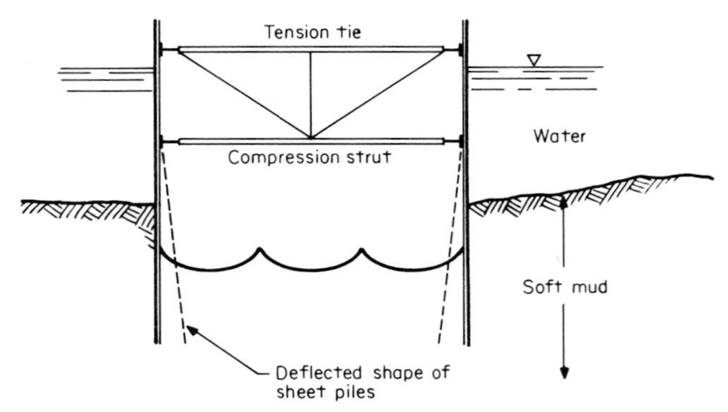

FIG. 7.14 Use of cantilevered concept for shallow cofferdam in weak material.

above that which otherwise would be required (see Fig. 7.14). Level B bracing must resist the force from the soil plus that of the tie. The sheet piles need to penetrate below the inside excavation only far enough to prevent run-in of sand, or a local shear failure of the soil underneath.

Since deflections are relatively large, the possible remolding effect on soft clays and, thus, an increase in active pressure, must be considered.

Temporary Support Piles and Bracing Frames (Steps 2 and 3)

These must provide sufficient lateral support to resist current and wave forces during the installation of the sheet piles. In calm water this can usually be done by the cantilever of vertical piles, using the bracing frame to establish a fixed-end condition and short moment arm. In strong currents or waves, batter piles or similar lateral support may be required.

The bracing frame generally consists of wales, struts, posts, and diagonals. Today, cofferdam bracing frames are almost always made of steel structural shapes and tubes.

Bracing frames are generally made up of a combination of welding and bolting. Bolts facilitate subsequent disassembly if required for removal. At intersections of struts and wales, sufficient stiffeners must be provided to prevent local distortion. Since field welds are very unsatisfactory due to wet conditions and vibrations of the elements, bolting is to be preferred for field connections. A typical framing arrangement is shown in Fig. 7.15.

Wales are subject to combined stresses, that is, normal loads applied horizontally and causing bending, plus bending due to dead weight in the vertical direction, plus axial compression from the ends and from diagonals. Standard structural design procedures provide methods for handling these combined forces.

Struts are subjected primarily to axial load as horizontal columns. They also have bending in the vertical plane due to dead weight, and any equipment supported on them. As columns, they must be supported in both the horizontal and vertical planes to keep the length to radius of gyration ratio (l/r) within allowable limits so as to prevent buckling.

Struts are also subjected to temperature loads. In cofferdams in waters, these are usually negligible since the top level (where temperature differentials are greatest) is generally free to expand and contract. However, they must be considered in land cofferdams where the walls are rigidly restrained.

In some cases, the lowest bracing frame must be installed below the level of predredging; that is, it is not practicable to predredge to a depth that will enable the entire frame to be set as a single prefabricated unit. In such a case, the third (lower) frame is made up separately and hung immediately beneath the upper prefabricated space frame. After further excavation inside, the frame can be lowered to grade.

For such a frame, having bracing in the horizontal plane only, the column effects (l/r) will usually control the size of the struts. The lower frame must be well supported, either by piles or from the upper frame. It should be truly horizontal and well blocked to the sheet piles. These precautions are important to prevent any tendency for the frame to rotate under load. Therefore, maximum

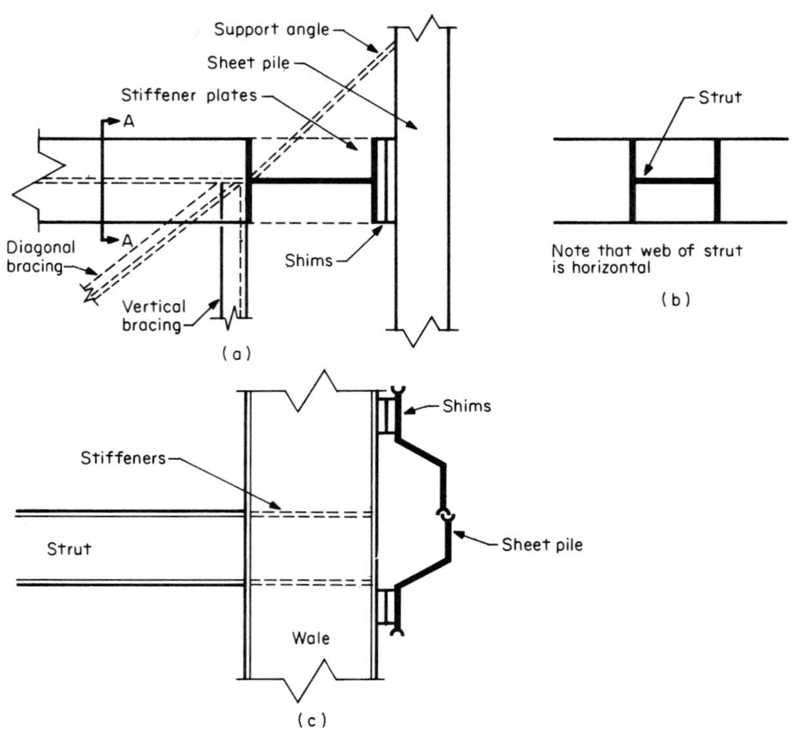

FIG. 7.15 Typical framing arrangement: (*a*) Elevation and (*b*) intersection of struts.

prefabrication of as complete a space frame as possible is to be preferred. To transport and place such a huge and heavy frame requires careful planning.

The space frame is usually best assembled on a barge. In extreme cases it may be assembled on skidways on shore, like an offshore platform jacket, and later skidded out onto a barge.

Alternatively, the space frame may be assembled on location, supported above water on large spud piles, and progressively lowered down as each level and its next set of vertical bracing are completed. In such cases, the spud piles, acting in cantilever, must be adequate to provide lateral support.

Finally, some of the bracing members may be made buoyant, especially if large-diameter pipe sections are used for struts.

If transported by barge, the frame, often weighing several hundred tons, can be picked by floating shear-leg crane barges or derrick barges and set over the support piles.

An alternative method is to float the space frame into exact location while still supported on the barge, then drive spud or support piles. The load is transferred to these support piles at high tide. At low tide the barge is extracted. Once again, consideration must be given to the lateral forces, including an accidental impact from the barge as it is moving out.

With very deep cofferdams, two or more prefabricated space frames may be set on top of one another and joined by pin piles (long pipe spuds running through vertical pipe sleeves). Funnel-shaped guides are used to help mate the top section to the lower.

Bracing by means of a series of circular compression rings is often an attractive solution because it eliminates cross-bracing which would interfere with excavation and subsequent pier construction. However, such bracing must be designed with particular care to prevent the possibility of buckling. Causes of buckling are the following:

1. Out-of-round tolerances in compression-ring prefabrication.
2. Uneven bearing of sheet piles on compression rings. Poor blocking.
3. Unbalanced soil loads.
4. Failure to consider each stage of excavation in the design: the ring which is currently the lowest at one stage may carry much more load at that stage than subsequently.
5. Accidental damage to a ring as by a bucket.
6. Installation at an angle with the horizontal.
7. Inadequate web stiffness leading to web buckling.

Because the mode of failure is sudden and transfers its load plus impact to an adjoining set, progressive collapse is possible and has occurred. Thus, each ring should be evaluated on the basis of its ability to arrest an adjacent local collapse.

Steel Sheet Piles (Steps 4 and 5)

Many different types are used. For cofferdams, where bending predominates, the Z sections, H sections, and O pipe sections predominate. Whatever section is adopted, the sheet pile should be sufficiently rigid and of sufficiently thick metal, say 3/8 in (10 mm) minimum, so as to withstand driving stresses and local twisting. Hence, the Z-27 is not normally suitable for cofferdams; the Z-32 and Z-38 are preferred. Figures 7.16, 7.17, 7.18, 7.19, and 7.20 illustrate some of the various types of commercially available sheet piles.

Use of 50,000 lb/in^2 (345 N/mm^2) yield steel is desirable to prevent local damage, increase interlock strength, and to provide greater bending resistance, hence, greater height between horizontal bracing levels. It is subject to much less damage and distortion than the steels of 36,000 lb/in^2 (248 N/mm^2) yield.

It is essential in water cofferdams (and most land cofferdams as well) that the sheet piles be continuously interlocked. To ensure this, they must be set accurately and driven progressively. By setting the corners first and working in both directions toward the middle, minor adjustments can readily be made.

As a control and guide to accurate setting, sheet-pile locations should be marked on the wale. An allowance should be made for interlock play [usually about ⅛ in (3 mm) per pile for Z piles manufactured in the United States.]

AISI standardized section designation	Nominal width, in	Weight, lb/ft² of wall	Section modulus, in³/lin ft of wall	Interlock tensile strength (usuable), lb/in
PZ 38	18	38.0	46.8	—
PZ 32	21	32.0	38.3	—
PZ 27	18	27.0	30.2	—
PDA 27	16	27.0	10.7	—
PMA 22	19⅝	22.0	5.4	—
PSA 23	16	23.0	2.4	3000
PSA 28	16	28.0	2.5	3000
PS 28	15	28.0	2.4	8000
PS 32	15	32.0	2.4	8000

Note: 1 in = 2.54 cm; 1 lb/ft² = 4.9 kg/m²; 1 in³/lin ft = 53.75 cm³/m; 1 lb/in = 175 N/m.

FIG. 7.16 Properties of standard sheet piling (see Figs. 7.17 and 7.18).

Sheet piles must be set against falsework or guides, giving enough support at a high elevation to hold them accurately in alignment and giving support against wind during the initial phases.

A partially set sheet-pile wall is very susceptible to wind damage, being a large, flat wall of little strength. For this reason, the wall may be tied to the upper

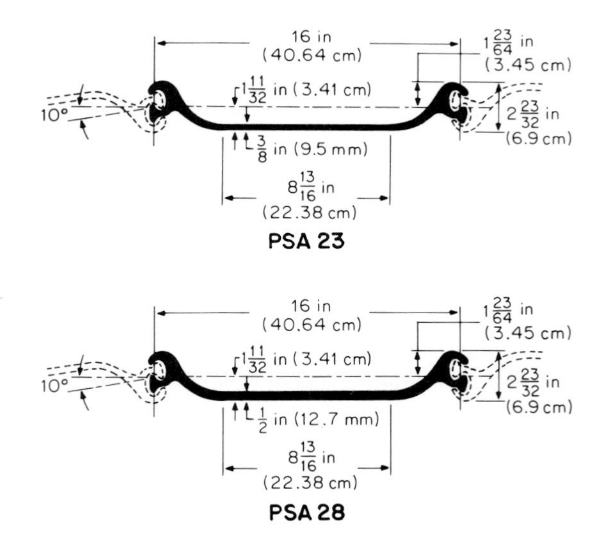

FIG. 7.17 Steel sheet piling.

guide (by temporary welding or bolting) to give it restraint in both directions. Also, it is not advisable to leave a large exposed panel of sheet piles undriven overnight or over weekends. Generally a panel should be set and at least partially driven before the end of the day's work (see Fig. 7.21).

Threading of sheet piles has historically been done by hand, but this resulted in many mashed fingers and was slow. New devices have been developed, both hydraulic and mechanical, by which the second sheet pile can be tied to the first at ground level, then run up and automatically engaged at the top.

In the case of multiple cofferdams and reuse of sheet piles, it is often practicable to handle, set, drive, and remove the sheet piles in pairs, using a short weld along the interlock at the top to mate each pair.

In driving, no sheet pile should have its top extending more than a few feet, e.g., 5 ft (1.5 m) maximum, below its neighbor at any time. Otherwise, it may tend to deviate from true alignment and the subsequent driving of adjoining piles will result either in high interlock friction or, worse, in driving out of the interlock.

Sheet piles are best driven by a vibratory hammer, at least as long as a satisfactory rate of penetration is being achieved (see Table 7.1). Vibratory hammers work well in silts and sands, satisfactorily in soft clays, and poorly in gravels, peat, and sandy clay hardpan. For these latter soils, either a diesel or steam hammer is used. The impact hammer should not be too large or it will lead to tip deformation. A driving head should be employed, properly grooved for the sheets selected.

In some very hard driving, e.g., when the sheet piles must toe into cobbles, etc., cast steel sheet-pile-tip protectors may be effectively used to prevent tip damage.

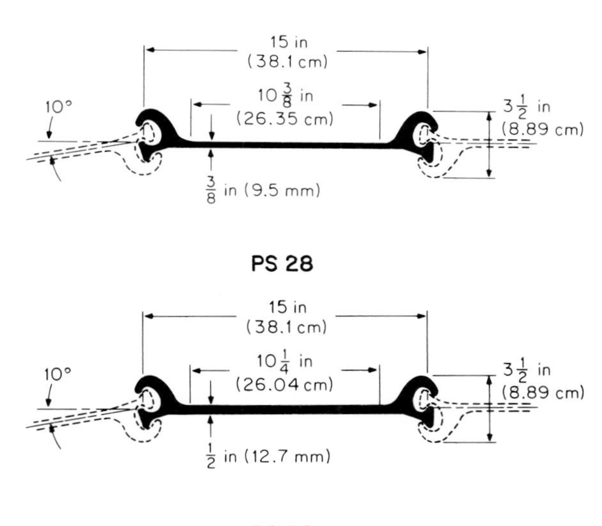

FIG. 7.17 Steel sheet piling (*Continued*).

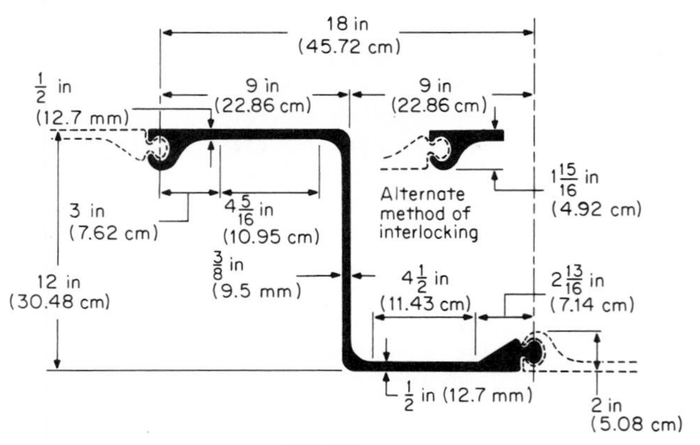

PZ 38

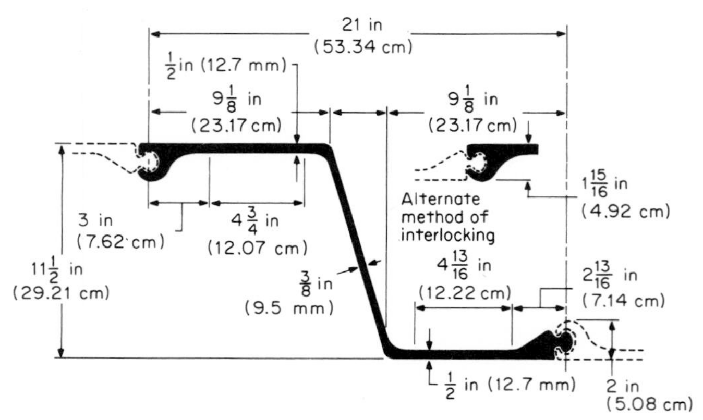

PZ 32

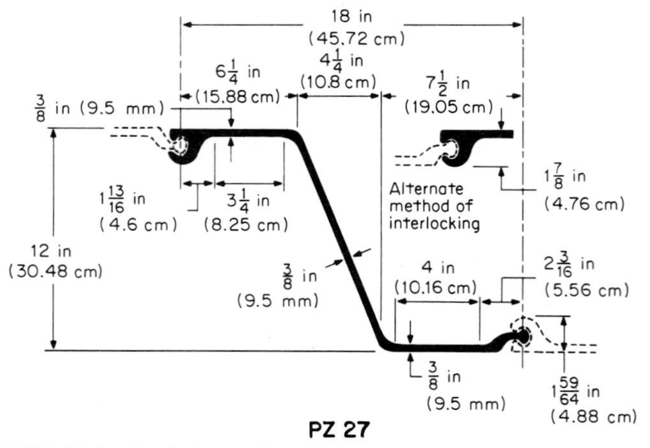

PZ 27

FIG. 7.18 Steel sheet piling.

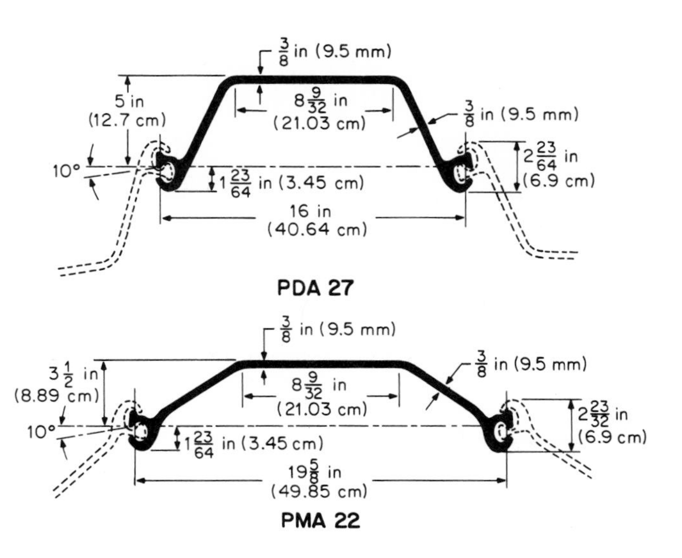

Sections That Will Interlock With Each Other				
PZ 38	PZ 32	PZ 27		PS 28
PDA 27	PMA 22	PDA 27	PMA 22	PS 32
PSA 23	PSA 28	PSA 23	PSA 28	

Note: The ball end of PZ 27 will interlock with the socket end of PZ 38 and PZ 32; however, the socket end of PZ 27 will not interlock with the ball end of PZ 38 and PZ 32.

FIG. 7.18 **Steel sheet piling (*Continued*).**

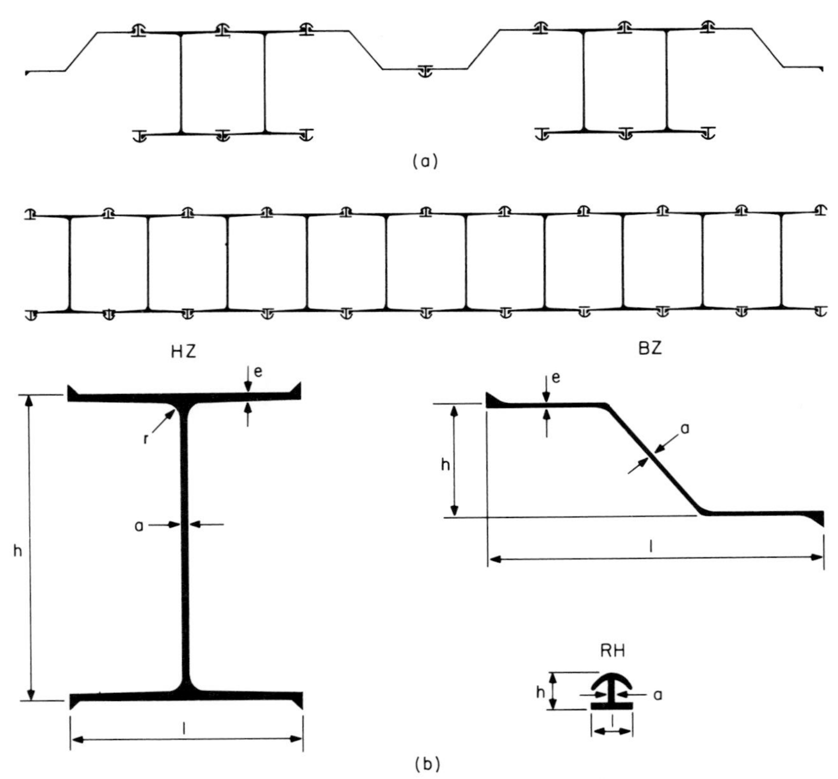

FIG. 7.19 Steel sheet piling. (*a*) A considerable increase of the modulus of the wall can be obtained by alternating a double BZ sheet pile with two HZ sections coupled together. (*b*) HZ sections continuously coupled together, with or without rear joint sections, constitute an extremely rigid wall capable of transmitting large vertical loads into the subsoil.

Tie and Block Sheet Piles (Step 6)

Ties are usually installed at the top because the sheet piles may not be bearing on the top wale after the cofferdam is fully dewatered. Rather, due to continuity effects, they will tend to move away from the top wale. To prevent vibration and to minimize bending moments in the sheets, provision of J-bolt ties is useful.

Bolts are superior to welds, since the latter often fracture under vibratory loads.

Excavation (Step 7)

Inside excavation is usually done with a clamshell bucket. It is essential to prevent hooking of the bucket under a brace; in addition to using care, temporary setting of one or more vertical sheets against the internal bracing, so as to act as guides, may be useful.

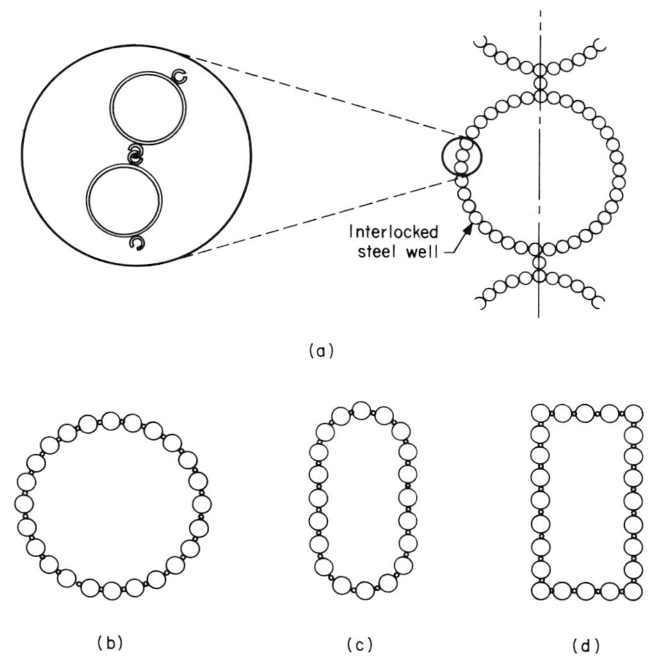

FIG. 7.20 Tubular steel sheet piles: (*a*) Interlocking; (*b*) circular plan; (*c*) elongated plan; (*d*) rectangular plan.

FIG. 7.21 Driving steel sheet piles in succession.

TABLE 7-1 Specifications for Impact and Vibratory Driver/Extractor

Based on information furnished by Manufacturer and Conmaco, Inc.

Vibratory driver/extractor

Eccentric moment, in-lb	Model	Manufacturer	Frequency, vehicles per minute	Horsepower	Amplitude, in	Maximum pull extraction, tons	Suspended weight, lb
6940	40E-3VT	Foster	700–1120	300	5/16–2	60	40,000
3600	ICE 812	Ice	480–1200	350	1/4–1	40	14,700
3500	40H-4	Foster	0–1600	460	5/16–1	30/45	18,000
2500	V-18	MKT	0–1600	224	1/2	40	14,000
1740	20H-4	Foster	0–1500	190	1/4–3/4	30	12,900
1462	V-14	MKT	1500–1850	140	1/4–1/2	40	10,000
870	IOE-1	Foster	955–1100	40	3/16–3/4	16	7,400
610	TH-2	Foster	0–1800	125	1/4–3/4	10	4,680

Impact extractor

Rated energy, ft-lb	Model	Manufacturer	Crane pull, maximum/minumun, tons	Blows per minute	Total weight, lb	Boiler horsepower, required ASME	Air required, ft^3/min	Inlet pressure, lb/in^2
12,000	HD-15	BSP	18/45	120	10,435	70	450	100
9,525	A-4	Nilens	/50	150	10,910	86	640	100
6,350	A-3	Nilens	/30	160	7,340	64	470	100
5,600	HD-7	BSP	8/20	150	3,750	—	250	100
3,615	P-14	Delmag	10/25	135	5,450	Diesel	—	—
1,640	1200-A	Vulcan	105/150	530	9,200	140	1,020	100
1,000	800-A	Vulcan	7/100	550	5,640	100	740	100
1,000	E-4	MKT	/100	400	4,400	30	550	100

In stiff clays it is hard to excavate under bracing, since a large ridge will be left. This can be knocked down by a diver and jet, or by swinging the bucket (taking care not to hit or snag the bracing).

After gross excavation, usually to 2 or 3 ft (60 or 90 cm) below grade due to the scallop-shaped bites taken by the clamshell bucket, the clay which is still adhering to the sheet piles, especially in the indented profiles, must be removed. This is usually done by using an H beam as a long chisel, augmented by a jet. This is run up and down along the sheet piling. The completeness of cleaning must be verified by a diver.

This cleaning is done to ensure that the tremie concrete seal will properly seal the entire inner perimeter. Obviously, in those relatively few cases where no seal is used, the sheet piles do not have to be cleaned.

Later, after piles have been driven, but prior to placing rock backfill or tremie concrete, silt and sediments are cleaned from the bottom by means of an airlift pump. This is effective in removing soft siltation material from the pockets left by the clamshell bucket.

Drive Bearing Piles (Step 8)

Unless the pilings were driven from floating equipment right after the predredging, they are driven now (see Fig. 7.22).

Since their tops or cutoffs are generally in or just above the tremie concrete seal, it follows that the heads have to be driven a substantial distance underwater. In the case of bridge piers, batter piles are often required, directed radially out on all four sides, as well as vertical piles.

FIG. 7.22 Driving long steel H piles with 200-ft (60-m) leads, Columbia River Bridge, I-205.

The piles should be carefully arranged so as to avoid intersection, both during setting and driving, and below ground.

Piles can be set through the bracing and supported in templates affixed to the top level bracing. There are three methods used to drive such piling: (1) with underwater hammer; (2) with a follower; (3) with above-water hammer, acting on a longer pile than required, followed by a cutoff and removal for reuse of the extra length.

With the underwater hammer, method 1, the difficulty lies with the physical dimensions of the hammer and leads and consequent interference from bracing and previously driven piles. If a hammer accidentally lands on a previously driven "high" pile, the hammer will be badly damaged. Hence, piles may have to be cut off before driving adjoining piles. If it hits on bracing, it may cause serious distortion.

Underwater steam hammers use air pressure in the hammer casing so as to exclude water. This air pressure reduces the efficiency of the hammer, particularly with depths greater than 60 ft (18 m).

The new hydraulic hammers (e.g., Hydroblok) do not use air and, hence, do not lose energy underwater.

The use of a follower, method 2, was discouraged or prohibited until recently because of bad experience with poorly made followers. Today's followers have a machined cast-steel driving head (socket) at their base and a properly reinforced head. They are fitted with guides so as to be supported in the leads and maintain axial alignment. Followers can be heat-treated after fabrication so as to anneal the welds and prevent brittle fracture under cyclic impact.

Both methods 1 and 2 encounter serious difficulties if any piles do not bring up to required bearing. Underwater splices are really not practicable for most piles, although they can be implemented with pipe piles and, in some cases, with prestressed concrete piles if proper detailing has been made beforehand. In any event, they are very costly and time-consuming.

Method 3 involves the handling and setting of a pile long enough to keep its head above water when its tip is at the designated elevation. Hence, if it becomes necessary to splice, the splice can be made above water.

Full hammer efficiency is available. The piles can be threaded past each other with minimum clearance required. Unless the total pile is prohibitively long, the setting and driving rate is significantly faster.

However, each pile must be subsequently cut off. In the case of batter piles, this may require some underwater cutting in order to set and drive adjoining piles. As many as possible will be cut after dewatering.

These long "cutoffs" can now be taken back to the fabricating yard and spliced on to the next piles. Thus, there is no cutoff loss, except on the last pier driven.

Batter piles generally extend under the edge of the cofferdam; hence, they may limit the penetration of the sheet piles. In some cases, it may be necessary to increase the plan size of the cofferdam so as to obtain adequate penetration prior to intersection with the batter pile.

Jetting is seldom done in cofferdam piers, but may occasionally be necessary with displacement piles. In at least one case, for bridge piers in sand, the use of jetting caused a loss in passive resistance for the sheets and a flow of sand underneath, resulting in collapse of the cofferdam even before dewatering.

Both jetting and vibrations can cause local liquefaction of the sand. Various steps to overcome this include placing of a lower set of bracing at the bottom of the excavation, so as to provide support in lieu of the soil's passive resistance, or placement of a thick layer of crushed rock at the bottom of the excavation, to prevent local liquefaction of the sand and permit ready escape of the pore water.

Leveling Course (Step 9)

A sand-and-gravel leveling course is used to bring the excavation to grade, ready to receive the tremie concrete seal.

This rock subbase also gives temporary support to the seal during placement and prevents intermixing of mud, sand, and concrete.

Underwater Concrete Seal (Step 10)

The purpose of the seal is (1) to prevent upward flow of water; (2) to act as a lower strut for the sheet piles; (3) to tie to the driven piles to resist uplift; (4) to provide weight to offset the differential head; and (5) to provide support for subsequent construction of the pier or intake structure. In many modern piers, the tremie concrete seal course is also designed to function as part of the structural footing block.

Underwater concrete may be placed either by the tremie (pipe) method or by the grout-intruded aggregate method.

For tremie concrete, the mix selected should be cohesive and highly workable. A typical mix is as follows:

Course aggregate—gravel, rounded, ¾ in (20 mm), maximum 55% of total aggregate

Fine aggregate—sand, coarse gradation, 45 to 50% of total aggregate

Cement—700 lb/yd^3 (415 kg/m^3) of type I or II

Water-reducing admixture

Plasticizing or air-entraining admixture

Retarding admixture (as required)

Water, to give a 6- to 7-in (15- to 18-cm) slump

Water/cement ratio, 0.45 maximum

Pozzolan, as a partial replacement for the cement, improves flow and reduces heat. The pozzolan should be ASTM Class N or F and should be tested for compatibility with cement and admixtures. Normally, a 15 to 20% replacement is used, although a higher percentage may be used, provided the delay in gain of strength is recognized.

Admixtures are commercially available with combined water/reducing, plasticizing, and retarding effects. Not all admixtures are compatible and suitable for underwater concrete. A trial batch of several yards, placed in a box or pit, may indicate, e.g., the degree of segregation.

Placement of tremie concrete is best initiated in a sealed tube or tremie pipe of 10- to 12-in (25- to 30-cm) diameter. A plate, with gasket, is tied to the end of the tremie pipe, which is then lowered to the bottom (see Fig. 7.23). The hydrostatic pressure holds the plate tight, so the pipe stays empty to the bottom. Obviously the empty pipe must have sufficient weight so as to not be buoyant. Concrete is then introduced to fill the pipe. With deep pipes, over 50 ft (15 m), it is desirable to first place ½ yd^3 (⅓ m^3) cement grout, then 1 yd^3 (⅔ m^3) of concrete from which the coarse aggregate has been omitted, then follow with the regular mix.

The upper end of the tremie pipe is capped with a hopper. Concrete should be delivered to the hopper, at atmospheric pressure, in a continuous stream, such as that obtained by pumping, conveyor belt, or a bucket with air-controlled closure devices.

The pipe is first placed on the bottom, empty and sealed as described. It is filled just above the balancing point. With a 10- or 12-in (25- or 30-cm) tremie pipe there is very little friction head loss. Then the pipe is raised 6 in (15 cm) or so off the bottom, the seal is automatically broken by the excess concrete head, and the tremie flows out. In normal tremie pours, the concrete exiting from the pipe flows up around the pipe to the surface, then slowly cascades down the surface.

The pipe must always be kept embedded in the fresh concrete, a distance of 3 to 5 ft (1 to 1.5 m) (see Fig. 7.24).

The joints in the tremie pipe should be bolted and gasketed. Otherwise, the flow of the concrete will suck in water and mix it by the venturi effect. This has caused disastrous washing out of the mix in some extreme cases.

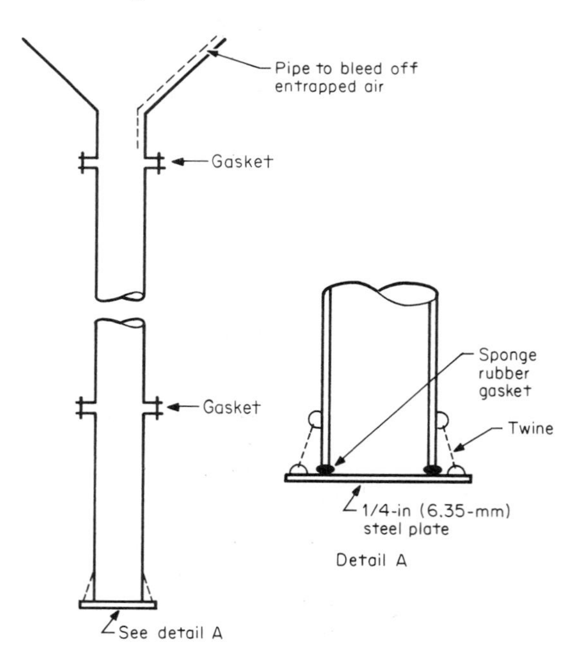

FIG. 7.23 Arrangement of tremie pipe for start of operations.

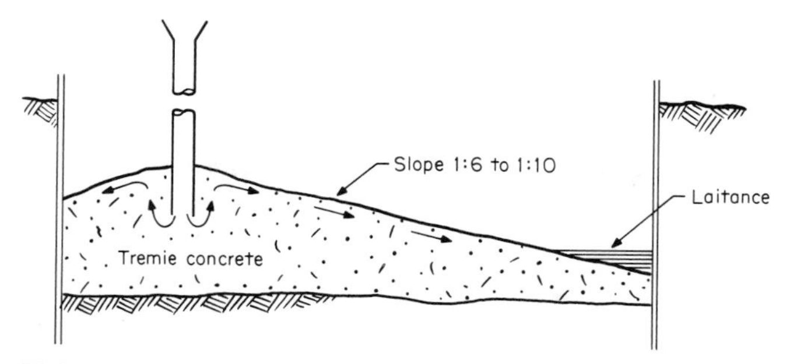

FIG. 7.24 Typical tremie pipe embedment.

If the tip of the pipe is accidentally raised above the surface of the concrete, the seal will be lost, the flow rate will increase, and water will be mixed with the concrete, causing severe segregation and laitance formation.

Any mixing of the tremie concrete and water leads to the formation of laitance (leached-out cement), a soft material of low specific gravity which rises on top of the concrete. Segregation can also occur in cases of severe washing. It is marked by a substantial increase in volume: the cofferdam fills with less concrete placed than calculated. Any yield greater than a 3 or 4% gain should be viewed as an indication of probable severe segregation.

The slope of the tremie concrete surface will be between about 6 to 1 and 10 to 1. Tremie concrete has been flowed up to 70 ft (21 m) from a single tremie pipe without excessive segregation; however, 30 to 35 ft (9 to 11 m) maximum is a more reasonable limitation.

In moving a tremie pipe, it should be raised clear of the water, resealed, and set back into the fresh concrete; in other words, a complete restart.

Dragging the tremie pipe through the fresh concrete inevitably produces segregation and excessive laitance.

Use of a rubber ball (volleyball) as a go-devil is widely practiced to start and restart a tremie pour, but it is not good practice. A volleyball is inflated with about 11 lb/in^2 (0.75 atm) air pressure. Once the ball is pushed deeper than 30 ft (9 m), it starts to collapse and loses its sealing ability. A solid "pig," whether a sack of straw or a gasketed wooden ball or cylinder, is much better.

While such may prove satisfactory in initial starting of a pour, in the restarting of a pour, after relocating the pipe or a loss of seal, the use of a go-devil of any type pushes a column of water down through the recently placed concrete, destroying it locally. Hence, the use of the sealing method is preferred.

The tremie-pipe layout and sequence should be such as to maintain acceptable flow distances and prevent cold joints (see Figs. 7.25 and 7.26). In many cases, this latter requires that a relatively high rate of pour be maintained. Typical rates are 50 yd^3/h (38 m^3/h) for small- and medium-sized cofferdams, 100 to 150 yd^3/h (76 to 114 m^3/h) for larger cofferdams. Retarding admixtures will be found very helpful.

As noted, tremie concrete will flow on a slope. Thus, the spacing of pipes and points of placement depends on depth of seal. The longer the flow distance, the greater the exposure of fresh concrete to washing. A typical flow distance limit

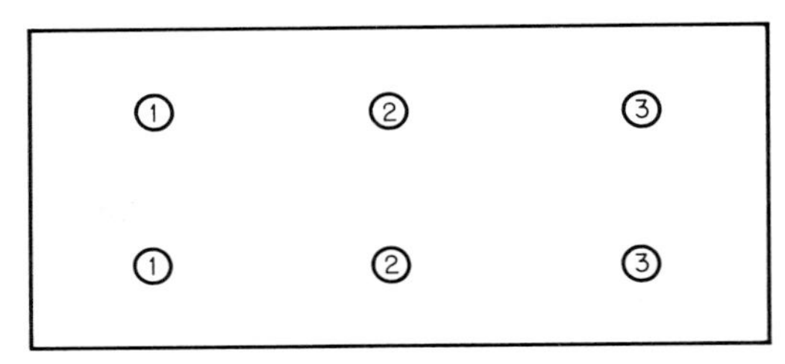

FIG. 7.25 Typical tremie pipe locations and sequence of concrete placement. Start placement through tremie pipes ①. Place to full height. Insert pipes at ② and continue placement until full height is reached. Insert pipe at ③ and complete placement. (Above typical for moderate seal thicknesses.)

is 25 to 35 ft (8 to 11 m), although satisfactory underwater concrete has been placed in thick seals with flow distances up to 50 ft (15 m) and more.

An airlift pump is usually operated in a far corner of the cofferdam to remove silt and laitance and to prevent its being trapped by subsequent concrete.

The tremie concrete surface will be irregular, with a mound at the location of the pipe and valleys in between pipe locations (see Figs. 7.27, 7.28, and 7.29). In attempting to fill the valleys near the end of the pour, great care must be taken to properly embed the tremie; otherwise, the concrete will just flow over the surface laitance. In many cases, therefore, it is best to just leave the valleys low. After dewatering, they can be filled with concrete placed in the dry, usually as a part of the footing block pour.

After the tremie seal has reached grade and initial set has occurred, a diver can remove the laitance by use of a jet. Later, after the cofferdam is dewatered, the laitance may have hardened so as to require a jackhammer to remove it.

FIG. 7.26 Layout of tremie pipes on one of main piers of Richmond-San Rafael Bridge.

FIG. 7.27 Surface of tremie concrete seal immediately after dewatering, Columbia River Bridge, I-205.

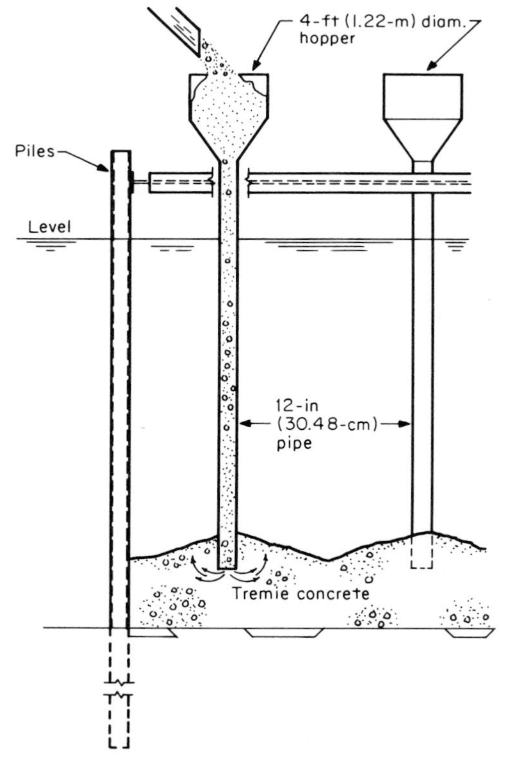

FIG. 7.28 Tremie method for underwater concrete.

FIG. 7.29 Unsatisfactory results with tremie concrete placement due to improper mix and procedure. Note segregation.

FIG. 7.30 Satisfactory results with tremie concrete placement using proper mix and procedures. (This is a second pier on the same project as that shown in Fig. 7.29, after corrective measures were instituted.)

Horizontal lifts are undesirable in a tremie concrete placement, as they will have seams of laitance on the joint surface. If a large cofferdam is to be subdivided, it should be done with a vertical bulkhead. After the first placement is made, the bulkhead can be removed and the second pour made. If the bulkhead is of precast concrete, the tremie will bond to it. Steel sheet piles may be used as an intermediate bulkhead and left in place.

The seal resists uplift by its weight and by its bond to the piling, which then resist uplift in skin friction against the soil. Typical bond values achieved are 80 lb/in^2 (0.55 N/mm^2), allowing the use in design of 20 lb/in^2 (0.14 N/mm^2). Between piles, the seal concrete must be thick enough to act as an unreinforced thick plate.

The net underwater weight of tremie concrete is about 140 lb/ft^3 (2250 kg/m^3) less 4% swell (over-yield) less 62.4 lb/ft^3 (1000 kg/m^3) of buoyancy, or about 70 to 72 lb/ft^3 (1120 to 1150 kg/m^3). Hence, it is relatively inefficient in resisting uplift, in that it may require additional excavation and longer sheet piles in order to make room for the required seal thickness.

In general, it must be assumed that full hydrostatic pressure will act on the underside of the concrete seal. The exception is when a positive cutoff is achieved by deep penetration of sheet piles and a filtered bleed system (dewatering system) has been installed below the seal. This is very difficult to assure in practice and, hence, is rarely used. When properly placed, tremie methods can produce homogeneous concrete of high quality (Figs. 7.30, 7.31, and 7.32).

In some cases of tremie-concrete seals on rock or hardpan, tie-down ground

FIG. 7.31 Placing tremie concrete in main pier, Columbia River Bridge, I-205. Concrete pumps delivering concrete to open tremie hoppers.

FIG. 7.32 High-quality cores of tremie concrete from San Mateo-Hayward Bridge.

anchors are drilled into the underlying material and anchored to or even prestressed against the seal.

The second method of placement of a tremie-concrete seal is by the use of grout-intruded aggregate. In this case, coarse aggregate [½ in (12 mm) and up] from which all fines have been removed is placed in the cofferdam. Vertical pipes for grout injection are installed on 5- to 7-ft (1.5- to 2-m) centers. Other vertical slotted pipes are installed to use as indicators.

Grout is now pumped through the pipes, so as to fill the voids and interstices in the rock with mortar. The grout must be very fluid and have high resistance to segregation. Usually the mix will consist of fine sand and cement in equal parts, plus an admixture designed to retard set, reduce water-cement ratio, promote fluidity, and prevent segregation.

If the seal thickness is too great for the first set of pipes [a typical lift is 5 to 7 ft (1.5 to 2 m)], a second set is installed, with its injection points set so as to enable the new grout to be injected into that which has been placed earlier, before the earlier grout has set.

The keys to successful placement of grout-intruded aggregate concrete are the cleanliness of the coarse aggregate (no fines, no silt) and proper control of the grout consistency. The rate of injection should be as rapid as possible.

The surface of a grout-intruded aggregate-concrete seal will require cleaning to remove the excess gravel, etc., much as does a tremie-concrete seal.

For grout-intruded aggregate, the sheet-pile interlocks must be fully engaged, and there must be no holes; otherwise, the fluid grout will flow out into the surrounding waters.

Concrete seals should normally be stopped about 6 in (15 cm) low, provided this can be done safely. The difference can be made up in the dry pour of the footing block. By stopping low, there will be less removal of humps and high points, in order to place the lower layer of reinforcing bars for the footing block.

Dewatering (Step 12)

During all previous steps, the water levels inside and out will have been equalized by a floodgate in the sheet piles, so as to prevent water flow through the interlocks and through the soils.

After the seal has gained sufficient strength (usually 3 to 7 days) the cofferdam is ready for dewatering. The floodgate is closed. It is often difficult to get the water level to lower initially, as water flows in through the interlocks as fast as it is pumped out. Later, after a head differential has been established, the sheet-pile interlocks will tighten.

Therefore, it is desirable to start dewatering with as much pumping capacity as possible. High-volume low-head (irrigation) pumps can be used to augment the high-head submersible pumps. Start at low tide when there is less total surface exposed (see Fig. 7.33).

If an initial start proves infeasible, there are a number of ways of sealing interlocks. Sand and cinders, sand and sawdust, and sand and manure have all been used. However, subsea vibration in current and waves may cause the material to fall out. Oakum caulking seams can be run by divers.

A final expedient is to drape weighted canvas over the outside face of the sheet piles. This will be sucked tight against the piling by the inflow.

FIG. 7.33 Dewatering cofferdam. Second Delaware Memorial Bridge.

Sheet-Piling Removal (Step 15)

Usually a vibratory hammer is most efficient in removal. Piles are often removed in pairs to facilitate their reuse.

It is often very difficult to pull the first pile. Once the first one is pulled, the others come more easily. For the first pile, there is double interlock friction and the general binding and stretching of the entire cofferdam.

Usually a pile in the middle of a side is easiest. Try driving (vibrating) down first to break the interlock friction. Sometimes a steam or diesel hammer must be used to drive the first pile down.

Impact extractors are made with powerful capacity to free a pile. Their clamping to the pile requires careful detailing (e.g., use of high-strength bolts); otherwise, they may rip out a section of steel from the sheet pile.

Another "last resort" is to rig jacks and break the binding by jacking one pile up against its neighbor.

Removal of Bracing (Step 16)

If possible, this should be designed to be lifted over the top of the partially completed structure. This is expeditious and facilitates reuse. This may not be feasible due to weight or interference, so removal in sections may be required.

The removal, in sections, is practical if well conceived and detailed during initial design. Generally speaking, provision must be made to relieve the built-in stress on a member to be disconnected before it is cut free. Bolted joints facilitate removal and cause less damage than burning.

Upon release of a member, the remaining forces must go elsewhere, for example, into the backfill and other bracing. Care must be taken to ensure that a second brace is not overloaded so as to fail in buckling. Blocking may be used to transfer the force to the structure.

When struts must pass through a massive portion of the new member, they are often best treated by casting them into the structure. Blockouts can be provided at the edges to permit cutting off back of the surface, if required, and filling the pocket with concrete. However, corrosion of submerged steel is extremely slow and minimal due to the nonavailability of oxygen.

For struts which penetrate thinner walls, sleeves may be provided so that the struts can be later removed by pulling through.

COFFERDAMS FOUNDED ON ROCK

In many cases, bridge piers and pump houses are founded directly on hardpan or rock. In such cases it may be extremely difficult to achieve sufficient penetration with the sheet piles to develop adequate shear resistance at the tip. Such penetration may be aided in specific soils by one or more of the following steps:

1. Use of hardened-steel tip protectors for sheet piles.
2. Continued driving at refusal with a relatively small hammer, so as not to damage the tip, but rather chip into the rock.
3. Predrilling (line drilling) along the line of sheet piles. One way to align the drill and piles properly is to first set and drive the sheet piles to a point just

above grade, taking care not to bend the tips by driving into rock. Then a 6- to 12-in (15- to 20-cm)-diameter bit is used to drill within the out-facing arch of each sheet or pair of sheets. In the case of H-section or pipe sheet piles, the drill may be run down the center.

When adequate penetration below the tip cannot be reliably and positively assured, then it is important to provide shear resistance at the lowest level possible by means of a bracing frame. This ensures the structural adequacy of the cofferdam, although it does not prevent run-in of sand under the tip.

Excavation at or below the tips should be carried out with some or all of the precautionary steps set forth in the next section.

Cofferdams which must penetrate overlying strata of hard materials or through boulders may present major difficulties. The sheet piles are easily deformed, split, or driven out of interlock.

In the case of an upper stratum of hard material, predrilling (line drilling) with a drill or high-pressure jet may permit penetration, especially if the piles are fitted with hardened-steel tip protectors.

If blasting is indicated (for example, when a coral or limestone layer must be penetrated), this is usually best done after the bracing frame has been set, but before setting the sheet piles. Blasting adjacent to sheet piles often causes distortion which later prevents driving to grade.

Boulders embedded in running sands present perhaps the worst difficulties. The hard boulders cannot easily be penetrated; they are extremely difficult to predrill because of sand run-in, and they do not always fracture by blasting. Some success has been had by the following methods:

1. Installing a third (or lower) set of bracing as deep as possible, just above the boulder elevation.
2. Keeping the water head inside the cofferdam above that outside, or filling the inside with bentonite slurry (drilling mud).
3. Injecting chemical grouts outside (silicates or polymers) to give cohesion to the sand. (This step is not always necessary.)
4. Progressively drive and excavate inside below the tips of the sheet piles. The driving should be performed with a relatively light hammer or vibratory hammer.

A steel spud pile may be driven down in the inward-facing arch formed by the sheet piles to displace a boulder into the excavated area. A high-pressure jet may be similarly used, although this is likely to aggravate the tendency for the sand to run in.

CELLULAR COFFERDAMS

Sheet-pile cellular cofferdams are utilized to dike off relatively large areas, where the use of internal bracing and/or tiebacks would be unsuitable due to length or interference with the new structure.

Cellular cofferdams are practicable for dewatered depths up to 80 ft (24 m) or so. They form a continuous gravity wall; hence, they impose a high bearing and overturning pressure and lateral shear on the soil beneath. For this reason, they are best used when founded on rock, hardpan, and dense sands.

The concept of the cellular cofferdam is that a circumferential ring wall is constructed progressively by setting and driving interlocking sheet piles to form a closed cell (see Figs. 7.34 and 7.35). The cell is then filled with granular material, such as coarse sand, sand and gravel, or rock. These fill materials, of course, develop tensile forces in the sheet-pile ring. The successful performance of this system depends on the internal shear strength of the fill, its friction against the sheet piles, and the tensile interlocking of the sheet piles.

When sheet-pile cells are constructed on sedimentary soils, the sheet piles penetrate below the internal excavation level, acting to seal off water flow beneath to some degree and to develop passive lateral resistance. Such a cofferdam tends to progressively develop the shape of a truncated cone and, as excavation proceeds inside, the cell may lean slightly inward due to consolidation of the soil under the eccentric loading.

If there is soft material in the sheet-pile cell (e.g., mud), it must be removed before the fill is placed. Otherwise, the soft material may develop high lateral fluid pressures against the sheet piles, equal to the surcharge pressure acting on it.

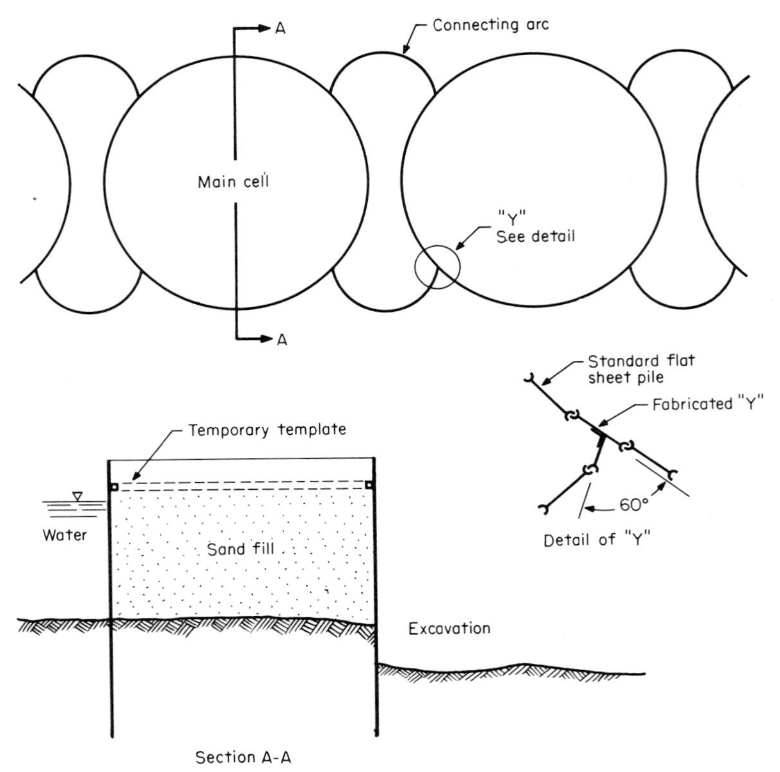

FIG. 7.34 Typical layout for steel sheet-pile cellular cofferdam.

FIG. 7.35 Steel sheet-pile cellular structure, Port of Portland, Terminal no. 6.

In order for the sheet-pile cell to function properly, the general geometry of the cell must be maintained at all horizontal cross sections, within limited tolerances. Further, the sheet piles must be stretched in tension by the fill in order to develop the necessary shear friction in the interlocks.

Finally, the entire ring of steel-sheet piles must be interlocked continuously.

To accomplish this, it is necessary that the sheet piles be accurately set. Therefore, falsework guides must be provided, consisting of at least two and preferably more ring wales. Their supporting falsework piles are usually steel pipe pile spud piles driven through sleeves in the lateral bracing of each ring wale.

The procedure is, therefore, to set the group of ring wales, then drive the vertical spud piles through the sleeves. Then the ring wales are raised to proper grade and pinned off.

Steel sheet piles are then set around the ring. The Y's (or T's) should be set first, carefully positioned in exact horizontal location and in vertical alignment. Standard flat sheets are then set progressively from the Y's toward a meeting point at center. The Y's and an occasional pile may be driven into the bottom a few feet only, during the setting, in order to enhance lateral stability. The junctions are always made at a central point between Y's so that if there is any stretching or "shaking out" to fit, it is done as far away from the stiff Y's as possible. To aid in accurate setting and joining, the position of each pile should be premarked on the top ring wale. It should be marked as the theoretical distance, center-to-center, of each interlock, plus a small increment, say ⅛ in (3 mm), to provide for stretch. The final fit-up should remove all slack, but not be so tight as to prevent a free-running entry of the last sheet.

Driving then proceeds in incremental fashion, starting at the Y's and working toward the centers. Each pile should be driven not more than 3 ft (90 cm) beyond the tip of its neighbor.

Excessively hard driving of sheets usually indicates inaccurate setting, which in turn develops excessive interlock friction at a later stage. (The field crew usually reports it as "encountering hardpan or obstructions"; and, of course, once in a while they are right!)

Vibratory hammers are ideal for driving the sheet pile in cells, since if excessive friction does develop, a group of piles can be raised back up and redriven in the same incremental iterative fashion.

Continued hard driving can drive sheet piles out of interlock, destroying the tensile capacity of the ring.

Most cellular cofferdams are circular cells, with short arcs connecting adjacent cells. In many cases, these short arcs can be set along with the adjacent circular cells, and thus set and driven vertically.

If, due to the need to stabilize each cell as it is set by filling with rock (as, for example, when working in a swift river), the arc has to be set later between two completed and filled cells, then the sheet piles will have to be set on a slight inward-leaning angle so as to keep the same arc distances at top and bottom.

Consolidation of fill within the cells helps to improve the internal friction and prevent liquefaction under earthquake. Recently it has become the practice to consolidate the fill by vibration. Such vibration is very effective in a cell, due to its confinement, provided the water can escape rather than just build up high pore pressure. In fact, in one case, intensive internal vibration caused the rupture of one cell where a layer of impermeable fill prevented escape of the displaced water.

Therefore, the construction of several vertical drains or wells, combined with internal vibration, will expedite consolidation.

In order for the sheet-pile cell to function most effectively as a gravity wall, it is desirable that it be fully expanded into a truncated cone before any backfill is placed behind it. Hence, backfilling should not start until the cell is filled.

One of the causes of many failures and partial failures of cells is the pushing of backfill out from shore, which in turn displaces mud against the arcs of connecting cells. The mud develops full fluid pressure under the surcharge, tearing the Y's from the adjoining cells.

Thus, the Y's are recognized as the weak points, that is, each Y and the three sheet piles connecting to it. Each leg of the Y is subject to high tension, combined with bending and moment in the flat portion. The details of the Y, its angle, and its method of fabrication are of primary importance. Many experienced designers/builders of cellular cofferdams use heavier-section steel, higher-yield steel, or both for these critical piles.

Whereas in the past many connectors were 90° T's, the present improved practice is to use 30° or 45° Y's, so as to minimize the moments.

Where cellular cofferdams are founded on rock, it is very difficult to seat the sheet piles so as to prevent large underflows of water. Among the methods used are the following:

1. Clean off to rock and preplace a graded layer of relatively low permeability sand and gravel into which the sheet piles are toed. This can control but not eliminate underflow.
2. Use of cast steel "protectors" on the tips of the sheet piles to permit toeing into rock.

3. Preexcavation of the rock surface so as to remove boulders, cobbles, ridges, and protuberances before the sheet piles are set.
4. Pouring a slab of tremie concrete inside the cell before filling it.
5. Pouring tremie concrete or placing sacked concrete on the outside of the cell. This is difficult to confine and is usually used as a last resort.
6. Placing an extensive blanket of clay on the outside of the cells.

The bursting effect of fill within the cells of cofferdams and the tensile stresses in the sheet piles may be reduced by draining the cell fill through weep holes in the sheets as the cofferdam proper is dewatered. Such slowly drained fill will also consolidate and have a higher angle of internal friction.

8

Temporary Earth-Retaining Structures

by Melvin Febesh
Robert S. Woolworth

The main purpose of temporary earth-retaining structures is to permit safe excavation of a hole into which a permanent structure can be built. The type of temporary structure and method of installation are dependent upon many factors, the principal ones being the type of soil encountered; the elevation of groundwater; the depth of soil to be supported; the proximity of existing structures which might be affected by the proposed excavation; and, if more than one system can be used to satisfy the projected conditions, their relative cost and time required for installation.

Special procedures and equipment have been developed in localized geographical areas to take advantage of local conditions, thus permitting economical installation of retaining structures under those special conditions. Contractors and engineers should be particularly careful in analyzing these special methods to be sure that the conditions at their site will permit these special methods to be used.

Temporary retaining methods can be discussed as two separate items: the structure in contact with the soil and the bracing system which holds the structure in place. There are also methods such as freezing and grouting, which change the soil into structural elements and can often be designed as self-supporting systems.

Before any system can be designed there are certain preliminary steps which should be taken and analyzed:

1. Analysis of soil-boring data. This would include determination of groundwater level, soil stratification, presence of underground obstructions (natural or created) and proximity of rock.
2. Survey of existing structures, including underground utilities.
3. Review of data available from libraries, utility companies, building department, U.S. Geological Survey, etc.
4. Examination of site and existing conditions including drainage of area, type of vegetation, etc.
5. Preparation of sketches to scale, indicating relationship of existing structures to proposed excavation, soil stratification, groundwater level, and proposed new structure.

A determination must be made of the amount of deformation that would be tolerable in the retaining structure so that no damage is caused to adjacent structures or utilities. This will indicate if underpinning is required of the adjacent structures, or if the retaining structure is to be designed to support the surcharge loads of these structures.

Total loads acting on the temporary retaining structure can then be determined and should include soil and water pressure and surcharge loads from traffic, construction equipment, material storage, and adjacent structures. A check should also be made of the base stability of the proposed excavation.

TYPES OF RETAINING STRUCTURE

Steel Sheet Piling

Steel sheet piling is rolled in a variety of shapes and weights with special joints to act as interlocks between sections of the sheeting. For use as temporary retaining structures, Z sections or deep-web sections with large section moduli are most commonly used (see Fig. 8.1).

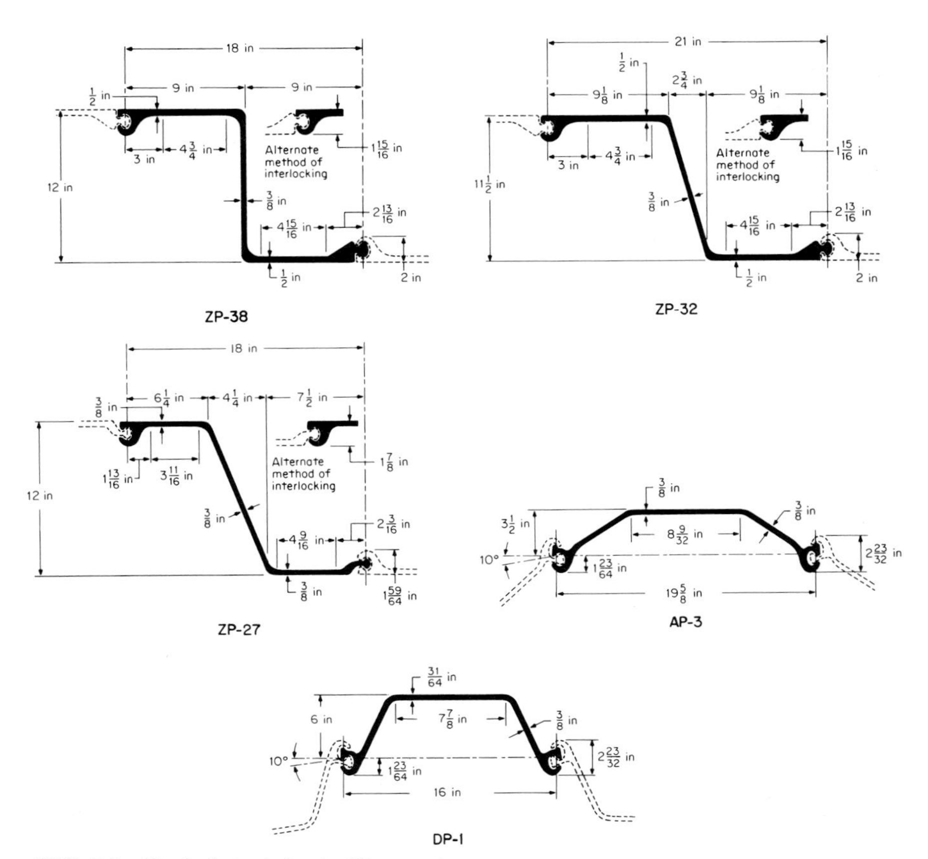

FIG. 8.1 Typical steel sheet-piling sections.

Greatest advantages are obtained in using steel sheeting when there is a high groundwater level and the soil to be retained cannot easily be predrained, or if it is desirable to maintain the groundwater outside the proposed excavation at its existing level.

The steel sheeting is generally driven to its final position with conventional pile-driving equipment before the excavation commences. To be most effective, the sheeting should be driven into an impermeable layer, or with sufficient penetration below the excavation subgrade to effectively eliminate flow of groundwater under the sheeting. Some leakage throughout the sheeting must be anticipated.

Steel sheeting loses its effectiveness when the interlocks are torn or if all sheets are not driven to the proper penetration. These conditions can be due to obstructions in the ground, hard layers of soil to be penetrated, or improper driving of the sheets. Loose sand or fill can undergo considerable consolidation during installation of the sheeting, so the use of this method should be evaluated when driving close to existing structures resting on these types of material.

Since material costs are relatively high, it is often desirable to pull out the sheeting after the temporary structure is no longer required.

Proper installation of the sheeting is important in ensuring the effectiveness of the system. A proper driving template should be set to obtain alignment of the sheets before driving. The top wale of the bracing system can often be used as the template. The sheets should be driven in "waves" or "steps"; i.e., the toe of any sheet should be driven no more than 5 to 6 ft (1.5 to 1.8 m) below the adjacent sheet, and the ball end of the sheet should be driven as the leading edge.

Soldier Piles and Horizontal Sheeting

This earth-retention system consists of vertical members, called soldier piles, generally 6 to 10 ft (1.8 to 3.0 m) on center, installed to below subgrade in advance of excavation, and the installation of horizontal members spanning between the soldiers as excavation proceeds.

Soldier piles are usually rolled sections (bearing pile or wide flange), but can be fabricated sections, pipe, or precast elements. They are usually installed with conventional pile-driving equipment, but can be installed in augered holes, set in pits, or by other means as required by site conditions. The horizontal sheeting is usually wood and is installed behind the flange closest to the excavation (inside flange). Steel or precast concrete planks can be used in lieu of wood sheeting. The sheeting can be installed on the inside face of the front flange and held in place by various methods such as clips, welded studs or bars, etc. (see Fig. 8.2).

There is some disagreement regarding the design of horizontal sheeting systems. Most designers agree that the full earth pressure is carried by the soldier piles and that the soil arches between the piles. The load on the sheeting is imposed by the soil under the arch and does not greatly increase as depth increases. This theory has continually proven itself on projects where 3-in (7.6-cm)-thick wood sheeting has been used for excavation depths of 50 to 60 ft (15 to 18 m). The New York City Transit Authority's *Field Design Standards* assume that all the load is carried by the sheeting and that the sheeting transfers the load to the soldiers; however, they permit a 50% increase in allowable stresses for the wood sheeting. In cases where the sheeting is installed behind the back flange of the soldier pile, it appears that the arching action of the soil is

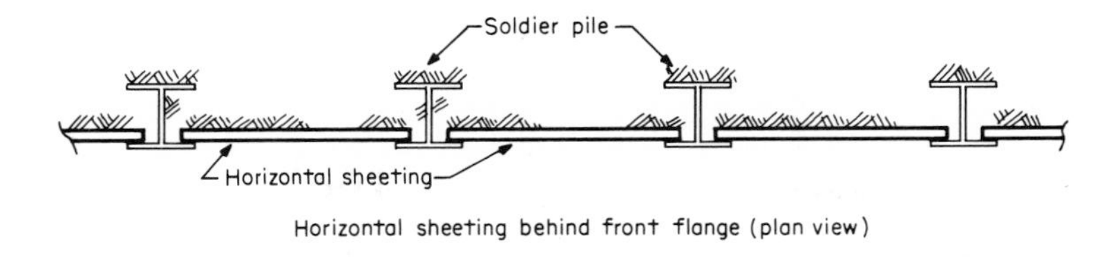

Horizontal sheeting behind front flange (plan view)

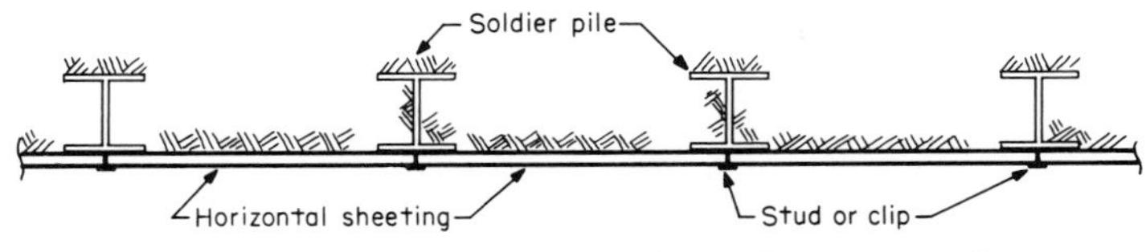

Horizontal sheeting in front of front flange (plan view)

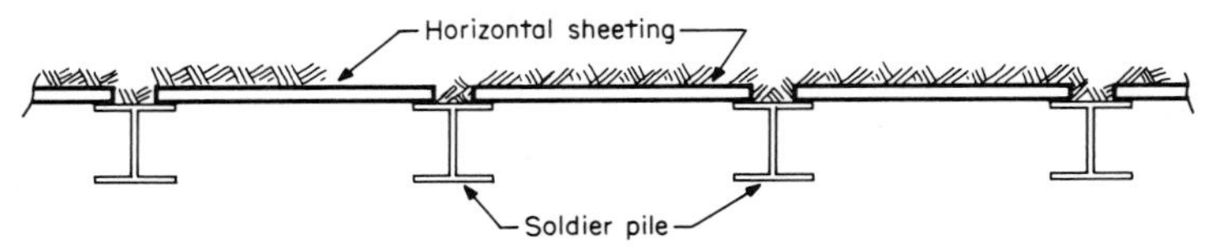

Horizontal sheeting behind back flange (plan view)

FIG. 8.2 Soldier piles and horizontal sheeting—typical details—plan.

destroyed and large movements outside the excavation may occur. In these cases the sheeting takes essentially the full soil pressure and should be designed accordingly.

When placing the sheeting, a space of 1½ to 2 in (3.8 to 5.1 cm) is normally left between sheets. These spaces are called *louvres* and are maintained by nailing short wooden blocks to the sheeting. These spaces permit packing of soil between the boards to replace any soil that might be lost during excavation for the boards. They also permit water to bleed through so that there is no buildup of water pressure. Should water entering between the boards carry material, the spaces are filled with salt hay which acts as a filter (see Fig. 8.3).

Vertical Wood Sheeting

Vertical wood sheeting is one of the oldest methods of ground support and under certain conditions is still the most economical. Most common uses are in excavations for utility or sewer trenches; however, for large excavations with depths of, say, 12 to 15 ft (3.6 to 4.6 m), this method proves to be very satisfactory.

Timber planks 2 to 3 in (5 to 7.5 cm) thick are generally used. The bottom edge is cut at approximately a 45° angle to aid in penetrating the soil. A timber bracing system is most commonly used. The sheeting is usually driven with light, air-driven hammers. Great care must be used during driving so that the sheets are not damaged. The sheeting should be driven in waves with tips penetrating only 2 to 3 ft (60 to 90 cm) below adjacent excavation levels. As excavation is carried to approximate bottoms of sheets, they are redriven an additional 2 to 3 ft (60 to 90 cm) below the excavation level and the cycle is repeated. At the

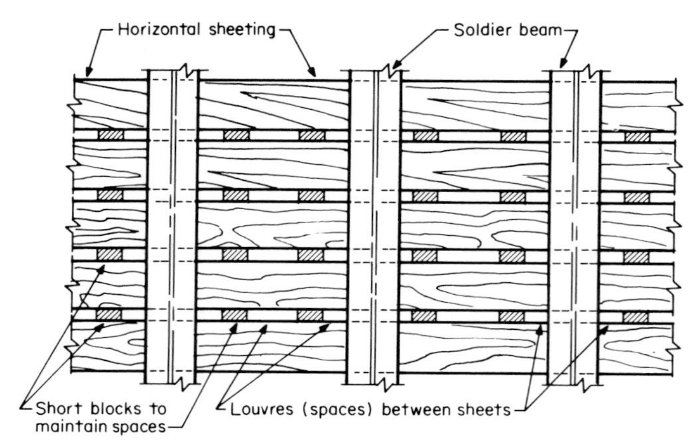

FIG. 8.3 Soldier piles and horizontal sheeting—typical details—elevation.

proper depth, bracing is installed. When groundwater is present, tongue-and-groove sheeting can be used to try to minimize infiltration of water (see Fig. 8.4).

Where ground conditions permit, excavation can be carried down to the level of the top wale, say 3 to 5 ft (0.9 to 1.5 m), before placing the top wale and then setting up the wood sheeting. The excavation, sheeting, and bracing then proceed as described above. Since the full earth pressure is carried by the sheeting, the rows of wales and braces are rather close together. This increases the cost of sheeting and excavation for deeper cuts.

Most designs of sheeting and bracing of shallow cuts are empirical, and many codes furnish tables indicating these designs (e.g., OSHA rules and regulations and New York City EPA specifications both give designs indicating sheeting thickness, wale and brace size, and spacing for various soil conditions and excavation widths and depths).

Vertical wood sheeting is often used when excavating for pits or deep footings below subgrade of general excavation. When excavating small holes, say 6 ft^2 (0.6 m^2), vertical wood permits rapid excavation at minimal cost. Bracing can be relatively light, since the soil arches around the holes and full earth pressure need not be supported. For larger holes, the arching action is lost and the bracing must be considerably stronger. It may then be advantageous to use lightweight steel sheeting. When excavating through soil with a high water table, the maintenance of the water table outside the sheets and the dewatering to excavation level on the inside create an unbalanced head with the possibility of a "blow" occurring under the sheeting. This possibility should be carefully considered before using vertical wood sheeting.

Augered Piles

It is quite common to install soldier piles in augered or predrilled holes, rather than by driving with conventional pile-driving equipment. In portions of the country where soil conditions permit the holes to be drilled and remain open, this method is probably more economical than driving. When this method of

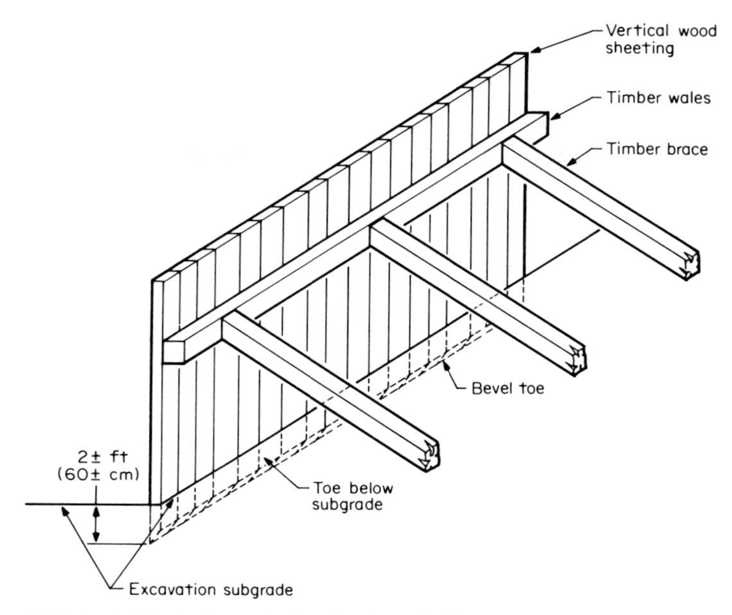

FIG. 8.4 Vertical wood-sheeting installation.

installation is specified or attempted in soil which is not suited for augering, the costs could be prohibitive and progress extremely slow. Where soil conditions require the use of temporary casing or drilling mud, some of the economy is lost. The presence of boulders or large obstructions may make the use of augering equipment uneconomical.

Using augered holes permits greater flexibility in design than when driving soldier piles. The holes can be filled with concrete (using reinforcing steel as required) to form concrete soldier piles, or steel sections can be installed, either fabricated or unfabricated. Where tie-back bracing is to be used, seats and/or sleeves for the tiebacks can be installed on the soldiers before placing in the hole. When placing structural shapes in augered holes, spacers are usually installed on the soldiers to help align them in the holes, and the holes filled with lean concrete—say one sack of cement per cubic yard (0.76 m^3) of concrete. This helps keep the soldiers in place as excavation proceeds and earth loads are imposed on the soldiers. As excavation proceeds, horizontal sheeting can be installed between the soldiers as with driven soldiers. In cohesive soils, the earth may arch between the soldiers without any support; however, it is good policy to install some type of support (wire mesh, chain-link fence, gunite, etc.) to protect workers from falling pieces of soil.

It is becoming increasingly common to see engineers specify use of augered holes for projects in urban sites to eliminate noise and reduce vibrations that are present when driving soldiers. One must be extremely careful when analyzing site conditions on such projects. Old foundations, boulders, etc. could seriously interfere with hole drilling and make this procedure extremely difficult, if not impossible.

Slurry Walls

The term *slurry wall* can mean any trench installed by the slurry method. However, for the purpose of our discussion, we will use the term to describe a reinforced concrete wall constructed by excavation utilizing the bentonite slurry method and backfilling with tremie concrete. Although we are here discussing temporary structures, the true economy of this type of construction is when the wall is also used as part of the permanent structure.

The general procedure for constructing a slurry wall is as follows:

1. Construct a "pretrench." This consists of building two guide walls, generally of reinforced concrete. The spacing between the walls is about 1 in (2.5 cm) greater than the width of the wall. The pretrench is constructed for the entire alignment of the slurry wall and helps guide the excavating equipment.
2. Fill the pretrench with a bentonite slurry. It is important that the level of the slurry be kept at approximately the original ground level and at least 2 ft (60 cm) above groundwater level.
3. Commence excavation (usually with a clamshell bucket), continually filling the trench with bentonite slurry to replace the excavated material. The length of wall dug at one time varies up to a maximum of 25 to 30 ft (7.6 to 9.1 m). Each section is called a panel. Several panels may be excavated at the same time as long as the space between panels does not permit bentonite to flow from one panel into the other.
4. After the panel is excavated to the proper elevation, the trench is cleaned of excess material and "end pipes" or steel beams are placed at each end of the panel to form a joint.
5. A prefabricated reinforcing steel cage is placed in the excavated trench.
6. Concrete is placed through tremie pipes, displacing the bentonite slurry. If end pipes are used, the pipes are withdrawn as the concrete receives initial set. If steel beams are used at the panel ends, they usually remain in place.
7. Continue with steps 2 through 6 until wall is complete. If end pipes are used, the adjacent panels require only one end pipe, since the joint has already been formed at one end. When hard soil or rock is encountered, chopping bits or churn drills may be required to remove this material.

The design and construction of slurry trenches are based on some theory and a lot of empirical relationships and experience. For our purpose it should suffice to know that slurry trenches do work and that they are an excellent tool in foundation construction.

Some things to consider when constructing slurry walls are the following:

1. When excavating a trench through fill or material with large voids, the bentonite slurry can run out through the voids and lead to a possible collapse of the trench. When this type of material is encountered, it may be necessary to alter the soil by compaction, filling, grouting, etc. to permit completion of the slurry wall.

2. If obstructions, boulders, etc. are removed during the trench excavation and these items extend outside the theoretical face of the wall, the void created during excavation will be filled with tremie concrete. If the excess concrete is on the inside face, it may be necessary to trim the wall as it is exposed during excavation.
3. Setting up a slurry plant is expensive, making the cost of slurry walls on small projects rather excessive.

The thickness of slurry walls is generally in the range of 2 to 3 ft (60 to 90 cm). This makes a rather stiff wall when compared to steel sheet piling or soldier piles. This permits larger cantilevers before installing top tiers of bracing and greater spans between bracing levels. New techniques are constantly being developed, such as installing precast panels in the excavated trenches and using post-tensioning, and should be considered for use on future projects.

BRACING SYSTEMS

The purpose of the bracing system is to maintain in place the structure which is in direct contact with the soil. The bracing system is generally installed in levels, or tiers, which are horizontal or (if the surface of the adjacent ground is sloping) parallel to the ground surface.

Deciding whether to use internal braces or tiebacks depends on several factors. Some soils are not suitable for using earth tiebacks. If rock is at exceedingly great depths, rock anchors are not economical. The geometry of the proposed excavation (depth of excavation in relation to its width) determine if cross-lot bracing or inclined braces bearing on footings at subgrade should be used. In cases where either tiebacks or internal braces may be used, the determining factor should be the cost and time of installation. In general, the cost of a tieback installation is greater than that of an internal bracing system. However, this is often offset by savings in the cost of excavation and constructing the perimeter walls. Note that tiebacks extend outside the excavation; if not within the property line, these may require special permits and/or easements.

Internal Bracing

Internal braces are usually HP, WF, or pipe sections. For high loads or long braces, pipe sections offer the advantage of requiring less lacing (intermediate bracing) than the other sections, but the higher cost of the pipe material often exceeds the saving involved. When installing cross-lot braces, one end is usually welded to the wale and the other end stressed with plates and wedges. When using inclined braces, the end bearing against the wale is usually welded to the wale and the other end is either cast into the footing (for light brace loads) or the load is transferred to the footing with plates and wedges. If deflections of the sheeting system are to be kept to a minimum, the bracing system can be prestressed by means of jacks.

When using internal braces, care must be exercised during the design of corners if the angle of the corner tends to "open up" and walk into the excavation.

Wales

Bracing systems using tiebacks or internal braces usually employ continuous horizontal members called *wales*. The wales are in contact with the earth-retention system and transfer the loads from that system to the braces or tiebacks. The main purpose of the wales is to permit the braces, or tiebacks, to be placed far enough apart so that they can be designed for high loads, minimize interference with construction operations, and permit work to be performed economically.

When designing wales, the span between points of support should be governed by the deflection of the wale as well as by the difference in cost of heavier wales as compared to additional braces or tiebacks.

Most deep excavations employ steel bracing systems. Wales are generally HP or WF sections. It is a good practice to choose as a wale a member which has a minimum 8-in (20-cm)-wide flange. This minimum flange width provides a large bearing area for braces or tieback brackets. Great care must be given to details of the connection between wales and braces (or tiebacks). Most failures in bracing systems occur at connections which are not properly designed or constructed. Stiffeners should be installed where required and brackets should be placed to resist the vertical component of brace or tieback load (see Fig. 8.5).

Berms

The vertical spacing between tiers of the bracing is critical to the stability of the entire retention system. As excavation proceeds, the earth below the excavation level provides the lateral support until the bracing level is installed. If the earth below the brace level is excavated on a slope with the toe of the slope at a level lower than the brace level, the wedge of earth that remains to support the sheeting is called a *berm* (see Fig. 8.6).

An analysis should be made of the entire system at each stage of excavation. The stability of the berm at each bracing level, before the braces are installed, may be critical and should be checked. The toe of the retention system should be checked at completion of excavation and bracing installation to verify stability of the toe. The slope of the berm should be analyzed to ensure its stability. In good soils slopes of 1 vertical to 1.5 horizontal are recommended. Poor soils might require slopes of 1:2 or 1:2.5. In poor soils it may also be necessary to install the bracing system in short lengths to maintain stability.

Tieback systems usually do not have the same problem with berm stability as internal brace systems. Ties are generally installed with the excavation being performed in horizontal lifts, eliminating the need for berms.

Tieback Systems

Tiebacks (or anchors) are a structural system which acts in tension and receives its support in earth or rock. The system consists of the earth or rock, which provides the ultimate support for the system; a tension member (or tendon) which transfers the load from the soil-retention system to the earth or rock; a transfer agent which transfers the load from the tendon to the soil or rock; and a stressing unit which engages the tendon, permits the tendon to be stressed, and allows the load to be maintained in the tendon (see Fig. 8.7).

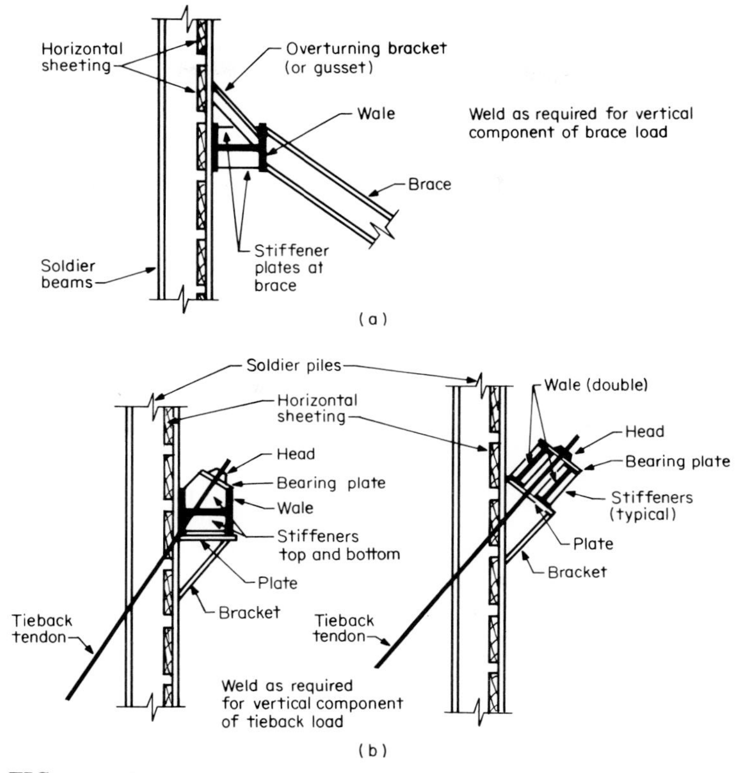

FIG. 8.5 Typical connection details: (*a*) internal braces and (*b*) tiebacks.

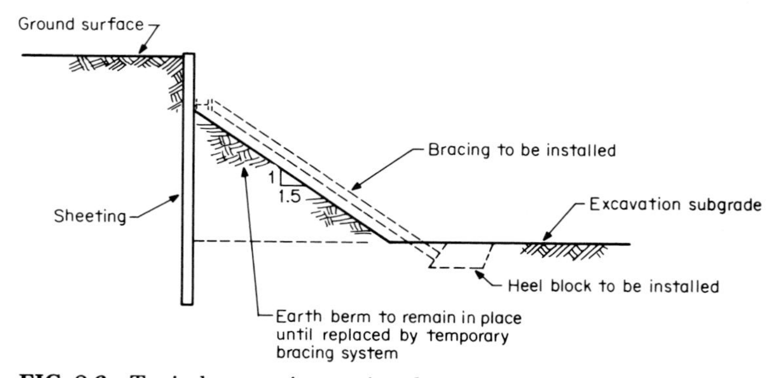

FIG. 8.6 Typical excavation section showing earth berm.

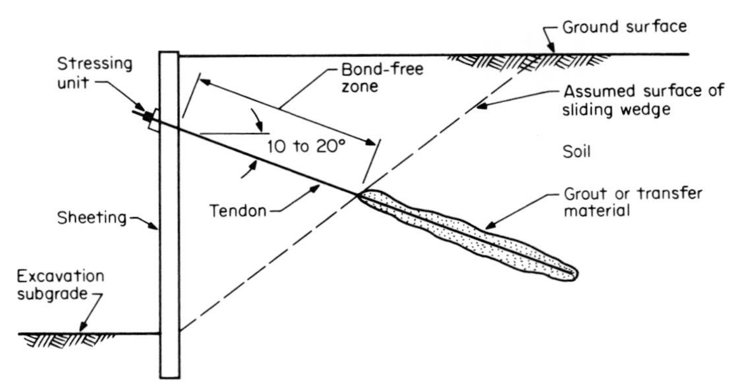

FIG. 8.7 Typical soil anchor.

Earth anchors are usually installed at an angle of 10 to 20° down from horizontal. This slight depressed angle aids in placement of the anchor and grout. If acceptable soil is not encountered at these levels, it is necessary to change the angle to engage the proper soil stratum. These relatively flat angles also have the advantage of not introducing large vertical loads into the earth-retention system, such as are introduced when placing rock anchors at a depressed angle of 45°. At this angle, the vertical load introduced at the bottom of the retention system is equal to the lateral load being supported.

In situations where excavations extend through soil into rock, with rock anchors restraining the sheeting and the excavation extending below the toe of the vertical members, great care must be exercised in protecting and bracing the rock face under the vertical members (see Fig. 8.8). Good construction practice would be to keep the vertical members a safe distance behind the face of the rock excavation and to line drill the rock face. The setback used is typically 3 to 4 ft (91 to 122 cm), but could be less in hard, competent rock; the decision should be made by someone experienced in rock mechanics. Excavation of the rock should

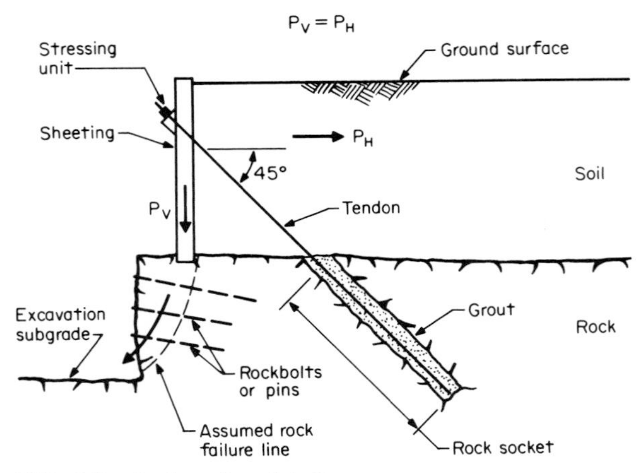

FIG. 8.8 Rock anchor detail.

be made in small lifts, and the rock bolted or pinned, as required, as each lift is blasted and removed.

The type of equipment and technique used to install the anchors are usually determined by the type of soil to be drilled through, the depth to the supporting stratum, and whether groundwater will be encountered. Where the soil is cohesive, large diameter, uncased shafts are drilled with truck- or crane-mounted drills. The bottoms of the holes are then belled out, the tendon and concrete placed, and the tie is stressed. Often the bell is eliminated and the length of shaft increased to develop the load in friction rather than by bearing.

Tendons

High-strength steel is generally used as the tendon. This permits the use of a high-capacity member in a small hole. Tendons can be high-strength strand [270 ksi (186 kN/cm^2)], high-strength rods, or wire.

Suppliers of tieback material furnish special heads or wedging assemblies to engage the tendons. These fall into two categories: friction connections and threaded connections. The friction type is used with strand or smooth rods while the threaded type is used with threaded bars or special fittings that can engage either smooth rods or strand. All types of heads that, to date, have been used by the author and are commercially available have been satisfactory and have achieved the desired results. The heads are part of a system which includes the tendons and a jacking assembly. The jacking system permits the tendons to be stressed to a predetermined load, then locking off the tendon at the desired load. The contractor should be advised that jacks for some of the systems are extremely heavy and require equipment to handle and place them. Some of the systems require off-site fabrication of the tendons. This can cause delay in placing the tendons if the tendon length is subject to field conditions. If the holes or sockets are unstable, this time lag could be critical and the hole may collapse while waiting for the tendon. It is to the contractor's advantage to use a system which is flexible and permits fabrication or adjustment of the tendons at the jobsite. This will permit placing and grouting the tendon as soon as the drilling of the hole is complete.

Laboratory tests and field experience indicate that when using strand or deformed bars, bond between the tendon and the grout is not critical. When using smooth rods or wire, some type of plate or nut should be attached to the bottom of the tendon.

Several types of mechanical anchors for soil are commercially available. In our experience with these anchors, we have not been successful in obtaining loads greater than 25 to 30 tons with any great consistency. As new and improved anchors of this type are developed, they should be investigated, since their installation could be quite economical.

Drilling for Anchors

During the course of drilling, if layers of soil are encountered which will not remain open without being supported, temporary casings are sometimes used to keep the holes open. This protection through the unstable layer increases the cost, but still permits the use of drilling equipment which is locally available.

Another common method used to install ties through unstable soil is with continuous flight, hollow-stem augers. This permits rapid drilling and, with the hollow pipe stem, allows the tendon to be placed and the grout installed through a completely protected hole. This method loses its advantage if boulders or very hard layers of soil are encountered. The drilling equipment used is very large, so it is not adaptable for operations in limited space. Holes can also be drilled and maintained open using drilling mud or slurry. Using this procedure, the holes are generally pressure-grouted.

The concrete or grout can then be placed by tremie methods, which are acceptable and provide the desired results.

When installing ties in large-diameter holes, centering devices are used. These keep the tendon approximately in the center of the hole, permitting complete encasement of the tendon with the concrete or grout. In small-diameter holes, centering devices are used if the tendon is stiff; however, if strand is used as the tendon, the irregular shape of the tendon makes it unnecessary to use spacers.

Common rock drills or air-tracks are often used to place earth anchors, and are the most common type of equipment used to install rock ties. This equipment permits installation of a continuous casing which can be withdrawn after the tendon has been grouted in place. Ordinary drills of this type can install casings up to 4½ in (11.4 cm) in diameter, while larger ones can handle casings up to 8 in (20 cm) in diameter. When used to install earth ties, the casings can be driven with a point or plug in the bottom, eliminating the necessity for cleaning the pipe. When withdrawing the casing, the plug can be knocked out or the point left in the ground. When used to place rock ties, the usual procedure is to drive the casing to the top of rock, attempting to seat the casing in the rock. Seating the casing in rock is quite important in creating a seal which prevents soil from entering the hole as the rock socket is being drilled. After the rock socket is drilled, it is flushed and cleaned and the tendon is placed and grouted. It is common to have large quantities of water entering the holes through seams in the rocks. Should this occur, the holes are filled with water and the grout placed by tremie methods.

Grouted Anchors

Most common materials used to transfer the loads from the tendon to the soil or rock are concrete, grout, and epoxy. Concrete is most often used where large-diameter holes have been drilled, say 12 to 18 in (30 to 45 cm) in diameter. Where small-diameter holes are drilled, grout or epoxy is used. For small-diameter holes in rock, most contractors use neat cement grout, use no additives or expanding agents, and use a grout with a low water-cement ratio. Best results are obtained when using a high-speed mixer which ensures a thorough wetting of the cement. These mixers usually provide a discharge pressure of up to 30 lb/in^2 (20.7 N/cm^2), normally sufficient to place the grout without the use of a grout pump. If pressure grouting is required, or if the grout must travel a long distance from place of mixing to point of placement, then a grout pump should be used.

Pressure grouting introduces larger quantities of grout into the hole, and when used with earth anchors will probably permit higher anchor capacities than if pressure grouting were not used. This technique requires special equipment and procedures, such as the use of packers which seal the hole and help maintain the pressure. Capacities of anchors installed with pressure grouting have reached several hundred kips with great consistency and reliability.

The cost of pressure-grouted anchors is high when compared with the cost of normally grouted anchors. The economy of this type of anchor should be investigated before using it.

Epoxy grouts come in the form of cartridges (sausages) which are placed in the holes after hole drilling has been completed. The steel rod, which is to be used as the tendon, is then connected to the drill rig, inserted into the hole, and turned with the drill rig. This procedure penetrates the cartridge casing and mixes the components of the epoxy. The advantage of epoxy grout is that it sets up faster than any other type of grout, permitting testing of the anchors in a relatively short time, and it can be used successfully under water.

Possible problems can occur in rock which fractures and breaks up during drilling and cleaning (blowing) operations. If the holes are enlarged, the cartridges have room to move and spin without the casing breaking. Under these conditions, the cartridge manufacturers recommend grouting the holes, then redrilling through the grout and using the epoxy. Here it is probably to the contractor's advantage to use the neat cement grout as the transfer agent, and use epoxy grout in more competent rock.

Most anchor designs indicate a "bond-free" zone, an area in which no transfer of load between the soil and anchor is desired during the stressing operation. The length of the bond-free zone varies according to the design criteria. However, the methods used to establish this zone are independent of the type of tendon or method of installation. The tendons in the bond-free zone can be wrapped, greased, painted, or sheathed in plastic tubes. The hole can then be filled with sand, lean concrete, or weak grout. Another method of establishing a bond-free zone is to fill the entire casing with grout, then to wash the grout out of the bond-free zone.

There are many hypotheses which have been advanced for anchor design, some primarily theoretical, some empirical; all require the application of judgment factors. Experience has shown that the only real proof of anchor capacity is by full-scale field test.

It is good practice to test every anchor to a load greater than the design before incorporating it into the bracing system. When earth ties are anchored in clay, an analysis should be made of the creep characteristics of the clay to ensure that significant movements will not occur during the time that the excavation is to remain open.

FREEZING AND GROUTING

Several earth-support techniques are used which act differently than the methods previously described. They do not employ sheeting, braces, or tiebacks, but change the physical condition and properties of the soil so that the soil itself acts as a gravity retaining wall, thus providing an earth-retention system.

Ground freezing is not a new method. It has been in use for over 100 years. The basic principle is very simple. It uses refrigeration to convert in situ pore water into ice. The thickness of the frozen wall of ice can be controlled by the number of freeze pipes installed, the temperature, and the length of time the freezing system is in operation before excavation commences. Freezing the soil also makes the soil mass impervious.

The most commonly used ground refrigeration system employs a coolant circulated through a manifold to a series of pipes which have been installed in drilled holes. The coolant freezes the soil and returns to the refrigeration plant for cooling and recycling.

Freezing can be accomplished in almost any type of soil. However, there are certain conditions which may limit its use or effectiveness. It is difficult to freeze ground where there is a considerable lateral flow of water. Where obstructions such as large boulders are present, the freeze pipes may not be installed straight and the increased spacing between the pipes may not permit the ground to freeze in that location. If the deviation in location is known in advance, the condition may be remedied by adding pipes in the enlarged spacing. If the deviation is not known, problems can occur when the excavation reaches this level of unfrozen soil. If steam or water lines are present in the ground in close proximity to the proposed wall location, they should be exposed and insulated in advance of ground freezing.

The cost of ground freezing is generally higher than the cost of more conventional ground-retention systems; therefore its use is limited to special situations. Costs of operating and installing a freezing system can be inflated by union requirements for the size of crews needed to perform the work; therefore thorough investigation of these requirements are essential on union projects.

After the need for freezing is completed and the system removed, the ground thaws and the soil is generally returned to its original condition.

Grouting, like ground freezing, is not a new method, the first documented use being about 200 years ago. Grouting consists of injecting material into the soil, usually in a predetermined pattern, with the object being to form an impermeable curtain (or mass) and/or increase the strength of the soil. Grouts may be divided into two types: cement-based and chemical.

To be effective the grout must penetrate the soil and fill the voids. It is therefore necessary for us to establish guidelines to assist us in determining which soils are groutable. An easy guide to use is that if a soil can be dewatered with well points, grouting will probably be successful. Chemical grouts generally have lower viscosity than cement grouts, and the gel time is easier to control; therefore it is more common to use chemical grouts for groundwater control. The successful use of grouting is dependent upon the soil and grouting technique. Very little information has been published on projects where grouting has not been successful. Grouting is relatively expensive, and with most grouting-specialty contractors unwilling to guarantee results, foundation contractors are generally unwilling to assume the risk of using grout instead of more conventional ground-support systems.

Most research in this country relating to grouting has been performed by the U.S. Corps of Engineers, and is related to grouting cutoff walls under dams. Their publications are readily available, as are various publications of the American Society of Civil Engineers on this subject.

EVALUATION OF SITE CONDITIONS

The first step in planning for and designing a sheeting and bracing system is to evaluate conditions at the site which will affect construction and performance of the system. Items to consider are: topography, soil conditions, on-site structures,

structures adjacent to or near the construction area, groundwater, and construction procedures and requirements.

Topography

Topography has obvious effects on construction activities. Flat or near-flat areas are relatively easy to work, while steep slopes along a retaining-structure alignment or even moderate slopes across alignment may require special consideration in planning site work. There are two situations in which topography may have a significant effect on design. One is where there is a significant difference in surface elevation on opposite sides of an excavation: it may be awkward to slope braces so as to have reasonably equal earth loads on both sides; or in cases with large differences, there may not be sufficient resistance readily available on the low side to restrain the loads from the high side. In the latter case, some auxiliary reaction must be developed such as reaction blocks, piles, or tiebacks.

The second situation is where there is a sharp rise in ground level a short distance beyond the excavation area proper. Digging a trench or large excavation on or at the base of a high slope may cause instability of the slope, resulting in extremely high pressures on a retaining structure. Such situations require a thorough investigation and study of conditions by a geotechnical engineer.

Soil and Groundwater

Soil and groundwater conditions are obviously matters to be considered, but all too often are not properly evaluated. Sands of different densities, soft and stiff clays, all have different characteristics with respect to difficulty of excavation and field measures needed to handle them. Add groundwater, alternating stratifications, and rock, and conditions can become quite difficult. Knowing what is there ahead of time permits preparing for it, but one must recognize that at least a few small surprises are likely no matter how carefully an area has been studied.

Obtaining adequate knowledge of subsurface conditions is critical to design, and if sufficient information is not made available along with plans and specifications for a project, then it must be obtained (by borings, test pits, seismic lines, or other means), and properly analyzed by a geotechnical engineer in order to plan a suitable sheeting and bracing system. Engineering analyses must evaluate not only lateral earth pressures, but also the stability of the bottom of the excavation, all as affected by soil strengths and groundwater pressures. And finally, not the least important is the matter of groundwater, both in terms of pumpage and in terms of how it may affect adjacent areas. Lowering a groundwater table for construction can sometimes cause settlements in surrounding areas, due either to migration of fines (locally) or increased overburden loads (at some distance), or may even affect nearby wells. Avoiding these potentially adverse effects requires planning.

Structures and Utilities

Surface structures—buildings, streets, or whatever—can be observed simply by a visit to an excavation site or along a pipe alignment. Where they are within the area to be excavated, the easiest method of removal is about the only consideration. Where alongside the excavation, a determination must be made of the type of foundation existing, whether it can withstand some limited deflection, or whether underpinning is needed. A compilation has been made of typical deflections near excavations relating to the type of retaining system, depth of excavation, distance to structure, etc. (see Figs. 8.9 and 8.10); from this an estimate can be made as to whether it will be necessary to consider underpinning. Where underpinning is needed, an experienced engineer should be called on to design and execute it.

Subsurface structures at a site may consist of shallow underground utilities, sewer and water mains, culverts and tunnels, or other buried structures. Some may simply be removed; some must be left in place; others may require temporary or permanent rerouting. Again, knowing what to expect minimizes unnecessary damage and can speed up the job if appropriate equipment is on hand to cope with the things encountered.

From the design aspect, subsurface pipe and structures which must remain in

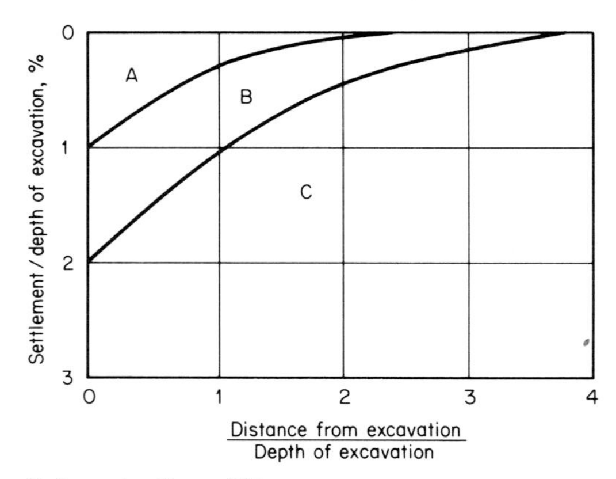

Relevant soil conditions:

Zone A—sand or firm to hard clay [$s > 500$ lb/ft^2 (24 kN/m^2)]

Zone B—very soft to firm clay [$s < 500$ lb/ft^2 (24 kN/m^2)] where there is a limited depth of clay below the base of the excavation, or where $\gamma H/s < 5$

Zone C—very soft to firm clay [$s < 500$ lb/ft^2 (24 kN/m^2)] where there is a significant depth of clay below the base of the excavation and where $\gamma H/s > 5$

NOTE: Good workmanship assumed. Poor workmanship (such as unfilled voids behind sheeting) will increase deflection.

FIG. 8.9 Settlement versus depth of excavation. *(From R. B. Peck, International Conference on Soil Mechanics and Soil Mechanics, 1969.)*

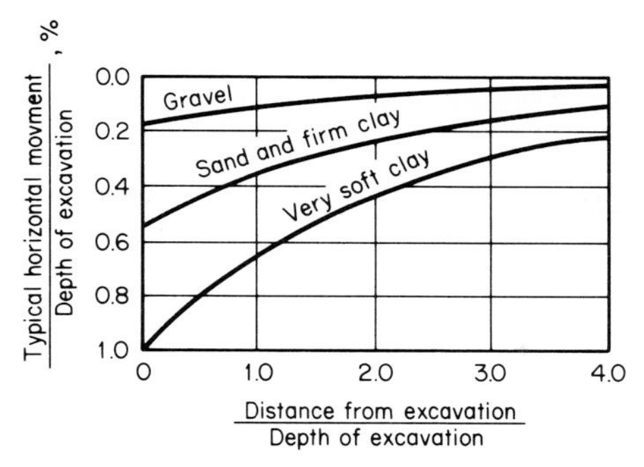

FIG. 8.10 Lateral displacement versus depth of excavation. *(From J. E. Crofts, B. K. Menzies, and A. R. Tarzi, Geotechnique, p. 165.)*

place can control the type of retaining structure to be used; for instance, vertical steel sheeting could not be used in such a situation unless soil conditions permitted leaving a gap in the sheeting which could be left open or blocked during excavation without adverse effects. Generally, pipes or conduits alongside an excavation will not be a problem and do not require special support, although deflections should be evaluated (see Figs. 8.9 and 8.10) to make certain that predicted movements will not cause damage. A tunnel or other large structure wall alongside may make conditions easier to deal with.

An important design consideration in sheeting and bracing design is surcharge loads (loads applied at the surface) alongside trenches. Such loads could be due to buildings or other structures immediately alongside the trench, construction or other vehicles and equipment, excavated material, or other stockpiled material. Designing for these loads results in heavier members and increased costs, so it is often desirable, where practicable, to keep such loads away. However, if these loads are applied without the retaining structure being designed for it, excessive deflections or failure can be expected to result.

Where structures alongside an excavation are underpinned, there is of course no surcharge load to be considered. Often it is possible and desirable to design the underpinning to serve as the retaining structure as well as the building foundation.

Rock

Excavations into rock may or may not require retention, depending on the nature of the rock and the joints and fractures present. Typically, retention involves rock bolts, rarely sheeting (except possibly skeleton sheeting in soft or broken rock). The presence of rock at depth may hinder the placement of sheeting and bracing down to that level if the rock is hard enough to prevent keying the toe of the sheeting into it. Where blasting is used, its effect on sheeting and bracing already in place (as well as on nearby structures) must be considered.

DESIGN

The proper evaluation of lateral earth pressures acting on sheeting is critical to safe design. It is generally desirable—and essential in situations involving deep excavations or important structures adjacent to excavations—to have conditions analyzed by an experienced and qualified geotechnical engineer. This section provides an introduction to the basics of sheeting and bracing design, and may provide sufficient data for the reader to analyze at least some simpler conditions, as well as to better understand what is involved in the more complex situations. It is assumed that the reader has a basic knowledge of static structural analysis and strength of materials.

Earth-Pressure Theories

As related to sheeting and bracing, the underlying concept of earth-pressure theory is that there are limiting conditions that develop in soil at the verge of failure (1) when it is allowed to expand laterally behind a deflecting vertical wall and (2) when it resists pressure applied to a vertical wall (see Fig. 8.11). One simplifying assumption used is that a plane surface represents the limit of the failing zone; analysis of the mechanics shows these surfaces will slope at 45° $\pm$ $\phi/2$, as shown in Fig. 8.11 (ϕ = internal friction angle of the soil).

With further assumptions (e.g., that soil pressures act parallel of the soil surface and that the presence of the wall does not affect shear stresses at the soil/wall interface), Rankine developed expressions for the active and passive earth pressures, respectively, for soils with both cohesion and friction:

$$P_a = \gamma Z K_a - 2c\sqrt{K_a}$$

$$P_p = \gamma Z K_p + 2c\sqrt{K_p}$$

P_a and P_p are active and passive earth pressures at depth Z below the ground surface, γ is the effective unit weight of the soil, c is the cohesive strength of the soil, and K_a and K_p are functions of the internal friction angle of the soil and the slope of the ground surface; for the typical case of level backfill, the expressions for these factors are:

$$K_a = \tan^2(45 - \phi/2)$$

$$K_p = \tan^2(45 + \phi/2)$$

where (45 $\pm$ $\phi/2$) is the angle of the failure plane from vertical.

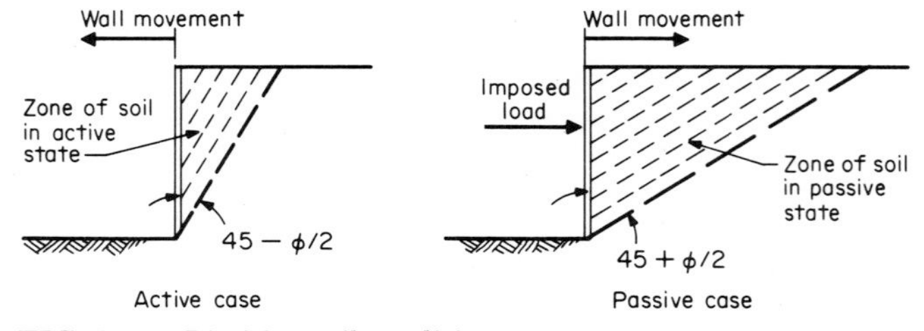

FIG. 8.11 Limiting soil conditions.

Coulomb thought it important to include the effect of soil friction (δ) on the wall. This friction is likely to develop a downward force on the wall as the soil slides down in the active case and an upward force as the soil in the passive condition tries to slide upward. With these considerations, his evaluation resulted in:

$$K_a = \frac{\cos^2 \phi}{\cos \delta \left[1 + \sqrt{\dfrac{\sin (\phi + \delta) \sin (\phi - \beta)}{\cos \delta \cos \beta}}\right]^2}$$

$$K_p = \frac{\cos^2 \phi}{\cos \delta \left[1 - \sqrt{\dfrac{\sin (\phi + \delta) \sin (\phi + \beta)}{\cos \delta \cos \beta}}\right]^2}$$

which give apparently more realistic values to the coefficients. With wall friction δ and the slope of the ground surface β equal to zero, these expressions reduce to those given above for Rankine.

The inclination of the failure plane for the Coulomb assumptions is determined from another complex expression.

Further evaluation and analysis have shown that the assumption of a plane surface for the failure plane is not realistic; it is really curved (approximating a log spiral). The error for the active coefficient is generally quite small, but it becomes large for the passive coefficient with a wall friction angle more than one-third the soil-friction angle. Caquot and Kerisel developed expressions for the log spiral values; the chart (Fig. 8.12) developed from those expressions provides a realistic basis for selecting active and passive coefficients in the ranges of normally encountered values for the various parameters.

Thus, for simple cases of horizontal ground surface and horizontal stratification, active pressures may be calculated from the Coulomb expression or from the log-spiral method (Fig. 8.12). The passive pressure coefficient should be calculated using the log-spiral method (Fig. 8.12) or can be calculated, somewhat conservatively, from the Coulomb equation with δ set equal to zero.

For complex stratifications, sloping strata, and other variations, reference should be made to texts which treat the topic in greater detail (NAVFAC DM-7.2, Lambe and Whitman, Winterkorn and Fang, Leonards, and others noted in Bibliography).

Cantilever Design (Granular Soil)

To demonstrate how these principles are applied in practice, along with the effects of water, and other items, a step-by-step example of the calculations for a cantilever wall is given, starting with Fig. 8.13, for all-sand soil conditions. This figure shows the basic conditions: a 12-ft (3.6-m) cut, 200 lb/ft^2 (9575 N/m^2) surcharge load up to the edge of the cut, sand having the properties shown, groundwater outside the cut 6 ft (1.8 m) below the surface, and at ground level inside the cut. In this example a steel sheet-pile wall is assumed.

The selection of representative groundwater levels for design should be done carefully; with tight sheeting, natural level is reasonable or slightly conservative; with pervious sheeting, although the seepage line into the excavation may be

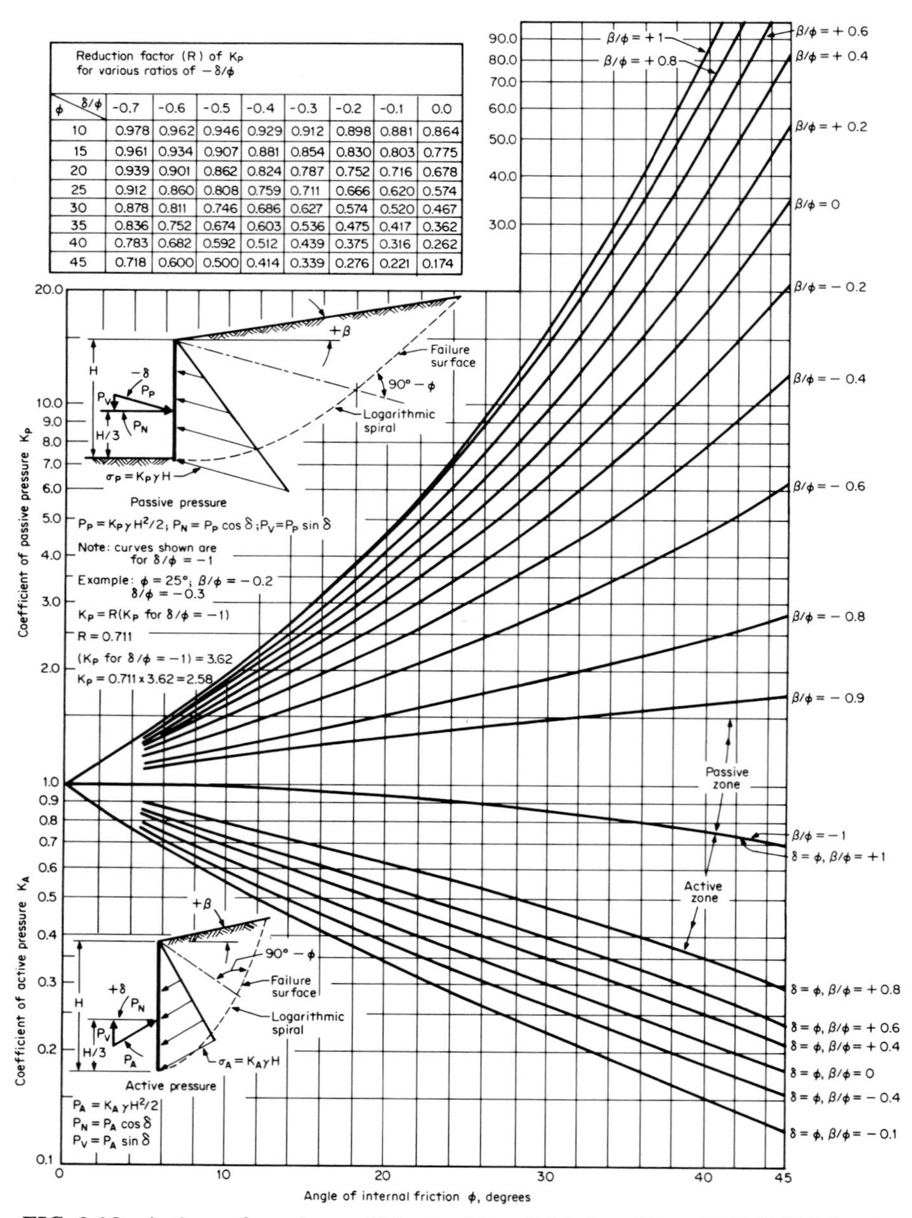

Reduction factor (R) of K_P for various ratios of $-\delta/\phi$

ϕ \ δ/ϕ	-0.7	-0.6	-0.5	-0.4	-0.3	-0.2	-0.1	0.0
10	0.978	0.962	0.946	0.929	0.912	0.898	0.881	0.864
15	0.961	0.934	0.907	0.881	0.854	0.830	0.803	0.775
20	0.939	0.901	0.862	0.824	0.787	0.752	0.716	0.678
25	0.912	0.860	0.808	0.759	0.711	0.666	0.620	0.574
30	0.878	0.811	0.746	0.686	0.627	0.574	0.520	0.467
35	0.836	0.752	0.674	0.603	0.536	0.475	0.417	0.362
40	0.783	0.682	0.592	0.512	0.439	0.375	0.316	0.262
45	0.718	0.600	0.500	0.414	0.339	0.276	0.221	0.174

FIG. 8.12 Active and passive coefficients with wall friction. *(From NAVFAC DM 7.2.)*

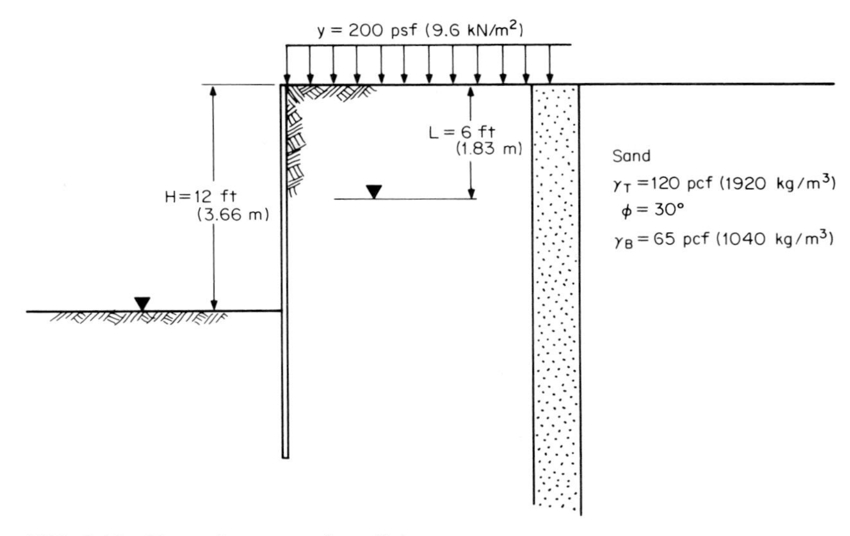

FIG. 8.13 Example: assumed conditions.

about at the bottom of the excavation, to account for seepage forces, the assumed external level should be a little above this. A fully dewatered site will have no water showing, but if the water level is at shallow depth below the excavation, it may still have a strong effect on passive and active pressures.

Figure 8.14 shows the basis for calculating vertical pressures from which lateral pressures are obtained. The vertical pressures must be separated into net (effective) vertical soil pressures and hydrostatic pressures; the lines are shown approximately to scale (numerical values are recorded at a few points). Also

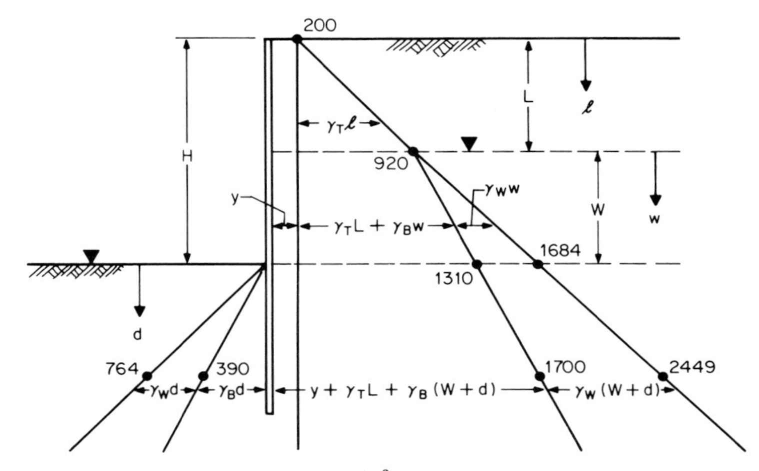

FIG. 8.14 Vertical pressures, lb/ft^2.

given are the expressions by which the vertical pressures can be calculated, at the respective levels in the diagram.

The minimum (active) and maximum (passive) pressures that could occur on each side of the wall are shown, approximately to scale, in Fig. 8.15. The total lateral pressures to be considered are the sum of the net vertical soil pressure times the appropriate coefficient of lateral pressure plus the hydrostatic pressure. Going to the Caquot-Kerisel charts (Fig. 8.12) with $\phi = 30°$ and β (ground slope angle) $= 0$, $K_a = 0.31$ with δ (friction angle of soil on the wall) $= 0.3\ \phi$.

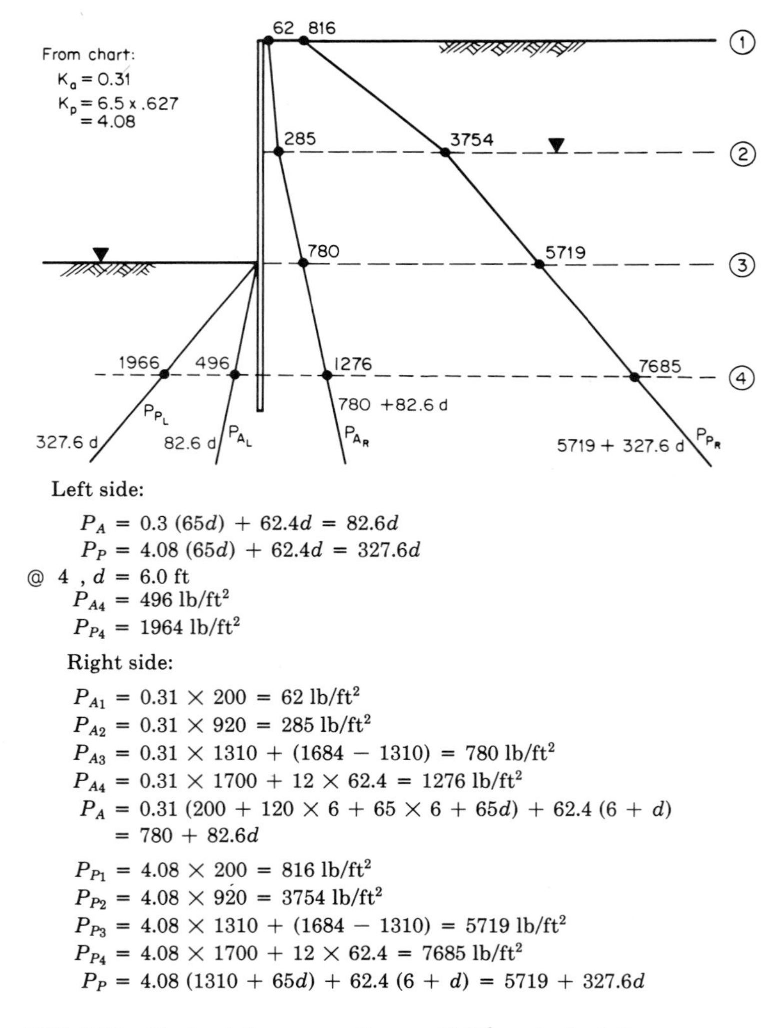

Left side:

$$P_A = 0.3\ (65d) + 62.4d = 82.6d$$
$$P_P = 4.08\ (65d) + 62.4d = 327.6d$$
$$@\ 4\ ,\ d = 6.0 \text{ ft}$$
$$P_{A4} = 496 \text{ lb/ft}^2$$
$$P_{P4} = 1964 \text{ lb/ft}^2$$

Right side:

$$P_{A1} = 0.31 \times 200 = 62 \text{ lb/ft}^2$$
$$P_{A2} = 0.31 \times 920 = 285 \text{ lb/ft}^2$$
$$P_{A3} = 0.31 \times 1310 + (1684 - 1310) = 780 \text{ lb/ft}^2$$
$$P_{A4} = 0.31 \times 1700 + 12 \times 62.4 = 1276 \text{ lb/ft}^2$$
$$P_A = 0.31\ (200 + 120 \times 6 + 65 \times 6 + 65d) + 62.4\ (6 + d)$$
$$= 780 + 82.6d$$

$$P_{P1} = 4.08 \times 200 = 816 \text{ lb/ft}^2$$
$$P_{P2} = 4.08 \times 920 = 3754 \text{ lb/ft}^2$$
$$P_{P3} = 4.08 \times 1310 + (1684 - 1310) = 5719 \text{ lb/ft}^2$$
$$P_{P4} = 4.08 \times 1700 + 12 \times 62.4 = 7685 \text{ lb/ft}^2$$
$$P_P = 4.08\ (1310 + 65d) + 62.4\ (6 + d) = 5719 + 327.6d$$

FIG. 8.15 Horizontal pressure diagram, lb/ft², and calculations.

For $\delta = 0$, K_a would be 0.33; a very small difference. For K_p, however, some conservatism is needed because of the very large effect of δ on K_p. Unless there are forces other than the basic soil pressures acting, or large deflections to be allowed, δ should be limited to no more than 0.3ϕ. At this value, $K_p = 6.5\ (0.627) = 4.08$. Then, following the basic relationship described above, the numeric values of the pressures can be calculated for various depths, as shown in Fig. 8.15, and expressions are developed which express the slopes of the pressure diagram below the bottom of the excavation, which are used in the required sheeting-penetration calculation.

The information used in the basic calculation is shown in Fig. 8.16. The acting soil pressures are represented by the area under the line *behvr*; line *beh* is the active pressure already calculated. The line *hn* represents the difference between the passive pressure on the outside of the excavation (line *gm*) and the active pressure (line *hq*) inside. Similarly, line *jr* is the difference between the passive pressure (line *js*) and active pressure outside the excavation. For this pressure diagram, line *behvr*, the sum of the horizontal forces, must equal zero, and the moments of the forces about any point must equal zero. A trial depth D is selected (typically equal to H) and Z is calculated by setting areas $(abehk) + (tor) = (kvt)$. This can be simplified a little by using the equivalent areas: $(abeqo) + (nvr) = (hnq)$. (See calculations under the diagram.) With this value of Z, take moments about point 0 (using the same areas); if the moment is other than zero, select a new D and repeat the calculation. The process is illustrated below Fig. 8.16. The calculations are completed by computing the maximum moment in the sheeting (at the point of zero shear) to select an appropriate sheeting section, and then increasing the required penetration of the sheeting by 20 to 40% to obtain a factor of safety of the order of 1.5 to 2.0.

The passive resistance indicated here assumes a static water condition. If there is an upflow into the bottom of the excavation, the effective weight of the soil must be reduced by the seepage pressure. Some of the calculations can be short-cut by a method suggested by Teng. The United States Steel *Steel Sheet Piling Design Manual* gives charts which, if used carefully, can save considerable effort for certain soil conditions. For higher retaining structures, the flexibility of the structural members can result in lower pressures and consequently lower moments in the members.

Restrained Sheeting (Cohesive Soil)

Calculations for braced or tied-back sheeting in cohesive soils are illustrated, starting with the conditions shown in Fig. 8.17. The same basic principles as given under earth-pressure theories apply, but their application involves different procedures because different soil characteristics are involved.

In cohesive soils, unless there is clear-cut evidence otherwise, the water table should be assumed at the soil surface, and total stresses are used in calculations, with $\phi = 0$. This applies to short-term excavations, without standing water. An excavation open for an extended period can mean other assumptions are needed with respect to water conditions, resulting in important differences in methods of computations, and consequently in results.

Figure 8.18 shows the lateral pressure profile developed from conditions given in Fig. 8.17. From the expression $K = \tan^2(45 \pm \phi/2)$, with $\phi = 0$, $K_a = K_p =$

1; then $P_a = \gamma Z - 2c$ and $P_p = \gamma Z + 2c$. At the surface a net negative pressure (tension) is calculated. For the purpose of these calculations, all tensile stresses are taken as zero: it is assumed the soil may develop a crack. At the lower levels, the pressure changes abruptly depending on soil cohesion. Below the excavation line, the net passive pressure (line jq) is the difference between the total passive pressure (line hn) on the excavated side and the total active pressure (line lt) on the unexcavated side.

Calculations for the sheeting and bracing start by taking the moments of the pressure diagram about point g, the level of the brace. The depth of sheeting penetration, d, is entered as an unknown. In this situation, tensile stresses above line ef are not included, and the very small positive pressure above line ef is ignored for simplicity. The calculated sheeting penetration is then 6.2 ft (1.9 m).

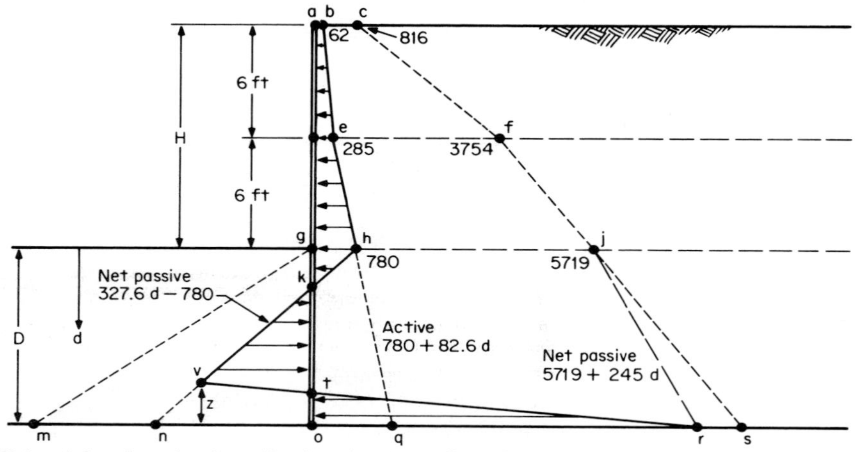

Using $(abeqo) + (nrv) = (hnq)$, solve for z. Start by assuming $D = H = 12$ ft, then:

$$6\left(\frac{62 + 285}{2}\right) + 6\left(\frac{285 + 780}{2}\right) + 12\left(\frac{780 + 780 + 82.6 \times 12}{2}\right)$$

$$+ \frac{z}{2}(327.6 \times 12 - 780 + 5719 + 245 \times 12)$$

$$= \frac{12}{2}(327.6 \times 12 - 780 + 780 + 82.6 \times 12) \qquad z = 1.69$$

Next, taking moments about point 0:

$$62 \times 6 \times 21 + \frac{6}{2}(285 - 62)\,20 + 285 \times 6 \times 15 + \frac{6}{2}(780 - 285)\,14$$

$$+ \frac{z}{2}(327.6 \times 12 - 780 + 5719 + 245 \times 12)\frac{z}{3}$$

$$= \frac{12}{2}(327.6 \times 12 - 780 + 780 + 82.6 \times 12)\frac{12}{3} \qquad \text{but } 73{,}254 \neq 118{,}138$$

Try $D = 10$ ft, $z = 0.71$, ΣM_o: $60{,}056 \neq 68{,}367$
$D = 9$ ft, $z = 0.40$, ΣM_o: $55{,}193 \neq 49{,}839$
$D = 9.4$ ft, $z = 0.563$, ΣM_o: $57{,}164 \cong 56{,}784$
use $z = 0.56$, $D = 9.4$

FIG. 8.16 Cantilever sheeting pressure diagram, lb/ft², and calculations.

To determine maximum moment in sheeting, calculate point of zero shear:

$$6\left(\frac{62 + 285}{2}\right) + 6\left(\frac{285 + 780}{2}\right) + \left(\frac{780 + 780 + 82.6d}{2}\right)d$$

$$- \frac{d}{2}(327.6d - 780 + 780 + 82.6d) = 0 \quad d = 8.00, (-3.24)$$

Then, taking moments about $d = 8.00$ ft.:

$$\Sigma M = 6 \times 62 \times 17 + \frac{6}{2} \times 223 \times 16 + 6 \times 285 \times 11 + \frac{6}{2} \times 495 \times 10 + 8$$

$$\times 780 \times 4 - \frac{8}{2} \times 409.8 \times 8 \times \frac{8}{3}$$

$$= 40{,}678 \text{ ft}\cdot\text{lb} = 488{,}141 \text{ in}\cdot\text{lb}$$

Required section modulus of sheeting, per foot of wall:

$$S_q = \frac{M}{f} = \frac{488{,}141}{20{,}000} = 24.4 \text{ in}^3$$

PZ27 has 30.2 in^3 per foot of wall and would be selected here.

FIG. 8.16 (*Continued.*)

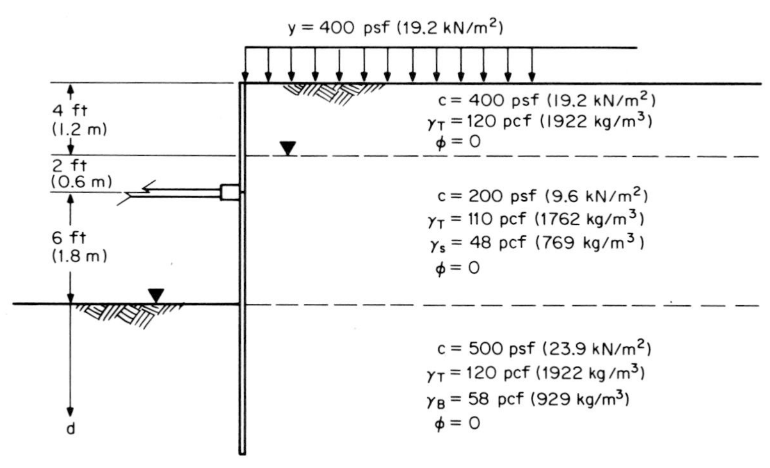

FIG. 8.17 Conditions: restrained sheeting.

Next, the horizontal forces are summed to calculate the reaction on the wall per foot of wall. Then the required sheeting penetration is found by adding between 20 and 40% to the calculated sheeting length as a factor of safety.

The sheeting section required is found by first calculating the maximum moment developed, which will be at the point of zero shear at or below the brace level (see Fig. 8.19). In this calculation, x is the distance below the top of the pressure diagram and is determined by a quadratic equation. The moment in the sheeting is then determined arithmetically. Selection of a suitable section of sheeting depends first on the material to be used. For timber, a 4-in (10-cm) thickness would be needed. If this is considered not feasible, then a steel section may be used; the PS32 sheet has an adequate section modulus.

The sizing of wales and braces follows simple static and structural analysis; in this case it is found that an 8-ft (2.4-m) spacing for braces puts too high a moment into the wale if timber is to be used; a 6-ft (1.8-m) spacing is structurally acceptable. The allowable cross-brace size is then quite small—4 by 6 in (10 by

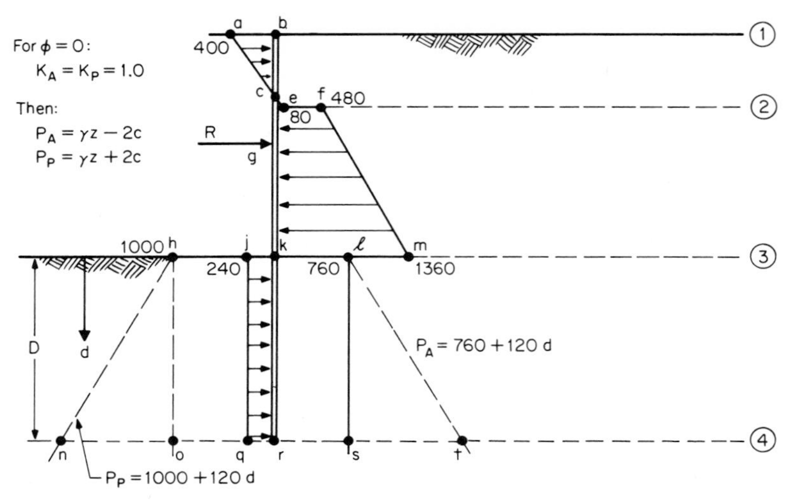

Left side:

$$P_{P3} = \gamma_z + 2c = 0 + 2 \times 500 = 1000$$
$$P_{P4} = 10 \times 120 + 1000 = 2200$$
$$\text{Net } P_P = (1000 + 120d) - (760 + 120d) = 240$$

Right side:

$$P_{A1} = \gamma_z - 2_c = 0 + 400 - 2 \times 400 = -400$$
$$P_{A21} = 4 \times 120 + 400 - 2 \times 400 = 80$$
$$P_{A22} = 4 \times 120 + 400 - 2 \times 200 = 480$$
$$P_{A31} = 400 + 4 \times 120 + 8 \times 110 - 2 \times 200 = 1360$$
$$P_{A32} = 400 + 4 \times 120 + 8 \times 110 - 2 \times 500 = 760$$
$$P_{A4} = 760 + 10 \times 120 = 1960$$

FIG. 8.18 Horizontal earth pressures, lb/ft^2—restrained sheeting.

structurally acceptable. The allowable cross-brace size is then quite small—4 by 6 in (10 by 15 cm); practicality often dictates it be the same size as the wale for simplicity of construction and ability to withstand accidental impacts during trench work. If a stronger (high-section-modulus) wale were used, perhaps a steel wide-flange section, then the spacing between braces could be extended.

It should be noted that in actual construction, during excavation to install the brace, the sheeting will be acting in cantilever manner and should be checked to see that the moment developed is not excessive. In this case, it can be seen by inspection that there will be no significant cantilever loads.

Multiple Brace Levels

The analysis given above for restrained sheeting can be expanded on to cover situations with more than one brace level. To illustrate this, consider an excavation planned for three levels of bracing (Fig. 8.20*a*). The first point requiring

Sum of monents about g

$$\Sigma M_g \circlearrowright\!+\; 8 \times 480 \times 2 + \left(\frac{1360 - 480}{2}\right)\frac{8}{2}\left(6 - \frac{8}{3}\right) - 240\, d \left(6 + \frac{d}{2}\right) = 0$$

$$d = 6.2 \text{ ft}$$

Sum of horizontal forces:

$$R - 8 \times 480 - \left(\frac{1360 - 480}{2}\right)\frac{8}{2} + 240\,(6.2) = 0 \qquad R = 4112 \text{ lb/ft}$$

Add 30% to d for safety factor, required penetration = 8.1 ft

To calculate maximum moment in sheeting, find point of zero shear at or below R by sum of horizontal forces:

$$R - x\,(480 + 110\,x) = 4112 - 480x - 110x^2 = 0 \qquad x = 4.31 \text{ ft}$$

Sum of moments of forces above X:

$$M = (4.31 - 2.00)\,4112 - 480\,\frac{(4.31)^2}{2} - 110\,\frac{(4.31)^3}{6} = 3573 \text{ ft lb}$$

Selection of sheeting by section modulus:

$$\text{Required section modulus } S = \frac{\text{Moment developed}}{\text{Allowable stress}} = \frac{M}{f}$$

$$S = \frac{3573 \times 12}{1000} = 42.9 \text{ in}^3 \text{ for wood}$$

$$\text{Section modulus for rectangular section} = \frac{bh^3}{12}$$

$$\text{calculate } h\text{: } 42.9 = \frac{12h^3}{12} \qquad h = 3.5 \text{ in}$$

4 in nominal thickness planking will serve satisfactorily.

(a)

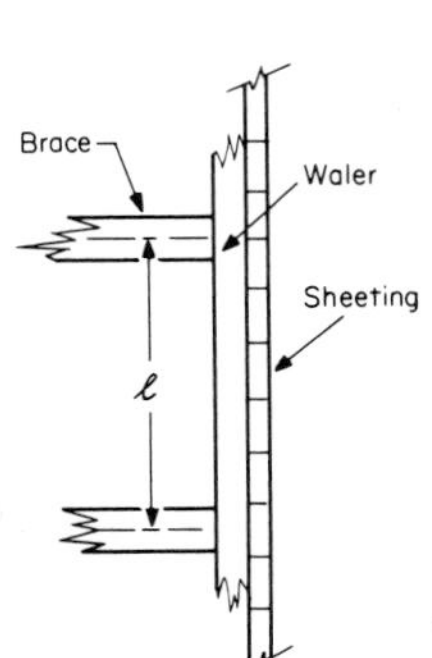

To select water size: Assume brace spacing $\ell = 8$ ft

Sheeting load R = 4112 lb/ft

$$\text{Maximum moment in waler } M = \frac{P\ell^2}{8} = \frac{4112 \times 8^2}{8} = 32896 \text{ ft-lb}$$

Required section modulus for waler (assume wood)

$$S = \frac{M}{f} = \frac{32{,}896 \times 12}{1000} = 395 \text{ in}^3$$

$$\text{For a square member } S = \frac{d^3}{6} = 395 \qquad d = 13.3 \text{ in.}$$

Try $\ell = 7$ ft then M = 25,186 ft-lb S = 302 d = 12.2 in

Try $\ell = 6$ ft then M = 18,504 ft-lb S = 222 d = 11.0 in

Use 12 in X 12 in timber, braces at 6 ft

Brace load = 6 X 4112 = 24,672 lb compressive force

At 1000 psi, need 25 in^2

Nominal 6 X 6 has 30.5 in. OK, but must check allowable stress vs. ℓ/d for member.

ℓ = unsupported span length

d = least dimension of member

See American Institute of Timber Construction Design Manual for allowable stresses

(b)

FIG. 8.19 Calculations for sheeting and bracing, lb/ft^2.

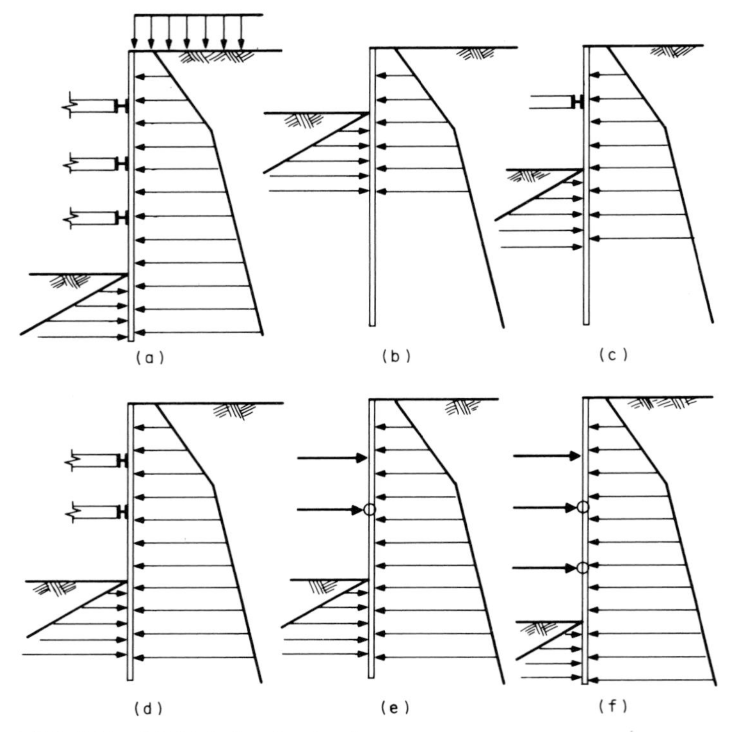

FIG. 8.20 Multiple bracing levels.

a computational check is the cantilever condition existing when the excavation for the first brace level has been completed (Fig. 8.20*b*). If we assume sheeting has already been driven to the required depth, there is no need to calculate required penetration, but the required section modulus must be checked as described in the previous section on cantilever design.

Next, a calculation is made with the first brace in place, and the excavation completed for the second brace (Fig. 8.20*c*). This calculation follows the steps given for restrained sheeting design. The following stage of excavation, with two braces in place and excavation made for the third level (Fig. 8.20*d*) requires either that the sheeting be considered a beam, and appropriate reactions be calculated, or, much more simply, that a hinge be assumed at the second brace level (Fig. 8.20*e*). The latter permits the calculation to follow the pattern of restrained sheeting design.

The final step, with three braces in place and excavation to final grade, can be solved in the same relatively simple manner by assuming an additional hinge at the third brace (Fig. 8.20*f*).

Moments in the sheeting must be calculated for each level of bracing to verify that the sheeting has adequate section modulus to carry the applied pressures. Efficient design would have approximately equal moments between each brace, that moment being the maximum which can reasonably be carried by the section modulus of the sheeting selected.

Soldier Piles and Lagging

Calculations for soldier piles and lagging (see Figs. 8.2 and 8.3), or any other sheeting system with full-height vertical members and horizontal sheeting use the same principles as the calculations described in the preceding sections, but with a few added considerations.

One item is the horizontal sheeting (lagging) itself, which spans between the vertical members (soldier piles), picking up the soil pressure and carrying it to the vertical members. Experience has shown that in general, a given lagging size can restrain soil where the typical pressure distributions used in retaining-structure calculations indicate the lagging should be highly overstressed. The explanation is that the soil mass arches between the soldier piles; the lagging deflects sufficiently to greatly reduce the pressure on it; this pressure is just enough to hold the soil which is acting as the arch in position. A possible version of the resultant pressure distribution is shown in Fig. 8.21*a*, where line 1 represents the average soil pressure (p) to be expected at the level being evaluated, line 2 is an idealized representation of the total pressures in the soil immediately behind the sheeting, and line 3 approximates the pressure actually reaching the lagging. Calculations for the required size of lagging are sometimes based on the pressure distribution shown in Fig. 8.21*b*. Figure 8.21*c* might be considered closer to fact, but the two figures result in the same maximum moment ($Pl^2/12$), so this is probably a reasonable value. No "correct" calculation has been established.

It should be noted that arching can be assumed in granular soils and stiff-to-hard cohesive soils. In soft cohesive soil, however, where the soil overburden pressure approaches 4 times the cohesive strength of the soil, arching will not develop, and the full active pressure of the soil will act on the lagging. This may

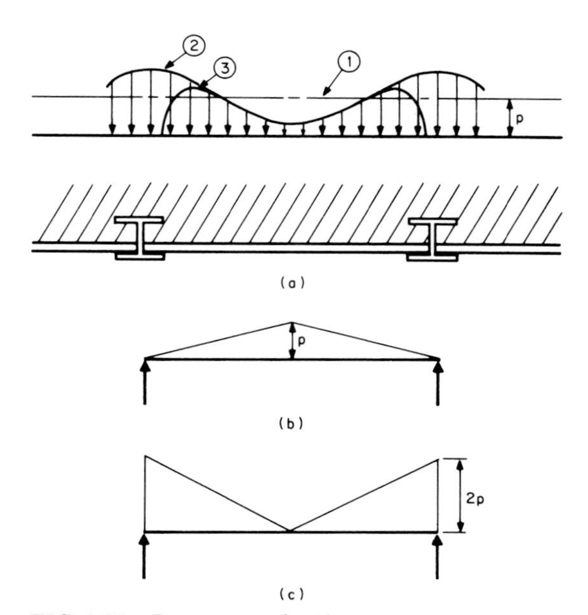

FIG. 8.21 Pressure on sheeting.

also occur where an excavation in firm cohesive soil is open for an extended period.

In calculations for the required size of soldier pile, the load on the pile must be taken as the full active lateral earth pressure over the spacing of the piles. With respect to resistance, however (there is no lagging below excavation level), the passive resistance available to the pile may typically be taken as the passive soil-pressure action on a surface approximately 3 times the width of the pile face. Again, this factor would not be appropriate in very soft clays, although a higher number could safely be used in very dense granular soil. And this should be applied only to normal sizes of piles; wide, drilled-in caissons, etc. should be evaluated as independent pile units (see NAVFAC DM-7.2).

In soldier pile and lagging configurations, it is common to have wales restraining the soldier piles to increase the spacing between braces or tiebacks. Size requirements for wales, braces, and connections are selected by normal structural design procedures.

Commonly Used Pressure Distributions

The difficulties and complications of developing soil-pressure diagrams for different situations, and the extended calculations resulting from them, has led to the use of simplified pressure distributions. These distributions have been developed and revised over a period of many years and are based to a large extent on actual field measurements as well as experience.

Diagrams illustrating the most current versions of these pressure distributions are shown in Fig. 8.22. The diagram for sand, Fig. 8.22*a*, covers all common situations: loose to dense sands and both braced and tied-back excavations. It is important to use an angle of internal friction representative of the material in situ; a loose sand will not have a high friction angle, a dense sand will.

For clays a greater variability must be taken into account, so different distributions are needed, as illustrated in Fig. 8.22*b* through *e*. For braced excavations, pressures have been measured to be higher than in tied-back excavations; when placing stress on ties during construction, it is important not to exceed the design load; the result could be to induce large lateral movements or even failure.

In calculations using these diagrams, the passive resistance in the bottom of the excavation should be calculated as previously described, subtracting the active pressure from the passive. The strut or tie reactions may be calculated assuming each support carries the load developed over half the distance to the next reaction; this is essentially the same as assuming hinges at each reaction (except the top).

Where there is a surcharge load alongside a trench, it should be taken into account as an added horizontal pressure pK_a, with

$$K_a = \tan^2(45 - \phi/2) \text{ for sands}$$

and

$$K_a = [1 - m(2q_u/\gamma H)] \text{ for clays}$$

where ϕ = internal friction angle of the sand
q_u = unconfined compressive strength of the clay
γ = unit wet weight of the clay
H = depth of excavation

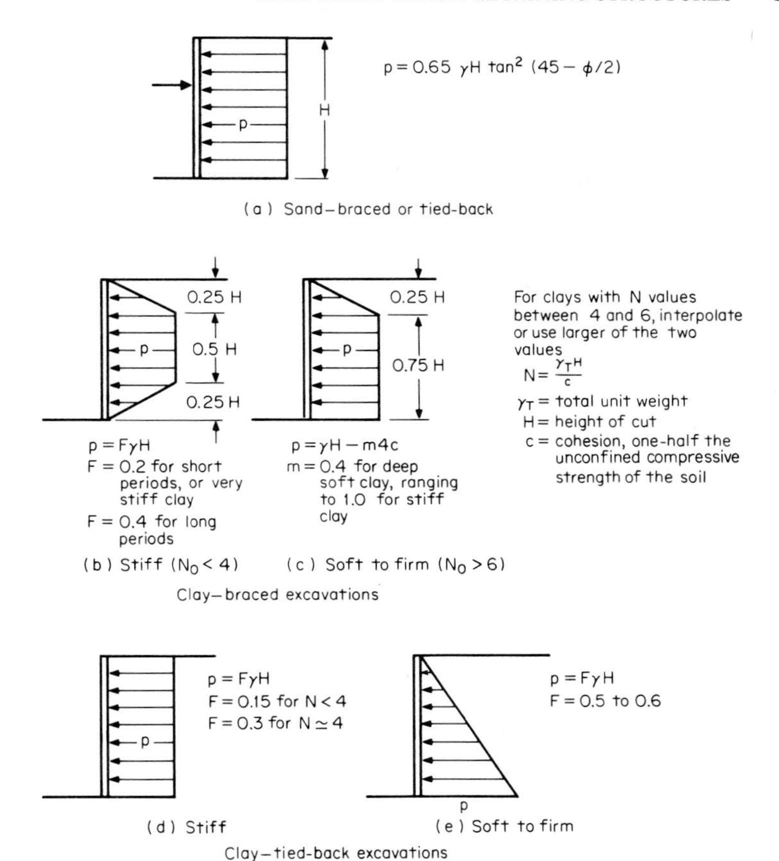

FIG. 8.22 Simplified pressure distributions.

p = average surcharge pressure
m = factor ranging from 0.4 for soft clays to 1.0 for stiff clays

The effect of a water table within the cut depth cannot readily be added to these pressure distributions except where water is at the ground surface and bouyant unit weights are used in calculating pressures. In cases where groundwater is present at an intermediate depth, it would be appropriate to go to full analysis. It should also be noted that the water table to be considered is that which will exist during construction, which may be different from that originally existing in the ground.

Base Stability

A factor important to check for in all excavations is bottom stability. In clays, it is basically related to shear strength; in sands, except for unusually low friction angles, the problem is primarily related to groundwater. Failure to take these factors into account is a primary cause of difficulties with braced and tied-back excavations.

For cohesionless soils, a simple verification of base/friction angle stability should be made as shown in Fig. 8.23, to see if it is permissible to terminate the sheeting at the bottom of the excavation. In general, any material with an internal friction angle over 25° will be stable in this configuration with respect to strength. Groundwater, however, can still have an important influence. Whenever groundwater (or the hydrostatic head under an impervious layer) is above the bottom of the proposed excavation in granular soils, there is a real potential for bottom heave, piping, or simply instability of the bottom. These problems can be averted by extending sheeting below the excavation and/or by some degree of dewatering (lowering head) so as to meet the criteria described below and depicted in Figs. 8.24 through 8.26.

When considering sheeted excavations in pervious soil, the most basic case is that with an "infinite" extent of uniform sand (see the upper diagram in Fig. 8.24); to qualify for this, the uniform deposit must extend a distance below the bottom of the cutoff wall at least equal to the width of the excavation. The chart defines the depth of penetration of sheeting required (with respect to the head of water above the bottom of the excavation) for various safety factors. An average safety factor of 1.0 is not adequate; it means that in some areas of the trench the bottom will have become quick.

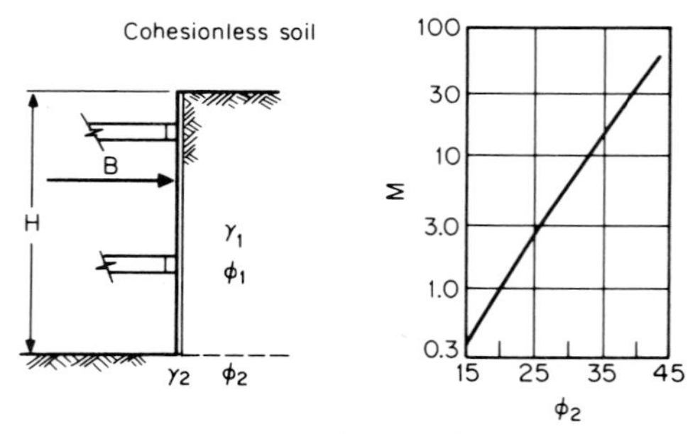

Factor of safety for stability of bottom of cut:

$$F_s = M \frac{\gamma_2}{\gamma_1}$$

γ_2 and ϕ_2 apply to material to a depth of $H/3$ or B below bottom of cut, whichever is less. For water at base of cut, use buoyant unit weight for γ_2. If sheeting stops at bottom of cut, groundwater must not be above that level or piping will occur at the bottom corners. With higher groundwater, deeper sheeting is required (see Figs. 8.24*b* and 8.24*c*), and/or groundwater level must be lowered.

FIG. 8.23 Bottom stability—cuts in sand.

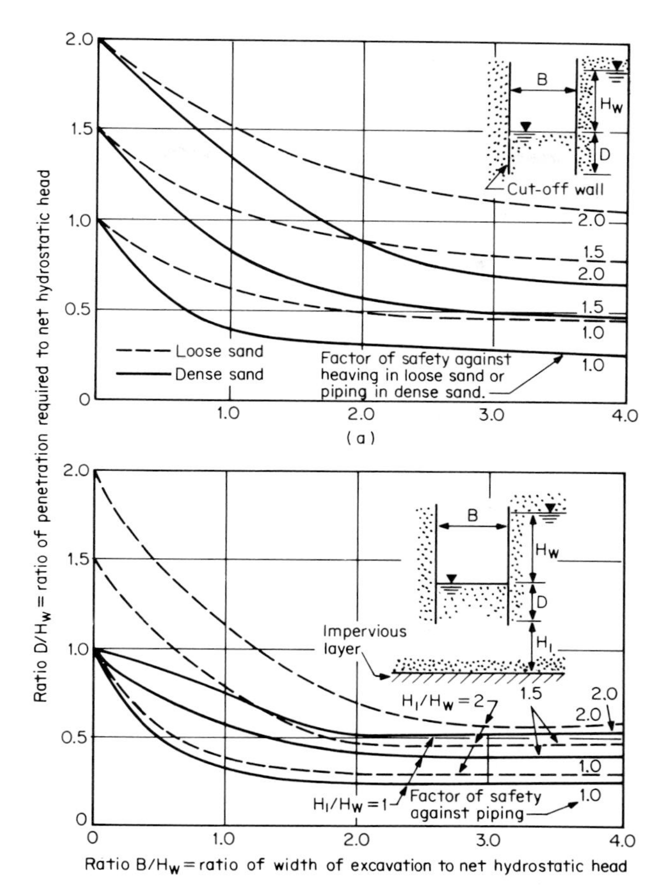

FIG. 8.24 Sheeting penetration versus piping in isotropic sand: (*a*) penetration required for cutoff wall in sands of infinite depth; (*b*) penetration required for cutoff wall in dense sand of limited depth. *(From NAVFAC DM7.1.)*

For locations where an impervious layer is at moderate depth below the bottom of the excavation, use the lower diagram of Fig. 8.24. In these figures, a layer with a permeability of less than one-tenth that of an adjacent material should be considered impervious. For impervious layers near the bottom of the excavation, see Fig. 8.25. For layers with limited differences in permeability, see Fig. 8.26.

For excavations in cohesive soils, bottom stability is affected by depth of cut, cohesive strength of the soil, and the thickness of clay below the bottom of the excavation. Charts and formulas are given in Figs. 8.27 through 8.30. Additionally, groundwater conditions below the excavation must be checked if there is a pervious layer present. For large excavations, the total weight of the soil between

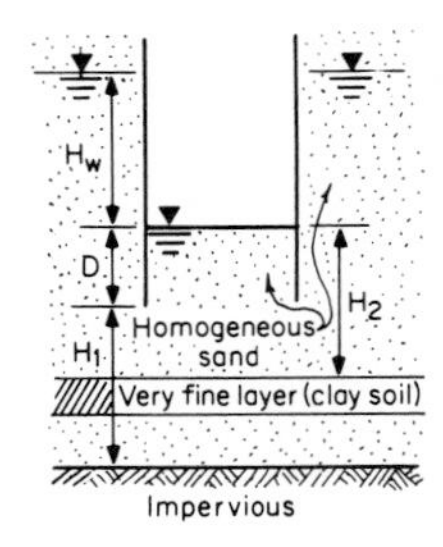

Fine layer in homogeneous sand stratum: If the top of fine layer is at a depth greater than width of excavation below cut-off wall bottom, safety factors of Fig. 8.24 apply, assuming impervious base at top of fine layer.

If top of fine layer is at a depth less than width of excavation below cut-off wall tips, pressure relief is required so that unbalanced head below fine layer does not exceed height of soil above base of layer.

If fine layer lies above subgrade of excavation, final condition is safer than homogeneous case, but dangerous condition may arise during excavation above the fine layer and pressure relief is required as in the preceding case.

To avoid bottom heave, γ_T X H_3 should be greater than γ_w X H_4

γ_T = total unit weight of the soil

γ_w = unit weight of water

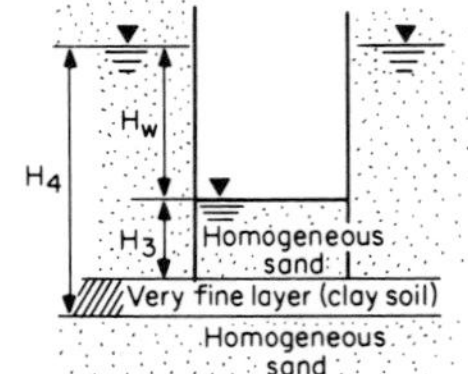

FIG. 8.25 Sheeting penetration versus piping in stratified sand. *(From NAVFAC DM7.1.)*

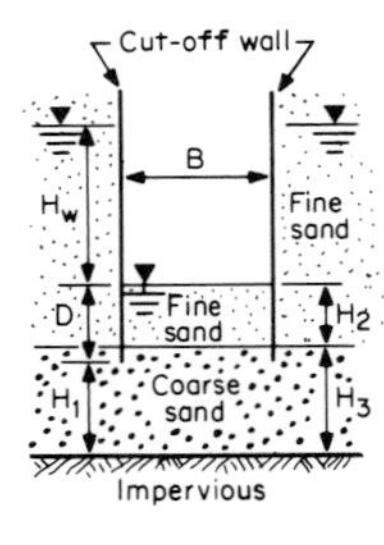

Coarse sand underlying fine sand: Presence of coarse layer makes flow in fine material more nearly vertical and generally increases seepage gradients in the fine layer compared to the homogeneous cross section of Fig. 8.24.

It top of coarse layer is at a depth below cut-off wall bottom greater than width of excavation, safety factors of Fig. 8.24 for infinite depth apply.

If top of coarse layer is a depth below cut-off wall bottom less than width of excavation, the uplift pressures are greater than for the homogeneous cross section. If permeability of coarse layer is more than than 10 times that of fine layer, failure head H_w = thickness of fine layer H_2.

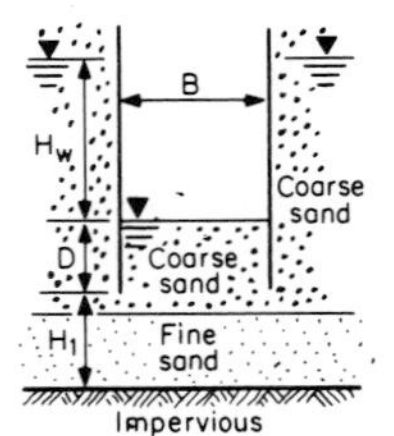

Fine sand underlying coarse sand: Presence of fine layer constricts flow beneath cut-off wall and generally decreases seepage gradients in the coarse layer.

If top of fine layer lies below cut-off wall bottom, safety factors are intermediate between those for an impermeable boundary at top or bottom of the fine layer using Fig. 8.24.

If top of the fine layer lies above cut-off wall bottom, the safety factors of Fig. 8.24 are somewhat conservative for penetration required.

FIG. 8.26 Sheeting penetration versus piping in stratified sand *(From NAVFAC DM7.1.)*

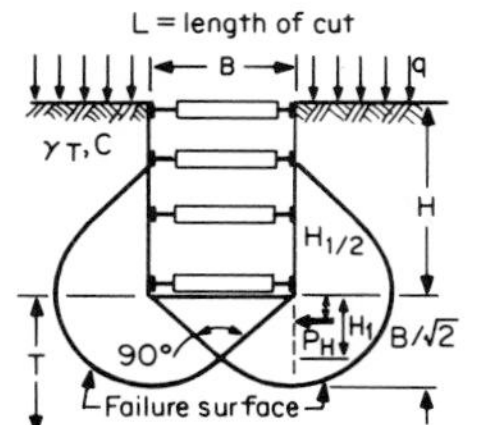

Cut in clay, depth of clay unlimited ($T > 0.78$)

If sheeting terminates at base of cut:

Safety factor $F_s = \dfrac{N_c\, C}{\gamma_T H + q}$

N_c = bearing-capacity factor, Fig. 8.28 which depends on dimensions of the excavation: B, L and H

C = undrained shear strength of clay in failure zone beneath and surrounding base of cut

q = surface surcharge

If safety factor is less than 1.5, sheeting must be carried below base of cut to insure stability.

Force on buried length:

If $H_1 > \dfrac{2}{3}\dfrac{B}{\sqrt{2}}$, $P_H = 0.7\,(\gamma_T HB - 1.4\,CH - \pi\,CB)$

If $H_1 > \dfrac{2}{3}\dfrac{B}{\sqrt{2}}$, $P_H = 1.5H_1\,(\gamma_T H - \dfrac{1.4CH}{B} - \pi\,C)$

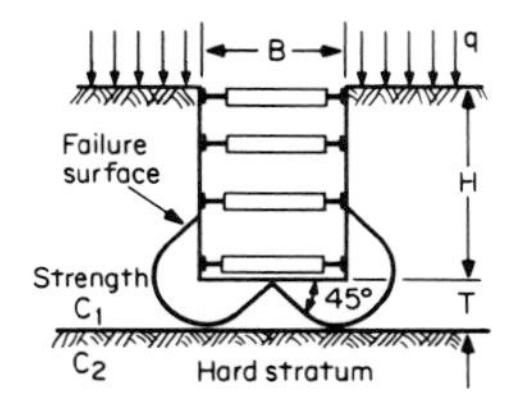

Cut in clay, depth of clay limited by hard stratum ($T \leqslant 0.7$)

Sheeting terminates at base of cut. Safety factor:

Continuous excavation $F_s = N_{CD}\,\dfrac{C_1}{\gamma_T H + q}$

Rectangular excavation $F_s = N_{CR}\,\dfrac{C_1}{\gamma_T H + q}$

N_{CD} and N_{CR} = bearing capacity factors from Figs. 8.29 and 8.30, which depend on dimensions of the excavation: B, L and H

Note: In each case friction and adhesion on back of sheeting is disregarded. Clay is assumed to have a uniform shear strength = C throughout failure zone.

FIG. 8.27 Stability of base for braced cut. *(From NAVFAC DM7.2.)*

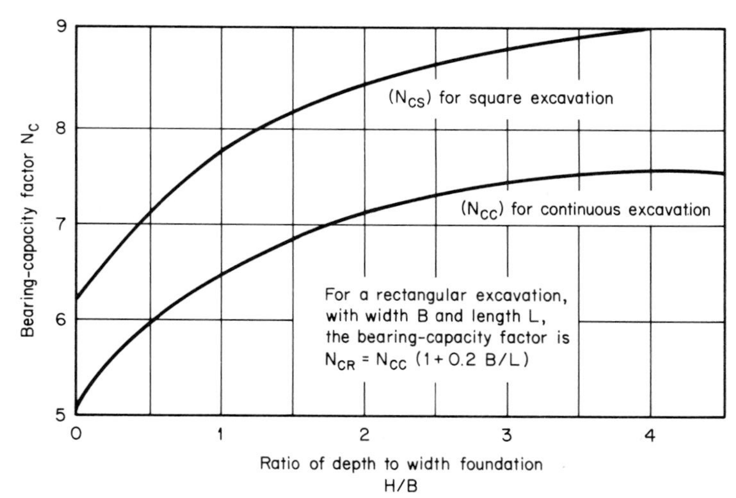

FIG. 8.28 Bearing capacity factors for deep uniform clays. *(From NAVFAC DM7.2.)*

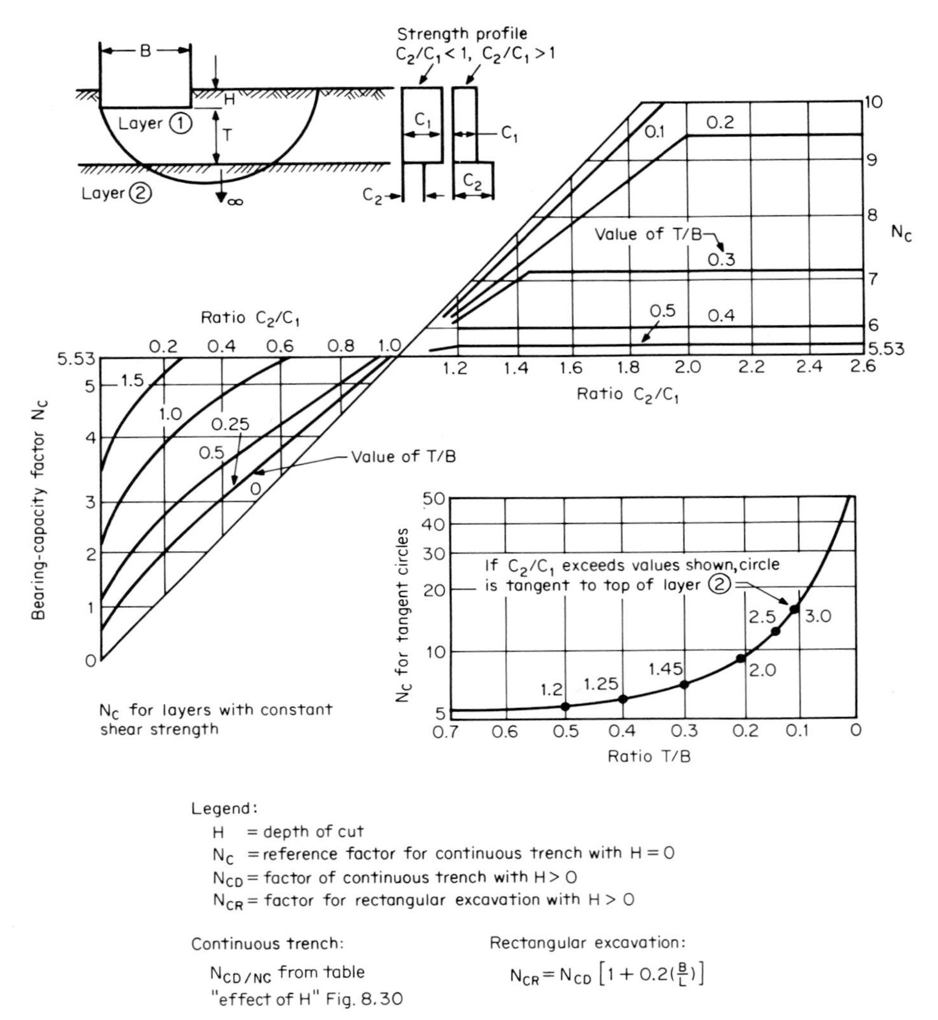

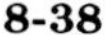

FIG. 8.29 Bearing capacity factors for two-layer cohesive soils ($\phi = 0$). *(From NAVFAC DM7.2.)*

the bottom of the excavation and the top of the pervious layer must be greater than the total hydrostatic head at the latter point. For narrower excavations, where the width is less than 1.5 times the thickness of the impervious material, advantage may be taken of the strength of the clay; total forces (excavation-wide, for a unit length along the trench) may be considered, then including vertical shear in line with the sides of the excavations. The pertinent relationships are shown in Fig. 8.31.

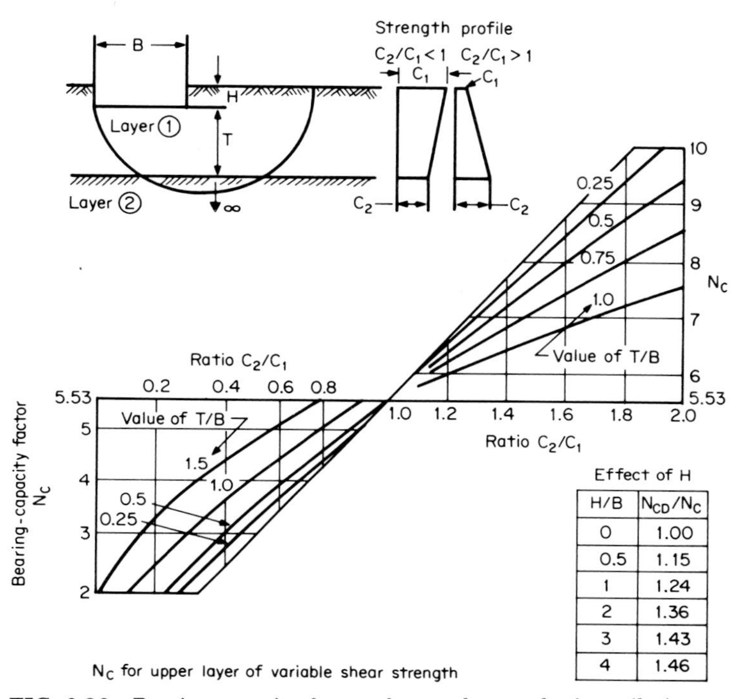

Effect of H

H/B	N_{CD}/N_C
0	1.00
0.5	1.15
1	1.24
2	1.36
3	1.43
4	1.46

FIG. 8.30 Bearing capacity factors for two-layer cohesive soils (ϕ = 0). *(From NAVFAC DM7.2.)*

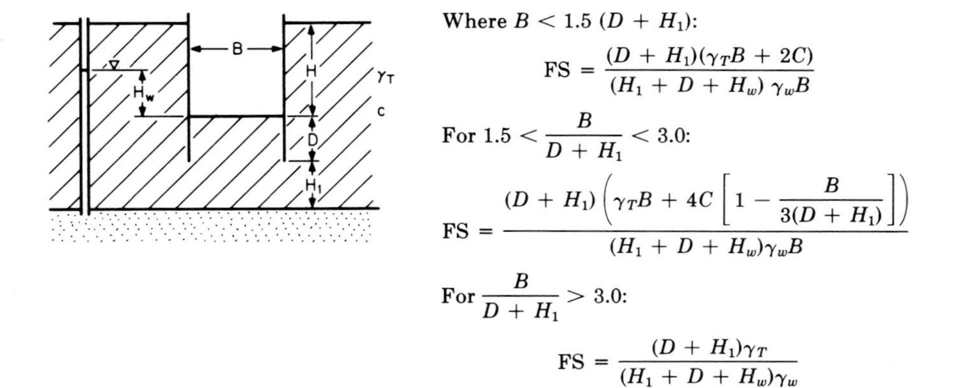

Where $B < 1.5\,(D + H_1)$:

$$\text{FS} = \frac{(D + H_1)(\gamma_T B + 2C)}{(H_1 + D + H_w)\,\gamma_w B}$$

For $1.5 < \dfrac{B}{D + H_1} < 3.0$:

$$\text{FS} = \frac{(D + H_1)\left(\gamma_T B + 4C\left[1 - \dfrac{B}{3(D + H_1)}\right]\right)}{(H_1 + D + H_w)\gamma_w B}$$

For $\dfrac{B}{D + H_1} > 3.0$:

$$\text{FS} = \frac{(D + H_1)\gamma_T}{(H_1 + D + H_w)\gamma_w}$$

FIG. 8.31 Excavation in clay with pervious layer below.

For wide excavations in deep cohesive soils, where sheeting restraint and/or passive reactions are within that same deposit, the overall stability of the excavation must be checked. This may be accomplished with stability charts, or may require a circular arc or wedge-type analysis (refer to NAVFAC DM-7.2, Taylor, or Lambe and Whitman in the Bibliography). In certain cases it may be appropriate to assume a failure surface which passes through a retaining member (such as a tieback), although more frequently the critical surface will encompass the entire wall and its restraining members.

Moment Reduction

When a retaining system consisting of sheeting restrained by one tie level and penetration into the subgrade is installed in a medium-dense or dense granular deposit, it has been found that the pressure on the sheeting, and consequently the moment in the sheeting (or other vertical support members), can be reduced in relationship to the flexibility of the sheeting. Using the chart in Fig. 8.32, an allowable reduction factor may be obtained for a given flexibility member (calculated per foot of wall). The flexibility number derives from the moment of inertia of the sheeting.

In summary, the steps are as follows:

1. Determine the maximum moment (per linear foot) by standard calculations for the assumed retaining structure configuration.
2. Select an appropriate maximum allowable unit stress in the sheeting material (f_s), note its modulus of elasticity (E), assume a type of sheeting, and identify its moment of inertia (I).
3. With the depth of excavation (H) and the sheeting penetration below the excavation (D), calculate the flexibility number (ρ), then determine the appropriate reduction factor from the diagram.

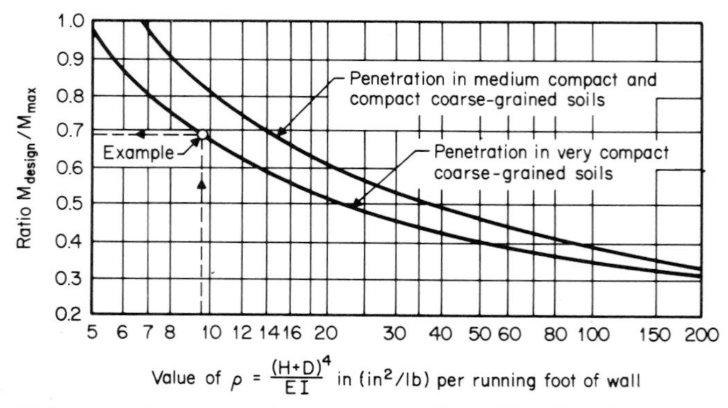

FIG. 8.32 Moment reduction factor. *(From NAVFAC DM7.2.)*

4. Apply the factor to the normally calculated maximum moment, then calculate the design stress for the sheeting. If it is not near the maximum allowable stress, select a different size of sheeting and recalculate until the most efficient sheeting section is found.

Factor of Safety

The selection of appropriate factors of safety is a critically important matter. Too low a factor, and failures or at least large movements occur, requiring replacement of the retaining structure or repair of buildings or other structures adjacent to the excavation—at considerable financial cost. Too high a safety factor means an uneconomical retaining structure that could be the reason for losing a bid. Just what is right depends on a number of factors, and experience is the best guide.

As a gross generalization, support for permanent structures should be designed with a safety factor of 2.0; temporary structures require a safety factor of about 1.5. There are specific situations where lower safety factors are acceptable; a few instances are cited in previous sections. But the selection of a typical safety factor must rest in part on consideration of the following:

1. Accuracy with which the soil characteristics are known. With a thorough investigation—frequent borings and laboratory testing of representative samples—it may be reasonable to reduce the indicated factors of safety by 10 to 20%. More commonly, boring information is limited and there are few, if any, tests; here a knowledge of an area would be needed to hold to moderate safety factors. It is hard to pick a high enough safety factor for situations where soil conditions are unknown or the reporting of them is unreliable. The only reasonable approach here is to go out and get enough information to allow the use of moderate factors of safety. It is always possible to use very low strength parameters, but this forces an ultraconservative (and high-cost) design.
2. The consequences of failure. Once identified as a factor, this is inherently obvious. So long as the safety of personnel is properly attended to, a relatively high risk factor is allowable on a trench across an open field. But an excavation near existing structures needs a conservative safety factor to prevent failure or detrimental movement.
3. Extreme soil conditions. Very soft clay should be considered a liquid; its low strength should not be relied upon. If soft to firm clay is stressed to its limit, it will start to yield and then, due to remolding, lose strength. This can occur whenever horizontal pressures exceed 4 times the cohesive strength of the soil. Very loose sands will not "set up" or arch; thus full active pressures must be taken into account at all points. Very dense granular soils may act as if they have a higher friction angle than is shown by laboratory test. Note that with deflection, sand may change density (increase or decrease), and it will then have somewhat different characteristics than it started with. Soil parameters and/or the safety factor must be adjusted accordingly.

4. Need to restrict deflection. A low safety factor will result in at least some deflections of the soil (and/or structures) behind the retaining wall. To restrict deflections to very low values, a safety factor of the order of 2.0 or higher is needed, along with prestressing of the retaining structure and its support.
5. The conservatism of the design criteria. The commonly used pressure diagrams (Fig 8.22) are based on a range of maximum recorded values; hence the factor of safety used may be quite low. Where calculations are primarily theoretical and results close to a realistic average, a higher safety factor is needed to account for locations where higher-than-average loads develop.

CONSTRUCTION PLANNING

Preparing to start a project which requires temporary ground support needs careful planning and coordination. Since speed of installation is usually critical, everything required for the installation must be available at the correct time. Thought must be given to alternate or corrective procedures which might be required if conditions at the site are different than anticipated.

The "lining up" of equipment and material often requires large storage or staging areas, probably not available at sites in urban areas. It is therefore of great importance to order equipment and materials from reliable suppliers who can be counted upon to make deliveries to the site at the proper times and who have sufficient resources available to supply additional material or equipment should job conditions require.

For most projects a time schedule in the form of a simple bar chart is sufficient to determine dates when materials and equipment will be needed, and these items should be ordered accordingly. When ordering materials which require long lead times for delivery, it may be advisable to order extra material (e.g., heads for tiebacks, high-strength wire) and have it available in case of need.

It is also advisable to obtain all permits required by local ordinances in advance of mobilizing material and equipment at the jobsite.

Planning should also include a determination of ramp or trestle locations and the planning sequence of bracing installation and removal. Planning should include coordination between the temporary earth-support system and the permanent structural system, ensuring that there is no interference (particularly when using internal braces). If tiebacks are to be used, necessary permission should be obtained from local authorities or adjacent property owners if the ties are to extend beyond the project property lines.

CODE REQUIREMENTS AND INSPECTION

Most local codes are not too specific about temporary earth-retention systems because they are temporary structures. Typical of this approach is the New York City Building Code. This requires a registered architect or engineer to file plans with the Building Department for all projects which extend more than 10 ft (3 m) below the legally established grades. These plans are to include details of sheeting, bracing, underpinning, etc. In addition, inspection and supervision of the installation is required by the architect or engineer. Upon completion of the installation, an affidavit is filed indicating that the installation is in accordance with the filed plans.

Typical of codes with more specific requirements would be the OSHA code. This code provides for design of sheeting and bracing, as well as inspection of excavation, sheeting, and bracing by "competent and/or qualified persons." This code also indicates angles of repose for various types of soil as well as providing tables indicating minimum requirements for sheeting and bracing of trenches (see Tables 8.1 and 8.2). The basic OSHA requirements (1980) for trenching are summarized below:

1. Banks more than 4ft (1.2m) high shall be shored or sloped to the angle of respose where a danger of slides or cave-ins exists as a result of excavation.
2. Sides of trenches in unstable or soft material, 4 ft (1.2 m) or more in depth, shall be shored, sheeted, braced, sloped, or otherwise supported by means of sufficient strength to protect the employee working within them (see Table 8.2).
3. Sides of trenches in hard or compact soil, including embankments, shall be shored or otherwise supported when the trench is more than 4 ft (1.2 m) in depth and 8 ft (2.4 m) or more in length. In lieu of shoring, the sides of the trench above the 4-ft (1.2-m) level may be sloped to preclude collapse, but shall not be steeper than a 1-ft (30-cm) rise to each 1/2-ft (15-cm) horizontal.
4. Materials used for sheeting and sheet piling, bracing, shoring, and underpinning shall be in good serviceable condition, and timbers used shall be sound and free from large or loose knots, and shall be designed and installed so as to be effective to the bottom of the excavation.

TABLE 8.1 Approximate Angle of Repose for Sloping of Sides of Excavations

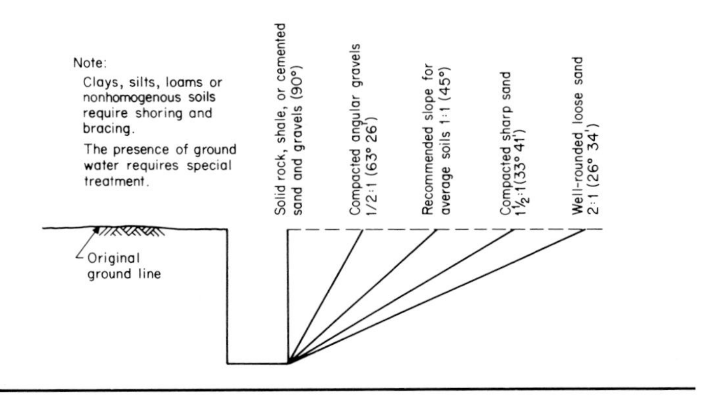

TABLE 8.2 Trench Shoring—Minimum Requirements

Depth of trench, ft	Kind or condition of earth	Size and spacing of members										
		Uprights		Stringers		Cross braces*					Maximum spacing	
						Width of trench, in				in	ft	ft
		Minimum dimension, in	Maximum spacing, ft	Minimum dimension, in	Maximum spacing, ft	Up to 3 ft	3 to 6 ft	6 to 9 ft	9 to 12 ft	12 to 15 ft	Vertical	Horizontal
5 to 10	Hard, compact	3 x 4 or 2 x 6	6			2 x 6	4 x 4	4 x 6	6 x 6	6 x 8	4	6
	Likely to crack	3 x 4 or 2 x 6	3	4 x 6	4	2 x 6	4 x 4	4 x 6	6 x 6	6 x 8	4	6
	Soft, sandy, or filled	3 x 4 or 2 x 6	Close sheeting	4 x 6	4	4 x 4	4 x 6	6 x 6	6 x 8	8 x 8	4	6
	Hydrostatic pressure	3 x 4 or 2 x 6	Close sheeting	6 x 8	4	4 x 4	4 x 6	6 x 6	6 x 8	8 x 8	4	6
10 to 15	Hard	3 x 4 or 2 x 6	4	4 x 6	4	4 x 4	4 x 6	6 x 6	6 x 8	8 x 8	4	6
	Likely to crack	3 x 4 or 2 x 6	2	4 x 6	4	4 x 4	4 x 6	6 x 6	6 x 8	8 x 8		6
	Soft, sandy, or filled	3 x 4 or 2 x 6	Close sheeting	4 x 6	4	4 x 6	6 x 6	6 x 8	8 x 8	8 x 10	4	6
	Hydrostatic pressure	3 x 6	Close sheeting	8 x 10	4	4 x 6	6 x 6	6 x 8	8 x 8	8 x 10	4	6
15 to 20	All kinds or conditions	3 x 6	Close sheeting	4 x 12	4	4 x 12	6 x 8	8 x 8	8 x 10	10 x 10	4	6
Over 20	All kinds or conditions	3 x 6	Close sheeting	6 x 8	4	4 x 12	8 x 8	8 x 10	10 x 10	10 x 12	4	6

*Trench jacks may be used in lieu of, or in combination with, cross braces. Shoring is not required in solid rock, hard shale, or hard slag. Where desirable, steel sheet piling and bracing of equal strength may be substituted for wood.

Note: 1 in = 2.54 cm; 1 ft = 0.305 m.

SOURCE: *Federal Register*, vol. 37, no. 243, Saturday, December 16, 1972.

5. Additional precautions by way of shoring and bracing shall be taken to prevent slides or cave-ins when excavations or trenches are made in locations adjacent to backfilled excavations or where excavations are subjected to vibrations from railroad or highway traffic, the operation of machinery, or any other source.
6. Minimum requirements for trench timbering shall be in accordance with Table 8.2.
7. Where employees are required to be in trenches 3 ft (90 cm) deep or more, ladders, extending from the floor of the trench excavation to at least 3 ft (90 cm) above the top of the excavation, shall be provided and so located as to provide means of exit without more than 25 ft (7.6 m) of lateral travel.
8. Bracing or shoring of trenches shall be carried along with the excavation.
9. Cross-braces or trench jacks shall be placed in true horizontal position, be spaced vertically, and be secured to prevent sliding, falling, or kick-outs.
10. Portable trench boxes or sliding trench shields may be used for the protection of employees only. Trench boxes or shields shall be designed, constructed, and maintained to meet acceptable engineering standards.
11. Backfilling and removal of trench supports shall progress together from the bottom of the trench. Jacks or braces shall be released slowly and, in unstable soil, ropes shall be used to pull out the jacks or braces from above after employees have cleared the trench.

BIBLIOGRAPHY

Canadian Foundation Engineering Manual, Foundations Committee, Canadian Geotechnical Society, Montreal, March 1978.

"Construction Safety and Health Regulations," Occupational Safety and Health Administration, Department of Labor, *Federal Register,* vol. 39, no. 122, Washington, D.C., June 24, 1974.

Crofts, J. E., B. K. Menzies, and A. R. Tarzi: "Lateral Displacement of Shallow Buried Pipelines due to Adjacent Deep Trence Excavations," *Geotechnique*, vol. 27, no. 2, p. 165.

Design Manual 7.1: Soil Mechanics, Department of the Navy, Naval Facilities Engineering Command, May 1982.

Design Manual 7.2: Foundations and Earth Structures, Department of the Navy, Naval Facilities Engineering Command, May 1982.

Diaphragm Walls and Anchorages, conference proceedings, Institution of Civil Engineers, London, 1975.

Dunnicliff, J.: *Geotechnical Instrumentation for Monitoring Field Performance,* NCHRP 89, Transportation Research Board, National Research Council, Washington, D.C., 1982.

"Earth Reinforcement—New Methods and Uses," *Civil Engineering,* vol. 49, no. 1, January 1979.

Field Design Standards, New York City Transit Authority, Civil Engineering and Architectural Division, New York, March 1, 1974.

Goldberg, D. T., W. E., Jaworski, and M. D. Gordon: *Lateral Support Systems and Underpinning,* vols. I, II, III, Federal Highway Administration Report no. FHWA-RD-75-130, Washington, D.C., 1976.

Ground Engineering, conference proceedings, Institution of Civil Engineers, London, 1970.

Lambe, T. W., and R. V. Whitman: *Soil Mechanics,* Wiley, New York, 1969.

Leonards, G. A.: *Foundation Engineering.* McGraw-Hill, New York, 1962.

"New York City Building Code," *The City Record,* vol. XCVI, no. 28917, New York, 1968.

"New York City Building Code," *The City Record,* vol. XCVI, no. 28917, New York, 1968.

"Open Cut Construction, Temporary Retaining Structures," seminar proceedings, Metropolitan Section ASCE, New York, 1975.

Peck, R. B.: "Deep Excavations and Tunneling in Soft Ground," International Conference on Soil Mechanics and Foundation Engineering, Mexico City, 1969, pp. 225–290.

Powers, J. P.: *Construction Dewatering: A Guide to Theory and Practice,* Wiley, New York, 1981.

Prentis, E., and L. White: *Underpinning,* 2d ed., Columbia University Press, New York, 1950.

Schuster, J. A.: "Controlled Freezing for Temporary Ground Support," *proceedings,* Rapid Excavation and Tunneling Conference, Chicago, 1972.

Steel Sheet Piling Design Manual, United States Steel, Pittsburgh, July 1975.

Taylor, D. W.: *Fundamentals of Soil Mechanics,* Wiley, New York, 1948.

Teng, W. C.: *Foundation Design,* Prentice-Hall, Englewood Cliffs, N. J. 1962.

Terzaghi, K., and R. B. Peck: *Soil Mechanics in Engineering Practice,* Wiley, New York, 1948.

Winterkorn, H. F., and H. Y. Fang: *Foundation Engineering Handbook,* Van Nostrand Reinhold, New York, 1975.

9

Underpinning and Shoring

by Rudi van Leeuwen
Fred Severud

Underpinning is the installation of temporary or permanent support to an existing foundation to provide either additional depth or an increase in bearing capacity.

When the foundation for a new structure is deeper than that of an adjacent existing structure, the installation of some kind of protection for the existing one is usually required. Underpinning may be required even though the new excavation is not directly adjacent to an existing structure. The soil conditions, the type of structure to be supported, location of the water table, or presence of rock above or below subgrade are some of the factors that dictate the type of support or underpinning required.

The location of the influence line of the new excavation in relation to a nearby structure has to be considered, taking into consideration the soil conditions, the water table and other temporary retaining structures, such as soldier pile sheeting or steel sheet piling installed between the excavation and the existing structure. Underpinning of interior column or wall footings may not be required when the exterior wall footing is continuously underpinned and provides protection of the interior.

Very frequently, underpinning is required because the structure originally was built on unsuitable bearing material, or a compressible layer (peat, organic silt, or a poorly compacted backfill or sanitary fill) underlies the bearing material. Such a structure may settle either during or after construction. In this instance, lack of boring information may be blamed for the damage to the structure and for the financial burden on owners to keep their buildings in a structurally healthy condition.

When an existing structure is supported on friction piles, which will be partially exposed during the excavation operation, the installation of underpinning may be necessary to support the weakened pile foundation.

Underpinning may also be necessary if the requirements of a structure have changed, e.g., when heavier floor loads are imposed or additional floors are added to an existing building.

A recent new use for underpinning (especially during the last 20 years) has been to provide extra basement space under existing buildings. In such applications exterior walls are underpinned and the interior column or wall loads are transferred through new columns resting on new footings, creating new additional floor space without altering the outside appearance of the structure.

One of the most frequent causes of settlement of existing buildings that results in the need for underpinning is the lowering of the water table due to tidal fluctuation, lowering of the groundwater by a water district, or the nearby construction of a subway tunnel or deep foundation with round-the-clock dewatering. As an example, this lowering of the water table can cause the top of timber-bearing piles to decay and will require remedial underpinning as indicated in Fig. 9.1.

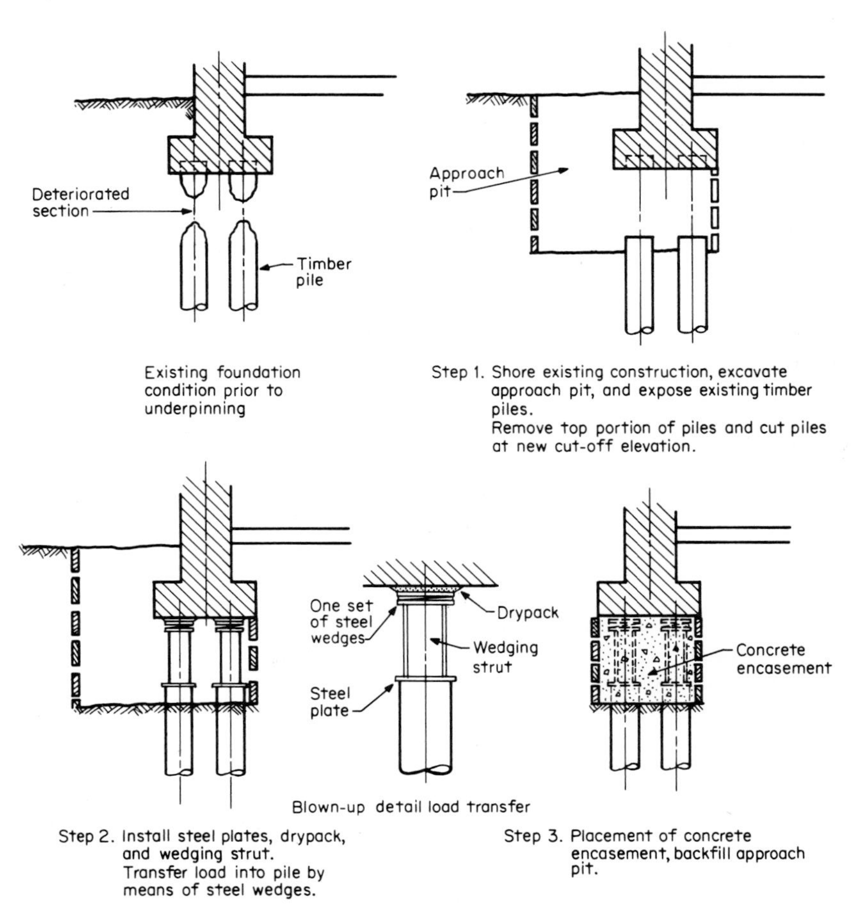

FIG. 9.1 Underpinning timber-pile foundations.

In certain instances, raising of the water table can effect a decrease in the bearing capacity of the soil, causing the structure to settle and again underpinning may be required to save the building.

To prevent any unexpected settlement of a structure, it is of the utmost importance that proper field investigations are performed, including borings, test pits, and soil tests. The cost of these tests is minor compared to the expenses of field investigation and survey of structural damage, design of remedial work, and the actual costs of the underpinning and repairs after the building shows signs of distress.

DETERMINING THE NEED FOR UNDERPINNING

When a structure starts showing signs of settlement or distress, it is of the utmost importance to establish level readings and offset readings. These readings should be taken by a professional on a daily, weekly, or monthly basis depending on the severity of the movements. Plotting these readings will indicate if the movements are decreasing or increasing and by analyzing the results, a decision can be made whether or not underpinning or other measures have to be taken to safeguard the structure and its occupants.

Prior to the start of the excavation for a new structure, it is advisable to have a professional examine all structures in close proximity to the construction site, even though no underpinning may be required. The examination will consist of a description of the structure with all its structural defects, backed up by photographs which will be part of the report. Most insurance carriers today insist on this type of report, which has to be submitted to them prior to the start of demolition or excavation operations.

SETTLEMENTS CAUSED BY UNDERPINNINGS

The most cautiously installed underpinning will be accompanied by some minor settlement of the structure. The settlement will vary with the type of underpinning system installed, the soil conditions encountered, the type of structure, and above all, by the quality of the work and supervision of the job. Most settlements can be held within ¼ to ½ in (6 to 12 mm).

It is very important to keep the settlement of a building uniform, because the difference in settlement from one point to another may cause structural damage.

Normally the excavation adjacent to the structure will be excavated to within 6 in (15 cm) above the bottom of the column or wall footings to facilitate the start of the approach pits. This may cause the unloading of the bearing soil, creating minor settlement.

The installation of a dewatering system in the new excavation may displace fines under the wall and column footings, creating a cause for additional settlements.

The installation of piles and caissons adjacent to existing structures may be the cause of movement and settlement, depending on the soil condition and the method of installation used. The installation of any type of bearing pile in sandy material, especially when driven with a vibratory hammer, adjacent to or in the vicinity of an existing structure requires precise monitoring of the structure before and during the pile installation.

The installation of lateral support for the structure itself and for the underpinning is very important. Any lateral movement of the structure will add additional loading to the bearing strata on which the underpinning is founded and thereby increase the total settlement. The lateral support can consist of cross-lot bracing, inclined bracing, or tiebacks in earth or rock. The vertical component of the tiebacks has to be added to the building load in each underpinning unit to determine the minimum bearing area.

Substitution of soldier piles or steel sheet-pile sheeting for underpinning does not prevent the structure from horizontal and vertical movement. In case of light structures, maintenance jacking may be used to prevent differential settlement, without the actual installation of underpinning.

Whatever method of underpinning is employed, it is always necessary to correct structural damage in walls and footing prior to the start of the underpinning, especially since the first step in an underpinning operation is to excavate under a footing. By doing so, the bearing capacity of the foundation will be decreased. In certain instances the use of temporary shores or needles is required.

In the case of a loose rubble foundation wall that has to be underpinned, it may be necessary to remove the rubble wall in short sections and replace it with a concrete wall. Alternatively, the rubble wall may be reinforced by repointing the joints with mortar or grout or by placing a concrete wall against the rubble wall and tying the walls together through the bonding action of the freshly placed concrete.

MINIMIZING DISTURBANCE DUE TO UNDERPINNING

In most instances underpinning can be installed without severe inconvenience to the occupants of the structure, especially if only exterior walls and footings are affected.

Some underpinning may be installed from within the basement of a structure when the locations of the existing boiler plant or other utilities do not obstruct the sinking of the approach pits. However, in many instances, structures have been underpinned without entering the building and with a minimum disturbance to the pedestrians on the sidewalks.

Performing the work entirely below the sidewalk and covering the areas of underpinning immediately on completion helps keep the disturbance to the public to a minimum.

UNDERPINNING METHODS

Different methods of support or underpinning are described below.

Temporary Support with Maintenance Jacking

Light structures (for example, wood frame garages) that fall within the influence line of an adjacent excavation and which do not warrant the expense of an underpinning installation may be supported on timber or concrete mats. When settlement occurs, the structure will be kept at the same level by means of mechanical or hydraulic jacks (see Fig. 9.2).

Bracket Pile Underpinning

When both the existing and future structures belong to the same owner, the use of bracket piles is very economical. (Most municipal building codes do not allow a building to be supported on a foundation that is located on someone else's property.) The steel bracket piles are driven or placed adjacent to the future structure in preaugered holes which are then backfilled with a lean sand-cement mix. The load is transferred from the structure into the pile through a steel bracket welded to the side of the pile. A combination of steel plates, wedges, and drypack is installed to ensure a tight fit between the structure and the bracket, as shown in Fig. 9.3.

This type of underpinning can be utilized for structures up to two stories high, depending on the weight of the building and the quality of the bearing material at subgrade of the new structure. The toe penetration of the piles is determined by the passive resistance below subgrade and by the vertical load distribution of the bracket pile. The spacing of the piles depends on the load distribution in the existing structure and also on the ability of the soil to arch from pile to pile for horizontal sheeting installation. The maximum spacing should not exceed 8 ft (2.4 m).

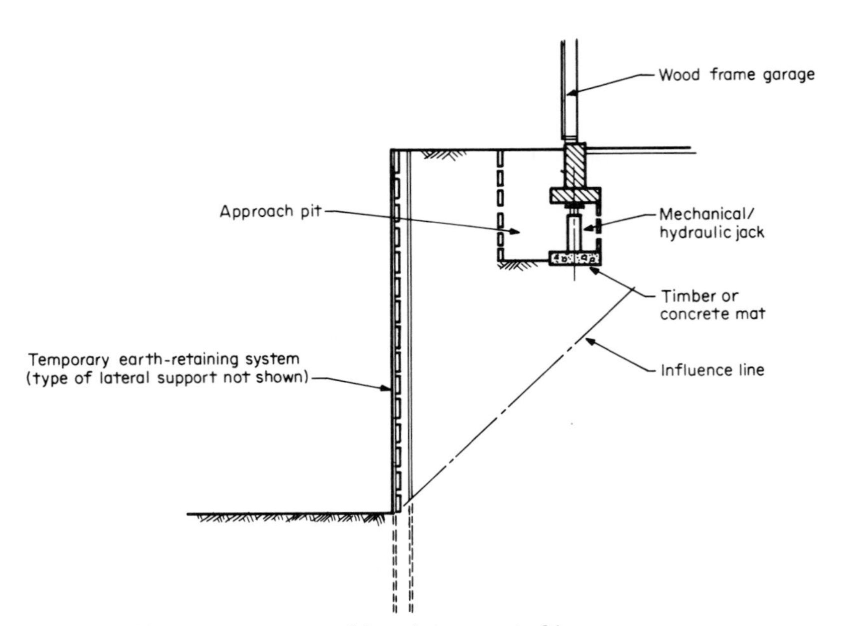

FIG. 9.2 Temporary support with maintenance jacking.

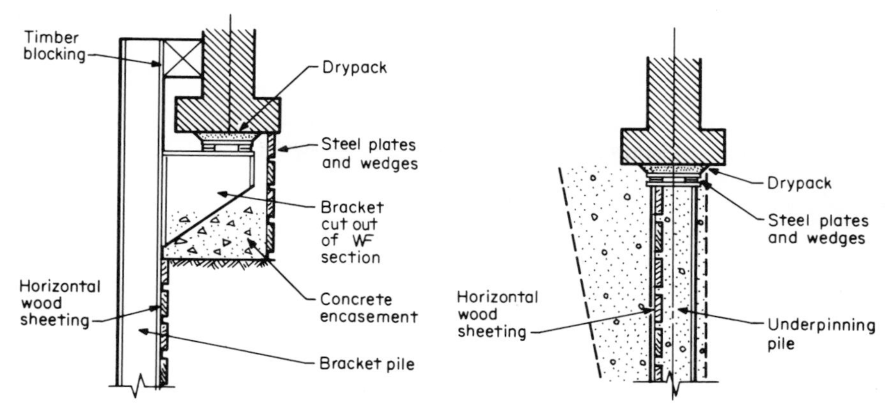

FIG. 9.3 Bracket pile detail.

FIG. 9.4 Slant-augered pile detail.

Slant-Augered Pile Underpinning

This method resembles that of the bracket pile underpinning and is most frequently used on the west coast for light one- and two-story structures. One advantage of this method is that the support system is located fully under the existing structure without infringing on the adjacent property, as shown in Fig. 9.4.

Augered holes at a slight angle (slant) are formed under the existing structure at intervals. The top portion of the hole is enlarged with a special tool to facilitate the placement of a steel beam in the hole, which will be set plumb. The pile is encased in a lean sand-cement mix. The building load is transferred by means of steel plates and wedges between top of pile and underside of foundation. In case the preaugered holes do not stand open due to ground-water or soil conditions, a bentonite slurry mix may be necessary to allow setting the steel beams and placing the sand-cement mix without great loss of ground under the existing structure.

Pit or Pier Underpinning

This method is the one most frequently used in the foundation industry. If the water table is above the level to which the new underpinning has to be extended, a dewatering system has to be installed prior to the start of the underpinning operation, so that all the pits can be sunk in the dry. If the excavation has to be performed partially in the wet, this not only makes the work more costly, but makes loss of ground unavoidable.

The piers may be continuous for the full length of an exterior wall, or an intermittent system of piers may be used. The building loads are transferred onto the construction piers by means of "drypack," a mixture of sand and cement with very little water that is rammed into the space between the existing wall or column footing and the hardened concrete, placed in the excavated pit. This method is more completely described subsequently.

Pretest Pile Underpinning

There are many occasions where the pretest pile method is the only solution for underpinning a structure: for example, in case the water table cannot be lowered to facilitate hand-dug pit underpinning, or when the bearing capacity of the soil at future subgrade is not sufficient to carry heavy concentrated loads and the tip elevation of underpinning has to be lowered to bear on deeply seated strata of hardpan or rock.

Pretest piles consist of steel pipe 10 to 24 in (25 to 60 cm) in diameter with a minimum wall thickness of 0.25 in (6 mm), up to 100 ft (30 m) in length, installed in 4- to 6-ft (1.2- to 1.8-m) sections. In most instances the pipe is cleaned and backfilled with structural concrete. The load is transferred by means of hydraulic jacks, steel plates, and wedges. A more detailed description of this method is presented subsequently.

Slurry Trench Wall

This new method of support was developed during the last two decades. It is used instead of installing a support system under a structure.

Originally the slurry wall was developed to eliminate the need for a temporary earth-retaining structure, such as sheeting and bracing or steel sheet piling. The wall is installed in a trench with sections 8 to 24 ft (2. 4 to 7 m) in length, excavated with a clamshell bucket with a Kelly bar or hanging from cables. A slurry trench wall may vary in width from 18 to 42 in (45 to 105 cm) depending on the needs of the structure; the depth is limited by the equipment being used. The excavated material is immediately replaced by a bentonite slurry mix. The lateral pressure created by the presence of the slurry prevents the walls of the excavation from collapsing. The reinforcing steel is placed through the slurry, after which concrete is tremied in place, displacing the bentonite mixture.

The advantage of a slurry trench wall is that it can be used as part of the permanent structure, diminishing the total cost of construction considerably.

Since the walls of the trench receive lateral support at all times, adjacent structures need not be underpinned. However, the slurry trench wall must be laterally supported during excavation for the new structure.

If a slurry trench wall is used in lieu of underpinning, and depending on the type of structure to be supported, the panel length should be reduced to 4 to 8 ft (1.2 to 2.4 m). The lateral support system has to take into account the surcharge loads introduced into the slurry wall by the structure sitting directly behind it.

PRECAUTIONS

The design of underpinning is far from an exact science. The conditions of the existing structure may be known at the time of the design, but the bearing capacity of the soil at subgrade may change from one pier to another. It is very important to design and prepare working drawings prior to the start of the underpinning operation. These plans will indicate what one hopes to find, but in many instances, especially with old structures without as-built record drawings, one has to be prepared to make drastic changes to cope with conditions encountered in the field.

During and after the installation of any type of underpinning, the following precautions should be taken:

1. During any excavation and sheeting installation, loss of ground is almost impossible to prevent, but it should be kept to a minimum by backfilling behind the boards where and when possible. Loss of ground will create voids behind the sheeting, which may instigate settlements of floor slabs or wall footings.
2. The building load must be transferred into the underpinning system. If the transfer is not performed properly, additional settlement will occur.
3. The installation of an adequate lateral-bracing system will prevent movement in the structure and in the new underpinning. A poorly designed bracing or tieback system will add to the total and final settlement of the structure.

It will be beneficial to the reader to review in greater detail two of the techniques mentioned before, namely, concrete underpinning piers and pretest piles.

UNDERPINNING PIERS

Procedures

When it is decided to sink underpinning piers under a structure, the following procedure should be followed, as illustrated in Fig. 9.5:

1. General excavation to within 6 in (15 cm) above the bottom of the existing wall or column footings.
2. An approach pit is excavated and supported by 2-in x 8-in (5-cm x 20-cm) dressed lumber. Minimum size of the approach pit is 3 ft x 4 ft (90 cm x 120 cm) to allow a laborer room to excavate and install the pit boards. The normal depth of an approach pit is 4 ft (120 cm) except where clearance requirements warrant deeper pits.
3. After the approach pit is completed, the excavation is extended under the wall or column footing and horizontal wood sheeting is placed.
4. The underpinning pit can now be extended to its proper depth by carefully excavating and sheeting of the shaft to the proper elevation.
5. Inserts are installed for an inclined bracing or tieback system.
6. All loose masonry and dirt are removed from the bottom of the existing wall or column footing.
7. Formwork is installed in the approach pit to facilitate the placement of the concrete pier up to within 3 in (7.5 cm) of the bottom of the existing footing.
8. Drypack is installed between the bottom of the existing footing and top of the new pier, in not less than 12 h after the concrete is placed.

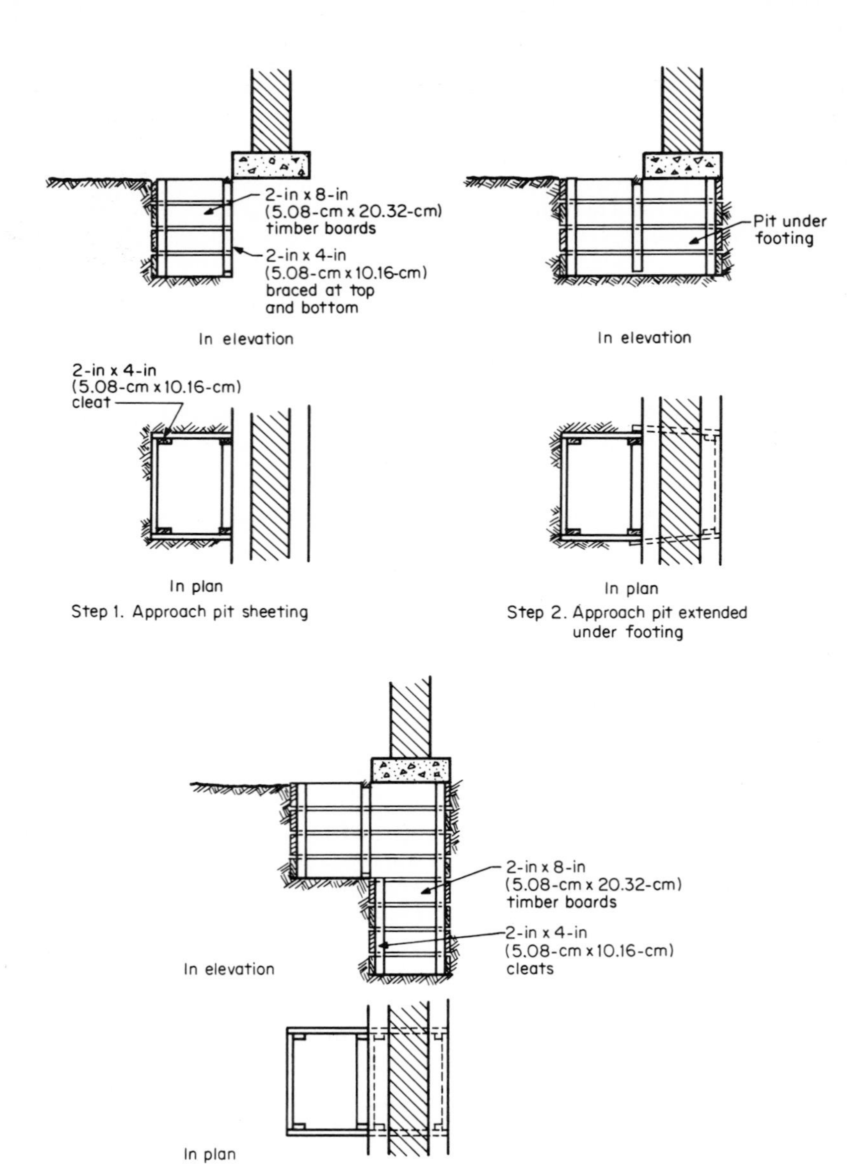

FIG. 9.5 Underpinning piers.

The excavated material is normally removed from the pit by means of buckets, utilizing a well wheel and manpower. However, if the piers are deep, mechanical equipment such as hoists are used.

If the soil is fairly cohesive, the excavation can progress in 2- to 3-ft (60- to 90-cm) lifts before the pit boards are installed; and in the case of very hard materials, only skeleton boards [one ring of 2-in x 8-in (5-cm x 20-cm) boards located at 16 in (40 cm) oc] will be required. However, in very fine, dry sand, because it runs freely when the excavation progresses before the next board can be placed, certain steps have to be taken, such as thorough wetting of the ground and/or the use of 2-in x 4-in (5-cm x 10-cm) lumber. It is advisable to use louvered boards at all times in order to enable backfilling behind the boards where necessary. (Louvered boards are boards placed with spaces left between them which can be packed with sand and salt hay in case the sand starts running after it dries up.) The use of untreated lumber for the installation of underpinning piers and sheeting has been discussed by engineers for many years. The lumber will eventually deteriorate, but will never lose its volume and will never be the cause for additional settlement of a structure. The only valid reason for using treated lumber is the presence of termites; otherwise it is a waste of the owner's money.

Sequence

The sequence of the installation of underpinning pits under bearing walls depends on the type and condition of the structure and on the soil condition, but in general no pits closer than 12 ft (3.6 m) on center should be opened simultaneously. The sequence of the installation of underpinning pits under column footings should not permit more than one pit to be opened at a time under the same column and pits under adjacent column footings should not be opened simultaneously.

In case an exterior wall footing has to be underpinned, but due to the light loads in the structure continuous underpinning is not required, intermittent underpinning piers may be installed, as shown in Fig. 9.6.

The sequencing mentioned above also applies here. The spacing between the piers depends on the loading in the structure and the soil condition. The interpier sheeting, as indicated in Fig. 9.6, has to support the soil behind the underpinning system. The thickness of the boards depends on the span between the concrete underpinning piers: 2-in (5-cm) boards up to 4 ft (1.2 m) and 3-in (7.5-cm) boards therafter.

In case the total amount of settlement has to be kept to a minimum, or pile driving will occur in the vicinity of the structure, it may be advantageous to install jacking pockets in each individual underpinning pier. Then if additional settlement occurs, mechanical or hydraulic jacks can be installed and the structure can be maintained at the same level.

Lintels

Some municipal building codes require continuous support of any exterior wall footing, and in some instances the structural integrity of the exterior wall footing does not allow intermittent underpinning piers. In these cases the installation of lintels between underpinning is required, as illustrated in Fig. 9.7.

There are many types of lintel constructions that can be used, but in case of

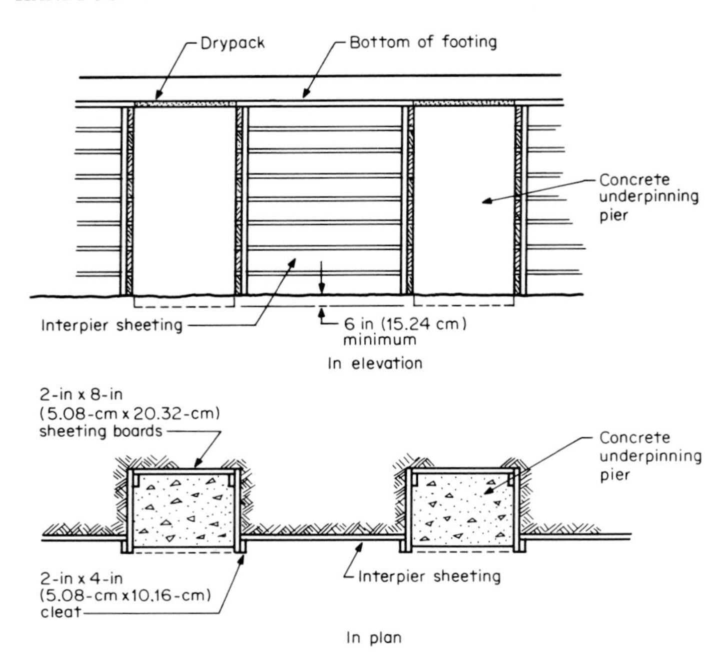

FIG. 9.6 Intermittent underpinning piers.

an emergency, a steel beam is the fastest solution. After the steel lintel is in place, drypack can be installed immediately between the steel girder and the bottom of the wall footing; permanent concrete encasement can be placed at a later date. If time is not of the essence, either of the two methods in Fig. 9.7 can be used.

Exposed Underpinning

In certain instances it is necessary to incorporate the underpinning into the new construction, especially where space is very tight, such as in construction of new tunnels and corridors. The underpinning should be installed in a manner such that the exposed face is clean and neat.

Fig. 9.8 indicates how the eventually exposed underpinning piers and the formwork are installed. After all the underpinning piers are fully completed, the general excavation is lowered while the interpier sheeting is being installed as indicated. When the excavation is completed, formwork is set between the underpinning piers, and the concrete for the fill-in wall is placed.

Overlapping Structures

Sometimes design criteria or lack of space make it necessary for a new column footing to project under the existing wall footing of an adjacent structure. When the adjacent structure requires concrete underpinning, a special method has to be utilized to facilitate this particular condition. The procedure of the underpinning installation is identical to that of a normal installation, except for the fol-

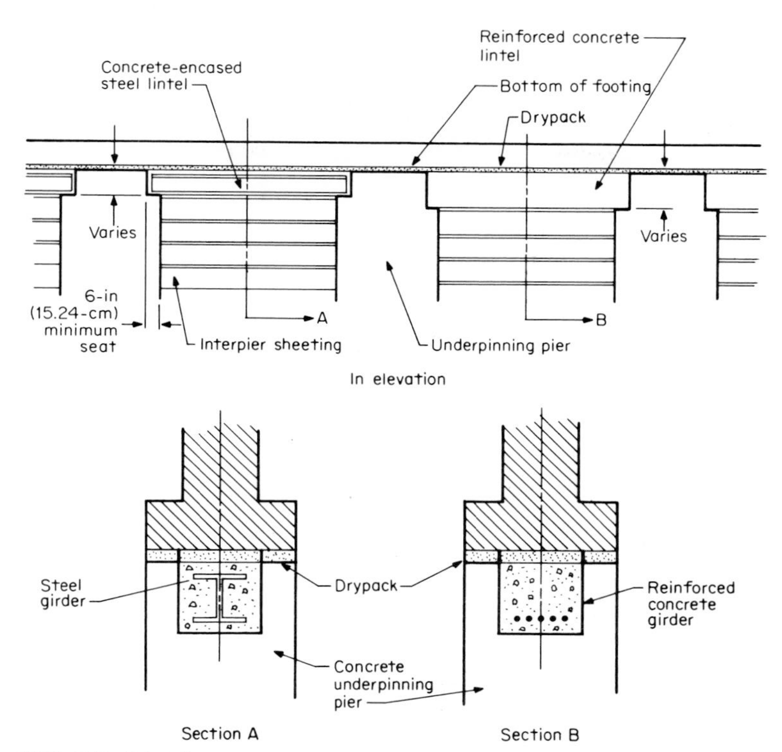

FIG. 9.7 Lintels.

lowing measures: The bottom of the excavation has to be lowered at least 12 to 18 in (30 to 45 cm) below the bottom of the future column footing beneath the existing wall, as shown in Fig. 9.9.

Concrete is placed up to the level of the bottom of the future column footing. Steel posts are embedded into this concrete mat. After the concrete has set, sand is placed and compacted up to the top of the future column footing. The underpinning pier is constructed from the compacted sand up to within 3 in (7.5 cm) of the bottom of the wall footing, and drypack is installed. After the general excavation has reached subgrade elevation, the sheeting within the area of the future column footing has to be stripped and the previously placed sand in the underpinning piers removed. The existing building load is now transferred through the previously placed steel posts into the soil. Formwork, reinforcing steel, and concrete for the new column footing can now be placed continuously under the adjacent structure.

Underpinning of Column Footings

The underpinning of column footings, individual or combined with wall footings, demands a great deal of attention due to the heavy concentrated loads that have to be extended to a lower elevation.

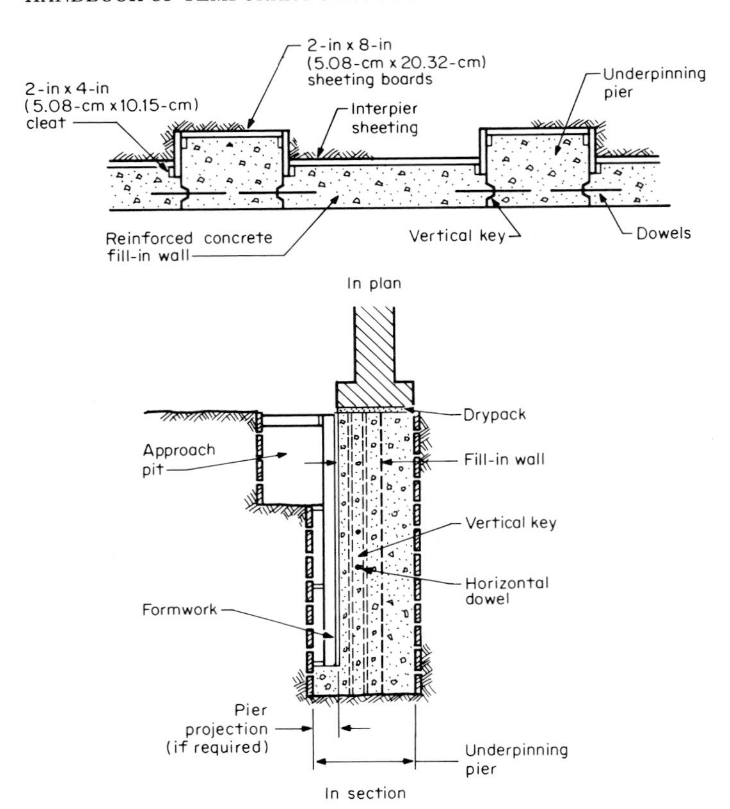

FIG. 9.8 Exposed underpinning.

The most difficult situation is the underpinning of a small individual column footing. In case the existing column footing is 5 ft x 5 ft (1.5 m x 1.5 m) or smaller, the only way to underpin the footing with concrete piers is to temporarily shore the column by means of inclined shores or needle beams supported on temporary mats, prior to the installation of the one or two underpinning piers required.

Fig. 9.10 indicates the underpinning of a 6-ft x 6-ft (1.8-m x 1.8-m) individual column footing with two alternates. Any loss of ground during the installation of the underpinning piers will cause immediate settlement. The construction sequence of the scheme with the three piers is 1, 2, and finally 3. The only time that the alternate, with only two piers, can be used is when the soil condition improves with depths and less bearing area is required for the underpinning piers than for the original column footing. The installation of the lintel is a necessity to prevent overstressing of the existing column footing.

Fig. 9.11 indicates the underpinning of a large column footing incorporated into the wall footing on both sides. To avoid entering the basement of the structure or living quarters of tenants, all piers are approached from the street side of the building. The procedure for the underpinning installation is as follows:

1. Install approach for pier 1.

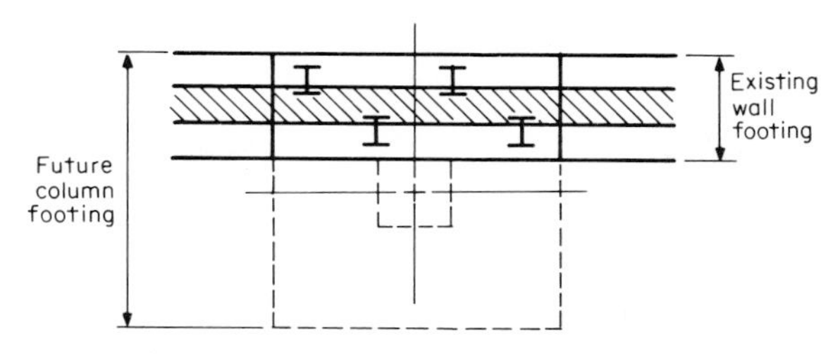

Plan—future column footing projecting underneath adjacent wall footing

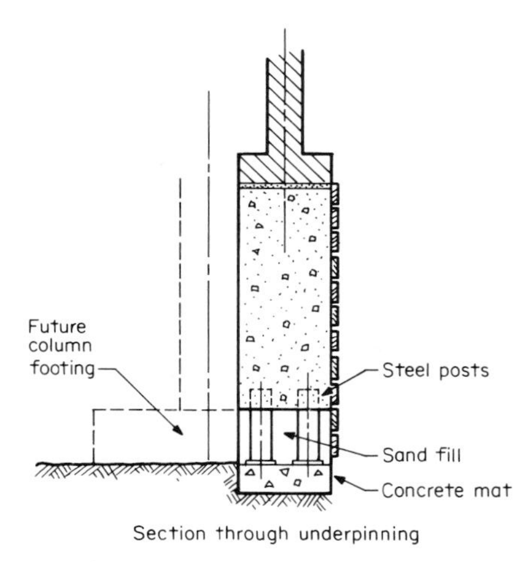

Section through underpinning

FIG. 9.9 Overlapping structures.

2. Excavate underpinning pit 1.
3. Place concrete in pit 1 to bottom of approach pit.
4. Place steel posts as indicated, transferring part of the column load into the posts by means of steel plates and wedges.
5. Excavate underpinning pit 2.
6. Place concrete in pit 2 and drypack pit 2.
7. Install formwork in approach pit to facilitate the placement of the concrete for pier 1.
8. Place concrete in remainder of pit 1 and drypack. The same procedure is used for the installation of piers 3 and 4.

In case only one approach pit can be installed, e.g., due to lack of available working space, the procedure will change slightly. After pier 2 is fully completed, the procedure will be as follows:

1. Excavate underpinning pit 3.
2. Place concrete in pit 3 up to the bottom of the approach pit.

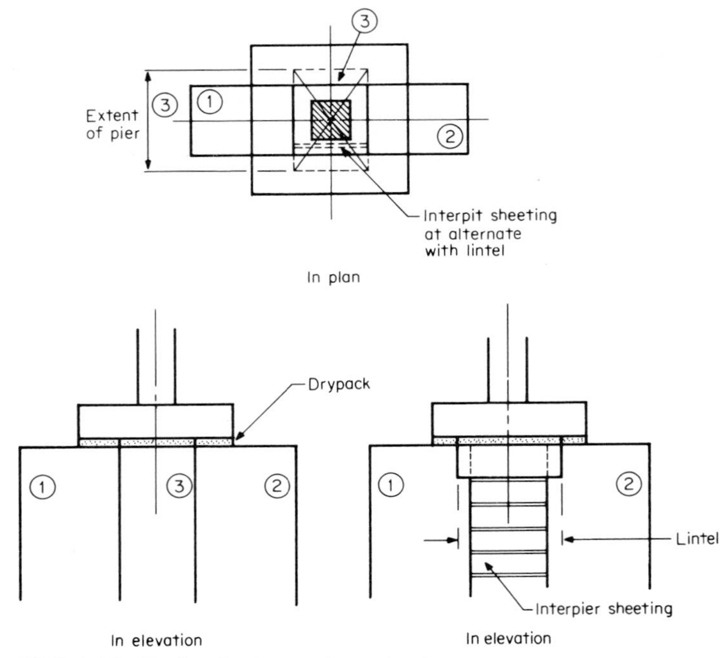

FIG. 9.10 Underpinning a column footing.

3. Place steel posts as indicated, transferring part of the column load into the posts by means of steel plates and wedges.
4. Excavate underpinning pit 4.
5. Place concrete in pit 4 and drypack.
6. Place concrete in remainder of pit 3 and drypack.
7. Install formwork in approach pit to facilitate the placement of the concrete for pier 1.
8. Place concrete in remainder of pit 1 and drypack.

In case the approach-pit excavation has to start from street level and the bottom of the footing is rather deep, the last procedure may be financially advantageous, due to reduction in the approach-pit excavation. However, consideration has to be given to the fact that the outcoming dirt from pits 3 and 4 have to be handled by workers over a larger distance in very cramped quarters, which may raise the overall cost of the operation.

Belled Piers

If ground conditions permit, it is possible to enlarge the bearing area of underpinning piers by belling-out the bottom section of the shaft. Belling-out of underpinning pits in dry, loose, runny silt and sand or in water-bearing sandy

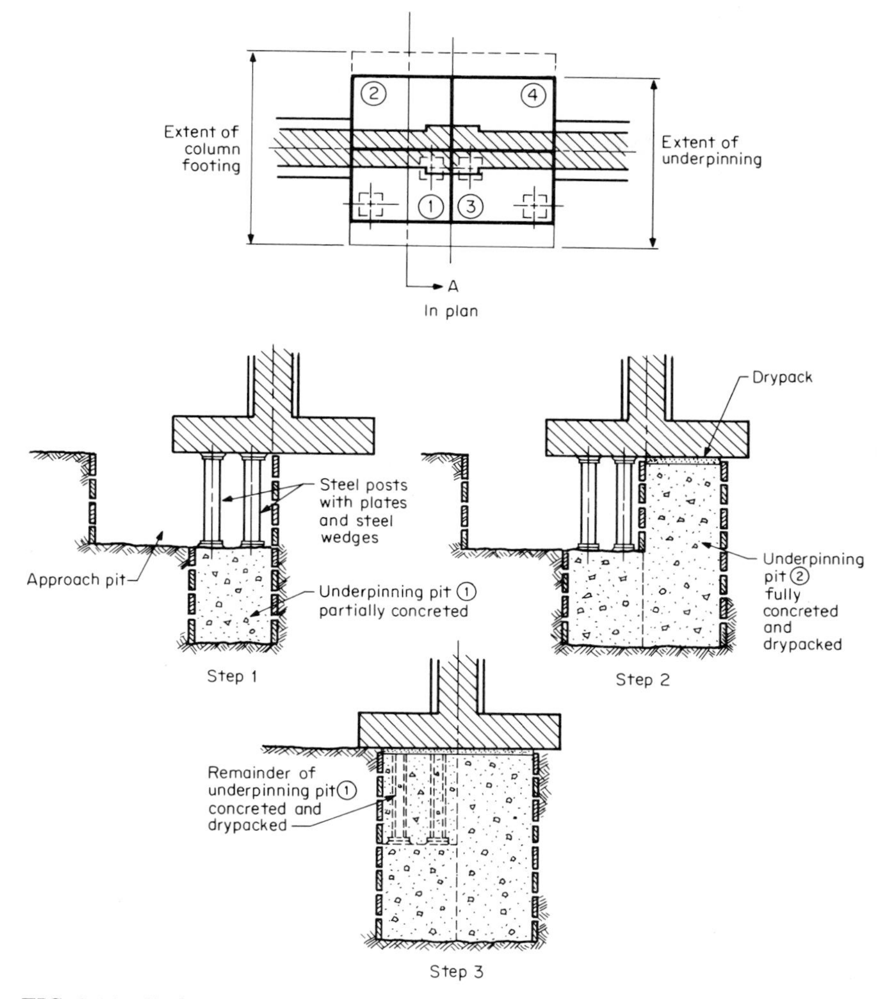

FIG. 9.11 Underpinning a column footing.

materials is very risky and will expose the structure to additional settlement due to loss of ground. However, in cohesive soils the belling-out of underpinning piers is technically feasible and will diminish the cost of the project due to lesser shaft excavation, especially when the underpinning piers are deep.

Fig. 9.12 indicates two different methods of using belling in underpinning an individual column footing. The actual column footing dimensions for the example are 8 ft by 10 ft (2.4 m by 3 m) and four piers of 4 ft by 5 ft (1.2 m by 1.5 m) each will accomplish the job. Total volume of pit excavation and concrete, assuming the depth of underpinning to be 15 ft (4.5 m), is 44.4 yd^3 (34 m^3). The first method shown on the figure indicates four piers, each with a 3-ft by 4-ft (90-

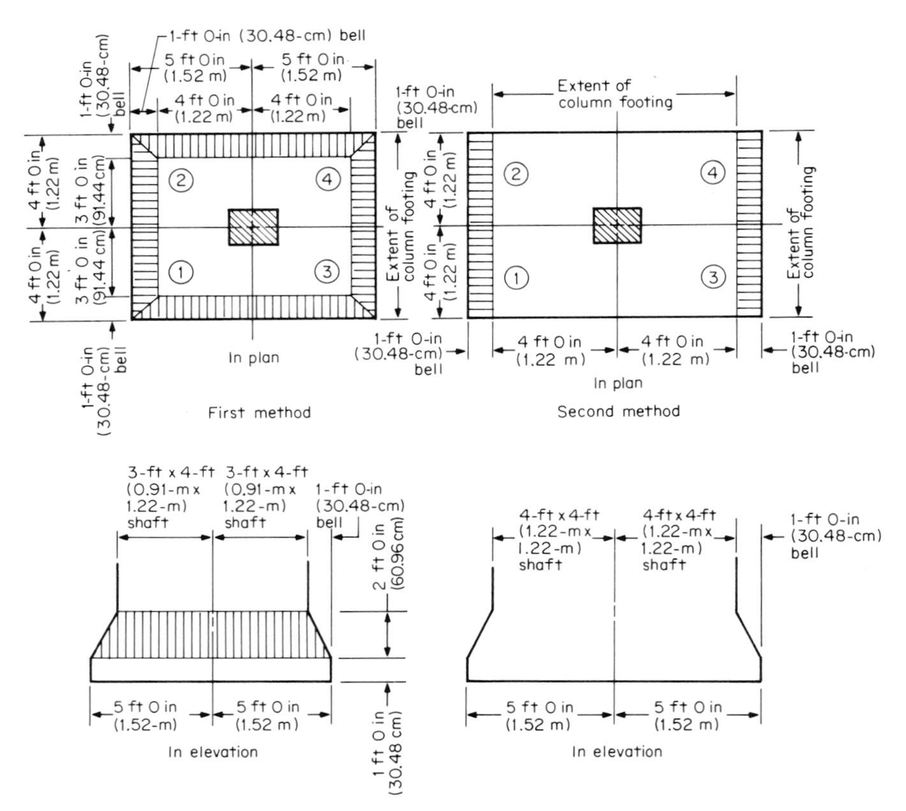

FIG. 9.12 Belled piers.

cm by 120-cm) shaft and a 1-ft (30-cm) bell in two directions. Total area in bearing is 80 ft^2 (7.4 m^2) with a total volume of pit excavation and concrete of 28.8 yd^3 (22 m^3).

Due to soil conditions, it may be very difficult to install bells on four sides in an excavated shaft and the second alternate shown in the figure may be the answer to the problem: four piers, each with 4-ft by 4-ft (120-cm by 120-cm) shaft and 1-ft (30-cm) bells on two of the four sides only. Total area in bearing is 80 ft^2 (7.4 m^2) with a total volume of pit excavation and concrete of 36.1 yd^3 (27.6 m^3).

If soil conditions allow, a third alternate (not shown on the figure, but similar to the second method) could be four piers, each with 3-ft by 4-ft (90-cm by 120-cm) shaft and 2-ft (60-cm) wide by 4-ft (120-cm) deep bells on two sides only. Total volume of excavation is 29.0 yd^3 (22 m^3). However, it has to be kept in mind that the soil condition in the area of belling has to be very hard or stiff to allow belling without causing loss of ground.

Belling of individual pits or continuing pits under bearing walls can reduce the total cost of underpinning. Care has to be taken that the belled-out portion of the underpinning does not interfere with future permanent foundations on the adjacent side.

PRETEST PILES

Procedures

The starting procedure for pretest piles is the same as the one used for the installation of concrete underpinning piers. The approach pits, however, will be deeper to facilitate the placement of steel pipe sections into the jacking pit. The length of the pipe sections will vary from project to project depending on field conditions. In case the water table is high and it cannot be lowered (by pumping, etc.), this condition will determine the depth of the jacking pit and the length of the pipe sections, which may vary from 3 to 7 ft (0.9 to 2.1 m).

In case boulders are suspected in the underlying strata, it may be advisable to install a cutting edge on the bottom section of pipe to prevent damaging the tip of the pile.

After completion of the jacking pit, a steel bearing plate is placed against the bottom of the footing, with drypack as a leveling device to compensate for the uneven bottom surface of the footing, as shown in Fig. 9.13.

The first section of pipe is placed in the jacking pit. After the pipe is set plumb, a temporary steel plate is set on top of the pipe. On this plate, two double-acting hydraulic jacks are placed. By applying pressure to the jacks, the pipe will be forced downward. It is of the utmost importance to monitor the structure during the jacking operation to avoid any uplift or damage to the structure, especially when the jacking progress is slow and high pressures are being applied to the jacks. After the jacks are extended to 75% of their stroke capacity, the pressure in the jacks is reversed and jacking dice are placed on top. This procedure is repeated until the next section of pipe can be installed.

The splice between two pipe sections can be accomplished by welding or with the use of external pipe sleeves. If the pretest pile is only exposed to vertical loading, a single weld pass will suffice. However, when the pile has to be watertight or is exposed to bending, a full-strength weld may be required. Splicing of the pipe section is very time-consuming due to the lack of adequate work space and the awkward location in the bottom of the pit where the weld has to be applied. To speed up the splicing operation, an external tight-fitting sleeve may be used. The sleeve is normally placed at the top of the pipe prior to lowering it into the jacking pit. If alignment, watertightness, or bending is a problem, the splice must be welded. The advantage of the sleeve is that most of the welding will be performed outside the confinement of the jacking pit and the welding between the sleeve and the next section of pipe is an easy down-hand weld instead of a full-penetration weld.

Before the next section of casing is placed, it may be advisable to remove the soil from inside the casing to reduce friction and thereby keep the loads in the jacks to a minimum.

If the soil is very soft, it may be advisable to add the next section and continue

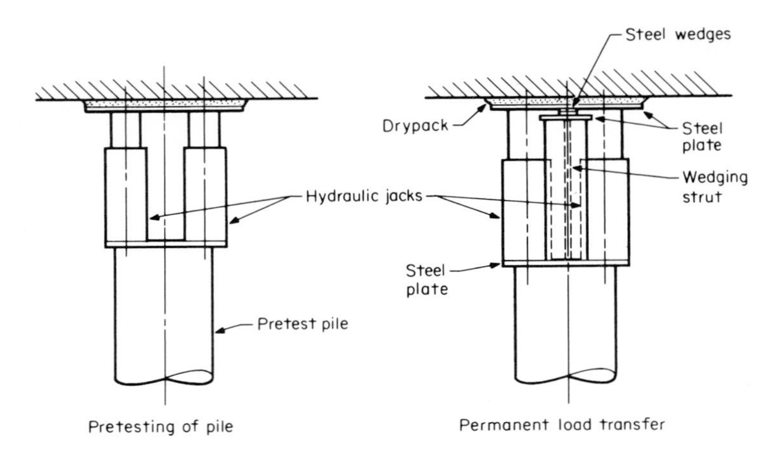

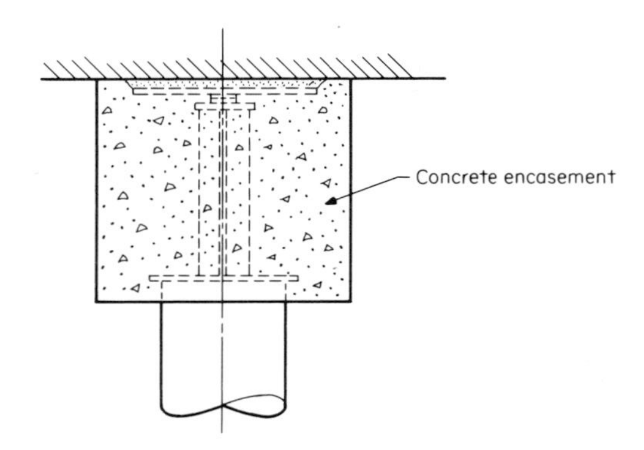

FIG. 9.13 Pretest piles.

jacking without cleaning the pipe. This procedure is repeated until the pipe is seated on the proper bearing strata or until the jacking loads get too high for the safety and integrity of the structure above.

Jacking through Soft Clay

Where piles are to be jacked through a stratum of soft clay, such as in the Chicago Loop area, it is good practice to place a compacted sand plug at the bottom of the first pipe section prior to the start of jacking. The sand plug will be forced upward slightly during the jacking of the pile to the hardpan layer, but it will prevent the soft clay from filling the pipe and thus little or no cleaning of the pile is required, which means a tremendous saving in time and money.

Overcoming Difficulties in Jacking

During the course of jacking the pipe, hard layers of soil mixed with gravel, cobbles, or even boulders may be encountered. The jacking pressures are getting too high and the conventional methods of removing the soil do not work. It will be necessary to drill or chop out the obstructions with special equipment, such as Bucyrus-Erie churn drills, to overcome the obstacles.

In certain instances, when jacking does not advance the casing due to the high frictional forces on the outside face of the pipe, it may be required to place an impact or vibratory hammer inside the jacking pit to advance the casing, but risks of settlement will increase.

Cleaning Out Piles

In most cases, jacked piles must be cleaned out and filled with concrete. There are different ways of cleaning out jacked piles, depending on the soil conditions and the total overall depth of the piles. Most methods are tedious and time-consuming, especially if the piles are long. The most common methods of cleaning are: orange peel buckets, pan-cake augers, mechanical augers, and air-lifting. It is of the utmost importance in areas where the water table is high to leave a plug of dirt above the tip of the pile at all times. Overexcavation inside the pile will result in loss of ground, which in turn will cause additional settlement.

Where the load on the pile is very light and the steel pipe alone is capable of carrying it, the dirt does not need to be removed from inside the pipe, saving the owner a considerable expense.

Pretesting

When the pile reaches a certain bearing stratum or the tip is at a predetermined elevation, the capacity of the pile has to be tested by applying 150% of the working load to the empty shell. If no substantial settlement occurs, the pile is ready for concreting. However, if the pile moves considerably [½ in (12 mm) to 1 in (25 mm)] under the 150% load over a period of 15 min, the jacking of the pile must be continued. When the pile passes this test, it is filled with structural concrete, and a permanent plate bearing firmly on the steel pipe and the concrete is placed on top of the pile. Two hydraulic jacks are set on top of the bearing plate in a manner that a wedging strut can be placed between them.

The pretest load is now applied to the pile (between 150 and 200% of the working load, depending on code requirements). The structure is carefully monitored for vertical movement and signs of distress in the areas of the pretest pile while the load is applied. If no substantial vertical movement occurs while the pressure in the jacks is maintained (taking into account the elastic shortening that will take place) a steel wedging strut with steel plates and one set of steel wedges is placed between the two jacks. The steel wedges are driven home until the pressure in the jacks drops. The pile is now fully completed, except for the concrete encasement around the top of the pile and the wedging strut.

By preloading the pile in the manner described above, the rebound of the pile after removing the jacks will be diminished from as much as ¾ in (19 mm) to between 1⁄16 in (1.5 mm) and 1⁄32 in (0.8 mm).

Bearing Capacity

The bearing capacity of pretest piles depends on the size of pile being used and the bearing material on which the tip of the pile is resting. The capacity may vary from 80 tons (73 t) for piles sitting on a sand and gravel stratum to 120 tons (110 t) and higher for piles bearing on sound rock.

It is very important to have a sufficient number of borings taken in the close vicinity of the structure to be underpinned with pretest piles. If the bearing stratum is fairly thin, it is possible to punch through it by overjacking or overexcavating the piles. The amount of additional pile length to be installed to reach the next bearing stratum, if present at all, could be very costly.

INFLUENCE LINES

The need for underpinning of a structure must be determined by an investigation of the relationship between future subgrade; the bottom elevation of the foundation of the existing structure; the distance between the existing and the new structure; the local soil conditions; and, in case the water table is above the future subgrade elevation, the manner of site dewatering.

The loads in the existing structure must be safely transmitted to a lower soil stratum capable of supporting these loads. The bottom elevation of the underpinning must be at a sufficient depth, so that the general excavation or the excavation for individual column footings or pile caps will not endanger the adjacent structure and deep enough so that the underpinning will not impose surcharge loads onto the future structure. To avoid undermining of installed underpinning during excavation, the bottom of the underpinning must extend at least 6 in (15 cm) below future subgrade (see Fig. 9.14). The meanings of influence lines and slope lines are illustrated in Fig. 9.15.

The angle of the slope line depends on the soil or rock conditions.

When a structure rests on rock or the underpinning is seated on rock, influence lines can normally be neglected. However, if the rock is of poor quality or is disintegrated, the underpinning has to be carried to sound rock or to an elevation determined by an influence line.

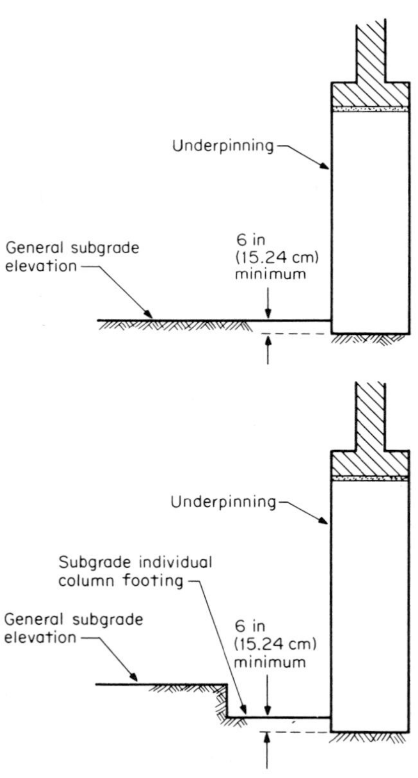

FIG. 9.14 Minimum depths of underpinning.

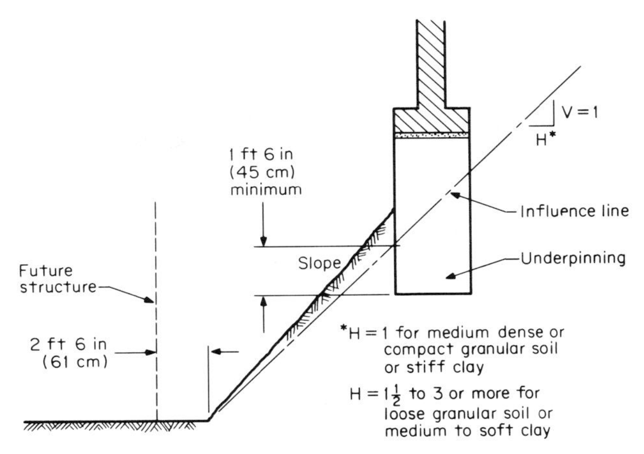

FIG. 9.15 Influence lines in soil.

If the subgrade is below the top of the rock, the quality of the rock face must be closely examined during rock removal. In case faulty rock is discovered, measures must be taken to prevent loss of bearing due to rock slides, by means of rock bolting and/or rock protection.

Figure 9.16 indicates a retainage system, i.e., soldier pile sheeting installed in front of an adjacent structure. Underpinning should be extended below the influence line as shown in the figure.

Figure 9.17 indicates underpinning seated on rock. The slope of the influence line depends on the soundness of the rock formation, which may vary from ver-

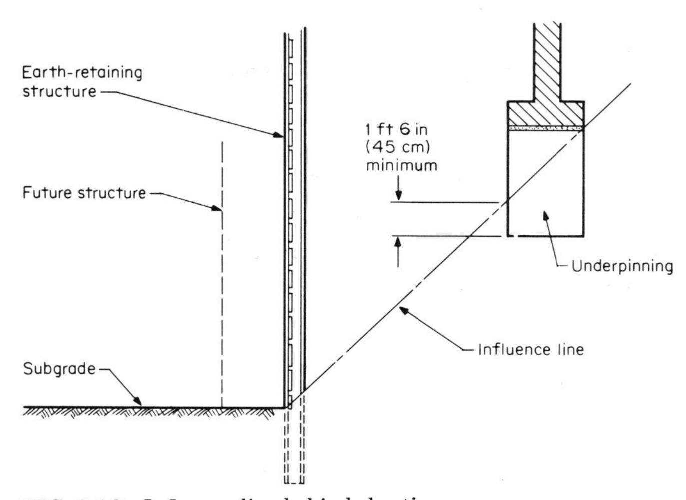

FIG. 9.16 Influence line behind sheeting.

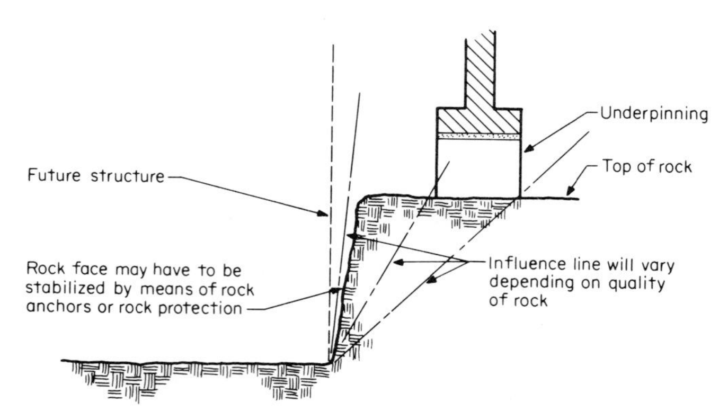

FIG. 9.17 Influence lines in rock.

tical in sound hard rock, to 1:1 in soft disintegrated rock. Rock anchors or rock protection may be required locally or for the full extent of the rock face, depending on the rock condition.

SAFETY PRECAUTIONS

Since, in most underpinning operations, workers are working under an existing building, often in pits dug into the founding layer, safety precautions are very important. Most municipalities include provisions regarding stability of trenches and pits. In most cases, sheeting of pit sides is required. In large pits or trenches, cross-bracing must be provided.

Some of the factors which affect stability of excavation sides are: type and density of soil, elevation of groundwater table, methods of excavation, nearby construction operations, other sources of vibrations, and surcharge loadings. The designer must always examine the applicable codes to be sure that the method chosen meets all code and life-safety requirements.

Type and Density of Soil

Loose sands and soft clays present the greatest design difficulties, since lateral pressures developed are the greatest in these soils. On the other hand, dense sand-gravel mixtures and stiff clays may be self-supporting at near vertical slopes for a considerable time, leading to limited loads on sheeting and bracing.

The presence of organic soil layers may lead to sliding failures of slopes and extremely high lateral pressures.

Elevation of Groundwater Table

This factor is especially important in sandy soils. Excavation below the groundwater table should be avoided, if possible, by dewatering before excavating.

If the excavation is in a tidal area, this fluctuation in the groundwater table may seriously affect dewatering operations, and another type of underpinning may have to be chosen.

Methods of Excavation

In many cases, underpinning excavations are done by hand, which allows sheeting and shoring to closely follow excavation.

If excavation is done by machine, extra care must be exercised that sheeting and shoring are not left behind by the speed of the digging operation. The weight of the machine and the vibrations it causes may, in themselves, cause a slope failure.

Nearby Construction Operations

Some operations which may lead to failures of slopes or high lateral forces on sheeting and bracing are vibrations and loads caused by: pile driving, heavy equipment, dewatering, blasting, and tunneling.

Other Sources of Vibrations

These include subways, railroad lines, highway traffic, and seismic loads.

Surcharge Loadings

The presence of slopes and/or surcharges adjacent to the excavation will also add considerable lateral loads. The effects of all these loadings must be anticipated in proper design of excavation supports.

CONSTRUCTION DOCUMENTS

Drawings for an underpinning operation are in the form of shop drawings. They must be very specific, indicating all the dimensions, materials, and a detailed description of methods to be used. Specifications are often given directly on the drawings as notes.

Usually the engineer of record will review drawings and specifications, but will not assume the responsibility of approving or disapproving them.

In most jurisdictions, the drawings require the seal of a professional engineer, and must be submitted to building officials for review. They will likewise not assume responsibility for the methods and materials used.

RESPONSIBILITY

In most projects, underpinning is a "design-build" operation. This means that the contractor who performs the work either has a professional engineer on staff or engages a professional engineer to do the design. In either case, the contractor assumes the risk that existing structures or utilities may be damaged by the operation. To minimize this risk, the PE needs complete information regarding the existing construction and the site subsoils. This will include a sufficient number of test borings and/or test pits, blueprints (and if possible, shop drawings) of the affected buildings, construction sequences, and any other information which could affect the underpinning operation. If as-built drawings are not available, the PE may need to perform an accurate survey of the buildings.

Before beginning the underpinning operations, a complete survey of existing structures (with photos, sketches, and reference measurements) is absolutely necessary. If this precaution is neglected, the team doing the underpinning may be held responsible for preexisting conditions.

When underpinning is indicated on structural drawings, the matter of responsibility should be carefully spelled out, and unless the structural engineer completely designs the underpinning, the method should not be shown on the drawings. The structural drawings should only indicate the extent of underpinning expected to be required.

INSPECTION

Since the contractor's professional engineer has the responsibility for the design of the underpinning, the PE should inspect it to be sure that it is being installed according to the design and to make sure that actual field conditions are as assumed. In some cases, changes may have to be made to the design, as when soils conditions and/or the existing construction is different from what was assumed.

The responsible PE who personally inspects the operation can spot these differences before they lead to problems and can make necessary changes expeditiously. This will prevent failures and also prevent unnecessary delays to the project.

10

Underground Construction Supports

by Louis J. Porcello
Allan G. Provost
Helmut G. Kobler

Contractors generally utilize a primary support system to control ground movement during the excavation phase of tunnel and shaft construction. Although this system is sometimes referred to as a temporary support, in most cases the primary support remains in place and is incorporated into the final support—usually a concrete lining—regardless of whether the support is in soil or rock. The final support is usually installed after completion of all excavation.

Notable exceptions to this method are:

1. Deep-shaft construction, where, in many parts of the world, the final concrete lining is traditionally advanced so as to remain within a few diameters from the bottom, thus minimizing the amount of temporary support required
2. Subway construction in soft ground, where cast-iron or fabricated-steel liners are utilized as the primary and final lining

3. Tunnels utilizing precast concrete liners installed behind a shield or tunneling machine, which remain as the final lining
4. Other tunneling methods, where an initial support medium, usually shotcrete, used alone or in conjunction with rockbolts, is later incorporated into the final lining.

Until recently, the presence of a temporary support system was sometimes disregarded by the engineer designing the final support. This procedure assumed an extra margin of safety provided by the temporary supports and the extra concrete between them, since the strength of the final lining usually was considered only for the portion inside of the temporary supports. In an attempt to curb the cost of tunnel construction, some designers have considered both support systems as one, specifying, for example, a minimum steel-rib section and spacing and minimum concrete thickness, regardless of whether the temporary steel supports were inside or outside of the minimum concrete line. This method, in effect, specifies the temporary steel-rib size, but allows the constructor to install heavier steel ribs, or resort to closer rib spacing, if ground conditions become worse than assumed by the designer for adequate support. In rockbolt-supported excavations (i.e., for subway stations in rock, mine openings, or powerplant chambers) the thickness of the permanent concrete roof can be minimized by considering the additional support offered by the owner-specified "temporary" tensioned and grouted rockbolts which remain in place after concreting. Thus in certain cases the owner will specify the primary supports, relieving the contractor from the task of designing them.

This text will assist the underground engineer in the design of "temporary"/primary supports when the owner does not specify them.

By necessity, the authors have limited each topic to a brief discussion and a simple design problem, but have referenced each method or theory well enough for the serious reader to follow it up in much greater detail than space allows in this chapter. The discussion of field practice for the installation of each type of support is, again, brief.

CALCULATING ROCK OR SOIL LOADS ON UNDERGROUND OPENINGS

The first and most difficult step in the design of an underground support system is to determine the magnitude of the rock or soil load acting upon the support system. (So as not to confuse the reader, it should be pointed out that the terms "rock load" and "soil load" are synonymous when dealing with loads on underground structures.) The reason for the difficulty in pinpointing an exact rock load are numerous, but are due in part to the fact that (1) the rock is not homogeneous; (2) it contains fractures, joints, and seams; (3) it is altered by weathering and chemical decomposition; and (4) its characteristics may be altered by the presence of water and the location of the water table in relation to the tunnel crown. Even an extensive core-drilling and investigative program may fail to reasonably identify and define the geostructural qualities of the material on which to base the design load calculations.

TABLE 10.1 Deere's Combined Table of Rock Loads and Classifications

Fracture spacing, cm			Rock load H_p, m		Remarks		Rock load H_p, m	Remarks
			Initial	Final				
100		1. Hard and intact	0	0	Generally no side pressure. Erratic load changes from point to point. Lining only if spalling or popping	1—Stable	0 to 0.5	
50	98	2. Hard stratified or schistose	0	0.25B	Spalling common	2—Nearly stable	0.5 to 1	Few rock falls from loosening with time
	95 90	3. Massive moderately jointed	0	0.5B	Side pressure if strata is inclined, some spalling	3—Lightly broken	1 to 2	Loosening with time
20	75	4. Moderately blocky and seamy	0	0.25B to 0.35C		4—Medium broken	2 to 4	Immediately stable; breakup after few months
10	50	5. Very blocky, seamy, and shattered	0 to 0.6C	0.25C to 1.1C	Little or no side pressure	5—Broken	4 to 10	Immediately fairly stable; later rapid breakup
5	25 10	6. Completely crushed		1.1C	Considerable side pressure. If seepage, continuous support required	6—Very broken	10 to 15	Loosens during excavations; local roof falls
	2	7. Gravel and sand	0.54C to 1.2C	0.62C to 1.38C	Dense Side pressure			
2	RQD*		0.96C to 1.2C	1.08C to 1.38C	$\sigma_h = 0.3\gamma\,(0.5H_t + H_p)$ Loose			
		8. Squeezing moderate depth		1.1C to 2.1C	Heavy side pressure	7—Lightly squeezing	15 to 25	High pressures
		9. Squeezing great depth		2.1C to 4.5C	Continuous support required	8—Moderately squeezing	25 to 40	
	Weak and Coherent	10. Swelling		up to 80 m	Use circular support. In extreme cases use yielding support.	9—Heavy squeezing	40 to 60	Very high pressures
		Terzaghi (1946): 1. For rock classes 4, 5, 6, 7 when above groundwater levels, reduce loads by 50% 2. For sands (7), $H_{p\,min}$ is for small movements ($-0.01C$ to $0.02C$); $H_{p\,max}$ for large movements ($0.15C$) 3. R is tunnel width $C = B + H_t$ = width + height of tunnel. I or circular tunnel, $H_t = 0$‡				Stini (1950) Note: Loads are for 5-m-wide tunnel For L meter wide tunnel $H_p = H_{p5\,m}(0.5 \text{ to } 0.1\,L)$		

*RQD is a percentage of core that is recovered in pieces more than 4 in (0.1 m) in length (per run of core).

‡The authors caution the reader that the designer, by setting $H_t = 0$ for a circular tunnel, can arbitrarily cut the design rock load in half. This may hold true for a homogenous soil which may distribute its load more uniformly on a circular opening (as compared to a horseshoe-shaped opening). If however, very block ground is encountered, the circular opening may offer no better load-distribution advantages over those of a noncircular opening.

Various geotechnical engineers have tabulated guidelines for estimating rock loads on tunnels. A widely used rock-load analysis in North America is Terzaghi's[1] table of loads for nine different rock conditions. Other rock and soils engineers, such as Stini[2] and Bierbaumer,[3] have also published guidelines for rock-load estimating. Deere et al.[4] published a combined table of five separate

	Rock load H_p, m Initial/Final	Side pressure, m Initial/Final	Invert pressure, m		
				A—Stable	Sound
	Little loosening			B—Unstable after long time	
Slightly broken	0/3 to 4	0/0	0	C—Unstable after short time	Sound stratified or schistose (some fissures?)
				D—Broken	Strongly fissured
Very broken	Loosening with time 3/11 to 13	0/1	1–2	E—Very broken	Fully mechanically disturbed
Extremely broken	Roof falls, loosening at time of excavation 5 to 10/11 to 15	2 to 4/2 to 6	4		Gravel and sand
				F—Squeezing	Pseudo-sound rock (properties change with time)
Soft, squeezing, or flowing; moderate depth	10 to 13/15 to 25	4/4	6		Some squeezing (genuine rock pressures), small overburden
Soft, heavy, squeezing; great depth	15 to 25/47 to 75	8/6	12	G—Heavy squeezing	Heavy squeezing; large overburden Swelling Silt, clay
Bierbaumer (1913) and others (ref. Bendel) (1948)				Lauffer (1958) Note: This classification is correlated with stand-up time	Rabcewicz (1957) Note: This classification has been used for evaluating feasibility of rockbolt types

rock-load classifications, with each compared to Deere's rock-quality designation (RQD) and fracture-spacing index. The combined table includes Terzaghi's original load table, but adds a tenth condition (for sand and gravel) and modifies the loads for circular tunnels. The combined tabulation is shown in Table 10.1.

The questions that initially confront the user of this table are how to identify the correct rock condition and whether to use the high or low end of the rock-load range for the rock condition once it is identified. The greatest error in support design would most likely result from improper assumption of rock load, rather than from an approximation of a particular stress-distribution method. It is therefore important that the designer of temporary supports acquire all available geotechnical information about the project. Many civil works projects are designed after extensive core boring and rock/soil testing is accomplished by a consultant engineering firm. These geotechnical reports (which are the basis of the final lining design in most cases) are public information and are available to all contractors bidding the work. Many of these reports contain a narrative section which may discuss similar projects performed in the area, highlighting the successes (or failures) of the construction procedures and support design. References to other reports also available to the interested tunneler are usually listed. In many cases, empirical field data from nearby tunnel projects, or projects in similar type rock or soil, will help the tunnel constructor in determining whether the support should be steel liner plate, steel ribs, shotcrete, rockbolts, or some combination of these, along with determining the size, thickness, or spacing of the support. The time spent in reviewing this information, or even the cost to employ a geotechnical consultant to recommend a safe but economical temporary support system, is usually less than the cost of a few days' production loss due to failure of poorly designed temporary supports.

Sample Problem (Worked in U.S. Customary Units Only)

Compute the rock load for moderately blocky and seamy rock on a 20-ft-wide × 20-ft-high horseshoe-shaped tunnel above groundwater level; rock density = 160 lb/ft^3. The following methods are used:

1. Terzaghi's method
2. Stini's method
3. Bierbaumer's method

A more detailed explanation of each of these methods follows:

1. Using condition no. 4 of Terzaghi's method:

$$\text{Initial load} = 0 \qquad \text{Side pressure} = 0$$

$$\text{Final rock load} = 0.25B \text{ to } 0.35C$$

where $B = 20$ ft
$C = 20 + 20 = 40$ ft

Therefore:

$$\text{Final rock load (min.)} = 0.25 \times 20 = 5 \text{ ft}$$

$$\text{Final rock pressure (min.)} = 5 \text{ ft} \times 160 \text{ lb/ft}^3 = 800 \text{ lb/ft}^2$$

Reduced by 50% (above water table, see note 1 in Table 10.1),

$$\text{Final rock pressure (min.)} = 400 \text{ lb/ft}^2$$

$$\text{Final rock load (max.)} = 0.35 \times 40 = 14 \text{ ft}$$

$$\text{Final rock pressure (max.)} = 14 \text{ ft} \times 160 \text{ lb/ft}^3 = 2240 \text{ lb/ft}^2$$

Reduced by 50% (above water table, see note 1 in Table 10.1),

$$\text{Final rock pressure (max.)} = 1120 \text{ lb/ft}^2$$

2. Using condition no. 4 of Stini's method, which is given in Table 10.1:

$$\text{Rock load for 5-m tunnel } H_{p5m} = 2 \text{ to } 4 \text{ m}$$

Adjust for the 20-ft (6.1-m) diameter by using Stini's equation

$$H_p = H_{p5m}[0.5 + 0.1\,(6.1)] = 1.11 H_{p5m}$$

Therefore

$$\text{Minimum load} = 1.11 \times 2 = 2.22 \text{ m} = 7.28 \text{ ft}$$

$$\text{Minimum pressure} = 7.28 \text{ lin ft} \times 160 \text{ lb/ft}^3 = 1165 \text{ lb/ft}^2$$

$$\text{Maximum load} = 1.11 \times 4 = 4.44 \text{ m} = 14.6 \text{ ft}$$

$$\text{Maximum pressure} = 14.6 \text{ lin ft} \times 160 \text{ lb/ft}^3 = 2330 \text{ lb/ft}^2$$

3. Using the "slightly broken" condition of Bierbamer's method:

$$\text{Initial rock load} = 0 \qquad \text{Side pressure} = 0$$

$$\text{Final rock load} = 3 \text{ to } 4 \text{ m}$$

$$\text{Minimum load} = 3 \text{ m} = 10 \text{ ft}$$

$$\text{Minimum pressure} = 10 \text{ ft} \times 160 \text{ lb/ft}^3 = 1600 \text{ lb/ft}^2$$

$$\text{Maximum load} = 4 \text{ m} = 13.2 \text{ ft}$$

$$\text{Maximum pressure} = 13.2 \text{ ft} \times 160 \text{ lb/ft}^3 = 2112 \text{ lb/ft}^2$$

A comparison of the three methods is given in the accompanying table.

	Terzaghi*	Stini	Bierbaumer
Initial rock pressure, lb/ft^2	0	N/A†	0
Minimum final rock pressure, lb/ft^2	800*	1165	1600
Maximum final rock pressure, lb/ft^2	1120*	2330	2112
Side pressure	0	N/A†	0

*Note that if Terzaghi's analysis were for a tunnel below the water level, minimum and maximum final loads would be 1600 lb/ft^2 and 2240 lb/ft^2 respectively, which would be much closer to Stini's and almost identical with Bierbaumer's analyses.

†NA means *not available.*

TIME AVAILABLE TO INSTALL SUPPORTS

An important factor that must be considered in the design of temporary supports is the time required to install them after the ground has been excavated. This time is referred to as standup time. (In his earlier papers on tunnel load analysis, Terzaghi referred to this time as bridge-action period.) A short standup time may dictate the type of support regardless of any structural design analysis to the contrary. For example, if a designer, through rock-loading and support-design analysis, determines that 31-lb (14-kg) wide-flange steel-rib sets, on 4-ft (1.2-m) centers, are adequate to support a given rock load, the designer must also determine if there is adequate standup time to excavate 4 or 5 ft (1.2 or 1.5 m) ahead of the lead support and install the new support before rock fallout starts to occur.

A qualitative definition of standup time was put forth by Lauffer[5] in 1958, who defined the active span to be supported as the smaller of either the tunnel diameter or the distance between the new face and the last-installed support. Seven types of rock masses were charted (see Fig. 10.1), with type A the best, type G the worst.

Standup time is important to the tunnel or shaft contractor because the longer the standup time, the easier it is for the work crew to install the supports in an organized manner, since they can be farther away from the working face and its inherent congestion. In drill-and-blast headings with steel-rib supports, the farther the ribs are from the working face, the easier it will be to position the machines used to drill the perimeter blast holes, and the result will usually be a tunnel with less overexcavation, or "overbreak"; contractors will thus be rewarded by removing less muck, and if a final lining is required, they can also reduce their overall concrete costs.

In summary, contractors utilize the standup time available to increase their advance rates and decrease their overbreak by more effectively utilizing crews and equipment.

The support-design methods outlined here do not consider the standup time allowed to install the supports. The two Commercial Shearing, Inc. texts,[6,7] as well as Szechy's text,[8] outline various techniques, such as compressed air, shield-driven tunneling, well-point pumping, or the installation of spiling or breast boards that are used by the tunneler to lessen ground movement in poor ground

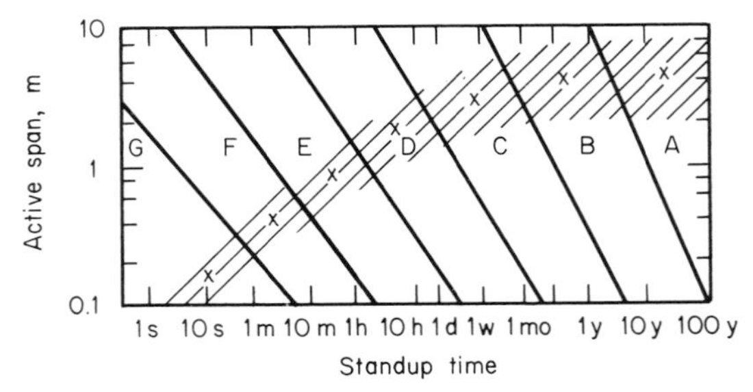

FIG. 10.1 Lauffer's standup time for seven different rock types.

until the temporary support can be installed. It will be up to the judgment of the support designer to determine if the ground conditions will permit the support to be installed safely prior to the occurrence of excessive ground movement.

EMPIRICAL ROCK LOAD AND SUPPORT DETERMINATION BY RSR METHOD

Wickham and Tiedemann[9] developed an empirical method for determining tunnel rock loads and support requirements, utilizing support data from 53 tunnel projects (primarily located in the western United States) and correlating the load and support requirement of each project with the characteristics of the rock being supported. Using three parameters, the characteristics of the rock being supported are graded to provide a rock structure rating (RSR).

The method first determines the value of parameters *A* and *B*, and then using the sum of *A* and *B*, determines the value of parameter *C*.

Parameter *A* considers the general area geology; its values are tabulated in Table 10.2.

Parameter *B* considers the joint direction and spacing in relation to the direction the tunnel is being driven; its values are tabulated in Table 10.3.

Parameter *C* considers the effect of groundwater inflow and joint condition as compared with the sum of parameters *A* and *B*; its values are tabulated in Table 10.4.

The RSR is the sum of these three parameters.

After calculating the RSR, the user can determine the rock load on the tunnel arch by referring to Table 10.5, or can utilize the rib-ratio/RSR relationship to

TABLE 10.2 Rock Structure Rating, Parameter *A*, General Area Geology

Maximum value = 30

	Geological structure			
Basic rock type*	Massive	Slightly faulted or folded	Moderately faulted or folded	Intensely faulted or folded
1	30	22	15	9
2	27	20	13	8
3	24	18	12	7
4	19	15	10	6

*The basic rock types are:

	Hard	Medium	Soft	Decomposed
Igneous	1	2	3	4
Metamorphic	1	2	3	4
Sedimentary	2	3	4	4

TABLE 10-3 Rock Structure Rating, Parameter *B* Joint Pattern—Direction of Drive

Maximum value = 45

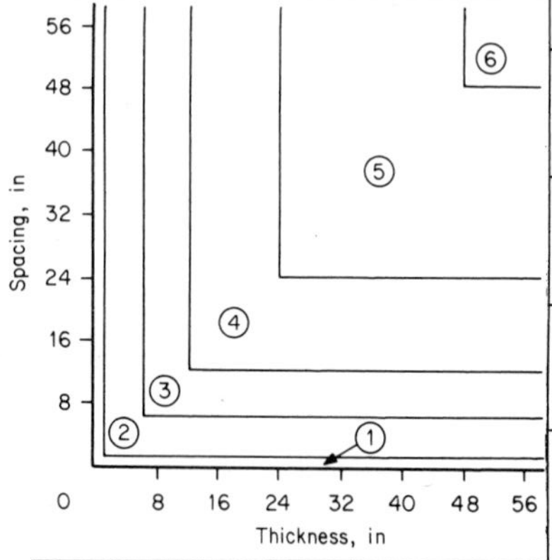

Spacing, in	Strike perpendicular to axis					Strike parallel to axis		
	Direction of drive					Direction of drive		
	Both	With dip		Against dip		Both		
	Dip of prominent joints					Dip of prominent joints		
Thickness, in	Flat	Dipping	Vertical	Dipping	Vertical	Flat	Dipping	Vertical
1 Very closely jointed	9	11	13	10	12	9	9	7
2 Closely jointed	13	16	19	15	17	14	14	11
3 Moderately jointed	23	24	28	19	22	23	23	19
4 Moderate to blocky	30	32	36	25	28	30	28	24
5 Blocky to massive	36	38	40	33	35	36	34	28
6 Massive	40	43	45	37	40	40	38	34

NOTES: Flat, 0–20°; dipping, 20–50°; vertical, 50–90°.

TABLE 10.4 Rock Structure Rating, Parameter C, Groundwater Joint Condition

Maximum value = 25

Anticipated water inflow, gpm/1000 ft	Sum of parameters $A + B$					
	13 to 44			45 to 75		
	Joint condition					
	Good	Fair	Poor	Good	Fair	Poor
None	22	18	12	25	22	18
Slight (<200 gpm)	19	15	9	23	19	14
Moderate (200–1000 gpm)	15	11	7	21	16	12
Heavy (>1000 gpm)	10	8	6	18	14	10

Joint Condition: Good = tight or cemented; Fair = slightly weathered or altered; Poor = severely weathered, altered, or open.

arrive directly at a tunnel rib size and spacing. The rib ratio is a ratio of the desired rib size and spacing, as it compares to a datum rib size and spacing designed for a tunnel of the same diameter subjected to a Terzaghi's[1] condition 7 load of 1.38 $(B + H)$ below the water table with a soil density of 120 lb/ft^3 (1920 kg/m^3). Theoretical spacing for this datum condition can be calculated by using Proctor and White's[6] table of load capacities for various rib sizes and tunnel diameters, assuming a maximum fiber stress of 27,000 lb/in^2 (18,620 N/cm^2)

TABLE 10.5 Correlation of Rock Structure Rating to Rock Load and Tunnel Diameter

Tunnel diameter D, ft	Rock load on tunnel arch W_r, kips/ft^2											
	0.5	1.0	1.5	2.0	3.0	4.0	5.0	6.0	7.0	8.0	9.0	10.0
	Corresponding values of rock structure ratings (RSR)											
10	62.5	49.9	40.2	32.7	21.6	13.8						
12	65.0	53.7	44.7	37.5	26.6	18.7						
14	66.9	56.6	48.3	41.4	30.8	22.9	16.8					
16	68.3	59.0	51.2	44.7	34.4	26.6	20.4	15.5				
18	69.5	61.0	53.7	47.6	37.6	29.9	23.8	18.8				
20	70.4	62.5	55.7	49.9	40.2	32.7	26.6	21.6	17.4			
22	71.3	63.9	57.5	51.9	42.7	35.3	29.3	24.3	20.1	16.4		
24	72.0	65.0	59.0	53.7	44.7	37.5	31.5	26.6	22.3	18.7		
26	72.6	66.1	60.3	55.3	46.7	39.6	33.8	28.8	24.6	20.9	17.7	
28	73.0	66.9	61.5	56.6	48.3	41.4	35.7	30.8	26.6	22.9	19.7	16.8
30	73.4	67.7	62.4	57.8	49.8	43.1	37.4	32.6	28.4	24.7	21.5	18.6

TABLE 10.6 Theoretical Spacing S_d, ft, of Typical Rib Sizes for Datum Condition

	Tunnel diameter, ft										
Rib size	10	12	14	16	18	20	22	24	26	28	30
4I7.7	1.16										
4H13.0	2.01	1.51	1.16	0.92							
6H15.5	3.19	2.37	1.81	1.42	1.14						
6H20		3.02	2.32	1.82	1.46	1.20					
6H25			2.86	2.25	1.81	1.48	1.23	1.04			
8W-31				3.24	2.61	2.14	1.78	1.51	1.29	1.11	
8W-40					3.37	2.76	2.30	1.95	1.67	1.44	1.25
8W-48						3.34	2.78	2.35	2.01	1.74	1.51
10W-49								2.59	2.22	1.91	1.67
12W-53										2.19	1.91
12W-65											2.35

in the rib flange. This datum spacing is shown for some typical sizes of ribs and tunnel diameters in Table 10.6.

Actual rib spacing for the tunnel being investigated would be determined by the equation

$$S_a = \frac{S_d}{\text{RR}} \times 100$$

where S_a = desired rib spacing for the tunnel under investigation
S_d = theoretical rib spacing for the datum tunnel
RR = rib ratio

Note that under no circumstances should S_a exceed 7 ft (2.1 m).

Rib ratio is determined from the empirical field data collected by Wickham and Tiedemann and can be calculated using the empirical relationship

$$(\text{RR} + 80)(\text{RSR} + 30) = 8800$$

Sample Problem (Worked in U.S. Customary Units Only)

Diameter of tunnel: 20 ft

General geology: Soft sedimentary, moderately folded

Joint pattern: Moderately jointed, with strike perpendicular to tunnel axis; dip 35° with direction of drive

Joint condition: Severely weathered (poor)

Anticipated water inflow: 300 gal/min per 1000 ft of tunnel

From Table 10.2:

$$\text{Parameter } A = 10$$

From Table 10.3:

$$\text{Parameter } B = 24$$

$$A + B = 10 + 24 = 34$$

From Table 10.4:

$$\text{Parameter } C = 7$$

$$\text{RSR} = A + B + C = 10 + 24 + 7 = 41$$

From the empirical relationship

$$(\text{RR} + 80)(\text{RSR} + 30) = 8800$$

the rib ratio is found to be 44.

From Table 10.6 for the 20-ft diameter, we have

$$S_d = \text{W8} \times 31 \text{ at 2.14-ft spacing}$$

$$S_a = \frac{S_d}{\text{RR}} \times 100 = \frac{2.14}{44} \times 100 = 4.87 \text{ ft}$$

Therefore, we can anticipate using a W8 × 31 at 5-ft spacing.

What is the rock load on the tunnel arch? From Table 10.5:

20-ft radius and RSR = 40.2 gives a rock load of 3.0 kips/ft^2

20-ft radius and RSR = 49.9 gives a rock load of 2.0 kips/ft^2

Therefore, by interpolation, the rock load for an RSR of 41 = 2.92 kips/ft^2.

SOFT-GROUND TUNNEL SUPPORTS

Pressed-Steel Liner Plate Supports

General

Pressed-steel liner plates are generally utilized in conventionally excavated soil or weak rock tunnels. Liner-plate manufacturers suggest that they may be utilized in tunnels up to 14 ft (4.2 m) in diameter without additional support. In larger tunnels, steel ribs can be utilized to provide extra support, but the liner plates then act as lagging and are normally of lighter gauge. The design portion of this section illustrates the use of liner plates as the sole primary tunnel support.

Design

Guidelines for design of cold-formed-steel tunnel-liner plate supports in soil are set forth in the AASHTO design manual, section 13.[10] This specification uses Marston's formula to determine the dead soil load. Additionally, live loads due to H20 loadings are tabulated. AISI publishes its own design manual[11] which adheres to the AASHTO criteria, but also tabulates live loads due to Cooper's E80 railroad loading (see Table 10.7). Live loads are generally considered as a negligible portion of the total load at depths of more than 10 ft (3 m) above the tunnel crown for H20 loads and 30 ft (9.1 m) for E80 loads. Marston's formula calculates the dead load P_d per foot (per 0.3 m) of tunnel as follows:

$$P_d = C_d WD$$

where W = moist unit weight of soil
D = diameter of tunnel
C_d = coefficient from Fig. 10.2 where H/D = ratio of overburden to tunnel diameter and ϕ = soil friction angle

In examining Fig. 10.2 note that AASHTO specifies friction angles for each soil type. Also note that the values of C_d become asymptotic at values of H/D higher than 10 to 15. AASHTO therefore allows us to disregard soil loads above 1.5 tunnel diameters in granular soils; above 2.25 tunnel diameters in saturated silt and clay; and above approximately 3.5 tunnel diameters in saturated clays. (The designer who is skeptical of such arbitrary assumptions of soil loading can, of course, use the loads recommended by Deere in Table 10.1.)

The AASHTO design procedure is as follows:

1. Determine the load on the tunnel (sum of live plus dead loads).
2. Initially determine the plate thickness required to resist the load thrust (joint or seam strength).

TABLE 10.7 Highway and Railroad Live Loads for Selected Heights of Cover

Height of cover, ft	Highway H20 loading including impact, lb/ft²	Railroad (Cooper's) E80 loading including impact, lb/ft²
2	N/A	3800
4	375	N/A
5	260	2400
6	190	N/A
7	140	N/A
8	110	1600
9	90	N/A
10	75	1100
12	N/A	800
15	N/A	600
20	N/A	300
30	N/A	100

SOURCE: *Handbook of Steel Drainage and Highway Construction Products,* 2d ed., American Iron and Steel Institute, 1971.

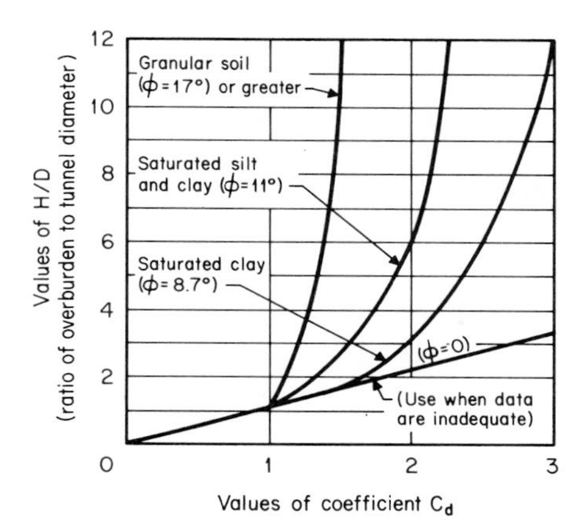

FIG. 10.2 Values of C_d for various types of soils; ϕ = friction angle of soil. (*From Sec. 13, Design Manual, American Association of State Highway and Transportation Officials, 1970.*)

3. Check the moment of inertia for minimum stiffness and adjust plate thickness if necessary.

4. Check the minimum plate area to resist buckling and adjust if necessary.

Sectional properties per inch of plate width are specified by AASHTO and are shown in Table 10.8. These properties are to be used in the four-step design procedure. Manufacturer's properties may vary slightly due to fabrication variances.

The seam strength required must be high enough to withstand the thrust on the plate multiplied by a safety factor of 3, where

$$\text{Thrust} = \frac{\text{total load} \times \text{diameter of tunnel}}{2}$$

The section must then be examined to determine if it can attain a minimum stiffness to allow for unbalanced loads due to imperfect installation procedures, i.e., local sloughing, grout pressures, or concentrated loads.

$$\text{Minimum stiffness} = \frac{EI}{D^2} \quad \text{lb/in (N/cm)}$$

where I = moment of inertia, in^4/in (cm^4/cm), from Table 10.8
E = modulus of elasticity for steel, 29×10^6 lb/in^2 (20×10^6 N/cm^2)
D = diameter of tunnel, in (cm)

Minimum required stiffness for

Four-flange plate = 111 lb/in (194 N/cm)

Two-flange plate = 50 lb/in (88 N/cm)

TABLE 10.8 Sectional Properties of Steel Liner Plate

Plate thickness		Ultimate design longitudinal joint (seam) strength, kips/ft		Effective area, in^2/in		Moment of inertia, in^4/in	
Gauge	In	Two flange	Four flange	Two flange	Four flange	Two flange	Four flange
14	0.075	20.0	N/A	0.096	N/A	0.034	N/A
12	0.105	30.0	26.0	0.135	0.067	0.049	0.042
11	0.119	N/A	N/A	N/A	0.076	N/A	0.049
10	0.135	47.0	43.0	0.174	0.085	0.064	0.055
8	0.164	55.0	50.0	0.213	0.105	0.079	0.070
7	0.179	62.0	54.0	0.233	0.114	0.087	0.075
5	0.209	87.0	67.0	0.272	0.132	0.103	0.087
3	0.239	92.0	81.0	0.312	0.150	0.118	0.120
¼	0.250	N/A	N/A	N/A	0.155	N/A	0.101
5⁄16	0.312	N/A	115.0	N/A	0.193	N/A	0.123
⅜	0.375	N/A	119.0	N/A	0.230	N/A	0.143

SOURCE: Sec. 13, *Design Manual,* American Association of State Highway and Transportation Officials, 1970.

If minimum stiffness cannot be attained for the initial section chosen, choose a heavier section accordingly.

In order to check for wall buckling, the liner's critical diameter D_c must first be determined.

$$D_c = \frac{r}{k}\sqrt{\frac{24E}{f_u}}$$

where r = radius of gyration of section, in/ft of length
k = constant varying from 0.22 for soils with $\phi > 15°$ to 0.44 for $\phi < 15°$ in U.S. customary units*
f_u = minimum specified ultimate tensile strength, 42,000 lb/in^2 (29,000 N/cm^2)
D_c = critical pipe diameter, in (cm)

If the actual tunnel-lining diameter is greater than the critical diameter D_c, then the buckling stress is

$$f_c = \frac{12E}{(Dk/r)^2}$$

If the actual tunnel-lining diameter is less than the critical diameter D_c, then the buckling stress is

$$f_c = f_u - \frac{f_u^2}{48E}\left(\frac{kD}{r}\right)^2$$

If f_c is more than the 28,000-lb/in^2 (19,300-N/cm^2) allowable yield strength specified by AASHTO, the new minimum required wall area A can be calculated by using the formula

$$A = \frac{\text{thrust} \times (\text{factor of safety for buckling} = 2)}{f_c}$$

and a new selection is chosen to meet the area requirement of this specification.

It should be pointed out that these formulas are based upon reasonable installation procedures to limit deflection or flattening of the completed liner plate to 3% of the nominal diameter. Also, if grouting pressures are anticipated to be larger than the computed external load, the grouting pressure should be used to calculate the section required.

Sample Problem (Worked in U.S. Customary Units Only)

Diameter of tunnel: 10 ft 6 in

Weight of soil: 140 lb/ft^3

*If designers are a bit wary of assigning such a wide-ranging value of k in calculating the liner's critical diameter, they may prefer to calculate the buckling strength by using Allgood's[7a] approach, which considers the soil's Poisson's ratio and the depth of cover over the structure.

Type of soil: granular, $\phi > 17°$

Liner plate type: four flange, 16-in length

Live load: AASHTO H20

Overburden height: 50 ft

From Table 10.7 we can see that an H20 loading has negligible effect on depths over 10 ft.

Solution

1. $H/D = 50/10.5 = 4.76$; from Fig. 10.2; $C_d = 1.4$
2. $PD = C_d WD = 1.4(140)(10.5) = 2{,}058 \text{ lb/ft}^2$
3. Thrust $= PD/2 = 2{,}058 \times 10.5/2 = 10{,}805$ lb/ft
4. Seam strength required = thrust × safety factor = $10{,}805 \times 3 = 32{,}415$ lb/ft

From Table 10.8, the minimum plate thickness for four-flange plate is 0.135 in (10 ga). The minimum stiffness for a four-flange plate is 111.

$$111 = \frac{EI \text{ minimum}}{D^2 \text{ (in)}}$$

$$= \frac{29 \times 10^6 I \text{ minimum}}{(126)^2}$$

$$I \text{ minimum} = 0.061 \text{ in}^4\text{/in}$$

which is larger than

$$I \text{ (10 ga)} = 0.055 \text{ in}^4$$

Therefore, use 8 gauge with

$$I = 0.070 \text{ in}^4$$

5. Check wall buckling.

$$\text{Critical diameter} = \frac{r}{k} \sqrt{\frac{24E}{f_u}}$$

where r = effective radius of gyration of four-flange, 8-gauge plate (use 0.43)
$k = 0.22$ for $\phi > 15°$
$f_u = 42{,}000 \text{ lb/in}^2$

Therefore

$$\frac{r}{k} \sqrt{\frac{24E}{f_u}} = \frac{0.43}{0.22} \sqrt{\frac{(24)29 \times 10^6}{42{,}000}} = 252 \text{ in}$$

Since actual tunnel diameter = 126 in < 252 in,

$$f_c = f_u - \frac{f_u^2}{48E} \times \left(\frac{kD}{r}\right)^2$$

$$= 42{,}000 - \frac{(42{,}000)^2}{48(29 \times 10^6)}\left(\frac{0.22 \times 126}{0.43}\right)^2$$

$$= 36{,}734 \text{ lb/in}^2$$

which is larger than 28,000 lb/in^2 allowable yield strength. Therefore

$$\text{Area required per foot} = \frac{\text{thrust}}{28{,}000} \times (\text{SF} = 2)$$

$$= \frac{10{,}805 \times 2}{28{,}000} = 0.7718 \text{ in}^2/\text{ft} = 0.0643 \text{ in}^2/\text{in}$$

This compares with area of 8 gauge:

$$\text{Area} = 0.105 \text{ in}^2/\text{in}$$

Thus, we can use 8-gauge plate, which meets all AASHTO criteria.

Installation of Liner Plates

In soft ground tunnels, the plates are generally installed close to the working face by the same crew that excavates the tunnel. The erection and bolting take place in a crowded and dirty environment using only the miners' muscles, aligning pins, and spud wrenches. At most, the only mechanical aid may be an occasional air-operated impact wrench. Flange holes are generally oversized to allow for alignment errors and the bolts and nuts have coarse, quick-acting threads, so they can be tightened even if coated with grime. The longitudinal joints or seams are alternately staggered to minimize weakness planes in that direction. The plates are 37$^{11}/_{16}$ in, or π feet (95.7 cm), in length; therefore, a 10-ft (3-m)-diameter tunnel lining will consist of 10 plates in its perimeter.

Figure 10.3 illustrates the installation of four-flanged liner plates on a California tunnel project. The second row of plates has a plate at spring line containing a threaded grout hole fitted with a pipe plug. Plates should be backpacked with pea gravel and then grouted to limit deflection to within 3% of nominal tunnel diameter. A certain percentage of plates should therefore be purchased with grout fittings. Backpacking normally starts at the invert, so that the completed ring does not sag or deflect under loads from the mucking or haulage equipment. It is difficult to backpack the sidewalls and arch until enough liner plate is installed so that the gravel or grout does not run toward the "free" end of the lining at the working face. In stiff soils it is sometimes advantageous to pack and grout in stages as the tunnel advances, and straw or excelsior is sometimes used to pack the free end of the rings to minimize grout losses.

Poor ground can sometimes be controlled by installing arch plates one ring

FIG. 10.3 Installation of pressed-steel liner plates. *(Commercial Shearing, Inc.)*

ahead of the remaining plates and by maintaining a sloping work face during the advance. If conditions warrant further arch support, commercially manufactured hydraulic poling plates, as illustrated in Figs. 10.4 and 10.5, may be utilized to contain the crown until plates can be installed. Some contractors utilize job-fabricated devices to control the crown in a similar fashion.

The engineer on a tunnel project utilizing liner-plate supports is responsible for certain functions to assure a safe and orderly advance as well as the efficient utilization of workers and materials.

1. The engineer should maintain an accurate inventory of liner plates delivered to the jobsite and update orders to compensate for loss or damage to the plates. This is especially important for elliptically shaped tunnels or for tunnels with curves in their alignment, which will require liner plates with more than one shape. The engineer should also maintain an adequate supply of nuts and bolts to compensate for normal losses.

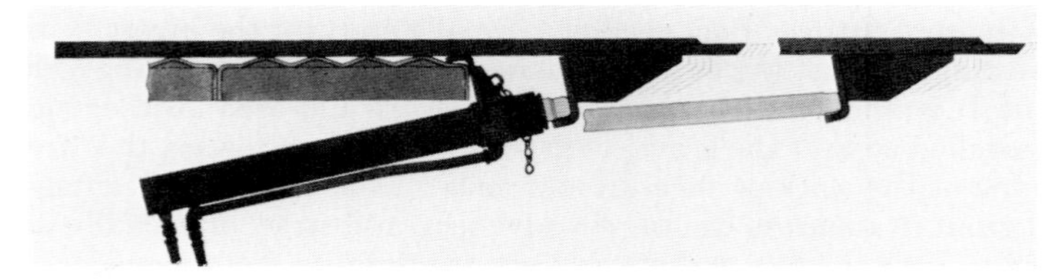

FIG. 10.4 Schematic diagram of a hydraulic cylinder shoving a poling plate forward from a liner-plate ring. *(Commerical Shearing, Inc.)*

FIG. 10.5 Using hydraulically actuated poling plates on a French tunnel project. *(Commercial Shearing, Inc.)*

2. Plates should be plainly marked and staged on flat cars prior to being sent to the heading. This minimizes confusion and delays by the erection crew. The proper number of plates with grout plugs should be marked and supplied when needed. Plates with grout plugs are more expensive than regular plates and should be used according to the engineer's or superintendent's schedule, not the miners'.
3. Lasers or other alignment aides should be checked frequently and advanced with the heading.
4. Accurate measurements of horizontal and vertical inside dimensions of the installed plates should be taken and recorded. Measurements within a few diameters of the face should be taken daily, farther back, at less frequent intervals or as required by field conditions.
5. A record of each quadrant grouted, along with its tunnel station, should be maintained to ensure that proper backpacking is being carried out.

Steel Tunnel Ribs

Design of Horseshoe-Shaped Steel Tunnel Ribs

One of the more widely used methods of designing a steel tunnel support is suggested by Proctor and White in chap. 11 of *Rock Tunneling With Steel Supports*.[6] This book is available at a reasonable cost from Commercial Shearing, Inc., and should be a part of the tunneler's library. Some assumptions used by Proctor and White are outlined below:

1. The rock load is assumed to be an inverted U-shaped load bearing on an inverted U-shaped tunnel crown. In order to simplify the calculations, only one-half of the arch is considered, as in Fig. 10.6.
2. Blocking points are spaced in a uniform radial pattern from arch spring line to crown. The uniform radial spacing results in a vertical load W at each

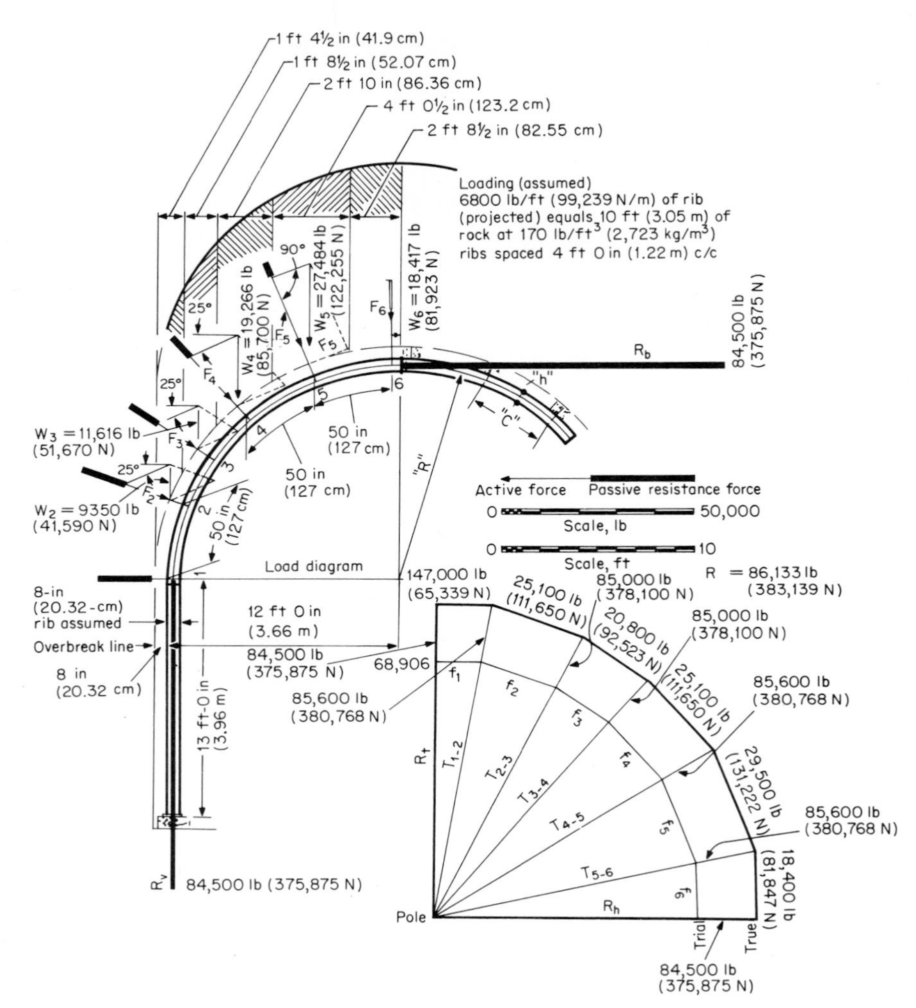

FIG. 10.6 Proctor and White's steel-rib loading and force polygon. *(Commercial Shearing, Inc.)*

blocking point, which is not uniform, but depends on the horizontal thickness of each rock block.

3. Vertical loads are resolved graphically into radial and tangential components, as illustrated in Fig. 10.7. Due to friction, the direction of the tangential component can never have a direction of more than 25° from the horizontal.

4. Moments in the curved rib are calculated by multiplying the axial force (T = thrust) in a particular section of the rib by h, or the distance between the

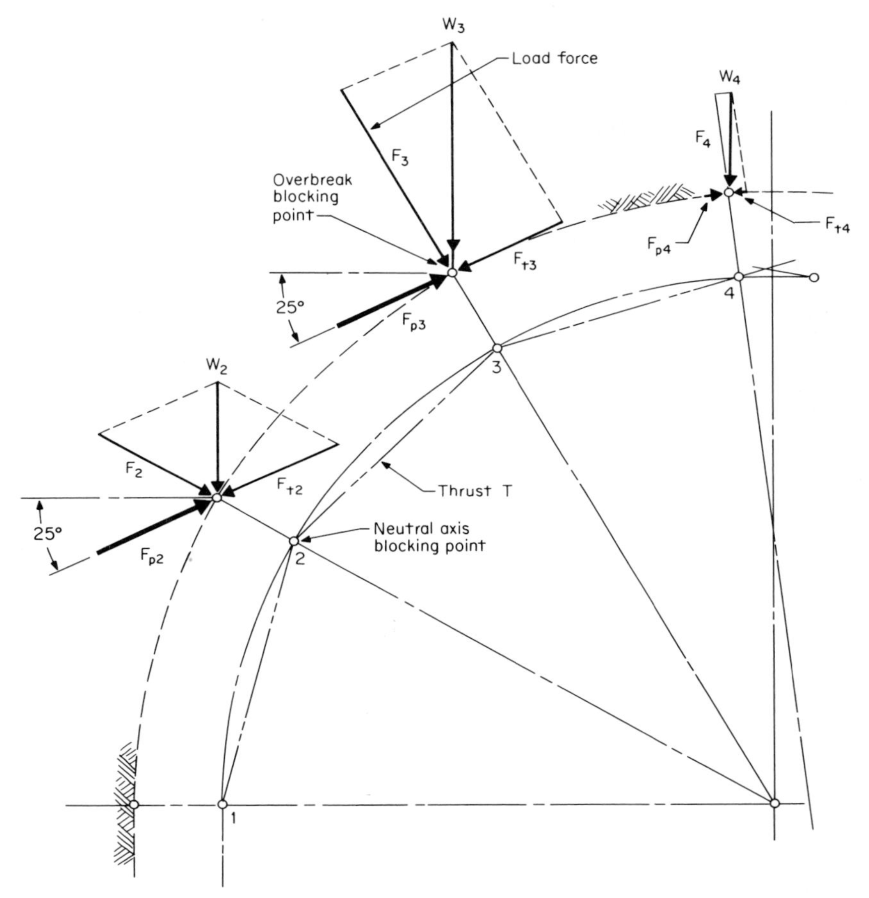

FIG. 10.7 Proctor and White's resolution of forces at overbreak blocking points. *(Commercial Shearing, Inc.)*

chord drawn between two adjacent blocking points and the neutral axis of the rib section, as illustrated in Fig. 10.8.

5. Since the arch approximates a curved, continuous beam, actual moments are somewhat less than $T \times h$, and reduction factors are given for the two-piece rib set as follows:

   a. Maximum moment, which is normally located near the arch splice, may be reduced by using a multiplier of 0.86.

   b. Moments at blocking points may be reduced by using a multiplier of 0.67. (This is the multiplier used to check moment in the straight portion, or leg, of the rib set.)

Expedient Design Analysis of Circular Steel Ribs for Shield-Driven Tunnels

Many analyses have been carried out for design of permanent circular tunnel linings. Szechy[8] reviews about 10 such analyses, but the only one in which he considers the hinge effect of bolted segments is the Hewett-Johannesson method[12] originally published in 1922. This analysis takes into account the weight of the segments, the vertical and horizontal pressures acting on the tunnel due to soil load, the effect of the water level in the soil, and the effect of compressed air if it is driven as a pressure tunnel. The analysis assumes segments of the cast-iron or fabricated-steel type, bolted rigidly in the longitudinal direction, and assumed as a four-hinged circle (as in Fig. 10.9*a*) in the transverse direction. None of these analyses addresses the condition faced by the tunnel contractor installing sectional steel ribs in a circular tunnel in soil.

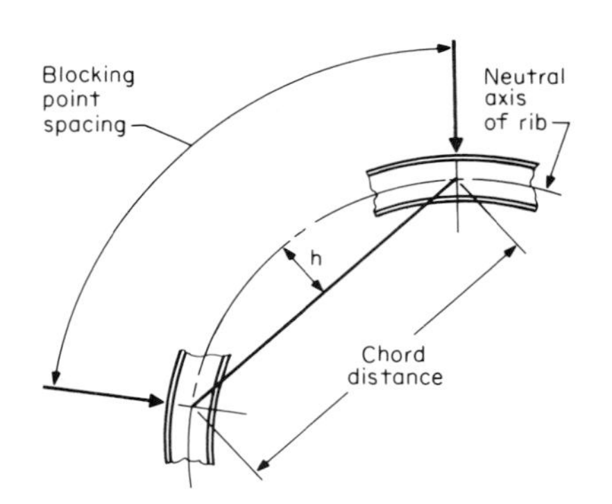

FIG. 10.8 Chord distance *h* between two blocking points on Proctor and White's steel rib.

Contemporary rib-and-lagged primary tunnel supports used in conjunction with shield or boring-machine excavation more closely approximate the sections shown in Figs. 10.9*b* and 10.9*c*, and are not rigidly connected to each other in the longitudinal direction. Shortly after the ribs emerge from the shield, they are generally jacked apart and shimmed with spacers, or "dutchmen," at jacking points so as to minimize voids (and earth movement) at the crown as soon as possible. The use of timber lagging does not allow the grouting of voids outside the primary support until after the final (con-

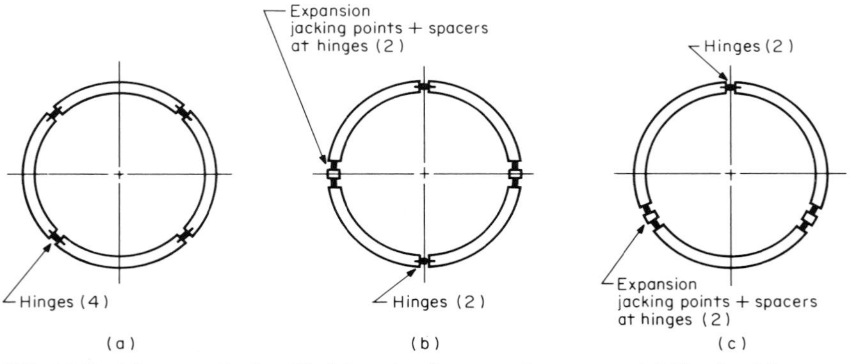

FIG. 10.9 Three methods of bolting circular tunnel supports: (*a*) Hewitt-Johanneson's four-piece bolted segments. (*b*) Four-piece rib set with expansion jacking points at spring line. (*c*) Three-piece rib set with expansion jacking points 30° below spring line.

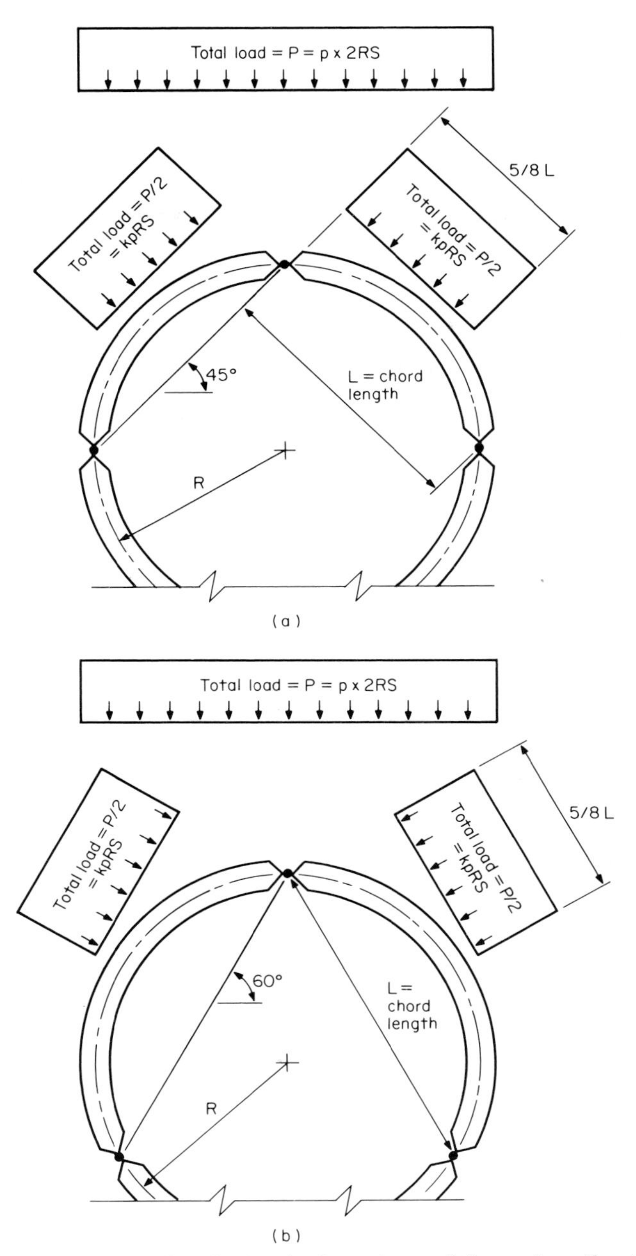

FIG. 10.10 Assumed load distribution for four-piece and three-piece rib sets: (*a*) Four-piece rib set at 90° per piece. Total load, *kpRs,* perpendicular to chord of arch piece and acting over five-eighths of chord length. (*b*) Three-piece rib set at 120° per piece. Total load, *kpRs,* perpendicular to chord of arch piece and acting over five-eighths of chord length. (P = total load; p = soil pressure, S = rib spacing, R = tunnel radius; K = earth-pressure coefficient.)

crete) lining is placed. Thus, some complex nonuniform loads may act on the ribs and lagging during the excavation phase, especially in soils where large boulders or voids may be encountered. Since the wide-flange rib is of uniform section throughout, the worst-case moment condition would be used to design the entire rib set.

The following expedient analysis assumes an uneven distribution of soil load (in this case, equal to the entire load as a function of tunnel diameter and soil pressure acting on only five-eighths of the length of the rib's chord distance) upon a four-piece rib set (Fig. 10.10*a*) and a three-piece rib set (Fig. 10.10*b*). The direction of the load is assumed perpendicular to the chord of the rib arch. No movement of the rib sections or their hinged ends is assumed; nor is passive soil reaction assumed as a result of rib movements or deformations; and the rib configuration is still assumed as circular, in spite of the jacking movement and the insertion of a spacer. The ribs are also considered slender enough so that moment and compression failure will occur before shear failure. The longitudinal force

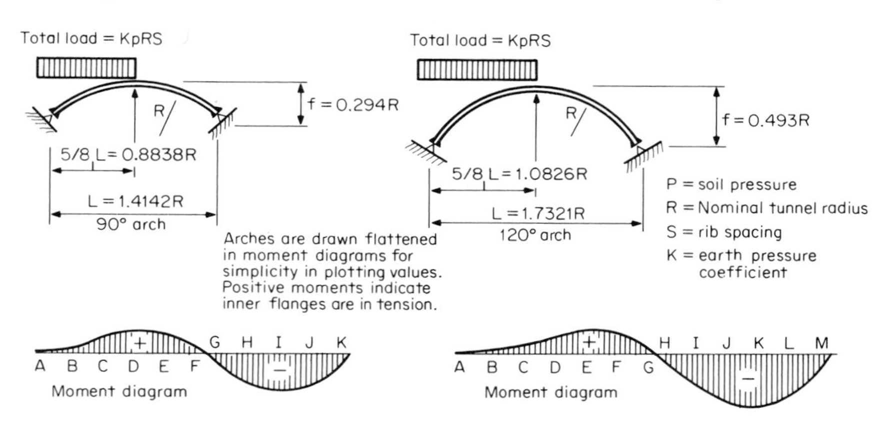

Location	Moment	Axial Compression
A	0	0.9557 KpRS
B	+0.0053 KpR^2S	0.8636 KpRS
C	+0.0139 KpR^2S	0.7766 KpRS
D*	+0.0229 KpR^2S	0.7252 KpRS
E	+0.0209 KpR^2S	0.6667 KpRS
F	+0.0079 KpR^2S	0.6663 KpRS
G	-0.0147 KpR^2S	0.7018 KpRS
H	-0.0369 KpR^2S	0.7283 KpRS
I*	-0.0443 KpR^2S	0.7338 KpRS
J	-0.0310 KpR^2S	0.7212 KpRS
K	0	0.6906 KpRS

*Denotes maximum moment.

(a)

Location	Moment	Axial Compression
A	0	0.6378 KpRS
B	Negligible	0.7647 KpRS
C	+0.0046 KpR^2S	0.6796 KpRS
D	+0.0155 KpR^2S	0.5940 KpRS
E*	+0.0233 KpR^2S	0.5254 KpRS
F	+0.0212 KpR^2S	0.4857 KpRS
G	+0.0097 KpR^2S	0.4848 KpRS
H	-0.0178 KpR^2S	0.5248 KpRS
I	-0.0465 KpR^2S	0.5624 KpRS
J*	-0.0599 KpR^2S	0.5761 KpRS
K	-0.0564 KpR^2S	0.5722 KpRS
L	-0.0389 KpR^2S	0.5510 KpRS
M	0	0.5138 KpRS

*Denotes maximum moment.

(b)

FIG. 10.11 Values of moment and axial compression for four-piece and three-piece arch ribs: (*a*) values for four-piece arch rib; (*b*) values for three-piece arch rib.

due to shield thrust is not considered in this analysis, nor is the torque reaction from a tunnel-boring machine.

The authors used Leontovich's[13] equations for two-hinged, constant-section circular bridge arches subjected to uniform loads over only a portion of their span. The total soil load is calculated by using the soil pressure multiplied by the full tunnel width (see Fig. 10.10). Each arch rib is then subjected to one-half the total load, shown as a uniform load acting on only five-eighths of the arch. An earth-pressure coefficient K is then used to reduce the load, since it is no longer acting in a vertical direction. This coefficient was derived from Coulomb's original earth-pressure theory (ca. 1776) used to reduce active soil loads acting on sloping retaining walls. (Figure 10.26 indicates values of K for various soil friction angles.)

A loading diagram for each type rib set, plus a moment and axial-load plot for each rib (flattened for ease of plotting), are indicated in Fig. 10.11.

Sample Problem (Worked in U.S. Customary Units Only)

Tunnel radius: $R = 6$ ft

Tunnel diameter: $B = 12$ ft

Rib set type: Four pieces at 90°

Rib properties: Section W4 × 13 lb/ft; $S_{xx} = 5.45 \text{ in}^3$; $A = 3.82 \text{ in}^2$

Rib spacing: $S = 3$ ft

Soil density: 120 lb/ft^3

Soil friction: $\phi = 30°$; $K_{4,\text{piece}} = 0.87$

The soil load is approximately Terzaghi's condition no. 7:

$$\text{Soil load} = 1.38B$$

Determine the maximum fiber stress at the maximum positive and negative moment locations.

Solution

Soil pressure: $p = 1.38B \times \text{soil density}$

$p = 1.38 \times 12 \times 120 = 1990 \text{ lb/ft}^2$

Moment constant: $KpR^2S = 0.87 \times 1990 \times (6)^2 \times 3 = 186{,}980 \text{ ft}\cdot\text{lb}$

Axial compression constant: $KpRS = 0.87 \times 1990 \times 6 \times 3 = 31{,}163 \text{ lb}$

1. At maximum positive location:

$$\text{Moment} = 0.0229\ KpR^2S = 0.0229 \times 186{,}980 = 4282 \text{ ft}\cdot\text{lb}$$

$$\text{Axial compression} = 0.7252\ KpRS = 0.7252 \times 31{,}163 = 22{,}600 \text{ lb}$$

$$\text{Maximum fiber stress} = \frac{M}{S_{xx}} + \frac{C}{A}$$

$$= \frac{4282 \times 12}{5.45} + \frac{22{,}600}{3.82} = 15{,}345 \text{ lb/in}^2$$

2. At maximum negative moment location:

$$\text{Moment} = 0.0443\ KpR^2 = 0.0443 \times 186{,}980 = 8283\ \text{ft}\cdot\text{lb}$$

$$\text{Axial compression} = 0.7338\ KpRS = 0.7338 \times 31{,}163 = 22{,}867\ \text{lb}$$

$$\text{Maximum fiber stress} = \frac{M}{S_{xx}} + \frac{C}{A}$$

$$= \frac{8283 \times 12}{5\ 45} + \frac{22{,}867}{3.82} = 24{,}224\ \text{lb/in}^2$$

Sample Problem (Worked in U.S. Customary Units Only)

What would be the maximum fiber stresses for a three-piece rib under the same conditions (except for $K_{3,\text{piece}} = 0.72$)?

Solution

Moment constant: $KpR^2S = 0.72 \times 1990 \times (6)^2 \times 3 = 154{,}742\ \text{ft}\cdot\text{lb}$

Compression constant: $KpRS = 0.72 \times 1990 \times 6 \times 3 = 25{,}790\ \text{lb}$

1. At maximum positive moment location:

$$\text{Moment} = 0.0233 \times 154{,}742 = 3606\ \text{ft}\cdot\text{lb}$$

$$\text{Compression} = 0.5254 \times 25{,}790 = 13{,}550\ \text{lb}$$

$$\text{Maximum fiber stress} = \frac{3606 \times 12}{5.45} + \frac{13{,}550}{3.82} = 11{,}487\ \text{lb/in}^2$$

2. At maximum negative moment location:

$$\text{Moment} = 0.0599 \times 154{,}742 = 9269\ \text{ft}\cdot\text{lb}$$

$$\text{Compression} = 0.5761 \times 25{,}790 = 14{,}858\ \text{lb}$$

$$\text{Maximum fiber stress} = \frac{9269 \times 12}{5.45} + \frac{14{,}858}{3.82} = 24{,}298\ \text{lb/in}^2$$

Commentary on Analysis

By inspection of Fig. 10.12, note that as the value of ϕ decreases, the values of K for the three-piece arch approach those of the four-piece arch—both being close to unity. Thus, the maximum fiber stresses will be greater for the three-piece arch than for the four-piece arch, because the moment multiplier at maximum negative moment location from Fig. 10.11*b* is approximately one-third greater than for Fig. 10.11*a*. This would appear to be contradictory, that a three-piece arch would be "weaker" than a four-piece arch, but remember the assumption of no transmission of moment across the bolted butt splices, i.e., a true hinged connection. It should also be pointed out that as the number of hinged joints increases, the maximum moments in each segment will decrease, but the ability of the structure as a whole to retain its circularity will be diminished.

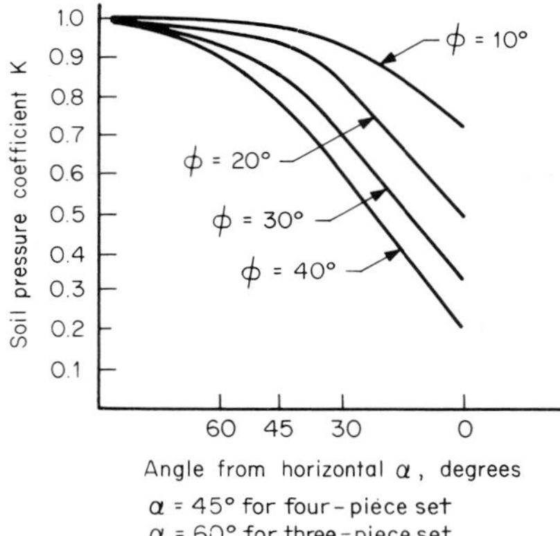

Soil friction angle ϕ, degrees	Loading 45° from horizontal (as in the case of a four-piece arch rib set)	Loading 60° from horizontal (as in the case of three-piece arch rib set)	Loading 0° from horizontal (horizontal load)*
10	1.00	0.94	0.71
20	0.94	0.84	0.49
30	0.87	0.72	0.33
40	0.80	0.60	0.22

*Values of the horizontal load factor compare closely with

$$k = \frac{1 - \sin \phi}{1 + \sin \phi}$$

also known as the active earth-pressure coefficient.

FIG. 10.12 Soil-pressure coefficient k versus angle from horizontal α for selected values of soil friction angle ϕ.

If even more unbalanced loads are anticipated, Table 10.9 can be utilized to calculate maximum positive and negative moments for a half-loaded condition and a three-eighths-loaded condition. On the other hand, if the load becomes more balanced, the maximum moments decrease and the design load would more closely approach a simple compression condition.

Installation of Steel Rib Supports

Horseshoe Tunnels in Rock

Steel rib sets are usually installed by the heading crew after the shotrock (muck) has been removed from the heading. Standup time determines the length of the round to be shot and the maximum distance from the new working face to which the new rib sets should be erected. After scaling off loose rock, the miners pre-

TABLE 10.9 Maximum Moment and Corresponding Axial Load for Half-Loaded Condition and Three-Eighths-Loaded Condition for Three- and Four-Piece Rib Sets

	Load KpRS acting on:			
	Half of arch		Three-eighths of arch	
	Four-piece set	Three-piece set	Four-piece set	Three-piece set
Maximum positive moment	$+0.0410\ KpR^2S$	$+0.0444\ KpR^2S$	$+0.0862\ KpR^2S$	$+0.0639\ KpR^2S$
Axial load at maximum positive moment	$0.6293\ KpRS$	$0.3443\ KpRS$	$0.3651\ KpRS$	$0.4377\ KpRS$
Maximum negative moment	$-0.0481\ KpR^2S$	$-0.0663\ KpR^2S$	$-0.0200\ KpR^2S$	$-0.0356\ KpR^2S$
Axial load at maximum negative moment	$0.6481\ KpRS$	$0.3787\ KpRS$	$0.4081\ KpRS$	$0.4077\ KpRS$

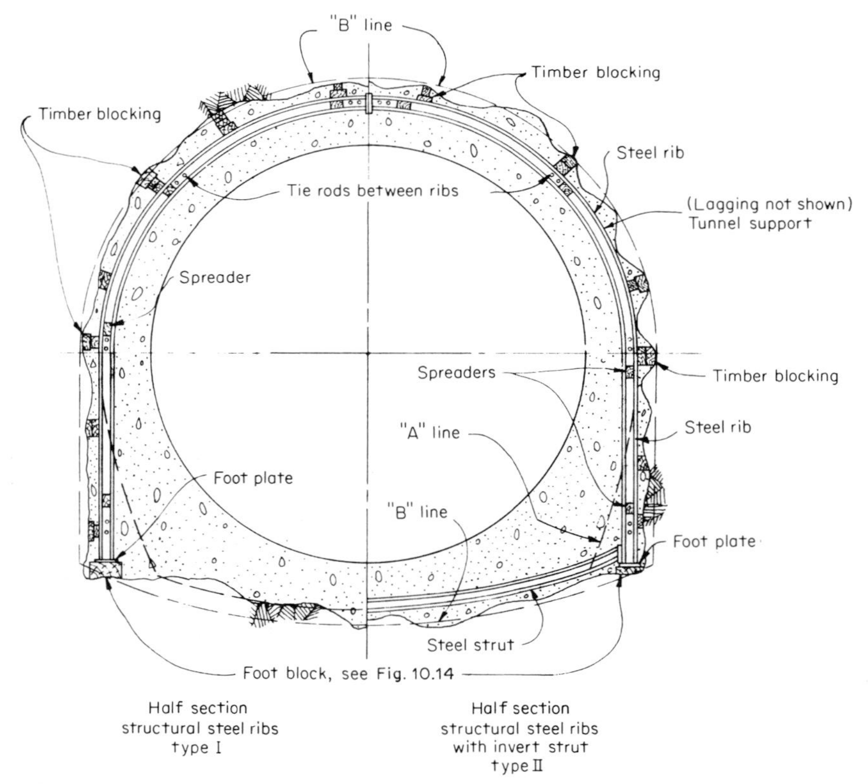

FIG. 10.13 A typical two-piece rib set for a small-diameter rock tunnel. The right half of the drawing illustrates an optional invert strut, to be utilized when side loads require it. *(Bureau of Reclamation, U.S. Department of Interior.)*

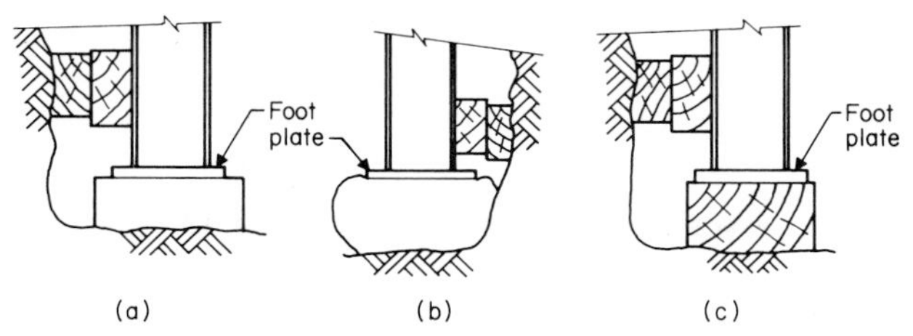

FIG. 10.14 Three methods of foot blocking: (*a*) precast concrete; (*b*) sacked concrete; (*c*) timber. *(Bureau of Reclamation, U.S. Department of Interior.)*

pare for rib erection by first digging to firm rock where the posts will stand. In order to avoid standing a lopsided set, foot blocks are installed to assure that the foot plates at the bottom of both posts are at the same elevation. Figure 10.13 illustrates a typical two-piece rib set, with an optional invert strut, for a small-diameter tunnel; Fig. 10.14 illustrates three methods of foot blocking. Once the foot blocks are leveled up, the miners will stand the posts. In smaller tunnels with light-steel ribs, the set is usually a two-piece unit with a single splice at the arch. If each half weighs less than 300 lb (136 kg), the miners can usually muscle the halves into place. For heavier sections, four-piece rib sets are commonly utilized, and the miners generally use the hydraulic booms on their drill jumbo, or a specialized erector to hold all the pieces up, until the splices can be bolted together. During this maneuver, it is important to insert the longitudinal tie-rods into their appropriate holes. In Fig. 10.15, Washington, D.C. miners are shown muscling an arch piece into position while inserting the tie-rod into its hole in the web.

FIG. 10.15 Assembling a rib set in a Washington, D.C. subway tunnel. *(Commercial Shearing, Inc.)*

Tie-rods are a necessary part of a rib-support system because they minimize longitudinal movement. Tie-rods are used in conjunction with spreaders between the rib sets. Spreaders are generally made of cut lumber, but they may also be made of steel pipe (with the tie-rod inside the pipe) or of angle iron. Timber spreaders are generally the most economical, and they also have the advantage that they can be trimmed to fit at the heading when adjustments to rib alignment and plumbness are necessary.A disadvantage is that they are subject to rot, or that they create voids in concrete, if the tunnel is to have a secondary lining. For these reasons, some owners specify that all or some of the timber spreaders must be removed prior to placing the secondary lining. The cost of removing the timber may, at times, offset its initial economy.

After the erection and assembly of the set, it should be blocked firmly against the rock surface according to the design requirements. The miners should be provided with precut timber of random thicknesses and lengths and an adequate supply of hardwood wedges to accomplish the blocking. Lagging (timber or steel channels) may also be installed at this time, if required to support the rock in the area between the rib sets. In order to speed up the tunnel drive, a portion of the blocking and lagging may be completed at the rear of the jumbo during the

FIG. 10.16 Steel ribs with wall plates and vertical posts in a Chicago water tunnel. *(Commercial Shearing, Inc.)*

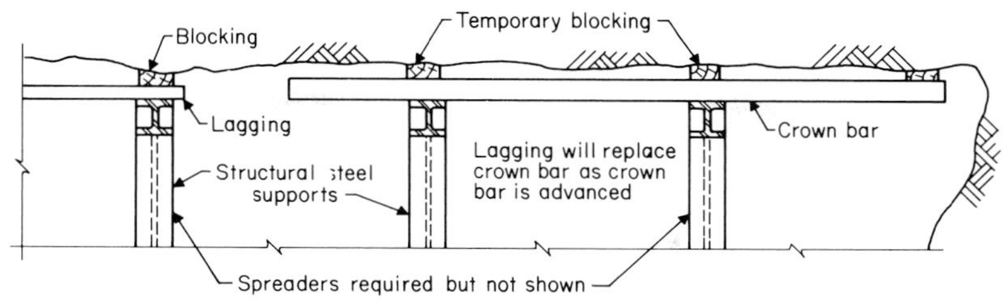

FIG. 10.17 Longitudinal view of crown bars used in a steel-rib-supported tunnel. *(Bureau of Reclamation, U.S. Department of Interior.)*

drill cycle. This option, of course, depends on the standup time available to the tunneler. Repairs to rib sets damaged by previous blasts should be carried out at this time, as well as reblocking or rewedging of any sets that have been loosened by the blasting or mucking operations.

Figure 10.16 shows steel rib arches with tie-rods and timber spreaders between the ribs, and also shows steel channel lagging placed outside the rib flanges to contain loose rock between the ribs. Notice also the horizontal wall plate beams between the arch ribs and the vertical posts. Many large-diameter tunnels are driven "top-heading-and-bench"—that is, the top half is first driven horizontally and the lower half is later "benched" as in a conventional quarry operation. In this type of excavation it is sometimes difficult to mate a post exactly to its respective rib as the bench is removed. The wall plate in this case distributes arch loads evenly to the posts. Installed during the top-heading drive, it also bridges the gap between bench excavation and post installation.

In ground with poor arch rock or a very short standup time, it sometimes is necessary to cantilever crown bars or spiling ahead of the leading rib set to control the arch.

Figure 10.17 shows crown bars which are shoved forward as the rib sets are advanced along with the tunnel heading. Figure 10.18 shows spiling extended ahead from the leading rib set and later trimmed to provide the proper clearance for the secondary lining. Spiling may also be drilled and inserted above and ahead of the working face.

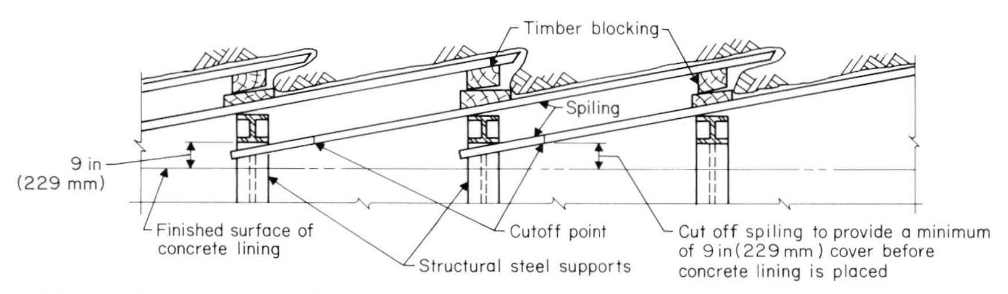

FIG. 10.18 Longitudinal view of spiling used in a steel-rib-supported tunnel. In this case the owner requires that the tail end of the spiling be cut off prior to placing the secondary lining, to provide a minimum cover of 229 mm (9 in). *(Bureau of Reclamation, U.S. Department of Interior.)*

The duties of the tunnel engineer in a steel-rib-supported tunnel are generally the same as outlined for a liner-plate tunnel—i.e., to provide adequate materials to the work crews, to assist them in proper tunnel alignment, and to monitor settlements or rib movements.

Steel-Rib Installation in Circular Tunnels

Shield-driven tunnels in soil

Steel ribs in shield-driven tunnels are generally installed at the rear, or "tail," of the shield by hand or with the aid of a mechanical erector. Figure 10.19 shows a mechanical erecting device on a tunnel-boring machine to be used on a Washington, D.C. subway project. A worker is shown near the transporter, which picks the support (in this case a segmental liner plate which will be utilized as a permanent support) off a flatcar and moves it forward into the tail-shield area (flatcar not shown). A rotating erector can be seen in the tail shield. The erector picks up each segment and rotates it into position, where it is bolted together. Generally a full circle of timber lagging is installed between each rib set to counteract the thrust used to propel the shield forward. Since the diameter of the rib set is slightly smaller than the hole bored by the shield, the ribs must be jacked at selected splice points to expand the rib set so it is in intimate contact with the soil. Figure 10.20 shows a miner inserting a spacer at spring line after completion of jacking. Figure 10.21 shows a completed circular tunnel. When curves are negotiated, care must be taken in cutting and installing the proper length lagging in its particular position around the ring.

FIG. 10.19 Rear view of a shielded tunnel-boring machine, showing a mechanical support transporter and a mechanical erector in the tail- or rear-shield area. *(Robbins Co.)*

FIG. 10.20 Expanding circular ribs on a California tunnel project. The hydraulic jack pushes out on reaction pads welded to the inside flanges of the ribs. *(Commerical Shearing, Inc.)*

FIG. 10.21 A circular rib-supported tunnel in Ohio. The four-piece rib sets are fully lagged and spacers are inserted at spring line on both sides of the tunnel. *(Commercial Shearing, Inc.)*

Circular rock tunnels bored by tunneling machines without shields

Optimally the contractor would prefer to install ring beams behind a shieldless tunnel-boring machine and avoid the inherent congestion near the work face, but standup time does not always allow this option. When ribs are installed close to the cutterhead, elaborate rib-transport and rib-erection equipment is generally supplied by the machine manufacturer as an integral part of the machine.

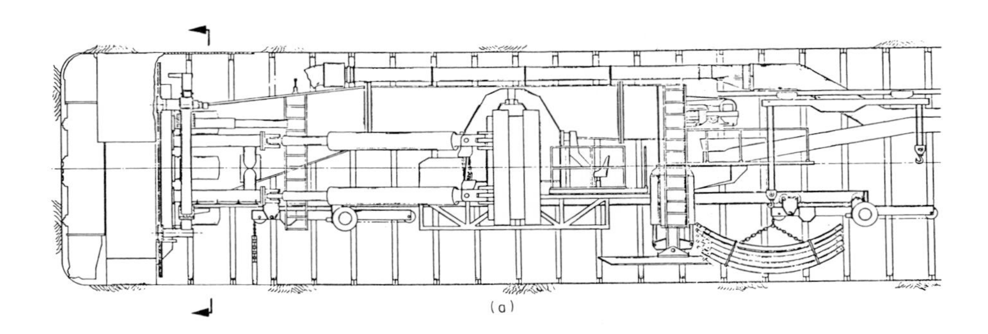

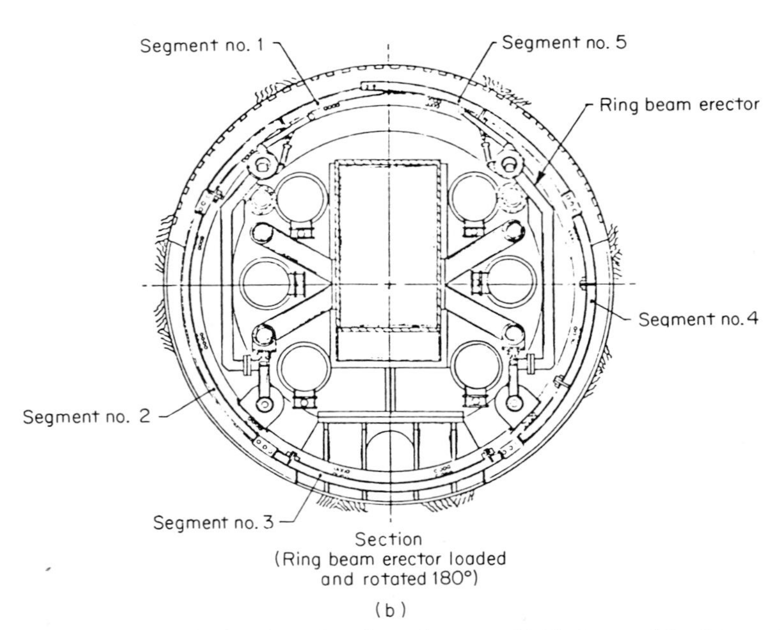

FIG. 10.22 (*a*) Longitudinal section through a nonshielded tunnel-boring machine with steel-rib installation close behind the cutterhead. *(Robbins Co.)* (*b*) Section through the erector. *(Robbins Co.)*

Figure 10.22*a* shows a longitudinal section through a large-diameter tunneling machine. A hoist and monorail picks an unassembled rib set off a flat car and transports it under the main body of the machine to a rotating erector located just behind the cutterhead. Figure 10.22*b* (section D-D) illustrates how the erector clamps on to the respective parts of the five-piece rib set and rotates them into position. A hydraulic expander, in this instance, is located at the crown splice.

ROCKBOLTS

A rockbolt consists of a bar, usually of high-strength steel, with one end anchored to the rock at the bottom of a drill hole and the other end restrained by a bearing plate at the collar of the hole on the rock surface. The bolt's anchor may be of the mechanical type (i.e., a steel shell which is expanded and forced against the sides of the drill hole), or it may be of the epoxy (resin) type which glues the rockbolt to the sides of the drill hole. By tensioning the bolt, the rock surrounding the bolt hole is subjected to a small amount of radial compression, and the bolt is considered an *active* support. Inversely, if the bolt is not tensioned when it is installed, it will later become tensioned as a result of rock relaxation and movement toward the shaft or tunnel, and this type of bolt is considered a *passive* support.

Bolt Design

Several approaches to rockbolt design are presented here. One approach uses the bolt-induced compression in the rock, which locks fractures and shears in the rock together and creates an arch of rock which protects the openings beneath it. Other approaches examine the weight of the rock mass, or of specific rock blocks within the mass, and then design a bolt pattern to resist the movement of the mass or the blocks. Regardless of the design approach, geotechnical engineers generally agree that a contractor must develop an installation and inspection program to ensure that the bolts and anchorages are capable of providing the design tension force. Installation programs include a hydraulically jacked pull test, or a torque-tension test, to check the integrity of the bolt and its anchorage. Inspection programs include selective testing of bolts installed in the past. If a regular inspection and corrective maintenance program cannot be carried out due to operation constraints of the construction procedure (i.e., the bolts can no longer be reached or are covered over), then the contractor should grout the bolts after their initial testing with a cement grout or resin.

Flat Roofs in Laminated Rock

This condition is commonly encountered in coal mining, where the rock overlying the coal seam is usually shale or sandstone. When the layer of rock adjacent to the roof of the opening is in a shale or other weak formation, it is usually bolted to an overlying layer of competent rock, such as sandstone. Obert and

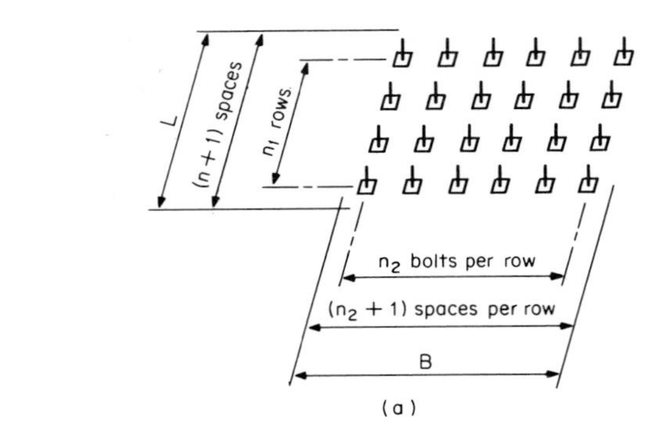

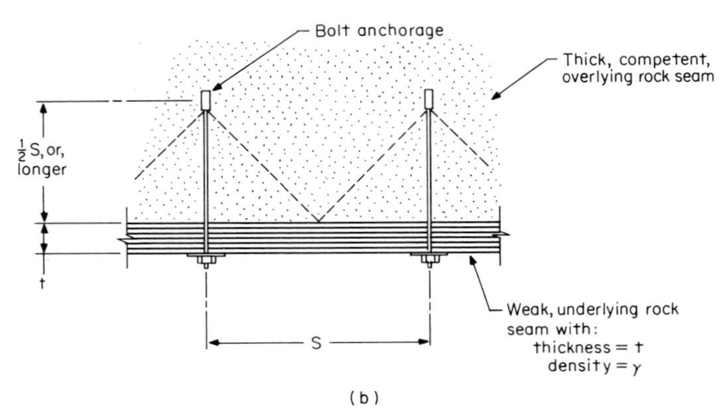

FIG. 10.23 Roof bolting a flat roof in laminated rock: (*a*) Bolting pattern. (*b*) Bolts supporting a weak underlying rock in a flat roof. *(From L. Obert and W. Duvall, Rock Mechanics and Design of Structures in Rock, John Wiley, 1967.)*

Duvall[14] consider reinforcement by suspension from a thick body of competent overlying rock (see Fig. 10.23*a*), where load per bolt W is

$$W = \frac{tL\gamma}{(n_1 + 1)(n_2 + 1)}$$

where n_1 = number of rows of bolts in a roof
n_2 = number of bolts per row
γ = unit weight of underlying rock seam, lb/ft^3 (kg/m^3)
t = thickness of underlying rock seam, ft (m)
B = width of opening, ft (m)
L = length of opening, ft (m)

This equation was based on a mine roof of infinite width and breadth, i.e., with no sidewalls to carry part of the lower seam's weight. However, the authors also examined the same seam with side and end support, and found, for a thick, self-supporting upper seam, that the weight per bolt approaches that for the infinite seam condition; i.e., the above equation is valid in both cases. To effectively transmit the bolt load from its anchor to the competent rock, the length of the bolt should at least equal the sum of the underlying rock seam thickness, plus one-half the bolt-spacing distance (see Fig. 10.23*b*). In most cases, bolt-spacing should not exceed 5 or 6 ft (1.5 or 1.8 m).

The roof-support designer should check with the bolt manufacturers for their recommended drill-hole size, type of anchorage, and bolt-tensioning procedures. It is also important to assure that the bolting crew is consistently developing the recommended bolt tension; and if not, to determine why not. (For example, the rock may be too soft for the type of mechanical anchor used, requiring a longer anchor or perhaps the use of resin anchorages.) Panek and McCormick[16] give a rough estimate of the amount of torque required to develop tension in a bolt by the equation (in U.S. Customary units)

$$\text{Tension (lb)} = C \times \text{torque (ft}\cdot\text{lb)}$$

where C = 50 for ⅝-in bolts
= 40 for ¾-in bolts
= 30 for cone-neck, or self-centering, headed bolts
= 60 for using hardened-steel washers between bearing plate and bolt head of nut

The recommended tensions for various grades and diameters of bolts are given in Table 10.10.

Rockbolting Circular Openings

This configuration is usually encountered in arched-roof, civil works construction.

Talobre's[17] approach utilizes a method of calculating radial stress around a circular opening to determine the load on an array of rockbolts. Figure 10.24*a* illustrates a somewhat circular opening in a rock formation which has relaxed and moved slightly toward the opening. Talobre assumes that a hydrostatic condition exists in the general area of the opening, i.e., where horizontal and vertical components of the residual stresses are equal. The shaded area represents an

TABLE 10.10 Tension Loads for Various Diameters and Grades of Steel Rockbolts

Bolt diameter, in	Stressed area at thread, in^2	Grade 30 bar yield load, lb*	Grade 55 bar yield load, lb*	Grade 75 bar yield load, lb*
⅝	0.226	—	12,400	17,000
¾	0.334	10,000	18,400	25.100
⅞	0.462	13,900	25,400	24,700
1	0.606	18,200	33,300	45,500
1¼	0.969	—	53,300	—
1⅜	1.555	—	63,500	—
1½	1.405	—	77,300	—

*It is customary to use 60% of the yield loads as the design working load of the bolt (Panek and McCormick[16]).

SOURCE: ASTM F432-76, copyright, ASTM, 1916 Race Street, Philadelphia, Pa. Reprinted with permission.

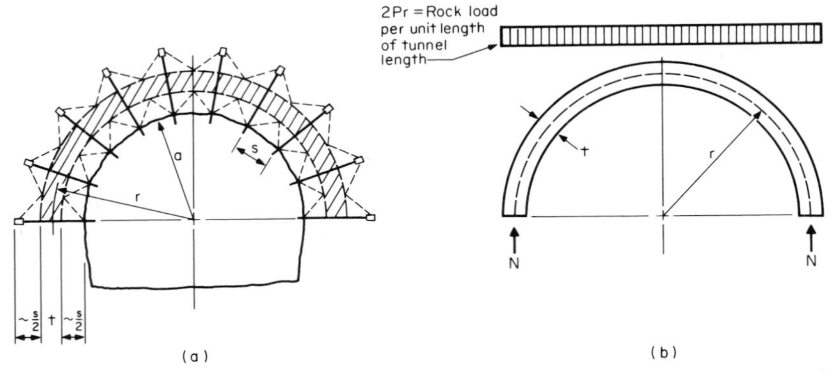

FIG. 10.24 Supporting a circular opening in rock with steel rockbolts: (*a*) Typical rockbolt pattern in arch. (*b*) Free body diagram of the supporting arch. *(From J. A. Talobre, "La méchanique des rockes et ses applications," 2d ed. Dunod, Paris, 1967.)*

arch, or vault, of radially compressed, consolidated rock. In the general case, the length of the rockbolt will be equal to the thickness of the arch t plus the rockbolt spacing s. The radius to the arch midpoint r is equal to the radius of the tunnel a, plus one-half the length of the rockbolt l, or

$$r = a + \frac{l}{2} = a + \frac{t + s}{2}$$

Figure 10.24*b* shows a free-body diagram of the top half of a 1-ft (30-cm)-long section of the arch. Thrust N at the spring line would equal

$$\frac{P \times 2r}{2} = P \times r$$

where P is the rock pressure around the opening.

Talobre then assumes that the circumferential or tangential unit stress at spring line equals thrust divided by the thickness of the arch, or

$$t = \frac{N}{t}$$

(For a fuller understanding of tangential and radial stresses around a hole, see the article "Stresses in Rock About Cavities" by K. Terzaghi and F. Richart, Jr., in the vol. 3, 1952–53 issue of *Geotechnique,* and see the appendix to that article by Richart.)

Talobre further assumes that this tangential stress is the major stress in a Coulomb-Mohr's circle, where from inspection

$$\frac{\sigma_r}{\sigma_t} = \frac{\sigma_{\min}}{\sigma_{\max}} = \frac{1 - \sin\phi}{1 + \sin\phi} = \tan^2\left(45 - \frac{\phi}{2}\right)$$

where ϕ = angle of friction of rock
σ_r = radial stress in arch
σ_t = tangential stress in arch

TABLE 10.11 Friction Angle ϕ for Various Rocks

Rock	$\phi°$ (intact rock)	$\phi°$ (residual)
Andesite	45	28–30
Basalt	48–50	
Chalk		
Diorite	53–55	
Granite	50–64	31–33
Greywacke	45–50	
Limestone	30–60	33–37
Monzonite	48–65	28–32
Porphyry		30–34
Quartzite	64	26–34
Sandstone	45–50	25–34
Schist	26–70	
Shale	45–64	27–32
Siltstone	50	
Slate	45–60	25–34

SOURCE: E. Hoek, "Estimating the Stability of Slopes in Opencast Mines," *Trans. Institution of Mining & Metallurgy,* London, 1970.

He then states that the load on the rockbolt is equal to the area covered by each bolt, multiplied by the radial stress. The bolt is then tensioned to twice the amount for a safety factor of 2. The approximate friction angles for various types of rock are shown in Table 10.11. Since the hydrostatic condition assumes that the rock around the opening has moved slightly toward the opening, the residual values of ϕ should be used in the calculations.

Sample Problem (Worked in U.S. Customary Units Only)

This problem is solved using trial-and-error assumption of t,s.

Tunnel radius: $a = 6$ ft

Arch thickness: $t = 4$ ft

Bolt spacing: $s = 4$ ft each way

Rock friction angle: $\phi = 35°$ (residual)

Rock load: Using Terzaghi's condition no. 3, the rock load can be determined using

$$H_p = 0.25 \times \text{tunnel diameter}$$

(or arch diameter in this case)

Rock density: 165 lb/ft^3

Rockbolt length: $l = t + s = 4 + 4 = 8$ ft

The radius of the supporting arch is:

$$r = a + \frac{l}{2} = 6 + \frac{8}{2} = 10 \text{ ft}$$

The rock load on the supporting arch is

$$P = 0.25 \times (10 \times 2)\ 165\ \text{lb/ft}^3$$
$$P = 825\ \text{lb/ft}^2 \text{ of tunnel}$$

The thrust on the arch is

$$N = P \times r = 825 \times 10 = 8250\ \text{lb/lin ft of tunnel}$$

The tangential (major) stress is

$$\sigma_t = \frac{N}{t} = \frac{8250}{4} = 2062\ \text{lb/in}^2$$

The radial (minor) stress is

$$\sigma_r = \sigma_t \times \tan^2\left(45 - \frac{\phi}{2}\right) = 2062 \times 0.27$$
$$\sigma_r = 558\ \text{lb/ft}^2$$

The area covered by each rockbolt is

$$4 \times 4 = 16\ \text{ft}^2$$

The total rockbolt load is

$$16\ \text{ft}^2 \times \sigma_r = 16 \times 558 = 8928\ \text{lb}$$

Stress the bolt to twice the load, or

$$8928 \times 2 = 17{,}856\ \text{lb}$$

From Table 10.10 choose a 1-in-diameter, grade 55 bar which can safely be tensioned to 60% of its yield load of 33,300 lb, or

$$33{,}300 \times 0.6 = 19{,}980\ \text{lb} > 17{,}856\ \text{lb}$$

(Check with the bolt manufacturer for recommended anchor type and recommended torque to develop approximately 18,000-lb tension.)

These calculations are a trial-and-error approach for arbitrarily assumed values of arch thickness t and bolt spacing s. If an unreasonable bolt load is calculated, then make new assumptions for t or s (or both) and recalculate.

Jaeger[19] states that rockbolt systems designed to cause average pressures on the surface of the rock cavity of between 10 and 20 lb/in^2 (7 and 14 N/cm^2) are usually sufficient to form a sustaining arch. Cording and Mahar[19a] point out where support pressures as high as 20 lb/in^2 (14 N/cm^2) were required to stabilize a 120-ft (86-m)-high cavern sidewall in rock with an excellent RQD, but also show where rock pressures of only 5 lb/in^2 (3.5 N/cm^2) were required in 20-ft (6.1-m)-wide openings in rock with a poorer RQD, and suggest that support pressures may be related to the size of the rock opening. A survey of a few underground caverns in North America would reveal that the average compression in the three major underground arches at the Churchill Falls Hydro Project[20] was specified at 12.5 lb/in^2 (8.62 N/cm^2); the three chamber arches at the NORAD Expansion Project[21] were compressed to an average of 8.5 lb/in^2 (5.9 N/cm^2); and a subway arch constructed in a major U.S. city is compressed to about 8 lb/in^2 (5.5 N/cm^2). In our sample problem we compressed 16 ft^2 (1.5 m^2) to 17,856 lb (79,420 N), resulting in an average pressure of 7.75 lb/in^2 (5.34 N/cm^2).

Bolting Across Joints

Haas[22] ran a series of tests on rockbolts acting across a shear plane and found that rockbolts are most effective across a shear plane when they are positioned so that they tend to elongate as the shear plane is subject to movement, least effective when shear plane movement tends to compress the bolts axially. In order to avoid trouble, the safest action that a tunneler should take in blocky ground is to try to avoid subjecting rockbolts to shearing action and to position the bolts so they do their work in tension. If this cannot be avoided, an analysis of each block and how it acts in relation to the rock opening should be carried out by the tunneler or geotechnical consultant.

Figure 10.25 illustrates our rock arch again, but this time considers the effect of a wedge of rock in the crown of the opening. The wedge of rock is estimated to be 13 ft (4 m) high by 12 ft (3.6 m) wide with a weight per linear foot of tunnel equal to 12,870 lb/ft (19,150 kg/m), acting in such a manner as to subject the bolts in the crown to simple tension. The weight of the wedge per 4-ft (1.22-m) bolt spacing would then be 12,870 $\times$ 4 = 51,480 lb (23,350 kg). By inspection, if the two bolts in the crown of the arch were lengthened [to approximately 14 ft (4.3 m) in this case], they would be able to hold up a 51,480-lb (23,350-kg) wedge prior to failure, because the yield strength of each bolt was 33,300 lb (148,120 N).

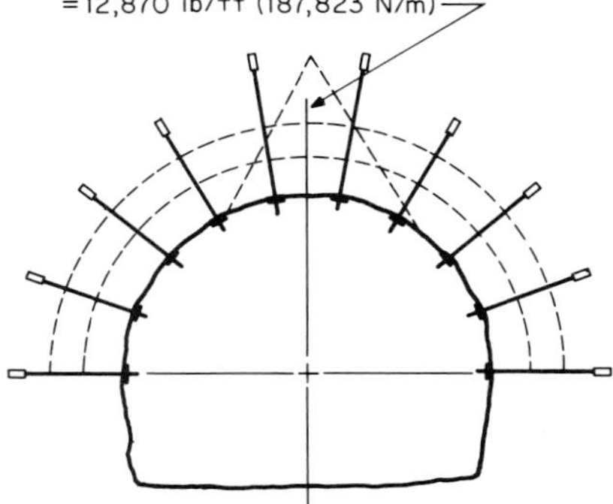

FIG. 10.25 Rockbolted arch in blocky ground subjected to a wedge failure in the crown.

Grouted Rockbolts

Bolts may be grouted after tensioning by either portland cement grout, or in the case of resin bolts, by using a fast-setting resin in the anchoring portion of the drill hole and a slower setting resin in the remainder of the hole to allow time for tensioning the bolt in the interim between each resin set. In examining the advantages of a fully grouted rockbolt system versus the same system without grout, we have arbitrarily chosen an 8-ft (2.44-m)-long bolt [less 1 ft (30 cm) for anchorage length] traversing a joint in the rock, which is dilating due to ground movement after bolt installation (see Fig. 10.26).

It can be seen that the extension of the ungrouted bolt will be 7 times that of the grouted bolt for the same force *T*. Thus, many geotechnical engineers consider rockbolt grouting a way to limit small rock movements after bolt installa-

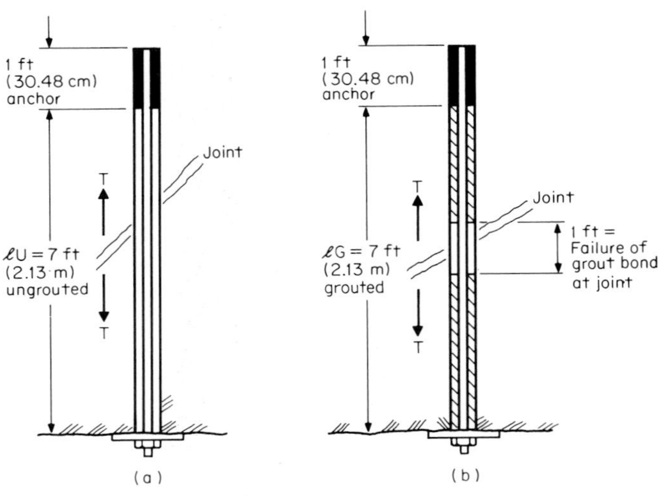

FIG. 10.26 Grouted and ungrouted rock bolts subjected to tensile loading across a joint: (*a*) Type U (ungrouted) and (*b*) type G (grouted).

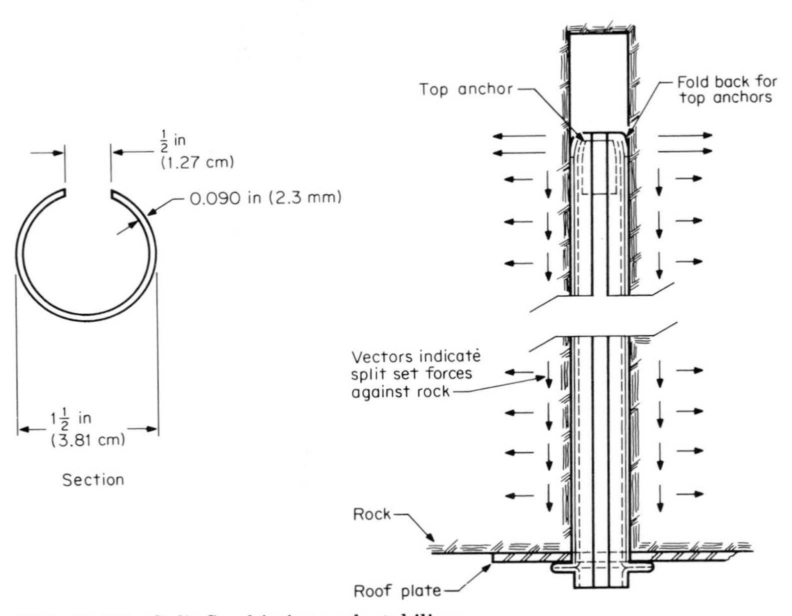

FIG. 10.27 Split Set friction rock stabilizer.

tion, movements which can slowly lead to an unstable condition in a shaft or tunnel.

An alternative active support device which eliminates the need for grouting is the Split Set (trademark of Ingersoll Rand Co.) friction rock stabilizer (see Fig. 10.27). The Split Set is a longitudinally split steel tube which is forced into a predrilled hole in the rock, adhering intimately to the rock surface of the drill hole by friction. The driving force, usually supplied by the rock drill used to drill the hole, loads the rock in compression. In addition to hard-rock support, it has been utilized successfully in rock too soft for mechanical anchors and in certain shales that cannot be reliably grouted with cement or resin. According to Scott,[23] testing has shown that it can develop an anchoring force of from ¾ to 2 tons/ft (22 to 58 kN/m) of borehole. An additional advantage is that if it is overloaded, it will slip slightly and still adhere to the borehole after the rock has completed its movement. Mechanical anchors, on the other hand, usually break or lower their load-carrying capacity once they are overloaded.

Passive Supports Related to Rock Bolting

Rockbolts require tension to do their work. Passive supports, on the other hand, are installed untensioned and do their work by taking up tension as the rock relaxes and the joints dilate. Two common types of passive supports are:

1. Perfo bolts (see Fig. 10.28*a*) are steel reinforcing bars grouted to the rock. The grout is inserted and held in the drill hole by a perforated sleeve until the reinforcing bar is inserted. For the most part, the perfo bolt has been supplanted by the untensioned rebar resin spiling.

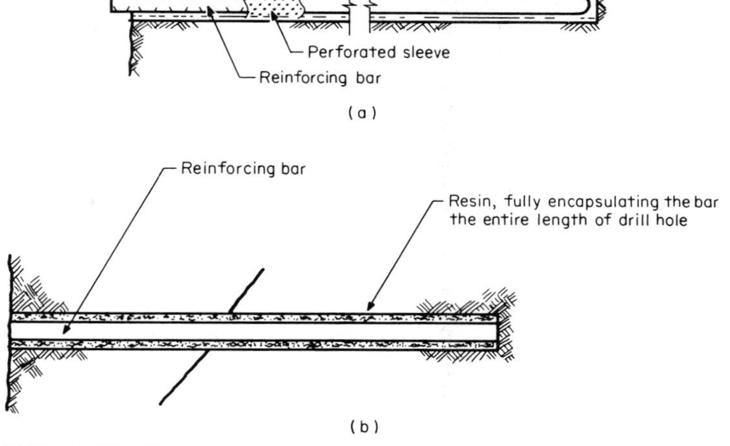

FIG. 10.28 Passive support devices related to rockbolts: (*a*) Perfo bolt. (*b*) Untensioned rebar resin spile.

2. Untensioned rebar resin spiles (see Fig. 10.28*b*) are steel reinforcing bars inserted in drill holes and are completely encapsulated with resin. The bars are in many cases driven in a fan-shaped pattern, pointing ahead of the tunnel or shaft working face. Manufacturers will recommend compatible drill-hole and bar sizes in order to install them efficiently by minimizing the amount of resin required.

When using passive devices of this nature, they should be installed as close to the working face as possible, and as soon as possible, to minimize initial joint dilation. Some geotechnical engineers claim that these devices, when installed at the working face as fast as possible, can be as effective as a tensioned-bolt system installed many hours after the ground is opened up. On many civil works projects, a combination of nontensioned bolting at the face, followed by tensioned bolting within one or two diameters of the face, is used effectively to limit ground movement.

Supplemental Materials

1. Bearing plates are used to transmit the reaction of the bolt's tension to the rock surface. As the rock becomes weaker, larger plates are used. Bearing plates should be thick enough to minimize "dishing," or deforming to a concave configuration, when the bolt is fully tensioned.
2. Roof mats, sometimes called "pans," or rock straps are made from 14- or 16-gauge sheet metal and are utilized to support weaker rock between the bolts.
3. Shotcrete is used on many occasions to support rock surfaces in between the bolt pattern. It can be utilized as plain shotcrete or can be reinforced with welded wire mesh.
4. Chain-link fabric is utilized, not as a supplemental support, but to protect workers and equipment from loose pieces of rock that may fall from the arch.

Installation of Rock or Resin Bolts, Spiles, Friction Rock Stabilizers

The first step the miner must take is to thoroughly scale loose rock from the new surface. In addition to providing a safe working environment for the driller, the scaling also uncovers sound rock with a surface flat enough to collar a drill hole, and later, flat enough to provide a good solid surface for the bearing plate.

Drilling a straight hole with the proper diameter at the far end (anchorage end) of the hole is important to the speed of the operation and the ability of the anchor to seat according to the manufacturer's recommendation. If hand-held jackleg drills are used, the miner usually is equipped with a set of drill steels in 2-ft (60-cm) increments. For example, an 8-ft (2.4-m)-deep hole would require a 2-ft (60-cm) starter steel to allow the miner good stability and leverage while collaring the hole, followed by longer drill steels, each with its own bits, to deepen the hole to 8 ft (2.4 m). The longest drill steel must have a bit diameter equal to the bolt manufacturer's recommended diameter, and the shorter drill steels must therefore have bit diameters slightly larger to allow clearance for the final bit. If, for example, three drill-size changes are required for the miner to arrive at a final

FIG. 10.29 Installing Split Sets with a hand-held jackleg drill in blocky ground. *(J. J. Scott and Ingersoll Rand Co.)*

FIG. 10.30 A specialized jumbo for drilling (left side) and installing (right side) Split Sets. *(J. J. Scott and Ingersoll Rand Co.)*

hole diameter of 1⅝ in (41 mm), the first or shortest steel would have a bit diameter of 1¾ in (45 mm), the second steel would have a bit diameter of 1$\frac{11}{16}$ in (43 mm), and the final, or longest, drill steel would have the 1⅝-in (41-mm) bit. It is therefore important to have the appropriate drill steels on hand, clearly marked to avoid delay.

The contractor may choose to utilize a drill jumbo to drill the bolt holes. The jumbo is generally equipped with power-fed drills having a higher energy output than jackleg drills, and it can usually drill holes 8 to 12 ft (2.4 to 3.6 m) long without a change of drill steel. The power feed is usually 2 to 4 ft (0.6 to 1.2 m) longer than the length of hole it can drill in a single pass; therefore, a power-fed drill designed to drill an 8-ft (2.4-m) hole in a single pass would have an overall length of 10 to 12 ft (3 to 3.6 m). Thus we can see that the 8-ft (2.4-m) long bolt in our sample problem would require a jumbo drill that would just barely fit in our tunnel with a 12-ft (3.6-m) diameter. A platform would also be required for the miners to torque or pull-test the bolts in the upper portion of the arch.

Figure 10.29 shows a miner driving a Split Set into the hole he has just drilled with his hand-held jackleg drill. Note how the blocky rock has spalled from under the bearing plates on the left side of the photo. Ungrouted rockbolts with mechanical anchors would be ineffective if the bearing plates were to become loose, whereas Split Sets or grouted bolts would still be effective in this situation. Note also how the chain-link fabric keeps loose rock from falling from the arch.

FIG. 10.31 Rockbolter on a boring machine. *(Robbins Co.)*

Figure 10.30 shows a drill jumbo drilling and installing Split Sets, while Fig. 10.31 shows a rockbolting drill mounted on a tunnel-boring machine. Mechanized rockbolting devices have become more commonplace in recent years as the demand has increased for safer and more efficient ground-support methods in high-speed tunneling projects as well as in coal, uranium, and other energy-related mining projects.

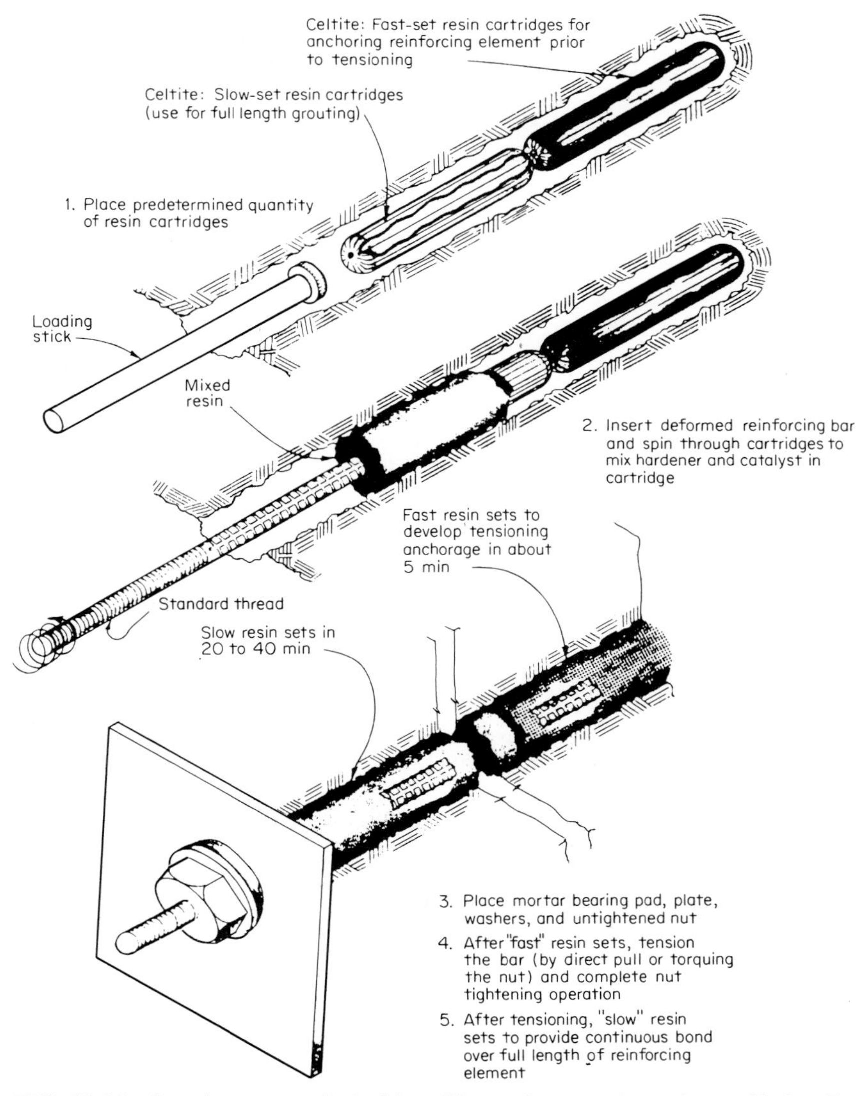

FIG. 10.32 Step-by-step method of installing resin-grouted, tension rockbolts. For vertical bolting, the manufacturer will supply parachute-shaped keepers to hold the resin cartridges in position until the bolt is inserted in the drill hole. *(Celtite, Inc.)*

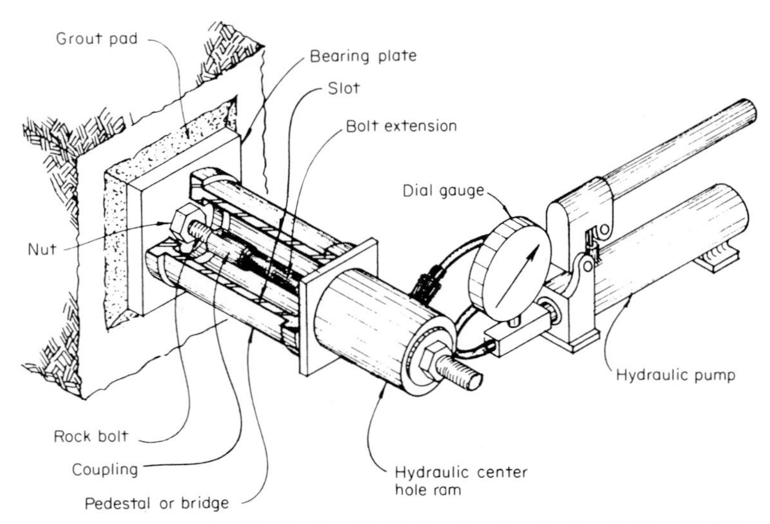

FIG. 10.33 A portable hydraulic device for tensioning bolts. *(Celtite, Inc.)*

Resin bolting has become very commonplace in civil works construction since the early 1970s. Prepackaged resin cartridges can be stored at the jobsite for relatively long periods of time until they are required at the heading. This convenience, as compared with the older method of preparing portland cement grout mixtures at the heading when required, has added to the popularity of resin grouts. Figure 10.32 shows the steps necessary to install and grout bolts using prepackaged resin cartridges. Resin manufacturers provide assistance in determining the proper amount of cartridges to fully encapsulate the bolt with grout. Care must be taken in fissured or seamy rock to insert extra resin to allow for losses to the fissures. Figure 10.33 shows how a hydraulic jack equipped with a center hole to accommodate the bolt, or a bolt-extension coupler, is used to provide the proper tension. Torque wrenches can tension the bolt faster than the jacking device, but a certain percentage of torqued bolts should be checked with the hydraulic ram to assure proper quality control.

Considerations for Construction Expedience in Rockbolting

1. Does the rockbolt length exceed the diameter of the shaft or tunnel, requiring a coupled connection (for both the rockbolt and for the drill steel used to drill the hole)? Does the diameter of either coupling require a drill-hole diameter larger than the manufacturer's recommended hole size at the anchor? This will require one to step down the size of the drill hole.
2. Can workers reach the bolts to torque them and later to test them (also by torquing)? A worker on a ladder can effectively apply only 75 to 100 ft·lb (100 to 135 m·N) of torque to a rockbolt. Two workers on a sturdy platform can attain a maximum torque of 400 to 500 ft·lb (540 to 675 m·N) with off-the-shelf torque wrenches. Air-operated impact wrenches tend to "wind and

unwind" very long bolts, without transmitting the proper torque to the anchor, thus requiring the workers to use either a hand-operated wrench or resin bolts with the thread on the "outside end" of the bolt.

3. Can changing the bolt-spacing length or required tension eliminate, or at least minimize, any of the aforementioned problems? Will the change require more rockbolts, thus offsetting the cost advantage? Will the change allow one to resume drilling and blasting sooner, or will it tie up the working face for a longer period of time?

SHOTCRETE

Shotcrete design is not much different than concrete design. The engineer in both cases assumes that through reasonable quality control in the selection, batching, and placement of materials, it is possible to assure a reliable design strength on which to base calculations. The shotcrete designer, however, can vary the early strength of the mix and can use this early strength to advantage in dealing with progressive rock deterioration in the tunnel or shaft. The authors will briefly discuss methods of assuring reasonable early and final design strengths of a shotcrete mix, and then focus on a few contemporary approaches in using shotcrete to support an underground opening in rock.

Mix Design

Prior to reviewing shotcrete design techniques, it is important to examine the strength-versus-time characteristics of coarse aggregate shotcrete mixes using commercial, fast-setting agents, commonly referred to as accelerators. Figure

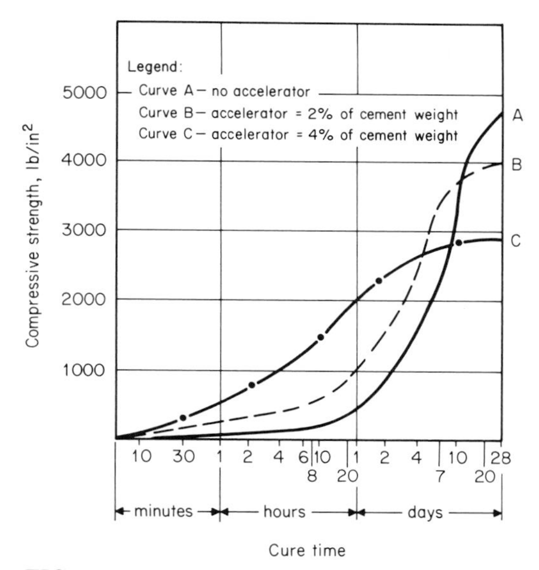

FIG. 10.34 Shotcrete strength versus cure time for a common mix with differing quantities of accelerator added.

10.34 shows strength versus time for three such mixes, with the proportion of each of the three mixes being the same, except for the amount of accelerator introduced to the mixes, expressed as a percentage by weight of the cement in the mix.

Note that for a constant cement content, the best 28-day strength is attained by using no accelerator in the mix. This, however, limits the early strength (4 to 8 h) of the in-place shotcrete, which the tunneler usually needs to prevent rock loosening in blocky ground. Note also that a large percentage (4% in our case) of accelerator may initially yield a high-early strength, but at a sacrifice of 28-day strength.

Prior to the start of excavation, the tunnel or shaft contractor should shoot test panels using various percentages of accelerator. A compromise must then be reached for the final mix design, which would provide an adequate early strength required to meet the anticipated ground conditions, but at the same time not reduce the 28-day strength to an unreasonably low value as compared with the final strength of a nonaccelerated mix.

A suggested mix, which is readily obtainable under normal field conditions, would have the following characteristics:

1. Aggregate ratio
  a. Sand—[¼ in (6 mm) minus]—60% by weight
  b. Coarse aggregate [¾-in (19-mm) maximum size]—40% by weight
2. Cement and accelerator
  a. Cement—650 to 700 lb/yd^3 (385 to 415 kg/m^3)
  b. Accelerator—1 to 2%, by weight, of cement quantity
3. Water
  a. Water-cement ratio of 0.38 to 0.45 by weight
4. Design properties
  a. Compressive strength at 8 h = 500 lb/in^2 (345 N/cm^2)
  b. Compressive strength at 7 days = 2500 lb/in^2 (1725 N/cm^2)
  c. Compressive strength at 28 days = 4000 lb/in^2 (2750 N/cm^2)

It should be mentioned that regulated set, reg-set, cements are being developed that promise to yield much higher early strengths with very little sacrifice of final strength. But for the most part, contractors are still using commercially produced portland cements and accelerators at this writing.

Design of Shotcrete Support Systems

Shotcrete can provide primary support in a rock tunnel or shaft, by itself or in conjunction with rockbolts or steel ribs. In many cases, it provides both the primary and permanent support to an underground opening. Rabcewicz[24] examined the hydrostatic forces acting on a shotcrete-supported opening according to the elasto-plastic stress distribution of Kastner.[25] He then used Fenner's[26] and Talobre's[17] equations to express the following relationship:

$$p_i = -C \cot \phi + [C \cot \phi + p_o (1 - \sin \phi)] \left(\frac{r}{R}\right)^{(2 \sin \phi)/(1 - \sin \phi)}$$

where p_i = required lining resistance
c = cohesion
ϕ = angle of friction
p_o = earth pressure, or γH, where H is overburden
r = radius of cavity
R = radius of protective zone

Assuming negligible cohesion, the equation is simplified to

$$p_i = p_o\,(1 - \sin\phi)\left(\frac{r}{R}\right)^{(2\sin\phi)/(1-\sin\phi)} = p_o(n)$$

where n is shown as a function of ϕ in Fig. 10.35.

When the radius of the protective zone equals the radius of the opening,

$$p_i = p_o\,(1 - \sin\phi)$$

and the cavity attains equilibrium without deformation. Examination of Fig. 10.35 indicates that using $r/R = 1.0$ would give the user the maximum p_i, or the thickest lining, while r/R values lower than 1.0 would give us low support requirements. The fact that one can calculate support requirement without regard to the size of the opening in rock has lead Hopper, Lange, and Mathews[27] to introduce another term, $\gamma(R - r)$, to the Fenner-Talobre equation as follows:

$$p_i = p_o\,(1 - \sin\phi)\left(\frac{r}{R}\right)^{(2\sin\phi)/(1-\sin\phi)} + \gamma(R - r)$$

where γ = unit weight of the ground, and all other terms remain the same.

By inspection, it can be seen that as the ratio r/R decreases, the value p_i decreases to some minimum value, then increases at a rapid rate. Some designers feel that if R is more than twice the value of r, the value of p_i may sometimes be overly conservative. This equation was utilized to explain the difference in sup-

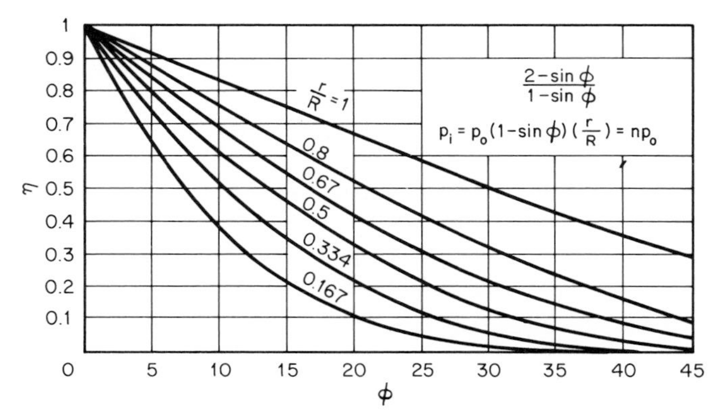

FIG. 10.35 Skin resistance p_i required to establish equilibrium of a cavity as a function of ϕ, angle of internal friction, and $p_o = \gamma H$. *(From L. V. Rabcewicz, Water Power, November/December 1964 and January 1965.)*

port requirements between the small pilot drift for a large-diameter highway tunnel and the main tunnel itself. Without the $\gamma(R - r)$ factor, the value of p_i would have been calculated as the same for either tunnel size. The authors claim that the value for R was compatible with the extensometer readings of the pilot bore, but in no way imply that this will always be the case. Mahar et al.[28] use the calculated value of p_i to determine the thickness t of the support system by the equation

$$t = \frac{p_i \times r}{f'_c} \times \text{FS}$$

where t = thickness of support system (shotcrete in our case)
p_i = calculated lining resistive pressure
r = radius of opening
f'_c = compressive strength shotcrete
FS = factor of safety (of 2 to 3)

Subsequent studies by the University of Illinois revealed the need for a more conservative design; thus the equation should be modified to

$$t = \frac{2p_i \times r}{f'_c} \times \text{FS} = \frac{4p_i \times r}{f'_c}$$

where the factor of safety is 2.

Deere et al.[4] examined the progressive failure of rock supported solely by shotcrete, due to fallout of key rock blocks, after the shotcrete has been installed

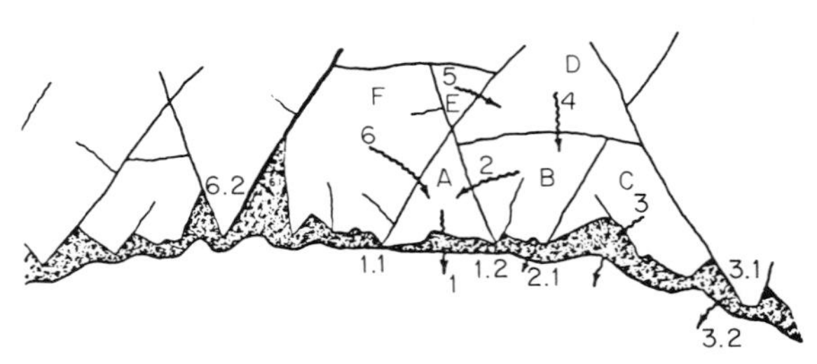

Step 1. Block A drops down, shearing through shotcrete along 1.1 and 1.2 and at each end of the block.

Step 2. Block B rotates counterclosewise and drops out, falling shotcrete in tension at 2.1.

Step 3. Block C rotates counterclosewise and drops out breaking rock-shotcrete bond at 3.1.

Step 4. Block D drops out followed by block E.

Step 5. Block F rotates clockwise and drops out, breaking poor bond between shotcrete and clay that was along weathered joints at 6.1 and 6.2.

FIG. 10.36 Progressive failure in shortcrete-supported rock.

and has reached partial strength. They suggest that the diagonal tension properties of the shotcrete multiplied by the perimeter of the rock block at the intersection with the shotcrete surface should equal or exceed the weight of the block. The progressive steps to failure are shown in Fig. 10.36. It should be pointed out that shotcrete can sometimes hold up far more weight than the diagonal tensile strength at the rock face would indicate through design computations, because in many cases the geometry of the rock blocks can key in a much greater weight. Of course, if the rock-block geometry is of poor structural configuration, shotcrete may not be effective.

Mahar et al.[28] suggest the following rules of thumb for using a thin membrane of shotcrete as the sole support in loosening ground: (1) the diameter of the opening should be less than 9 m (30 ft); (2) rock joints should be rough and clean; and (3) rock blocks should be less than 1.5 m (5 ft) in size, with no large wedges apparent.

We have isolated one such block and show it in Fig. 10.37 as trying to rotate counterclockwise into a tunnel around an axis $A - B$. The only resistive force is provided by a layer of shotcrete with thickness t and a certain diagonal tension strength f_v (commonly referred to as "shear strength"), acting at 45° to the plane of the shotcrete over a diagonal thickness of $1.41t$. Deere uses $f_v = 4\sqrt{f'_c}$ as the diagonal tension strength of the concrete or shotcrete.

Heuer[29] examines the long-term requirement of shotcrete to support the slowly increasing swelling-and-squeezing pressure on a tunnel lining. He suggests that

$$t_e = \frac{\text{LF}}{0.85\phi}\left(\frac{p_i r}{f'_c}\right)$$

where t_e = effective thickness of shotcrete lining
$\text{LF}/0.85\phi$ = load-reduction and capacity-reduction factor for which a value of 2 is recommended for temporary shotcrete support and 2.5 to 3.0 for a final lining
r = radius of tunnel
p_i = average radial pressure
f'_c = normal unconfined compressive strength of shotcrete

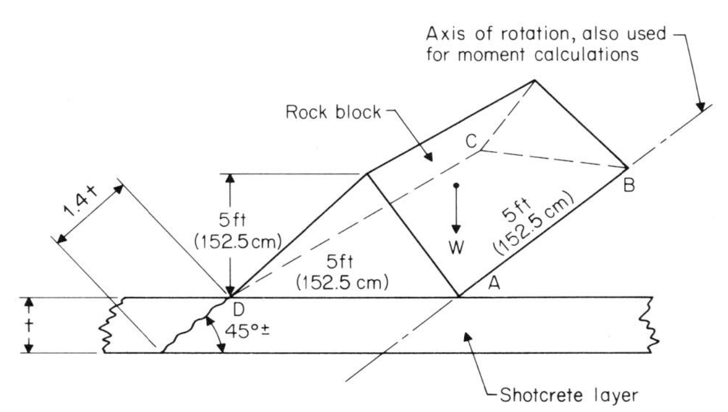

FIG. 10.37 Shotcrete layer preventing a block of rock from falling out due to rotation around axis A-B, based on Cecil's[32] theory.

The effective thickness t_e is somewhat less than the nominal thickness of the applied coating of shotcrete. Heuer suggests that t_e is 2 in (5 cm) less than the nominal thickness for machine-bored tunnels, or 4 in (10 cm) less in the case of drilled and blasted tunnels.

Therefore, for temporary shotcrete support in a drilled and blasted tunnel, with slow swelling pressures anticipated, the nominal amount of shotcrete required would be

$$t_{\text{nominal}} = 2\frac{pr}{f'_c} + 4 \text{ in (10 cm)}$$

Fernandez-Delgado et al.[32a] found, through both field testing and large-scale laboratory studies, that the bond between the shotcrete and rock is quite significant in determining its effectiveness as a support in tunnels with rock blocks that may in time loosen and bear against the lining. In rock with poor adhesion character-

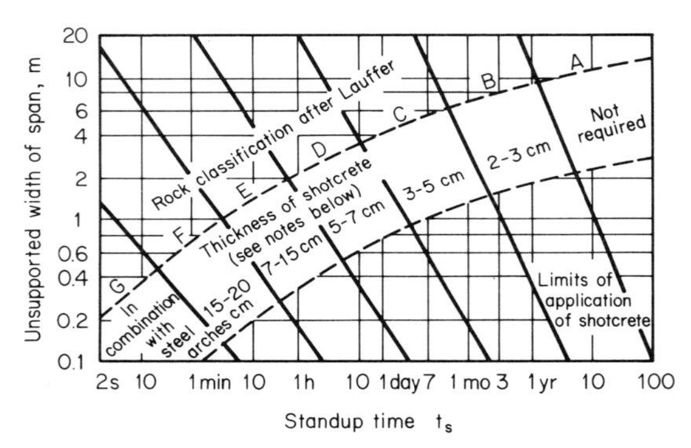

NOTES:

B. Alternatively, rockbolts on 1.5- to 2.0-m spacing with wire net; occasionally reinforcement needed only in arch.

C. Alternatively, rockbolts on 1.0- to 1.5-m spacing with wire net; occasionally reinforcement needed only in arch.

D. Shotcrete with wire net; alternatively, rockbolts on 0.7 to 1.0-m spacing with wire net and 3-cm shotcrete.

E. Shotcrete with wire net; rockbolts on 0.5- to 1.2-m spacing, with 3- to 5-cm shotcrete sometimes suitable; alternatively, steel arches with lagging.

F. Shotcrete with wire net and steel arches; alternatively, strutted steel arches with lagging and subsequent shotcrete.

G. Shotcrete and strutted steel arches with lagging.

FIG. 10.38 Lauffer's chart according to Linder.[30]

istics, additional shotcrete may have no significant effect after a certain thickness is reached.

For the most part, shotcrete design is based on empirical data obtained from previous successes (or failures) in the area of shotcrete-supported tunnels. Here we again refer to Lauffer's chart (Fig. 10.38) used by Linder[30] to compare rock quality types A through G with the shotcrete thickness required to support a tunnel alone, or in conjunction with some other support device.

Sample Problem (Worked in U.S. Customary Units Only)

The rock load can be determined using Stini's condition no. 7, Terzaghi's condition no. 8, or Lauffer's E-F.

Angle of friction: $\phi = 25°$, cohesion negligible, joints rough and clean

Radius of plastic zone: $R = 50$ ft

Radius of tunnel: $r = 9$ ft (horseshoe-shaped)

Rock density: $\gamma = 165$ lb/ft^3

Compressive strength: $f'_c = 5000$ lb/in^2 for 28-day strength
$f'_c = 800$ lb/in^2 for 8-h strength

The maximum rock block size is a 5-ft-high pyramid with a 5-ft × 5-ft × 5-ft base.

1. Find the necessary shotcrete thickness using the Rabcewicz method with $r = R$ and a Stini condition no. 7 load of 70 ft based on an adjustment to an 18-ft tunnel diameter.

$$p_o = \gamma H = 165 \text{ lb/ft}^3 = 11{,}550 \text{ lb/ft}^2 = 80 \text{ lb/in}^2$$

from the chart in Fig. 10.35, for

$$\phi = 25° \qquad \frac{r}{R} = 1 \qquad n = 0.58$$

$$p_i = p_o(n) = 80 \times 0.58 = 46.4 \text{ lb/in}^2$$

$$\text{Illinois/Mahar's } t_{28 \text{ day}} = \frac{4p_i r}{f'_c} = \frac{46.4 \times 9 \times 12}{5000} \times 4 = 4.0 \text{ in}$$

$$\text{Heuer's } t_{28 \text{ day}} = 2\frac{p_i r}{f'_c} + 4 \text{ in}$$

$$= 2\frac{46.6 \times 9 \times 12}{5000} + 4 = 6.0 \text{ in}$$

2. Find the shotcrete thickness for the same loading using the Hopper-Lang-Mathews method:

$$\frac{r}{R} = \frac{9}{50} = 0.18$$

from the chart in Fig. 10.35, for

$$\phi = 25° \qquad \frac{r}{R} = 0.18$$

$$n = 0.05$$

$$p_i = \gamma H(n) + \gamma(R - r) = 165 \times 70 \times 0.05 + 165 \times (50 - 9)$$

$$= 7342 \text{ lb/ft}^2 = 51 \text{ lb/in}^2$$

$$\text{Illinois/Mahar's } t_{28 \text{ day}} = \frac{51 \times 9 \times 12}{5000} \times 4 = 4.4 \text{ in}$$

$$\text{Heuer's } t_{28 \text{ day}} = \frac{2 \times 51 \times 9 \times 12}{5000} + 4 = 6.2 \text{ in}$$

3. Find the necessary shotcrete thickness to support a 5-ft × 5-ft pyramid of rock as in Fig. 10.37 using the tensile strength of 8-h-old shotcrete.

$$\text{Weight } W = 165 \times 5 \times 5 \times \frac{5}{2} = 10{,}312 \text{ lb}$$

(It should be pointed out that we are analyzing the forces due to the weight of the block only. Additional forces caused by increasing ground pressure on the opening or by swelling seams are not analyzed here.)

The moment due to weight W about axis $A - B$ is

$$10{,}312 \times \frac{-5}{2} = -25{,}781 \text{ ft}\cdot\text{lb}$$

$$f_v = 4\sqrt{f_c'} = 4\sqrt{800} = 113 \text{ lb/in}^2$$

The resistive force of shotcrete on sides

$$DC = AD = BC$$

will counteract the weight of the block.

The resistive force of each side is

$$60 \text{ in} \times t \times 1.41 \times 113 \text{ lb/in}^2 = 9560\, t \text{ lb per side}$$

The moment of the shotcrete resistive force is

$$\text{Force } AD\left(\frac{5}{2}\right) + \text{force } CD\left(\frac{5}{2}\right) + \text{force } CD(5)$$

$$= 9560t\left(\frac{5}{2}\right) + 9560t\left(\frac{5}{2}\right) + 9560t(5) = 95{,}600t \text{ ft}\cdot\text{lb}$$

For the block equilibrium around axis $A - B$

$$95{,}600t - 25{,}781 = 0$$

$$t = 0.27 \text{ ft} = 3.23 \text{ in}$$

So, we can choose from a variety of solutions for our requirement to temporarily support a tunnel opening solely by a shotcrete coating. It appears that 3 or 4 in (7.6 or 10 cm) of shotcrete would be sufficient for early protection of the tunnel from deterioration due to progressive block failure, while 6 in (15 cm) of shotcrete would protect us from moderate squeezing and swelling after the tunnel has been opened for some period of time. Lauffer's chart for condition F/E would recommend approximately 15 cm (6 in) of shotcrete, plus marginal use of rockbolts.

Empirical data from nearby projects could also provide significantly more accurate design basis than this approach. In many civil works tunnels, rock chambers, or shaft projects, the owner or engineer will specify the type of shotcrete support required, with particular emphasis on early and 28-day strengths and the lining thickness. Temporary bolts or steel ribs may also be specified to be installed in conjunction with the shotcrete.

Wire Reinforcement in Conjunction with Shotcrete

Sample Problem (Worked in U.S. Customary Units Only)

Advocates for use of mesh could use a rough approximation of its value in supporting the 5-ft × 5-ft × 5-ft block in the previous calculation. Referring again to Fig. 10.37, assume we had embedded 6 × 6, W_3/W_3 wire mesh in the 3.23 in of shotcrete. The area of the wire would be equal to 0.06 in^2/ft each way.

If the wire has a working strength, still in its elastic range, of 28,000 lb/in^2, it would provide a resistive force of

$$0.06 \times 28{,}000 = 1{,}680 \text{ lb per foot of length}$$

The additional resistive moment around axis *A-B* would be

$$1{,}680 \text{ lb} \times 5 \times \frac{5}{2} + 1{,}680 \times 5 \times \frac{5}{2} + 1{,}680 \times 5 \times 5 = 84{,}000 \text{ ft}\cdot\text{lb}$$

Since we had earlier calculated that 3.23 in of plain shotcrete would provide 25,781 ft·lb of resistive force, the wire mesh would be equivalent to 84,000/25,781 × 3.23, or 10½ in of 8-h-old shotcrete!

Going one step further, the tensile strength of 28-day-old shotcrete would be

$$f_v = 4\sqrt{5000} = 283 \text{ lb/in}^2$$

and, by inspection, we can see that the same wire mesh would be equivalent to 10½ × (113/283) or 4¼ in of 28-day-old shotcrete.

This would suggest that wire mesh would greatly assist in providing the tensile strength necessary in blocky ground that the shotcrete cannot provide on its own.

Opponents of using the wire mesh would argue that to expect any shotcrete (whether reinforced or not) to be reasonably effective, we must have an intimate bond between the shotcrete and the rock surface, and that the wire mesh reduces this bond by causing voids in the shotcrete behind the mesh. An uneven rock surface could conceivably create voids between the wire and rock, and even greater voids would result if the field crew were only a little bit careless in the mesh installation and the shotcrete application. The opponents would also argue that in order to be effective, shotcrete should be applied as soon as possible after the ground is opened up. The time required to attach the mesh closely to the rock surface (usually by using many short rockbolts) would detract from the shotcrete's effectiveness, plus expose the workers to unnecessary risk while installing wire mesh.

Some designers arrive at a halfway point in this argument by first installing a coat of nonreinforced shotcrete immediately after opening up the rock, then installing wire mesh and a second coat of shotcrete. This provides a safer working environment for the meshing crew and a smoother surface upon which to install the mesh. This method has been successfully utilized on many projects.

A recent development is to introduce steel wires, or "fibers," into the shotcrete mix to increase the strength of the coating. Although there are still some field problems associated with plugging of the mixing and application machinery and excessive rebound of the fibers during application, there have been some dramatic increases observed in the strength of the coating. Alberts,[33] using 1.3 to 1.4% fibers (by volume of mix), reports a 50% increase in tensile strength and 180% increase in flexural strength as compared with a plain shotcrete mix.

Further testing by Moran[34] with shotcrete mixes containing 0.5 to 1.5% fibers by volume reveal that although the compressive strength may not increase substantially over nonfiber mixes, the toughness index, or the ability of the fiber mix to carry a load after cracking and deformation, is dramatically increased over the toughness of a nonfiber mix.

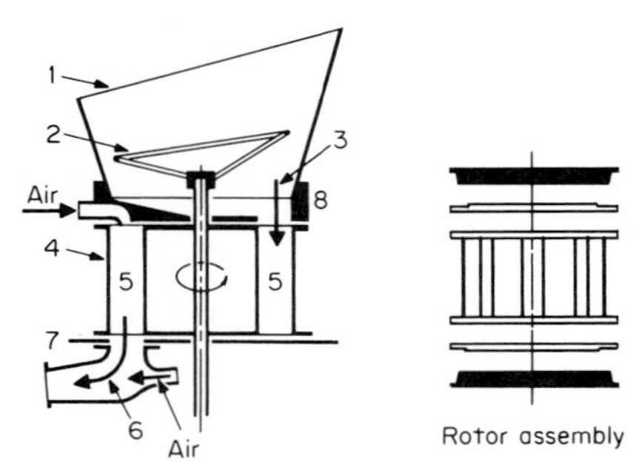

FIG. 10.39 Operating principle of the dry-shotcrete gun. The dry or damp material being gunned is loaded into the pressureless hopper (1) where a rotary stirring spider (2) loosens and pushes it toward the feed hole (3). Driven by an electric, air or diesel motor, the hermetically sealed rotor (4) rotates between the clamping plate (8) of the hopper and the cover plate (7) of the base. The rotor contains compartments (5), open at both ends, into which the material falls when it passes under the feed hole, and out of which it falls when it passes over the discharge port (6) on the opposite side, into the path of the compressed air current. *(Allentown Pneumatic Gun, Inc.)*

Field Practice

Shotcrete may be applied by the wet or dry process.

In the wet process, the mix is delivered as a low-slump concrete and is generally pumped to a nozzle. At the nozzle, compressed air is injected into the stream of flowing concrete and propels it toward the rock surface. Wet shotcrete may also be applied by loading the mix into a sealable chamber and then forcing the mix through the nose and nozzle by pressurizing the chamber with compressed air. In the wet process, the accelerator will be in liquid form and will normally be introduced to the mix at the nozzle.

The dry process generally utilizes a mixture of dry cement and damp sand and aggregate, which is introduced into rotary guns (see Figs. 10.39 and 10.40). The guns propel the damp mix through a hose to the nozzle where the proper amount of water is introduced into the stream of material. Accelerator may be introduced to the mix either as a liquid at the nozzle or as a dry powder as the mix is loaded into the hopper of the rotary gun. The nozzleman is the key to an efficient and high-quality shotcrete program. Some owners require that the contractor's nozzleman must have creditable prior experience on similar shotcrete projects and may also require a qualified backup nozzleman.

Figure 10.41 illustrates a nozzleman applying a wet shotcrete coating to a rock surface. Prior to shotcreting, he should clean the rock surface of all dust and dirt from blasting operations to assure a good bond between the rock and the coating. He is shown positioning the nozzle so that the stream of shotcrete hits the rock in a perpendicular direction. He also positions the nozzle so that it is 3 to 5 ft (0.9 to 1.5 m) from the rock surface. Some contractors mount the nozzle on a remote-controlled boom equipped with a swivel head. In this case, the nozzleman would operate the boom and water controls from a position at the rear of the nozzle, where he would be exposed to less shotcrete rebound and possible rockfalls. The contractor should provide the nozzleman and other members of the shotcrete crew with face-mask filters, eye shields, and protective skin ointment to protect them from the caustic effects of the cement and accelerator.

As in any concrete operation, care in supplying the sand, aggregate, cement, and additives under strict quality-control standards as well as good placing procedures will result in a high-quality shotcrete. A cautionary note: Deficient design or field application can prove more dangerous than no shotcrete at all,

FIG. 10.40 A dry-shotcrete rotary gun with the hopper clamping plate folded back, exposing the rotor assembly. The gun operator must assure that foreign objects or extra large chunks of aggregate do not enter the rotor pockets, or plugging will surely occur. *(Allentown Pneumatic Gun, Inc.)*

FIG. 10.41 Application of wet shotcrete over wire mesh reinforcement in a large-diameter rock shaft. Proper manipulation of the nozzle will minimize rebound and will result in a dense, high-quality coating.

because the coating will cosmetically hide potentially treacherous rock conditions that would otherwise be detected.

SHAFT SUPPORTS

The shaft contractor uses the same materials as the tunnel contractor (i.e., plates, ribs, bolts, etc.) to provide temporary support while sinking the shaft.

Design

Liner plates may be utilized as the sole support in small-diameter earth shafts. The American Iron and Steel Institute[11] provides a chart (Fig. 10.42) to determine the equivalent fluid pressure acting on a liner-plate shaft. As in tunnel liner-plate design, thrust on the plate, when utilized as the sole support, is determined by the equation $T = PD/2$, where P is the radial pressure and D is the shaft diameter. The thickness of the plates may be increased in stages as the shaft deepens, to meet the increasing thrust on the seam.

Earth pressures on shafts may also be calculated using triangular, trapezoidal, or rectangular load distributions as in soldier-pile or cofferdam construction, covered elsewhere in this book.

The designer should consider the additional load due to groundwater pressure when utilizing liner plates. However, if ribs and timber lagging are used, the water will drain through the support, and its effect on support loading should not be considered.

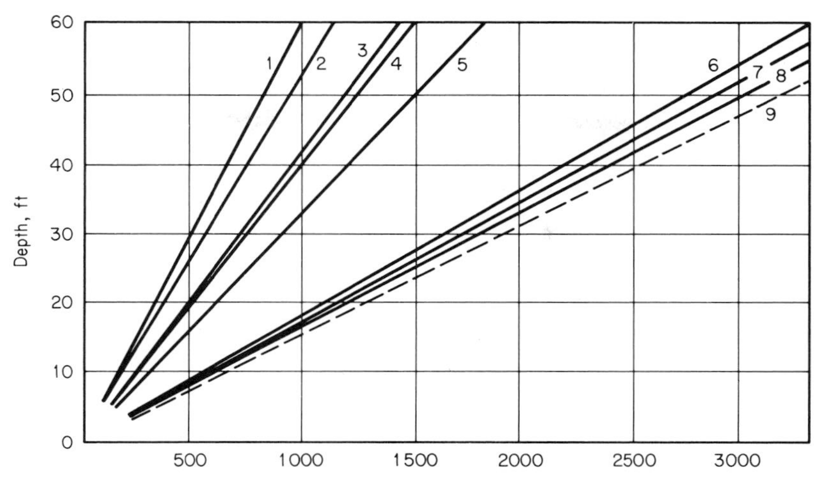

1. Clay: Lumpy and dry
   Earth: Loose and either dry or slightly moist
2. Earth: Fairly most and packed
3. Earth: Perfectly dry and packed
4. Clay, sand, and gravel mixture
5. Drained river sand
6. Earth: Soft flowing mud
7. Clay: Damp and platic
8. Earth: Soft, packed mud
9. Hydrostatic pressure of water

FIG. 10.42 Equivalent fluid pressure for caisson construction. *(American Iron and Steel Institute, Handbook of Steel Drainage and Highway Construction Products, 2d ed.)*

FIG. 10.43 Collaring a 32-ft finished diameter mine shaft in Colorado, using steel ribs with 36-ft outside diameter liner plates. *(Harrison Western Corp.)*

FIG. 10.44 A 90-ft deep, center-strutted shaft in Chicago overburden: (*a*) Circular ring beams and lagging are supported by tie-rods hung from a rectangular structural steel bearing set. (*b*) Corner of the bearing set is supported by a drilled and concrete-filled bearing pile. *(S. A. Healy Co.)*

Field Practice

The contractor usually builds a collar or bearing set to support the liner plates or ring beams in a vertical direction and prevent them from settling as the shaft is deepened.

In Fig. 10.43 a contractor is shown constructing a collar for the first 12 ft (3.6 m) of a steel-rib and liner-plate shaft that will extend an additional 70 ft (21 m) to sound rock. Concrete will be placed between the outside of the liner plate and the surrounding soil. The concrete collar will provide a bearing surface to counteract the weight of the ribs and plates as the shaft is deepened. Note the tie-rods and pipe spreaders on the left, as well as the hardwood wedges between the ribs and plates. The outside diameter of the liner plate ring is 36 ft (11 m). Pouring a 2-ft (60-cm)-thick permanent concrete lining inside the plates will result in a 32-ft (9.75-m) inside diameter shaft.

Some contractors hang the temporary earth support from a steel bearing set. Figures 10.44*a* and *b* illustrate a rectangular bearing set, resting on concrete bearing piles, used to support a large ring beam and timber-lagged earth shaft. The bearing piles may be drilled to rock, or to a competent earth strata. Note also in Fig. 10.44*a* that the timber lagging is between the rib flanges, but is wedged tightly against the outside flange to minimize voids.

FIG. 10.45 View from the bottom of a deep shaft under construction in Idaho. *(J. S. Redpath Corp.)*

Contractors sinking deep mine shafts in rock generally advance a permanent concrete lining along with the excavation, keeping it within a few diameters of the working face. Figure 10.45 illustrates a deep rock shaft under construction. The camera is pointing upwards from the bottom of the shaft. Split Sets and chain-link fabric provide temporary ground support until the concrete lining is placed. The forms are immediately above the temporary support. Note also the bottom of the multiple deck staging with its wells or openings for the muck-removal buckets. The staging allows miners to install temporary and permanent supports concurrent with the excavation. The staging is equipped with its own independent hoisting system to allow more flexibility to the concurrent operations.

Shafts are generally sunk to provide access for constructing a horizontal tunnel. At the shaft bottom, the contractor will cut a hole through the temporary shaft support in order to provide a tunnel portal. Bracing must be provided to redistribute the vertical and tangential shaft loads around the tunnel portal. Figure 10.46 illustrates a tunnel portal at the bottom of a shallow ring beam and lagged earth shaft. Vertical posts will redistribute the vertical loads acting on the shaft lining, while the tangential loads may be redistributed by a lintel beam, or

FIG. 10.46 A shallow Chicago earth shaft with a tunnel portal at its bottom. *(J. Lattyak, MSDGC.)*

by doubling up on the shaft ring beams at the elevation of the portal. Also note that due to the shallowness of the shaft, the contractor chose to drive vertical timber lagging *outside* the ring beams as the shaft was deepened.

REFERENCES

1. Terzaghi, Karl: "Rock Defects and Loads on Tunnel Supports," sec. I, in R. V. Proctor and T. L. White, *Rock Tunneling With Steel Supports,* Commercial Shearing and Stamping Company, The Youngstown Printing Company, Youngstown, Ohio, 1946, Revised 1968.
2. Stini, J.: *Tunnelbaugeologie,* Springer, Vienna, 1950.
3. Bierbaumer, A.: *Die Dimensionierung Des Tunnel-bauerworks,* Engleman, Liepzig, 1913.
4. Deere, D., R. B. Peck, J. E. Monsees, and B. Schmidt: "Design of Tunnel Liners and Support Systems," DOT Contract 3-0152, University of Illinois, February 1969.
5. Lauffer, H: "Gebirgsklassifikation für den Stollenbau," *Geologie Und Bauwesen,* vol. 24, no. 1, pp. 46–51, 1958.
6. Proctor, R. V., and T. L. White: *Rock Tunneling with Steel Supports,* Commercial Shearing and Stamping Company, Youngstown Printing Company, Youngstown, Ohio, 1946, Revised 1968.
7. Proctor, R. V., and T. L. White: *Earth Tunneling With Steel Supports,* Commercial Shearing, Inc., Youngstown, Ohio, 1977.

7a. Allgood, J.: "Summary of Soil-Structure Interaction," Tech. Rep. no. R771, U.S. Naval Civil Engineering Laboratory, National Tech. Informational Services, Springfield, Virginia, 1972.

8. Szechy, K.: *The Art of Tunnelling,* Akademiai Kiado, Budapest, 1970.
9. Wickham, G., and H. Tiedemann: "Ground Support Prediction Model (R.S.R. Concept)," Tech. Rep. no. 125, Jacobs Associates, for Advanced Research Projects Agency of the U.S. Department of Defense, Washington, D.C., 1974.
10. *Design Manual,* sec. 13, American Association of State Highway and Transportation Officials, Washington, D.C., 1970.
11. American Iron and Steel Institute: *Handbook of Steel Drainage and Highway Construction Products,* The Lakeside Press, New York, 1971.
12. Hewett, B. H. M., and S. Johannesson: *Shield and Compressed Air Tunneling,* McGraw-Hill, New York, 1922.
13. Leontovich, V.: *Frames and Arches,* McGraw-Hill, New York, 1959.
14. Obert, L., and W. Duvall: *Rock Mechanics and the Design of Structures,* John Wiley & Sons, New York, 1967.
15. Panek, L.: *Combined Effects of Friction and Suspension in Bolting Bedded Mine Roofs,* U.S. Bureau of Mines, Washington, D.C., 1962.
16. Panek, L. A., and J. McCormick: chap. 13 in Panek and Williams, *Mining Engineering Handbook,* Society of Mining Engineers, 1973.
17. Talobre, J.: *La Mecanique des Roches,* Dunod, Paris, 1957.
18. Hoek, E.: "Estimating the Stability of Slopes in Opencast Mines," *Trans. Institution of Mining & Metallurgy,* London, 1970.

19. Jaeger, C.: *Rock Mechanics and Engineering,* Cambridge University Press, London, 1972.

19a. Cording, E., and J. Mahar: "Index Properties and Observations for Design of Chambers in Rock," *Engineering Geology,* Elsevier Scientific Publishing Company, Amsterdam, 1978.

20. Benson, R., R. Conlon, A. Merrit, P. Joli-Coeur, and D. Deere: "Rock Mechanics at Churchill Falls," *Symposium Trans. American Society of Civil Engineers,* New York, January 1971.

21. Provost, A., and G. Griswold: "Excavation and Support of the NORAD Expansion Project," Paper 11, British Tunnelling Society, *Tunnelling '79,* London, March 1979.

22. Haas, C.: "Rock Bolting to Prevent Shear Movement," *Symposium Proceedings,* M&E Series 79-08, SMA-American Institute of Mining Engineers, New York, February 1979.

23. Scott, J.: "Testing of Friction Rock Stabilizers," *Proc. SME-American Institute of Mining Engineers,* New York, March, 1977. ("Split Set" is a patented trade name of the Ingersoll-Rand Company and James J. Scott.)

24. Rabcewicz, L: "New Austrian Tunnelling Method," and parts I and II, *Water Power Magazine,* November and December 1964, January 1965.

25. Kastner, H.: *Statik Dei Tunnel Und Stollenbaues,* Springer, Berlin, 1965.

26. Fenner, R.: *Untersuchungen Zur Erkenntis Die Gebirgsdrucks,* Glückauf, 1937.

27. Hopper, R., T. Lang, and A. Mathews: "Construction of Straight Creek Tunnel, Colorado," *Proc. North American Rapid Excavation and Tunneling Conference,* 1972.

28. Mahar, J., H. Parker, and W. Wuellner: "Shotcrete Practice in Underground Construction," U.S. Department of Transportation, Rep. UILU-ENG-75-2018, Urbana, Illinois, 1975.

29. Heuer, R.: "Selection/Design of Shotcrete for Temporary Support," publication SP45, *Use of Shotcrete For Underground Structural Support,* American Society of Civil Engineers-American Concrete Institute, 1973, pp. 160-174.

30. Linder, R: "Spritzbeton Im Felshohlraumbau"*Die Bautechnik,* October 1963.

31. Muller, L.: "Removing Misconceptions on the New Austrian Tunnelling Method," *Tunnels and Tunnelling Magazine,* London, October, 1978.

32. Cecil, O.: "Correlations of Rockbolt Shotcrete Support and Rock Quality Parameters in Scandinavian Tunnels," University of Illinois thesis, Urbana, Illinois, 1970.

32a. Fernandez-Delgado, G., E. Cording, J. Mahar, and M. VanSintJan: "Thin Shotcrete Linings in Loosening Rock," *Proc. Rapid Excavation and Tunneling Conference,* Atlanta, June 1979.

33. Alberts, C.: "Steel Fiber Shotcrete," report for Stabilator AB, Stockholm, on its part in the North Research Project, 1976.

34. D. Moran: "Steel Fibre Shotcrete—A Laboratory Study," *Concrete International,* January 1981.

ACKNOWLEDGMENTS

The authors wish to thank the following individuals for supplying information, photographs, illustrations, and other valuable assistance in the preparation of this chapter:

J. Farrell, D. Pierce, P. Mertineit, B. Coulter, and M. Bertoldi, Harrison Western Corporation, Denver
E. Glibbery, Morrison-Knudsen Company, Inc., Boise, Idaho
T. King, Tosco Corporation, Denver
H. Henson, Colorado Highway Department, Denver
G. Wickham and H. Tiedemann, Jacobs Associates, San Francisco
P. Tilp and K. Schoeman, U.S. Bureau of Reclamation, Denver
H. Jacoby and L. Nicolau, Grow Tunneling Corporation, New York
E. Cording and R. Heuer, University of Illinois, Urbana
A. Corbetta, CF&I Steel Corporation, Pueblo, Colorado
G. Malina and D. O'Connor, S. A. Healy Co., McCook, Illinois
G. Greenfield, Celtite, Inc., Cleveland
N. Dahmen, The Robbins Co., Seattle
J. Lattyak, Metro Sanitary District of Greater Chicago
R. Hendricks and, B. McKinstry, J. S. Redpath Corp., Tempe, Arizona
W. Roberts, Allentown Pneumatic Gun, Inc., Pennsylvania
W. Harrison, Frontier-Kemper, Evansville, Indiana

11

Roadway Decking

by Bernard Monahan

Constructors learn their profession by slowly applying theoretical concepts to the real world of urban construction. This is particularly true of temporary structures. This subject matter is particularly vulnerable to the often-heard criticism that public agencies and responsible professionals are immune to change.

Methods for decking over streets are frozen into the subconscious of the experienced engineer. Experience is a marvelous teacher; just as the home builder who erects a foundation at the ocean's edge learns the meaning of tides and wave action, so too learns the construction engineer who attempts to span a major urban intersection. The variables are great and often defy rational analysis, not because of our inability to define the probable forces and resulting stresses, but more often because of our inability to weigh all of the possibilities in a manner that is safe for the vehicle operator or pedestrian using the surface and also for the construction worker performing daily duties. The surface decking system must be integrated into the general support system that is required for lateral stability.

Bridging may be fabricated out of timber, concrete, or metal. Future generations may develop the use of fiberglass or other lightweight products to the point where they may be substituted for timber, concrete, or metal.

The use of fiberglass sections or plates to span street openings has not yet been recognized as a medium in which additional research would result in savings of life-cycle costs for subway or other subsurface construction. Perhaps future students and practitioners will recognize fiberglass as a viable alternative. For the present, however, this chapter will provide the details that an engineer must evaluate regardless of the element used for the actual spanning of the opening.

An earlier chapter has dealt with the spanning of a simple trench; this chapter will deal with the decking of an entire street in an urban metropolis such as New York or London.

The typical cross section will be simplified in order to clarify the basic requirements. The chosen cross section is as shown in Fig. 11.1.

Figure 11.2 shows a city street decked over for underground construction.

The construction engineer must generalize the design criteria and then address each segment of the design as a special situation.

The need is simply that the engineer is to design a floor-framing system that is capable of supporting pedestrians, vehicles that are normally found traversing urban crossways, and heavy construction equipment. The engineer, given this responsible charge, must come up with a practical design that will accomplish this and withstand the forces of nature as well. Decking systems must also be capable of withstanding impact and vibration loads resulting from various construction methods such as blasting.

DECKING DESIGN

A decking or flooring system might be envisioned as in Figs. 11.3 or 11.4. The decking beams and the decking itself are considered to act as simply supported beams. They are supported by girders which in turn are supported by columns that transmit the loads to the foundation. A proper analysis starts with the imposition of a design load which should include an allowance for impact and a safety

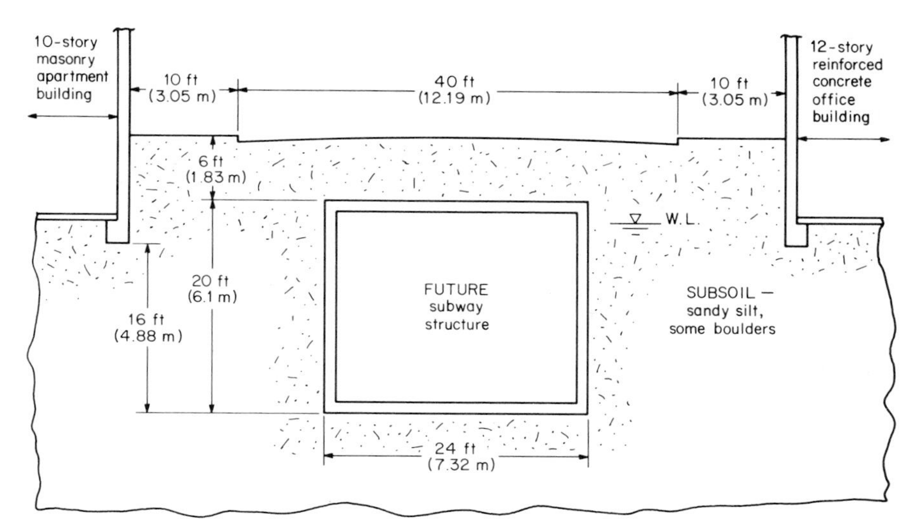

FIG. 11.1 Typical cross section.

FIG. 11.2 Timber decking systems must be designed for construction equipment as well as for public use. *(Thomas Crimmins Contracting Co.—Photo Collection.)*

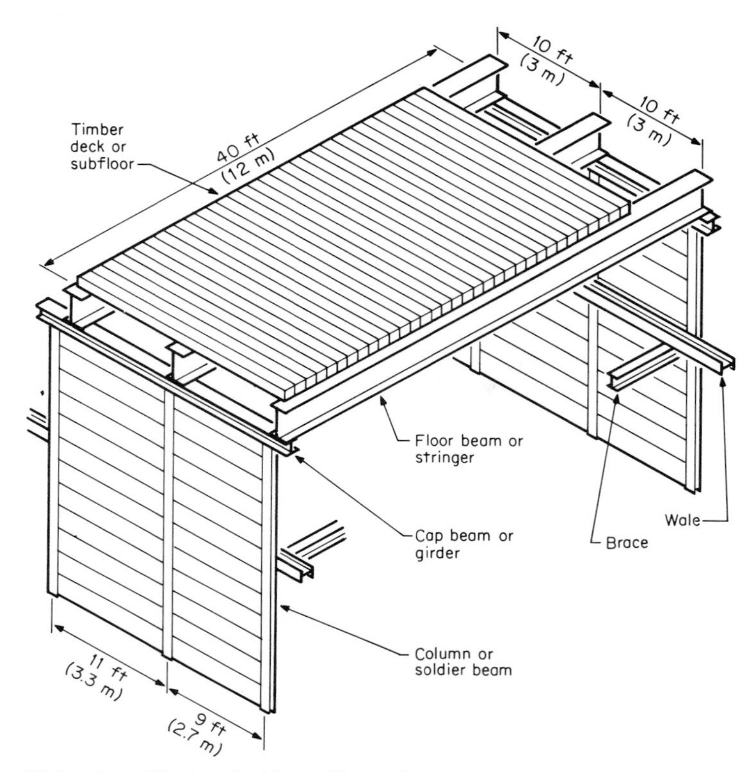

FIG. 11.3 Timber decking with cap beam.

factor to cover unknown variables in the construction procedure and in the makeup of the structural members.

A typical live-load analysis might assume a loaded transit or ready-mix truck (Fig. 11.5) with its wheels positioned at the midspan or a combination of worst conditions. Two trucks might even be positioned side by side while they await discharge instructions. Usually the live load is computed by assuming a uniform load of 200 to 300 lb/ft^2 (9500 to 14,400 N/m^2) or by positioning an anticipated loading in a worst condition.

DECK SUPPORTS

Each column or soldier beam will support one-half of the designated loads for the segment of roadway decking that is located between column supports.

A settlement or load analysis must be performed on the pile sections to determine the validity of assuming that the pile can support the design load. The analysis can consist of a load test or judgment based on past experience, or on an accepted criterion such as the *Engineering News-Record* formula.

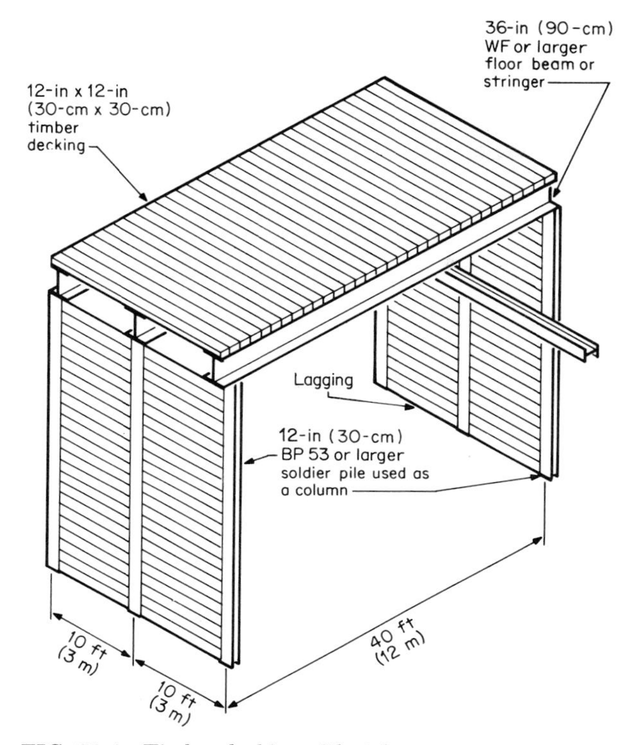

FIG. 11.4 Timber decking with stringers.

Each case is different; however, most probably a high driving resistance will convince most engineers that sufficient penetration has been achieved. The more exasperating condition is the case in which a soldier beam is driven and intercepts a boulder or stone which prevents its further penetration. The column is obviously unsatisfactory; it must be freed; however, the choices to achieve this freedom are limited and almost always time-consuming.

If the engineer is sure that the obstacle is natural and not a manufactured utility such as an electrical duct bank or a water main or some other in-service utility that is not shown on the available plans, then a drill steel might be used to sound the depth of the obstacle. It might be that the borings failed to locate a natural ridge in the substrata rock formation. However, more likely it is a sizable boulder of 2 ft (60 cm) or more in diameter. A pit could be excavated alongside the soldier pile, and hand drilling and chipping techniques might be used to free the tip of the pile. Frequent short piles will force a contractor to predrill the anticipated soldier-pile locations. This may simply consist of drilling to subgrade with a 2- to 3-in (5- to 7.5-cm) auger or boring tool and then redrilling holes that are known to contain obstacles with a well-driller's rock bit.

Blasting techniques are rarely used in the circumstances just described because of the proximity of surface utilities and the need for heavy charges when

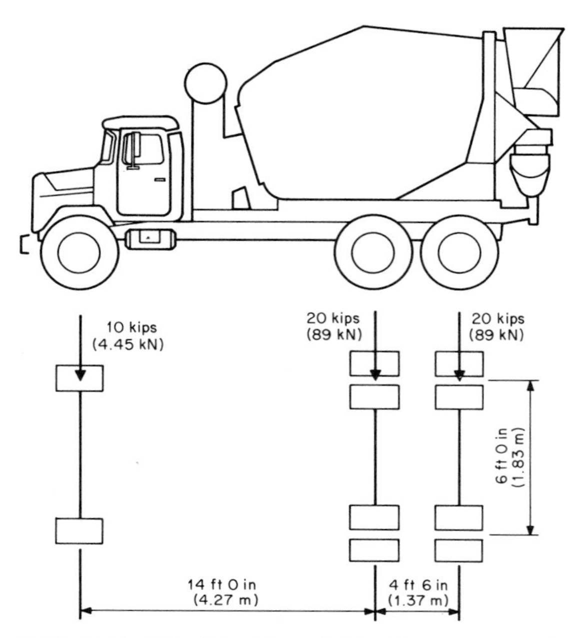

FIG. 11.5 Wheel loading of 15-yd^3 concrete truck [wheel load = 10 kips (4.45 kN) per wheel; resultant of two rear wheels = 20 kips (89 kN); front axle load = 20 kips (89 kN); rear axle load = 40 kips (178 kN) each].

no relieving face exists. A more common approach is to drive the soldier beams and to underpin those cases (hoping they are few) in which the pile is hung up on an obstruction. This case requires that the decking system be completed and the short pile be resupported, usually by means of welding a spliced section to the bottom of the previously obstructed pile. The method, as well as the use of temporary supports, must be judged on the basis of conditions at the time that resupport is necessary. Usually the waler support can be welded to the soldier piles and thus act as a needle-beam support to assure a bridging of the soldier pile during underpinning procedures.

TIMBER DECKING

At this point, let us assume that the columns have been installed satisfactorily and that the contractor is now ready for the actual decking sequence. Figure 11.3 indicates the use of the girder or cap beam. Its use is voluntary; if the column spacing is the same as the decking-beam spacing, say 10 ft (3 m), then each cross beam could sit directly on a column or soldier pile (Fig. 11.4). Remember that the column-support spacing is mandated by the lateral pressures, and if a typical timber lagging is to be installed, then a 10-ft (3-m) spacing will probably result. The omission of a cap or girder beam requires that the column or soldier beam have a fairly close tolerance, say within 6 in (15 cm) of design location. Those soldiers which fall outside of the tolerance will then require a cap beam that spans a minimum of two bays or three soldiers.

The question of cap or girder beam having been answered, the engineer must now choose a decking or floor beam that will support the given loads and whose deflections will not be excessive.

Figure 11.6 gives a typical segment. (Note that bending, shear, and deflection calculations would dictate the use of a W33 x 240 beam for this case.) Deflection of support beams is a factor in assessing hanger requirements. Excessive movements may cause leaks in water lines and shorts in electric ducts.

The important points to remember in analyzing a specific decking requirement are that this is to be a heavily trafficked urban thoroughfare and that any assumptions should always be made on the conservative side of the judgment coin. Timber decking systems must also be designed for the support of construction equipment as well as for public use.

The resulting deck is as shown in Fig. 11.4 and is drawn to show the optimum case of the deck and floor beams sitting directly on the evenly spaced column

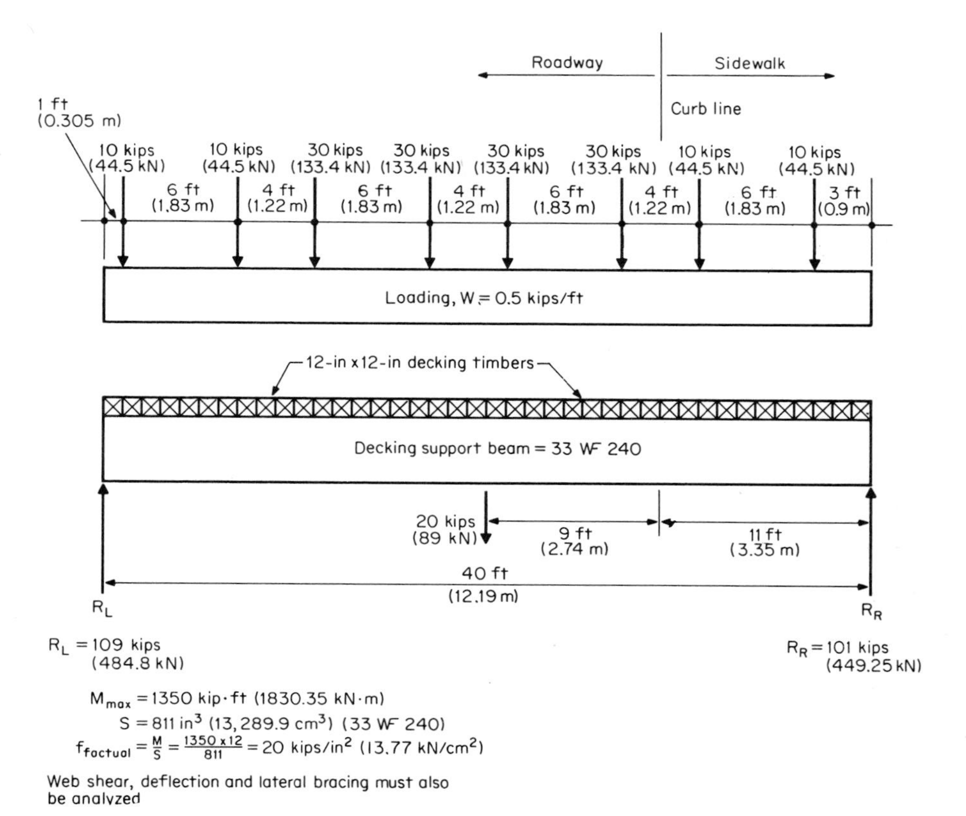

FIG. 11.6 Decking support beam analysis.

supports. In this circumstance the girder or cap beam may be deleted. Figure 11.7 shows the various other members of the decking support, such as braces and tie-rods.

The connections between columns and floor beams are normally bolted to prevent a gradual "walking" of the floor beam off the head of the column. The tops of the columns or the ends of the floor beams should be secured against rotation when subjected to lateral forces. Normally a spacer or tie-beam placed on each end at the third points will be sufficient to prevent rotation on spans of less than 50 ft (15 m).

The floor beam must be checked for buckling and stiffener plates added at vital areas. Usually stiffener plates are required at the supports because of the heavy load concentrations.

Having designed the floor or decking beam, the engineer may wish to give some thought as to how this 33-in (84-cm) beam section can be installed without disrupting the normal traffic patterns. The answer is, of course, that there must be a great deal of inconvenience. However, the hardship to the public is usually minimized by following a standard procedure. One typical sequence is as follows:

1. Install soldier piles (columns).
2. Reroute gas lines and other utilities that either interfere with the floor-beam installation or would cause a hazard to the construction crews.
3. Excavate and vehicular-plate the cross or slot trenches that are sufficient to receive the floor beams, which are to span the entire street below the surface.
4. Close street; reroute traffic, preferably during off hours; and install floor beams.
5. Reinstall vehicular plates and reopen roadway.

Specific periods or hours are usually prescribed for installation of floor beams or any operation that is expected to interrupt the normal traffic flow. In certain

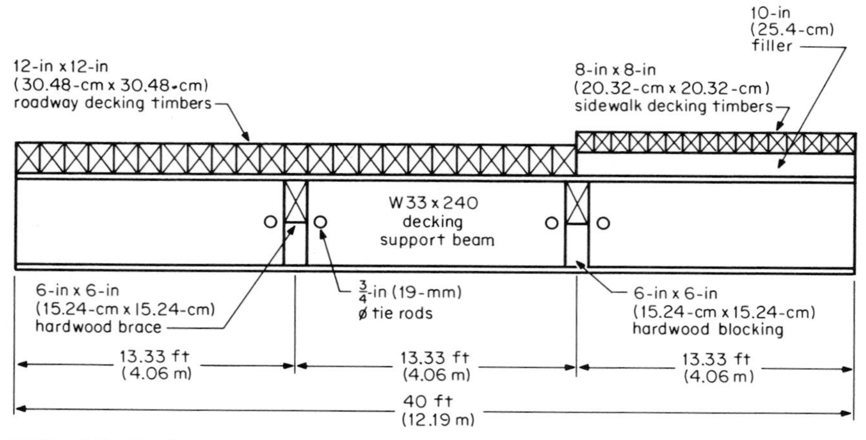

FIG. 11.7 Decking cross section.

business districts it might require a 10 P.M. to 6 A.M. operation with noises muffled to prevent complaints of loss of sleep.

Having completed the design and installation of the floor-beam-support system, one must consider the decking or surface design. The design must be simple, for the deck sections will be subjected to all of the known design variables and to some others that were not envisioned. The decking must be durable, for it will be in service for at least 3 years. The decking must be portable, for it must be remembered that the decking is only a cover or roof that enables a contractor to install the actual structure below it while the city continues its daily routine. Finally, the decking must be simple in order to avoid errors in loading and installation. (See Figs. 11.8 through 11.11.)

One system that fits all of these requirements is a timber-decking sequence. Usually a good grade of 12-in x 12-in (30-cm x 30-cm) timber is a simple beam that is durable and portable. It is relatively light and can often be moved into position by three workers. The timber beams are standard and may be interchanged; the friction factor of wood is relatively good when compared with steel or other metals. Complaints due to slipping or falling are usually minor and are not greater than those of a normal busy roadway. Disadvantages are that wood is a natural material that is often subject to the discontinuities of nature. A timber beam may develop a fault line through an internal knot or flaw after having been installed, and a heavy loading may cause cracking or actual failure. Repetitive wetting and drying cycles may weaken the resiliency of the section and may lead to failure. All things considered, timber has proven itself to be a dependable

FIG. 11.8 Twenty-foot decking timbers placed parallel to traffic flow. *(Thomas Crimmins Contracting Co.—Photo Collection.)*

FIG. 11.9 Decking system being installed along a curb. Timbers are perpendicular to traffic pattern due to subsurface requirements. (Normally roadway decking timbers are 20 ft in length and are placed parallel to the direction of traffic.) *(Thomas Crimmins Contracting Co.—Photo Collection.)*

material for decking. Timber beams tend to redistribute stresses to adjacent members when stressed. Timber is not fragile or brittle; it is rather a flexible material that seems always to meet our expectations even under adverse conditions.

The following definitions and design procedures for a timber deck are introduced at this point.

Structural Lumber: The term "structural lumber" refers to material that has been graded for strength in accordance with nationally recognized standards and to which specific values have been assigned for the various strength properties.

Design Procedures: Lumber used structurally in buildings, bridges, and other structures is subject to the same design procedures and basic formulas that are applied for other kinds of structural materials. Lumber, however, is not an isotropic material and this condition must be recognized in proper design.

Use of Actual Dimension: Calculations to determine required sizes of members must be based on actual or net dimensions.

Allowable Stresses: Allowable bending stresses determined for single-member uses are increased 15%. Allowable unit stresses for normal loading duration may be increased as follows when duration of the full design load

FIG. 11.10 Ten-foot roadway decking timbers that span between two support beams are utilized to provide access hatches for construction operations. *(Thomas Crimmins Contracting Co.—Photo Collection.)*

does not exceed the periods indicated: 15% for 2 months, as for snow; 25% for 7 days, as for loads during construction; 33⅓% for wind or earthquakes; 100% for impact.

Precautions must be taken to prevent the timbers from gradually walking off of the floor-beam flanges. A welded seat angle or metal runner is often used to prevent this type of serious failure.

A 10-ft (3-m) timber-deck beam weighs approximately 500 lb (225 kg), and if it is dropped to the floor of an excavation, it becomes a deadly missile. Carelessness of this nature is negligence and will often result in crews refusing to continue to work and in stop-work orders from governing agencies. A realistic contractor realizes that safety and good workmanship pay high dividends, and for this reason specially skilled workers are normally used when placing decking beams in crowded urban centers.

This discussion, so far, while general, has actually pointed toward the 10-ft (3-m) span that is most commonly found in the New York metropolitan area. New York has a laissez-faire attitude toward trucks and their allowable wheel loads. The streets of New York are probably the busiest in the world and they are often found to contain vehicles that would punish other cities beyond our wildest speculation. New Yorkers care for their streets and property, but they realize that business seems to prosper under a laissez-faire approach. Overloaded vehicles are common in the city and trucks are rarely if ever weighed. The result of this knowledge is that a 10-ft (3-m) span is considered a maximum for a timber deck in New York City. In other cities, this is not the case: truck loads are more reasonable and enforcement is such that a 12-ft (3.7-m) span is allowable and considered a safe practice. Concrete trucks are limited to 10 yd^3 (7.65 m^3) in most other cities, rather than the 15 yd^3 (11.5 m^3) allowed in New York City.

All the previously mentioned design considerations are pertinent to the installation of either a 10- or 12-ft (3- or 3.7-m) span type of deck. The engineer must always ascertain what type of loading can be expected under normal conditions and what is the maximum anticipated loading.

PRECAST DECKING

A recent innovation has been the precast concrete slab. A 5-ft x 10-ft (1.5-m x 3-m) section is prefabricated and shipped to the jobsite. The slab when positioned is more durable than the 12-in x 12-in x 10-ft (30-cm x 30-cm x 3-m) timber beam system; however, it is also more costly. The concrete slab weighs 3 times the weight of wood, and the design normally requires a heavier support system. Larger support floor beams require deeper sections and result in a further lowering of utilities that are normally hung from the underside of the floor-beam system.

The minimum thickness of a concrete panel is usually 9 in (22 cm), with the preferred thickness being 12 in (30 cm).

There is an old adage that "you cannot argue with success." The subway contractors and designers have compared alternate systems on every contract that has ever been let. The results have always been that the most reliable and cost-effective system has remained the timber-beam system. It is being challenged by the precast or concrete slab system; however, the added weight of the concrete system poses a formidable extra cost every time a labor foreman is instructed to lift a slab at an access opening. The 12-in x 12-in x 10-ft (30-cm x 30-cm x 3-m) timber is easily removed and replaced, whereas the concrete slab requires the use of a mechanized crane for its removal and replacement.

UTILITIES

It is worthwhile for the reader to consider the standard conditions, if indeed such exist, that face the subway contractor. The intersection contains utilities running in all four directions. Some of these utilities are close to the surface and will interfere with the proposed decking system. Other utilities, such as gas, are automatically rerouted outside of the excavation, for it is feared that an undiscovered leak or break would cause a serious explosion.

Sewer and steam must take a high priority, for a gravity sewer must be maintained at grade and a steam line requires condensate pump pits at low points in the line. Electric, telephone, and water are usually routed in a manner that allows the normal construction operations to proceed. That is, the duct bank might require a removal of the duct encasement, and the water line might be made satisfactory by utilizing hangers and anchors and securing the line to the underside of the floor beam system.

FIG. 11.11 Excavation progresses by removing decking timbers. Spacer timbers prevent lateral movements in decking beams. *(Thomas Crimmins Contracting Co.—Photo Collection.)*

FUTURE TRENDS

What are some of the innovations that are on the horizon? What might future engineers expect in the twenty-first century? The answers to these or any similar questions fall into a hypothetical concept and we cannot foresee the future. However, certain generalizations might be logically presumed.

Tunneling methods might become so advanced and cost-effective that the massive open-cut type of excavation might become obsolete and be replaced by a tunnel which is installed at a level below that of the utilities. This would enable the contractor to concentrate on the shafts for debris removal rather than the open-pit type of problem.

The central city might become less populated and the need for constant access along and on top of subway structures might become less desirable. This would enable the contractor to provide pedestrian access along sidewalks and to fence in the entire excavation during the construction period. This type of construction is in the same category as the large office building basement construction. Neither of these two descriptions is really innovative; both are practiced today where the conditions are suitable. However, where a new subway is to be placed in a metropolitan setting, the old but tested methods will be slow to give way to innovation. The uses of aluminum mats and fiberglass panels in place of timber decks are likely possibilities.

DESIGN CONSIDERATIONS FOR BEAMS AND COLUMNS

Design of Wood Beams: Investigation of the strength and stiffness requirements of a wood beam under transverse loading should take into consideration the following factors: bending moment induced by the load, deflection caused by the load, horizontal shear at supports, and bearing on supporting members. Any one of these four factors may control the design, although deflection of a temporary beam is usually not a matter of safety and a limit is applied only where appearance or comfort is important.

Design of Simple Solid Columns: The most common form of wood column is a simple solid member of rectangular cross section. A critical factor in the design of such columns (or braces) is the slenderness ratio, or l/d for rectangular sections and l/r for round and other sections. In determining the slenderness ratio, l is the laterally unsupported length measured along the longitudinal axis and is the distance between supports which prevent lateral movement in the direction of the least dimension d or least radius of gyration r. The l/d ratio for a simple solid rectangular column should not exceed 50.

For short solid columns where the l/d ratio is in the order of 10 to 12 (and sometimes higher depending on the species and grade of lumber), the allowable load is governed by the axial compressive strength of the timber without reduction for slenderness.

Treatment: For pressure-impregnated preservatively treated wood, no reduction in allowable stresses is necessary. For wood pressure impregnated with a fire-retardant treatment, a 10% reduction in the allowable stresses is recommended.

REGULATIONS AND RESPONSIBILITIES

The installation must conform to the requirements of the Occupational Safety and Health Act (OSHA) of the federal authorities, as well as the area requirements of the various states and municipalities. The building code usually contains many of the minimum standards that are considered acceptable and is beneficial in guiding the engineer in design requirements. Safety to the public is paramount in determining the minimum limits for installing decking or other structures that are designed and installed to protect the public during construction operations.

The inspection of construction installations is the responsibility of the governing agency. This task may be delegated to an inspection agency; however, the basic responsibility remains with the owner and the design engineer. Usually, the contractor is required to propose installation details that have been prepared by a responsible engineer in the contractor's employ, and these details are then examined by a professional engineer in the employ of the consultant. It is often stated, within the project specifications, that temporary structures are the responsibility of the contractor and that the examination of these proposed details by the engineering consultant does not relieve the contractor of this responsibility. In practice, accidents usually result in damages that are shared by all parties.

DRAWINGS AND SPECIFICATIONS

Drawings for temporary decking or sheeting should be equal in detail and completeness to the contract drawings. Specifications for temporary facilities should be provided to the contractor and should clearly outline the minimum requirements for acceptable installations.

COSTS

Estimating the cost of various decking installations will be dependent on the contractor's experience regarding previous installations. The basic ingredients of any cost estimate are labor, materials, and equipment. Labor and equipment costs are dependent on time or productivity factors, whereas materials are simply a bill of materials multiplied by the unit costs of each item.

A standard approach would be to estimate the productivity of a work crew during an average week on the assumption that the crew will complete a segment of decking during that week. For analysis purposes, let us assume that the crew will complete 200 lin ft (60 m) of decking 50 ft (15 m) in width during a 5-day period. The night shift will install four decking beams per night, or 20 per week. The day shift will install timber decking on the previously installed beams. The cost of this crew and equipment might approximate $30,000 per week on a straight-time three-shift basis, and if their productivity was as stated previously, then the unit cost of labor and equipment would be $3/ft^2$ ($32/m^2$). The cost of materials might be twice this, or $6/ft^2$ ($65/m^2$). A cost of $9/ft^2$ ($97/m^2$) would thus be arrived at; a slower productivity rate or a higher crew cost would increase the labor and equipment factor. Cost estimates vary with union restrictions and traffic patterns in urban centers.

12

Construction Ramps, Runways, and Platforms

by Bernard Monahan

The planning of access is basic to the completion of every construction project. City traffic must coexist with construction endeavors and trenches must be covered during peak traffic periods. Trucks must be provided ramps or runways to load and unload materials. Cranes require platforms within the construction project in order to maintain clear streets for material deliveries and to provide safe reach for the crane.

The purpose of this chapter is to present some of the pertinent considerations that must be evaluated when designing a support system for construction access.

DEFINITIONS

A ramp is a sloped surface which joins different levels. It is utilized on a construction project to provide access between a foundation or basement and the street level.

A runway is a strip of level road which forms a track for wheeled vehicles. It is a narrow extension out into the construction site.

A platform is a raised horizontal surface of wood or steel used by cranes and vehicles to bring materials into construction projects.

DESIGN AND RESPONSIBILITIES

The designs of ramps, runways, and platforms are often left to the contractor. The consulting engineer will often require that plans for temporary structures be submitted for examination. The responsibility for the design of the temporary structure remains with the contractor, the logic being that only the contractor could be aware of the problems and needs of the construction crews. Government authorities tend to agree that access and supplies during construction operations are best planned for by the contractor's personnel. The contractor as a major participant in the building team is willing to accept this charge and often will retain or employ professional engineers to plan for access to the construction project.

In order to properly plan for access, the constructor must segregate the job into building blocks and realistically grasp the magnitude of the problem. A high-volume traffic access might be approached on a different level from a rarely used secondary emergency approach.

The physical conditions of ramps, runways, and platforms might be considered similar in that each structure must be designed for the worst condition; however, a single-usage design might be supplemented by specific controls, such as a flagman, whereas a highly trafficked approach might require wider access and higher safety factors.

The engineer responsible for the design of temporary ramps, runways, and platforms must understand the physical conditions that could lead to failure of the structure. Only by having a clear understanding of the weakest link in the support chain can the professional plan to overcome the condition that could cause collapse.

The cost of construction is increasing; the cost of a construction failure is excessive and avoidable. Good training and quality control can reduce the cost of failures in terms of human life and monetary considerations. It is up to the construction industry team to provide the expertise required to meet the challenge for safe and efficient access passageways.

The basic human factor in the construction industry is the civil engineer, employed or retained by contractors, design consultants, and by government. The civil engineer's responsibilities include the design and supervision of the installation of ramps, runways, and platforms. Too often so-called professionals will delegate the responsibility for important structures to construction craftsmen rather than to trained professionals. This leads to sloppy practices that may cause failures.

VEHICULAR RUNWAY PLATES

The most common type of ramp or runway covering is the steel plate that is used to support vehicular traffic where trenching is required. Figure 12.1 shows an intersection where utility relocations required extensive plating for temporary runway access.

Figure 12.2 gives the reader an idea of the variation in utilities that might be found in a midtown thoroughfare. Each utility excavation must be analyzed independently and in union with other utility excavations.

Temporary runways are often required to expedite the handling of materials during construction. These projects require the contractor to install shoring to support cranes and delivery trucks. In some cases the trucks carrying concrete to the site are the determining factor in designing a roadway support system. In other cases it is a special crane-handling situation that determines the design values for the runway support system.

FIG. 12.1 Street plates are utilized to temporarily cover excavations. *(Thomas Crimmins Contracting Co.—Photo Collection.)*

FIG. 12.2 Street excavations sometimes require that the entire street be covered with vehicular plates which are normally 1 in × 5 ft × 10 ft. *(Thomas Crimmins Contracting Co.—Photo Collection.)*

In either case, rolling loads can be dangerous, for the typical transit ready-mix concrete truck may weigh 50 tons (45 t) when fully loaded. The problem becomes even more serious if consideration is given to the fact that the loading is repetitive and the friction loadings in a horizontal direction can cause serious failures. The usual method of securing street plates to roadways is to drive a series of 6-in (15-cm) spikes into the pavement surface. This method is effective in low-traffic areas where the trench plating is left in place for prolonged periods. However, in high-traffic areas where the runway plate is repeatedly removed and replaced, the spikes create a failure plane which may lead to a serious vehicular accident.

The use of steel plates supported on horse-head supports is commonly seen at intersections where trenches tend to branch out in order to follow various utilities that might require relocation. Typically, utilities such as sewers and steam lines require the right of way, for they must be installed to grades suitable for drainage or, in the case of steam lines, condensation pits. Other utilities must be diverted around these priority installations. The trenching system at a busy intersection may lead to a platforming or covering of most of the right of way. A safe method of covering an intersection is imperative and, of course, is mandated by law.

Figures 12.3 and 12.4 are presented to show the proper securing of the vehicular plate to prevent slippage.

Sheeting and bracing of the sides of the trenches serve a twofold purpose in that they prevent the failure of earth and provide vertical support for the street plating system; 6-in (15-cm) spikes are usually used as fasteners to prevent the

FIG. 12.3 Vehicular plates must be secured to prevent movement due to the vibrations caused in the plate from the flow of traffic. *(Thomas Crimmins Contracting Co.—Photo Collection.)*

FIG. 12.4 Horizontal movement of vehicular street plates may be prevented by the proper use of spikes, wedges, and asphaltic edging. *(Thomas Crimmins Contracting Co.—Photo Collection.)*

lateral sliding of plates off of their supports. The supports are generally oversized 8-in x 8-in (20-cm x 20-cm) or 12-in x 12-in (30-cm x 30-cm) timber because of its resistance to rolling as well as its structural characteristics.

Runway support systems have long been relegated to a relatively minor role, with the philosophy that the system was beyond the planning control of the design engineer and belonged to the artisan for the skilled employment of materials that were capable of supporting vehicular traffic. This philosophy is no longer practical and it is now up to the engineering profession to provide design standards for street runways and other vehicular-traffic hazards. Standards require professional controls of quality and performance and this can only be accomplished through additional restrictions imposing professional inspections of field installations.

TIMBER TRESTLE BRIDGES

The timber trestle bridge is one of the simplest types of bridges. Spans are usually limited to 25 ft (7.5 m) using timber stringers. The stringers rest on trestle bents and abutments. Figures 12.5 and 12.6 illustrate the various components of timber trestle bridges.

The first priority is to classify the loading limitations that are expected. An error in judgment regarding weight limitations can be avoided by clearly marking the safe capacity of the bridge. Many construction engineers overlook this standard requirement; they design and construct a bridging system with a certain crane as the design model. A new or rented crane requires a structural-support revision that all too often is never carried out. A bridging failure often leads to loss of life and serious injury. The design is usually limited to a single lane approach by a concrete truck (Fig. 12.7) or by a crane. The driver or operator must be limited as to the rate of speed while on the vehicular bridge and must be cautioned regarding sudden stops.

A major hazard regarding the construction of temporary bridging is the tendency of spiked timbers to loosen. A daily engineering inspection of a heavily trafficked bridge is recommended during the initial break-in periods. Loose spikes and timbers must be tightened and secured. Main supporting stringers must be tightened and secured. Main supporting stringers must be reviewed regarding bearing surfaces. Stringers have a tendency to move with the load and will often creep off the cap support. Overlapping stringers provide adequate support, but a bolted or spiked connection will often be preferable. A substantial end dam or heel will prevent this tendency for horizontal shifting.

The thread and the decking can often be combined. Common sizes of components of timber bridges are shown in Table 12.1.

A 4-in x 12-in (10-cm x 30-cm) plank laid on the flat will usually provide the required strength for construction equipment. Bridge footings require special attention; on permanent structures they are often driven to rock in order to provide a secure support.

On temporary structures the tendency is to overlook the need for adequate bearing for the footing timbers. The 4-in x 12-in (10-cm x 30-cm) planks used as footings are really only leveling sills, and if a site investigation leads the engineer

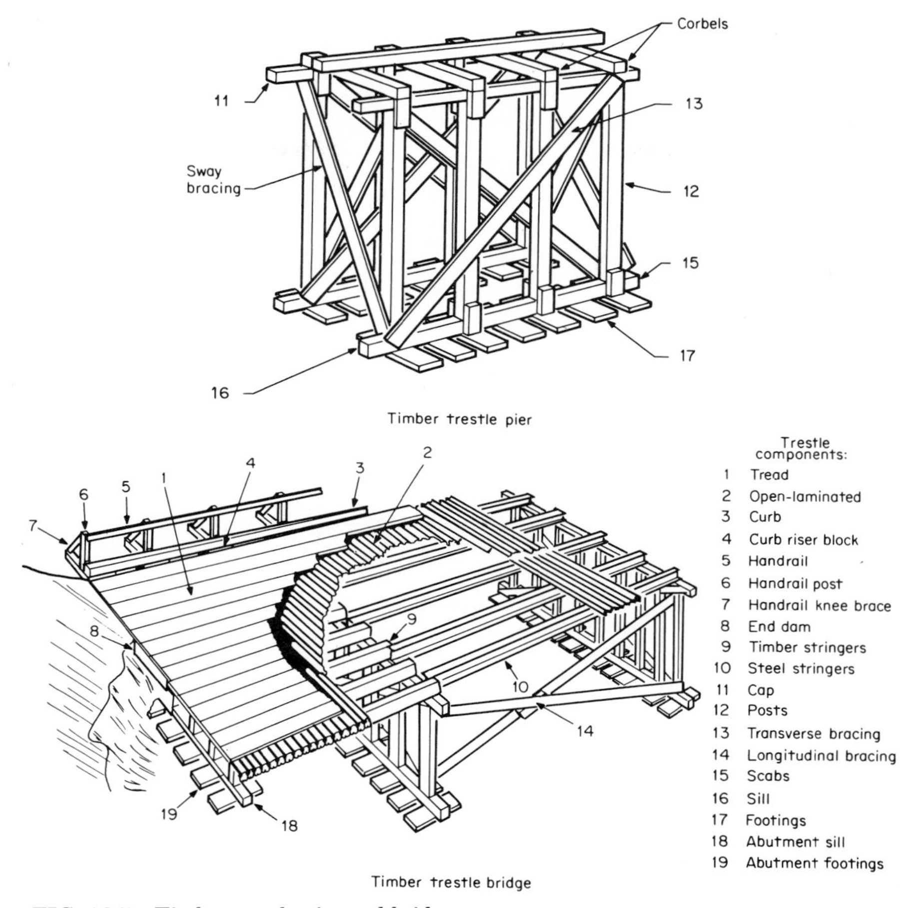

FIG. 12.5 Timber trestle pier and bridge.

to suspect the bearing capacity of the underlying soils during rains or other adverse weather conditions, then a separate soils analysis is in order. Investigations of this nature must be based on borings and other specific data.

Usually a temporary structure will not have the benefit of a detailed soils analysis and the construction engineer will be required to visually judge the proper bearing capacity. Sands might fall in the 3 ton/ft^2 (30 t/m^2) capacity area while a silty clay or organic meadow mat might be incapable of providing any worthwhile support without excessive settlements. Piles may have to be driven or a raft or floating support system may have to be designed. These are judgments that must be made on site observations and investigations.

In most cases the stringer design is the most critical member of the bridge; however, a check should be made of the capacity of the posts and the cap beams.

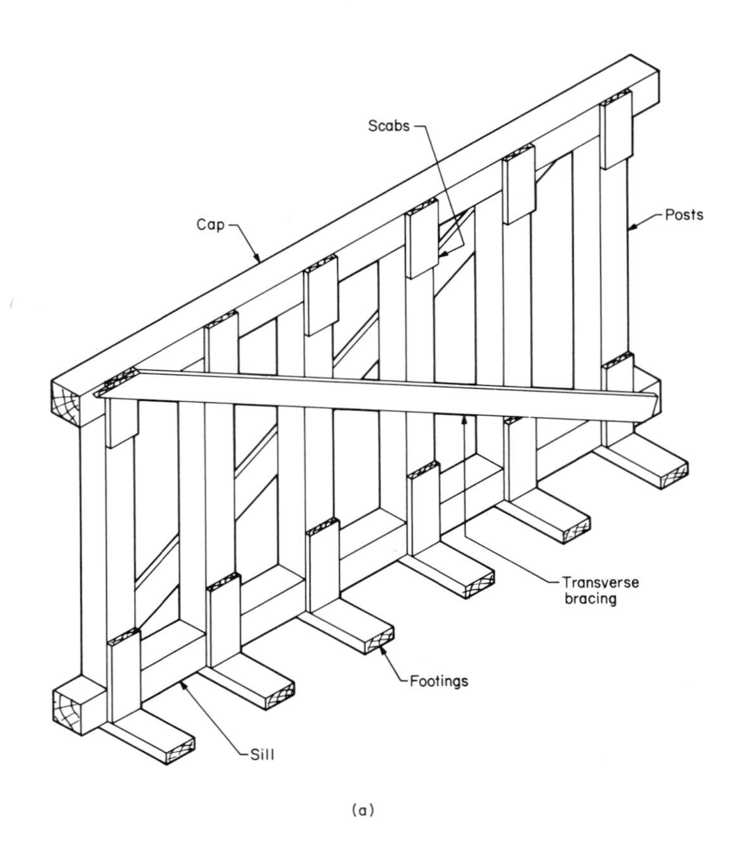

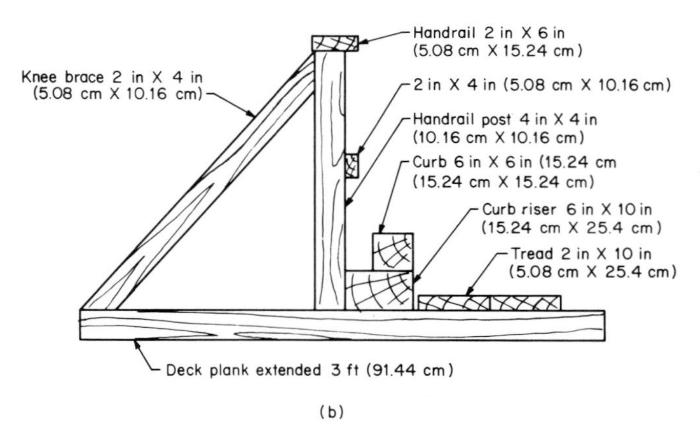

FIG. 12.6 (*a*) Typical timber trestle bent and (*b*) curb and handrail system.

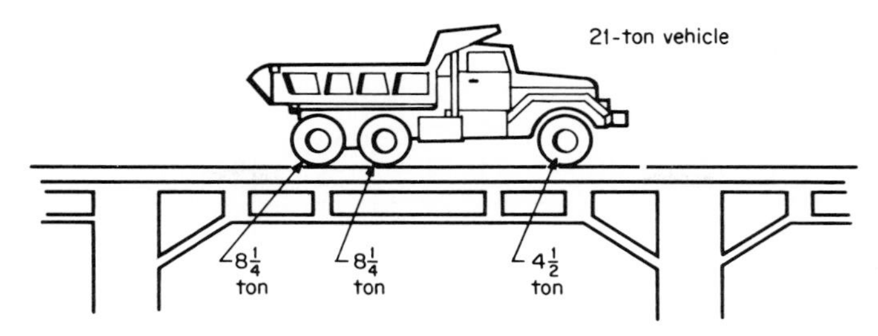

FIG. 12.7 Effect of vehicle on bridge depends on: (1) gross weight of vehicle; (2) weight distribution to axles; (3) speed at which vehicle crosses bridge.

A 12-in x 12-in (30-cm x 30-cm) post is not normally designed to carry a load of more than 36 tons (32 t). The maximum unsupported length of the post should be checked for buckling. Gross bracing of bents is important, for it prevents a side-sway force from causing a collapse of the bent.

PREFABRICATED BRIDGES

The Bailey-type bridge panel provides a rapid means of bridging streams and excavations for various construction equipment. It can be assembled in different ways for various spans and loadings. It is a through truss bridge supported by

TABLE 12.1 Bridge Components of Timber Trestle Bridge

No.	Bridge components	Common sizes or reference
1	Tread	2-in planking
2	Open-laminated deck	4-in planking
3	Curb	6 in x 6 in
4	Curb riser block	6 in x 10 in
5	Handrail	2 in x 4 in
6	Handrail post	4 in x 4 in
7	Handrail kneebrace	2 in x 4 in
8	End dam	4 in x 12 in
9	Timber stringers	12 in x 12 in
10	Steel stringers	12 BP 53
11	Cap	12 in x 12 in
12	Posts	12 in x 12 in
13	Transverse bracing	4 in x 12 in
14	Longitudinal bracing	4 in x 12 in
15	Scabs	4 in x 12 in
16	Sill	12 in x 12 in
17	Footings	4 in x 12 in
18	Abutment sill	12 in x 12 in
19	Abutment footings	4 in x 12 in

NOTE: 1 in = 2.54 cm.

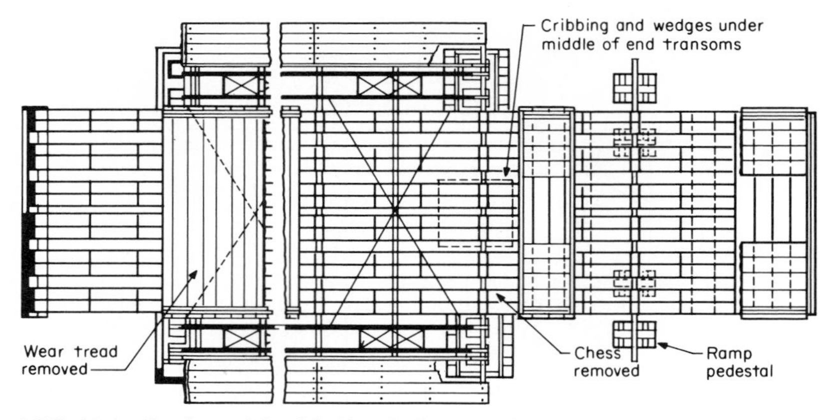

FIG. 12.8 Steel panel fixed bridge, Bailey type (plan).

two main trusses formed from 10-ft (3-m) steel sections called panels (Figs. 12.8, 12.9, and 12.10). The roadway decking can be either timber or steel. The reliability of this type bridge is unequaled. It requires a minimum of maintenance and the standard components may be reused under many conditions.

Acrow Corporation of America supplies a similar type of system as shown in Figs. 12.11 through 12.14.

A standard panel-bridging system has various single-lane and double-lane widths and includes several types of decking for different traffic-loading requirements. Timber or steel decking can be provided; the former for temporary use and the latter for temporary, semipermanent, or permanent use. A special feature of a panel bridge is that footwalks are cantilevered outside the main trusses and are therefore safe from roadway traffic. Footwalks of 3- to 4-foot (0.9- to 1.2-m) widths can be provided capable of accepting a live load of up to 120 lb/ft^2 (5750 N/m^2). The footwalks themselves may be steel, chequerplate, or timber.

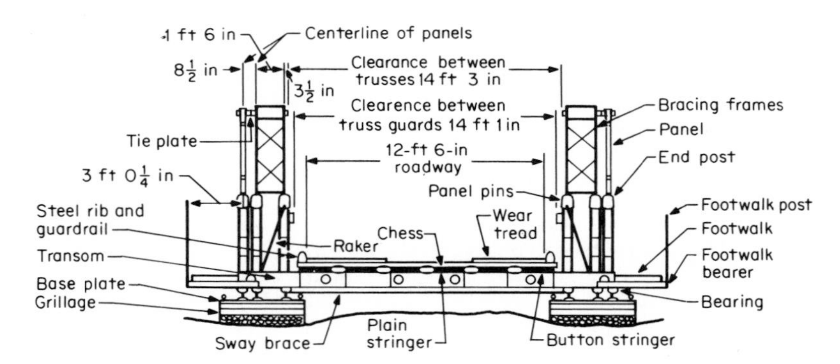

FIG. 12.9 Steel panel fixed bridge, Bailey type (end view).

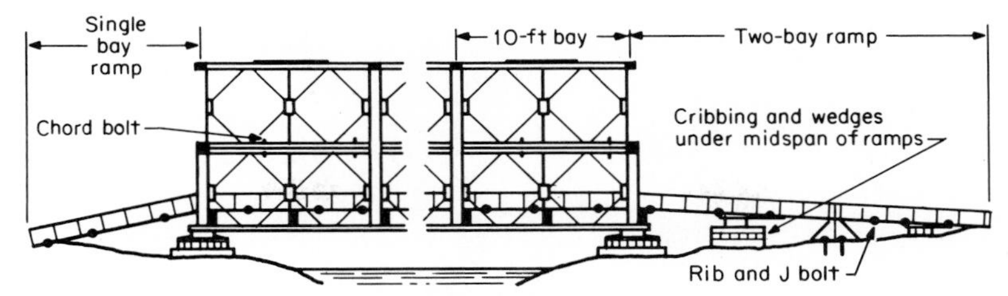

FIG. 12.10 Steel panel fixed bridge, Bailey type (elevation).

Sloping bridges or sections should not be steeper than 1 on 10 for normal vehicular traffic. Typical widths of bridges are as shown in Fig. 12.14.

Prefabricated bridges are available and engineers and contractors are well advised to consider their implementation whenever bridging is necessary.

FOUNDATION RAMPS

Office building foundations normally require the use of ramps for access during construction.

Shallow basements (one and two levels) are normally approached by means of earth fill. Deeper excavations might require timber or steel ramps (Fig. 12.15). The basic rule is that ramps must not exceed a 15% slope for the most powerful

FIG. 12.11 Acrow panel bridges are suitable for various spans.

FIG. 12.12 110-ft two-lane permanent Acrow panel bridge purchased by Morris County, New Jersey.

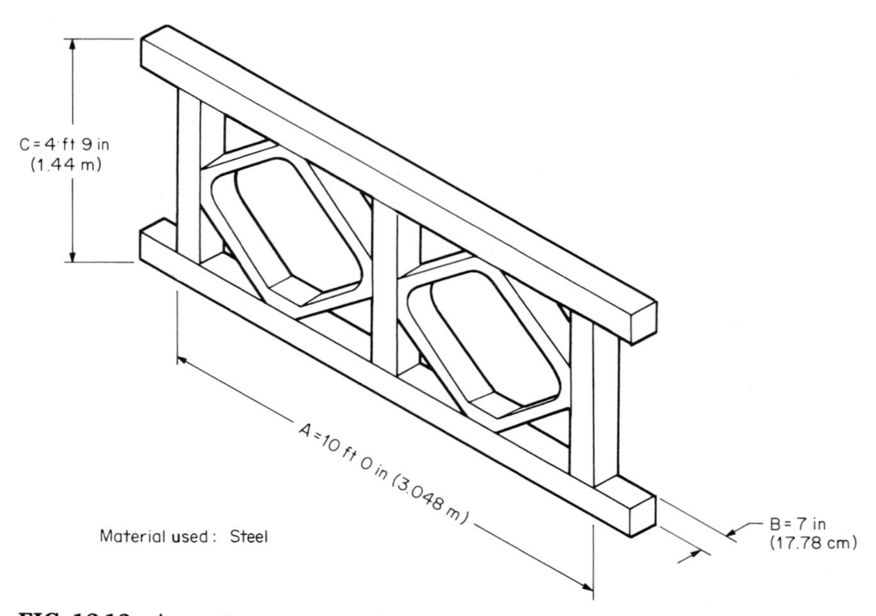

FIG. 12.13 Acrow truss components.

Acrow Panel Bridge is supplied in four roadway widths.

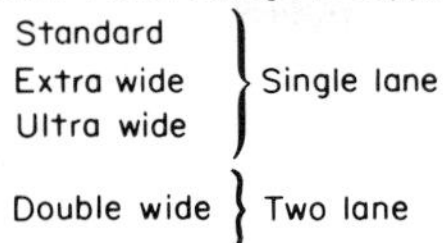

Roadway widths and clearances are shown below.

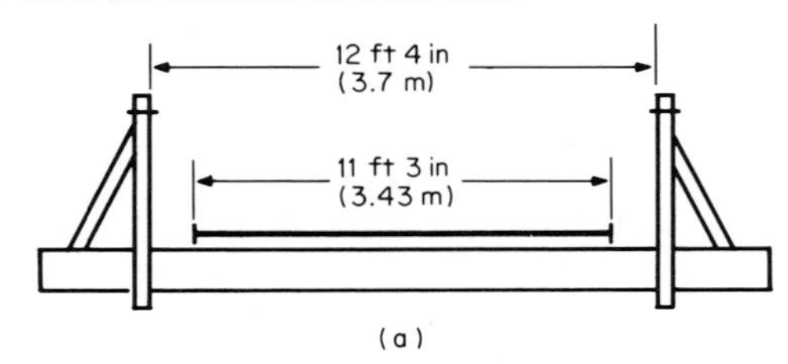

(a)

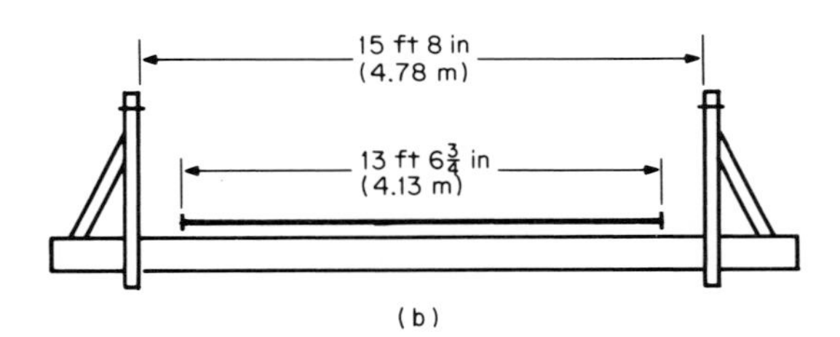

(b)

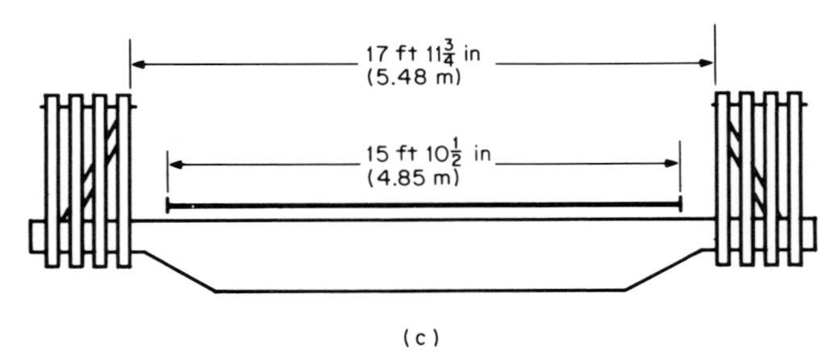

(c)

FIG. 12.14 Acrow bridge dimensions: (*a*) standard; (*b*) extra wide; (*c*) ultra wide.

FIG. 12.15 Construction ramp for a large building foundation. *(Thomas Crimmins Contracting Co.—Photo Collection.)*

10-wheel trailers. Errors in the slope can be very costly with an expensive installation that requires winches to assist trucks in and out of the cut. This, of course, defeats the beneficial advantages of a construction ramp.

Timber and steel ramps are preceded by earth or fill ramps. The excavating machines must dig their own ramp as the work progresses; it is only after the excavation has reached subgrade that a steel or timber ramp can be constructed.

One-Basement Structure

In this case excavation is less than 15 ft (4.5 m). Earth or fill ramps are normally excavated by the equipment as it progresses toward subgrade. Earth ramps may be totally unsatisfactory from a traction standpoint and for this reason they are often topped with loose building bricks. The rough edges of the brick form high spots for the truck tires to grab on when utilizing the ramp access. The brick topping should form a blanket at least 12 in (30 cm) in depth in order to minimize repairs. The topping must be maintained in order to preclude the forming of potholes. A ready supply of brick or even stone aggregate is necessary in order to assure continuous operation. The existence of fine silts or clays on the ramp will lead to a loss of traction for the trucks. These materials, when wetted, are slow to dry and are known to form slipping planes.

The key ingredient of a good earth ramp is drainage. The surface must be easily drained with materials that are heavy enough to resist erosion during heavy rains. Rutted roads make poor transportation and an unusable ramp can cause serious construction delays.

Heavy snows or freezing rains will prevent the utilization of ramps, as is the case with all roads. A ramp, however, is limited in length, and most contractors will find it advantageous to maintain an ice- or snow-free ramp. This is accomplished by removing large accumulations of snow with a machine, possibly a front-end loader, and by blowing the remaining snow off to the side of the ramp using compressed air flow pipes. These are 6-ft (1.8-m) long ¾-in (19-mm) pipes hooked to the air lines and having a hand-operated shutoff valve. A ready supply of salt or calcium chloride will often provide the final determining factor as to whether a contractor operates during cold or icy conditions.

Two-Basement Excavations

Two-basement excavations [approximately 30 ft (9 m)] are similar to one-basement excavations. However, the need for prior planning of access is more urgent. An error in judgment could require considerable effort to revise the access.

Figure 12.16 indicates some of the typical features of a timber ramp with steel stringers. The surface decking will usually consist of 4-in x 12-in (10-cm x 30-cm) rough planking which rests on a series of steel stringers interwoven with 12-in x 12-in (30-cm x 30-cm) timbers. For simplicity and uniformity, the steel stringers are normally 12 in (30 cm) in depth. Figure 12.17 indicates the typical ramping detail, which is, of course, similar in style and appearance to the temporary bridging discussed earlier.

The steel stringers are utilized to support the actual wheel loads of the vehicles utilizing the ramp. Great care must be exercised to route the vehicle in a manner that is safe for the designed trespass; 8-in x 8-in (20-cm x 20-cm) guide

FIG. 12.16 Construction ramps are usually fabricated using a combination of steel and timber. *(Thomas Crimmins Contracting Co.—Photo Collection.)*

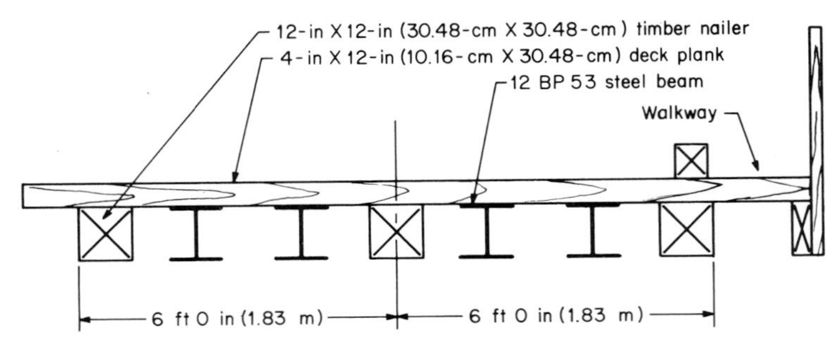

FIG. 12.17 Typical vehicular ramp cross section.

blocks or curbs are often utilized to maintain the vehicular access in a path that is fully supported by the stringers.

Walkways are usually positioned along the edge of the ramp, and they are provided with secure rails, all in accordance with OSHA requirements.

A ramp is a sturdy bridging system that must meet stringent quality controls. The ramp must be planned simply and it must be constructed in a secure fashion in order to meet the requirements of daily use. Like all bridges, it must be secured in place; the likelihood of horizontal movement is great due to the laterally induced forces. The standard method of resisting these thrusts is to secure the toe and the cap of the ramp against movement. Figure 12.18 indicates a typical cap detail.

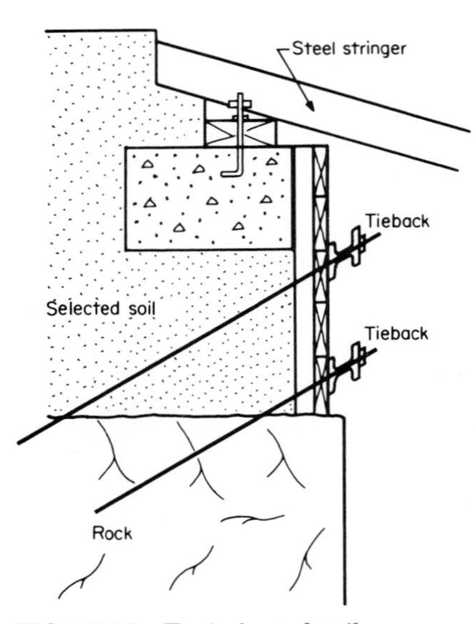

FIG. 12.18 Typical cap detail.

The design of the cap footing will depend on the bearing value of the surface soils and must be carefully examined; often it is necessary to continue the footing until firm bearing is established, possibly even to rock. However, most often it is advantageous to increase the pad size in order to lower the required bearing value.

A typical construction sequence is shown for the building of a ramp in the photographs in Figs. 12.19 through 12.24.

RUNWAYS AND PLATFORMS

Construction runways are normally utilized on facilities where the construction proceeds in a horizontal rather than a vertical direction. The runway is built on top of the new construction, utilizing the walls and grade beams as supports.

FIG. 12.19 Ramp, stage I. *(Thomas Crimmins Contracting Co.—Photo Collection.)*

FIG. 12.20 Ramp, stage II. *(Thomas Crimmins Contracting Co.—Photo Collection.)*

FIG. 12.21 Ramp, stage III. *(Thomas Crimmins Contracting Co.—Photo Collection.)*

FIG. 12.22 Ramp, stage IV. *(Thomas Crimmins Contracting Co.—Photo Collection.)*

FIG. 12.23 Ramp, stage V. *(Thomas Crimmins Contracting Co.—Photo Collection.)*

FIG. 12.24 Ramp, stage VI. *(Thomas Crimmins Contracting Co.—Photo Collection.)*

Figures 12.25 and 12.26 give some important details regarding the construction of a crane runway and platform over a newly constructed sewage channel and over concrete girders, respectively.

The positions for lifting must be carefully shored, for it is when the crane is swinging over a single outrigger that the maximum stresses are induced into the shoring members. Turnoffs or additional spans may be added to provide for special lifts.

Figures 12.27 through 12.32 show various parts and other aspects of working platforms.

Platforms are sometimes required in order to provide access for structural-steel erection. The placement of timber mats and shores may be utilized to strengthen existing or new construction.

The width of the support system is determined on the basis of the widest vehicle use. A crane runway must be wide enough for the placement of outriggers, while a truck runway need only be the standard 12.5 ft (3.8 m) clear. The length of a ramp or runway will be determined by the physical characteristics of the project, such as its depth or its size.

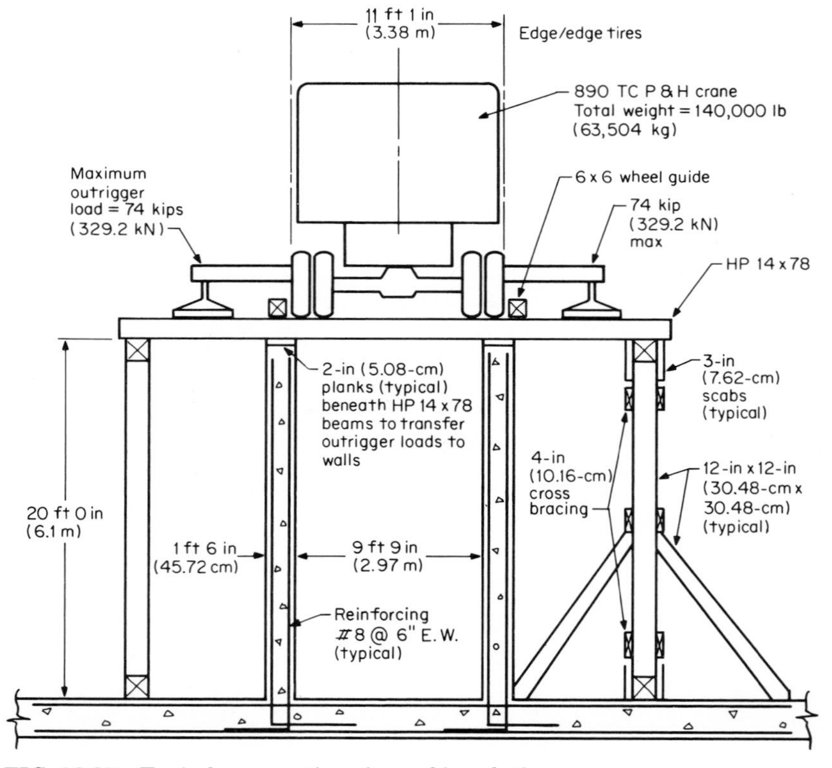

FIG. 12.25 Typical cross section of a working platform.

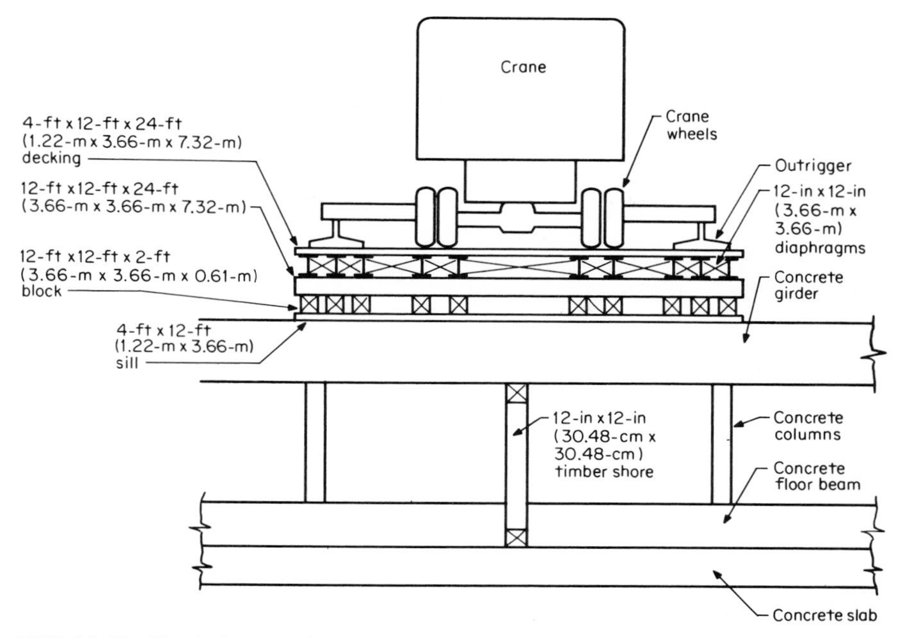

FIG. 12.26 Typical crane platform.

FIG. 12.27 Construction runway provides access for cranes and trucks. *(Thomas Crimmins Contracting Co.—Photo Collection.)*

FIG. 12.28 Timber runways are required for reinforcing and for concrete deliveries. *(Thomas Crimmins Contracting Co.—Photo Collection.)*

FIG. 12.29 Concrete grade beams may be strengthened by means of intermediate shores. *(Thomas Crimmins Contracting Co.—Photo Collection.)*

FIG. 12.30 Timber mats being placed to provide crane access for the erection of steel. *(Thomas Crimmins Contracting Co.—Photo Collection.)*

FIG. 12.31 Shoring of concrete deck in order to provide support for crane platform. *(Thomas Crimmins Contracting Co.—Photo Collection.)*

FIG. 12.32 Timber mats placed on sills. *(Thomas Crimmins Contracting Co.—Photo Collection.)*

The City of New York requires that a crane permit be obtained from the city before establishing a crane within the city's limits. The firm seeking the crane permit must have a professional engineer's certification that the site has been examined and found to be suitable for supporting the crane during its intended use. The engineer's examination must include an appraisal of the probable bearing value of the subsoil, a statement regarding the presence of vaults under the bearing pads, and calculation showing the intended use and loadings.

The bearing value of the soil is not to be assumed greater than 1500 lb/ft^2 (71,820 N/m^2) unless the engineer has specific knowledge regarding the nature of the subsurface soils. Borings and soil-testing programs are not usually carried out for simple temporary crane installations; therefore, most engineers rely on a visual inspection of the soil bearing value and assign a value of 1500 lb/ft^2 (71,820 N/m^2) unless the subsoil appears to be incapable of providing even this minimum level of support.

A standard rubber-tired crane has four outrigger pads whose bearing surfaces are 4 ft^2 (0.37 m^2) each for a total of 16 ft^2 (1.5 m^2). A loading of 60,000 lb (27,180 kg) per outrigger is common, and therefore a footing must be provided to distribute this loading over a larger area. Typically each outrigger will be placed on a timber crib consisting of 12-in x 12-in (30-cm x 30-cm) beams bolted together to form a mat of approximately 40 ft^2 (4 m^2).

The question of subsurface vaults is often more difficult to solve, for urban construction sites may have had a history of four or five earlier buildings whose drawings are not known and whose construction may have included a sidewalk vault for storage. These nineteenth-century vaults are usually massive brick arches supported by piers and they are normally of a greater strength than the minimum assigned to street bearing levels. However, in the interest of public safety, it is mandatory that a responsible engineer investigate and determine the safe bearing values before a crane is permitted to function in a crowded metropolitan area like New York City.

ROADS AND DRIVEWAYS

Access to a construction site is usually no serious problem. Trucks and other vehicles can operate on the natural ground with a minimum of difficulty. Dust can be alleviated by sprinkling with water, a temporary expedient that requires frequent repetition, or by the application of an asphalt oil palliative. Heavy and frequent traffic requires more extensive measures.

Of primary importance is drainage. Water must be removed from the roadway as fast as it accumulates, by means of crowning the surface, sloping, and side ditches or drains. The stabilized soil itself is not suitable as a wearing surface and must be protected. A common method is the road-mix method in which asphalt emulsion or cut-back asphalt is mixed with the surface material and the layer compacted by traffic. This provides an excellent wearing surface that is easy to construct, is watertight, can be repaired easily, and can be removed with minimum effort when no longer needed. Road-mix asphalt can be applied to almost any soil except clay. Another method is the application of a seal coat. A seal coat consists of the spray application of a liquid asphalt to the compacted and graded roadway surface, usually followed by the application of a thin layer of sand or find gravel. The asphalt may be cut-back liquid asphalt, asphalt emulsion, or hot asphalt cement.

If access roads are to become part of the permanent installation, it will be necessary to construct them in accordance with the project specifications.

RAILROADS

Rail facilities are necessary in tunnel construction as well as for moving bulk and heavy materials and construction components. Panelized track units are used for sidings and industry spurs, one advantage being the speed of construction. Another advantage for the temporary constructor is their high salvage value upon completion of their temporary duty. Several railroads have centralized fabricating plants in which they build their own track panels. A recent development is a prefabricated standard-gauge steel tie-and-track section 39 ft (12 m) long. The sections are shipped to the site by rail or truck, can be handled by a small crane, are easily laid and joined with a minimum of skilled labor, and can be salvaged and reused upon completion of their service.

Rails are secured to the steel ties by pairs of steel jaws inserted in holes drilled in the ties on each side of the rail seats. The joint is secured by a tapered steel key or wedge. Similar panels are also fabricated of rail with wood ties. Load-carrying capacity depends on the weight of the rail used and the spacing of the ties. For example, the steel tie units recommended for infrequent light traffic include 70-lb (32-kg) or 90-lb (40-kg) relay rail with the steel ties on 36-in (90-cm) centers. A heavy unit, designed for heavy and frequent traffic, would consist of 100-lb (45-kg) or heavier relay or new rail with the ties spaced 24 in (60 cm) or 30 in (75 cm). A 39-ft (12-m) panel consisting of 90-lb (40-kg) rail with ties spaced at 30 in (75 cm) weighs about 4700 lb (2130 kg).

Panels are laid on the prepared subgrade, ballasted, raised to grade, and the ballast is tamped with either hand or power tools in the usual manner.

Tunnel construction usually requires temporary rail lines to remove muck away from the heading and to move supplies into the tunnel. Tunnel tracks are frequently narrow gauge, some as small as 24 in (60 cm). Panel tracks are frequently used in underground work because of their convenience. Special steel and wood composite ties are available for use in machine-made circular cross-sectional tunnels. Fitted with wood end blocks, the ties are designed to fit the contour of the tunnel. When reused, the blocks can be modified to fit the different diameter tunnels.

13

Falsework/Shoring

by Colin P. Bennett
*Robert T. Ratay**

*The contribution of the Scaffolding, Shoring, and Forming Institute, Inc. to this chapter is acknowledged.

This chapter is intended to provide basic examples of the various types of shoring equipment available to the construction industry. The chapter also gives examples of efficient use, advantages and disadvantages, and special considerations related to shoring methods. All types of shoring can be efficient and cost effective if matched to the proper job. Matching the most efficient shoring system to its most suitable application is the responsibility of the shoring layout design engineer.

All shoring erection must follow the shoring layout specifically and must be erected properly and in accordance with the layout. The reader should keep in mind that safety requirements are of primary importance regardless of the type of shoring being used.

Discussions of proprietary shoring systems are purposely avoided, so the following sections are intended as a general guide to the principles and procedures of safe shoring design and utilization. The manufacturer or agent should provide specific information on the characteristics and use of equipment. Under actual field conditions the manufacturer's design tables and instructions should be consulted and followed for the type of equipment in use.

STEEL-FRAME SHORING

In the late 1930s and the 1940s welded-steel-frame scaffolding made its appearance as a more efficient and simpler replacement for the older, tube-and-coupler metal scaffolds. The concept of a welded frame taking the place of three or more

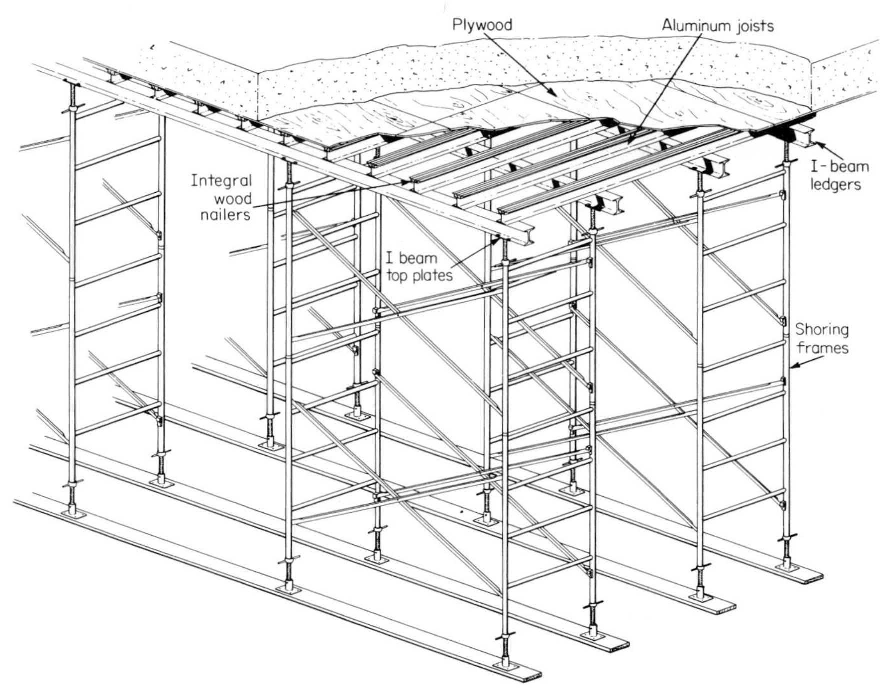

FIG. 13.1 Typical shoring/formwork section with steel-beam ledgers and aluminum joists.

pieces of tubing connected to each other by loose scaffold couplers changed the scaffolding practices in many areas of the world. Line drawings of tube-and-coupler scaffolding and frame-type shoring are shown in Figs. 13.2 and 13.3, respectively.

It was not until the 1950s that these labor-saving scaffolding "frames," as they generally were called, began to be used to provide vertical falsework/shoring support to horizontal formwork for slabs, beams, and other similar concrete construction. A frame width of 5 ft (1.524 m) became standard with 2 and 3 ft (0.61 and 0.91 m) widths available on a more limited basis.

The earliest applications used wooden ledgers (stringers) seated directly on the top, or "header," bars of the frames. This method was not satisfactory or efficient since releasing the shoring loads after concrete setting required the lowering of frames to break the bond to the concrete, as shown in Fig. 13.4. For heights of one or two frames this was a passable but awkward process; for higher work requiring multiple tiers of frames (lifts) it was very difficult to safely release the threaded screw legs at the base of the scaffold.

Attendant problems were soon discovered concerning the wooden ledgers. Unless 4-in (10-cm) wood was used, the ledgers were laterally unsupported and could not be loaded to their full strength. Also, the header bars were insuffi-

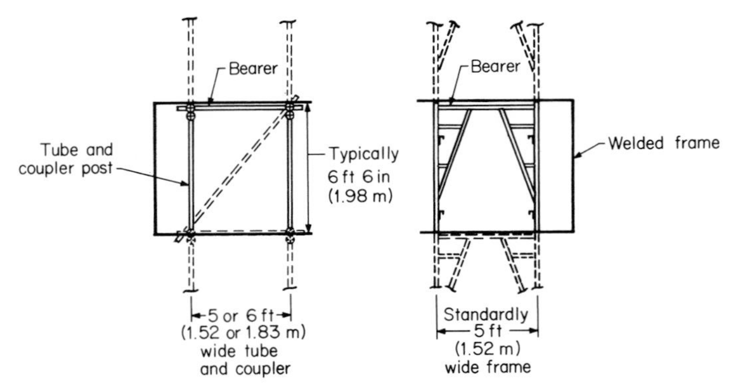

FIG. 13.2 Cross section through tube-and-coupler- and frame-type shoring.

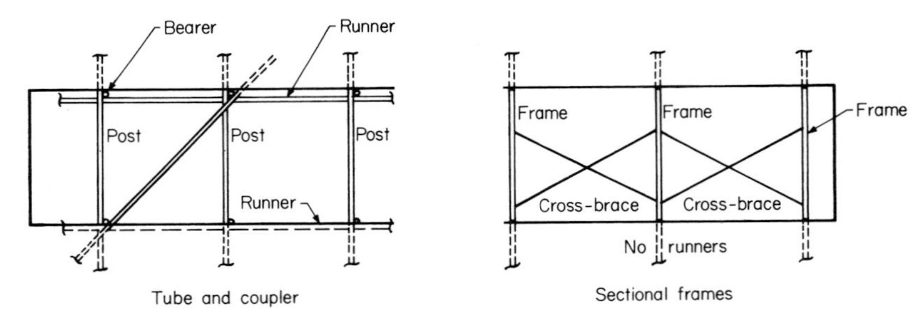

FIG. 13.3 Side view of tube-and-coupler- and frame-type shoring.

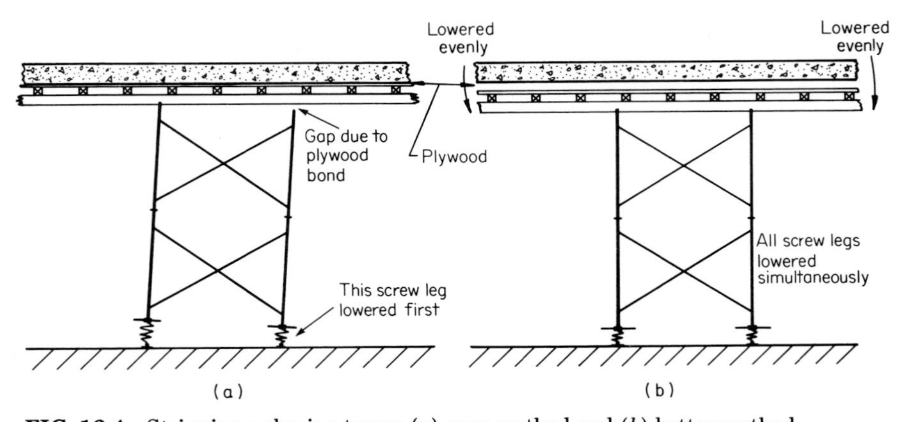

FIG. 13.4 Stripping a shoring tower: (*a*) poor method and (*b*) better method.

ciently strong to efficiently develop the load capacity of the frame legs. Efforts to reinforce the header bars of the frames did not eliminate the collateral problems of the local crushing of the lumber at bearing points on the tubular members. Reinforced or not, practical limitations resulted in the inability of the header bars to carry much more than the strength of *one* leg of the frames, i.e., half-capacity leg loading. This resulted in requirements of up to twice as many frames as theoretically necessary to support any given concrete load. Fifty percent inefficiency with unnecessarily high costs was the result. Many concrete contractors were therefore reluctant to try this and were unwilling to make additional investments to purchase 4 × 8 or 4 × 10 in (10 × 20 or 10 × 25 cm) ledgers which loaded the frames more efficiently. They were accustomed to using 3 × 4, 4 × 4, and 4 × 6 in (7.5 × 10, 10 × 10, and 10 × 15 cm) joists and ledgers in conjunction with wood or metal single-pole shores. The scaffolding industry was, and still is, a rental-oriented one, but logically, rental of wooden joists and stringers is impractical.

However, some scaffolding manufacturers did make small I beams (I4 × 7.7) available for rental as an efficient substitute for 4 × 8 and 4 × 10 in (10 × 20 and 10 × 25 cm) wood ledgers. While these became popular with many contractor and did expand the use of frame shoring, there still remained physical limitations and drawbacks such as necessary lateral overlapping of the beams that made positions on the frame header bars inefficient by having to position the ledgers away from the frame legs. (See Fig. 13.5.) Also, careless placement of such

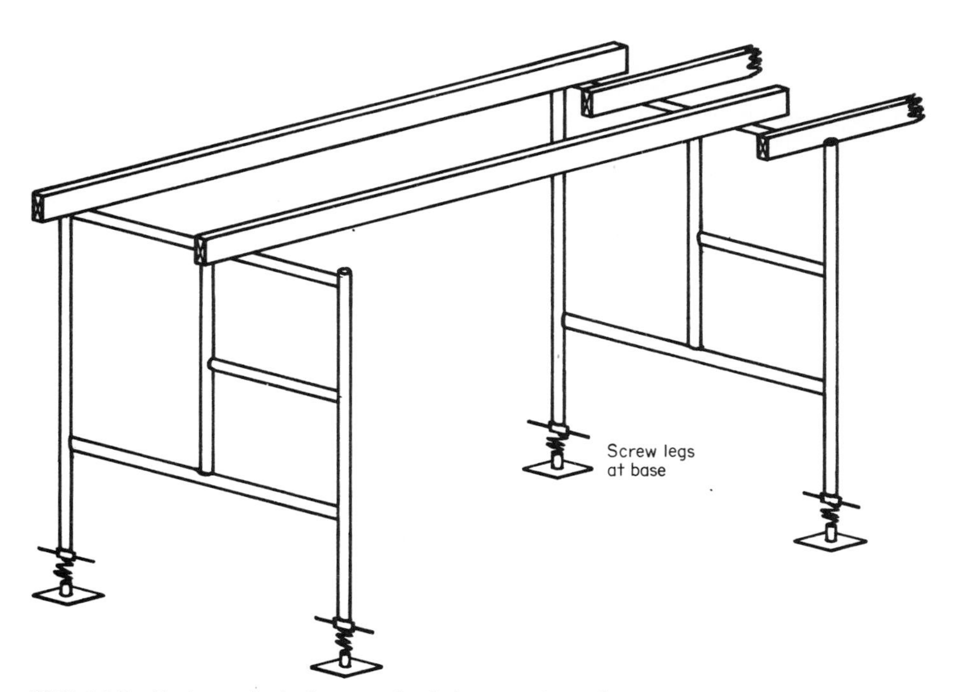

FIG. 13.5 Early method of supporting ledgers on frame bars.

beams subjected the header bars to bending stresses and thus again failed to achieve optimum leg-load capacities of up to 5000 lb (22,240 N) per leg.

The solution

To realize the frames' full potential, the shoring loads had to be applied *concentrically* to the frame legs. Accordingly, the next development was a historic one for both the construction industry and the scaffold manufacturers: the design and manufacture of specialized shoring accessories for use in conjunction with the standard manufactured scaffolding frames. The components were U heads or channels (Figs. 13.6 to 13.8) to support and locate stringers and ledgers; adjustable threaded screw legs of up to 24-in (60-cm) open adjustment [although 12 to 16 in (30 to 40 cm) was normal], base plates with nail holes, and shorter diagonal bracing as small as 2 ft (60 cm) between frames. As these components became more plentiful, huge numbers of standard access scaffolding frames for shoring became available having safe load capacities between 3000 and 5000 lb (13,344 and 22,240 N) per leg. Thus, shoring concrete with scaffold frames became an everyday event instead of an occasional one. (See Fig. 13.9.)

Heavy-Duty Shoring Frames

The next step was another logical one: design frames *especially for shoring.* While doing this, why not also make them stronger?

Some premature efforts were made to manufacture frames with legs of nominal 1.9- or 2-in (4.8- or 5-cm)-OD tubing. Unfortunately, the resulting load capacity of 5000 to 7000 lb (22,240 to 31,136 N) per leg was not much greater than those obtainable with standard scaffolding frames. Frames generally had become standardized to use 1⅝-in (4.13-cm)-OD tubing with wall thicknesses of up to 0.125 in (3.2 mm).

Subsequently, the heavy-duty shoring frame made its logical appearance with tubular legs of 2⅜ or 2½ in (6.0 or 6.35 cm) OD and with design shoring load capacities of 10,000 lb (44.480 N) or more per leg. Figure 13.10 shows typical heavy-duty shoring towers of various configurations.

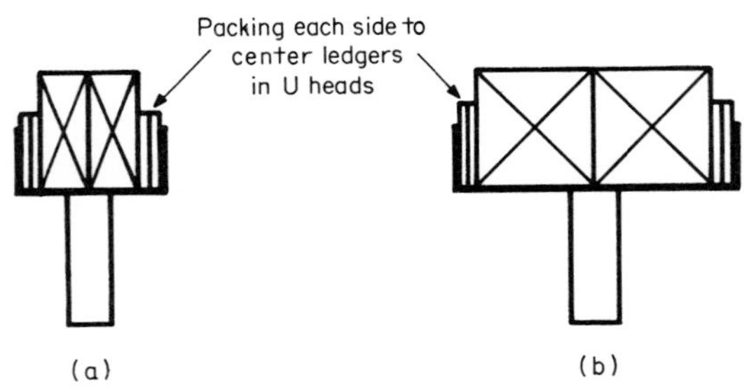

FIG. 13.6 Ledgers in U heads: (*a*) two 2-in (5-cm)-wide ledgers in 4-in (10-cm) U head; (*b*) two 4-in (10-cm)-wide ledgers in 8-in (20-cm) U head.

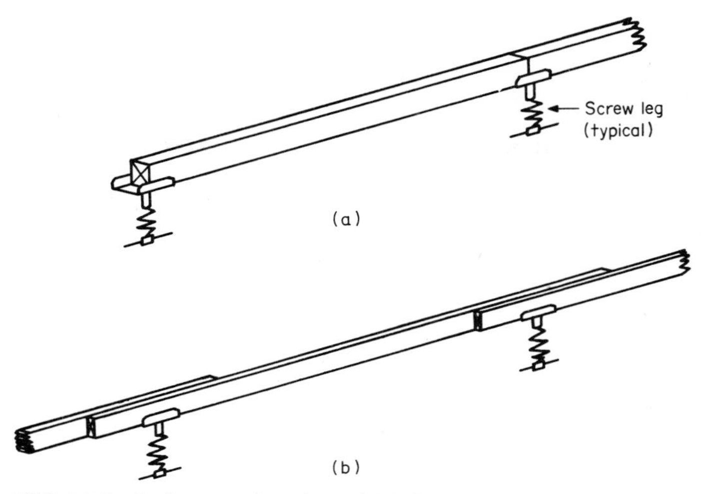

FIG. 13.7 Ledger continuations: (*a*) ledgers butted in U heads and (*b*) ledgers lapped at U heads.

The advantages and disadvantages of heavy-duty shoring frames were as follows:

Advantages	Disadvantages
Only one-half the number of frames for a given shoring condition	Weight increased to 70 lb (31.75 kg) for a 4 × 6 ft (1.22 × 1.83 m) high frame
Use of standard cross braces	Short supply while inventories were being enlarged
Simpler design procedures	
Sturdier accessories for rough handling	Market saturation (mid-1970s) took many years

From then on, the staple of the industry were frames with capacities of 10,000 lb (44,480 N) per leg. The 5-ft (1.524-m) width of the standard "access" scaffolding frames was changed to the more convenient modular 4-ft (1.22-m) width by 5 or 6 ft (1.524 or 1.83 m) high sizes; these quickly became the "bread and butter" items of shoring construction. Also, the shoring accessories were made of thicker and stronger materials to accept the larger vertical loadings involved with these heavy-duty frames. Larger U heads and screw legs and the availability of preexisting brace sizes made this type of shoring one of the most versatile and *adaptable* tools that the building contractor had ever known.

Another radical development was the telescoping tube used inside the frame leg. Its application was similar to that of a single-pole shore with a pin and pinholes for rough adjustment: it had an adaptor with a short length of screw thread for fine adjustment. These telescoping tubes were named "extension" legs or

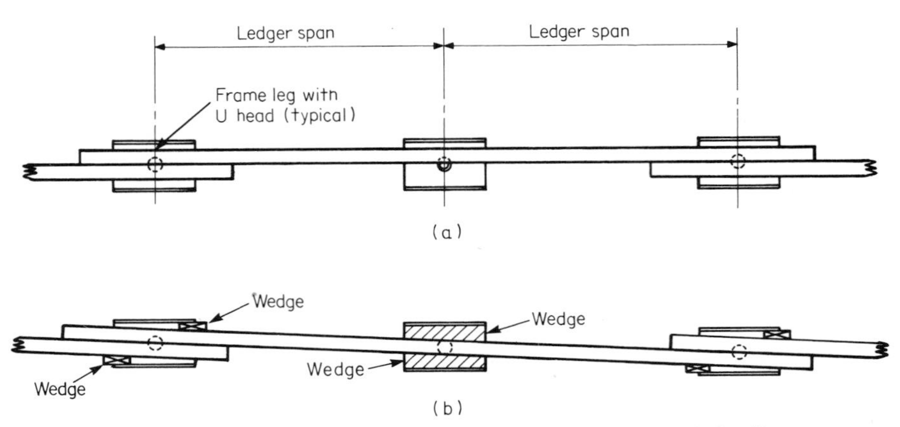

FIG. 13.8 Ledger lapping: (*a*) ledgers lapped this way give eccentric loading at center leg; (*b*) correct way to lap ledgers.

tubes and had working lengths of up to 5 ft (1.529 m). With their rows of pinholes they were quickly christened "piccolos."

Today, there are many more types of equipment available for shoring, covered later in this chapter. However, the workhorse of the concrete construction industry is still the 4 × 5 or 4 × 6 ft (1.22 × 1.524 or 1.22 × 1.83 m) heavy-duty welded shoring frame.

Almost all makes of this frame are furnished with some form of quick-acting mechanical locking mechanism that enables fast attachment of the cross bracing to the frames. However, wing nuts, hand-turned on threaded studs welded to frame legs, are still being used by some.

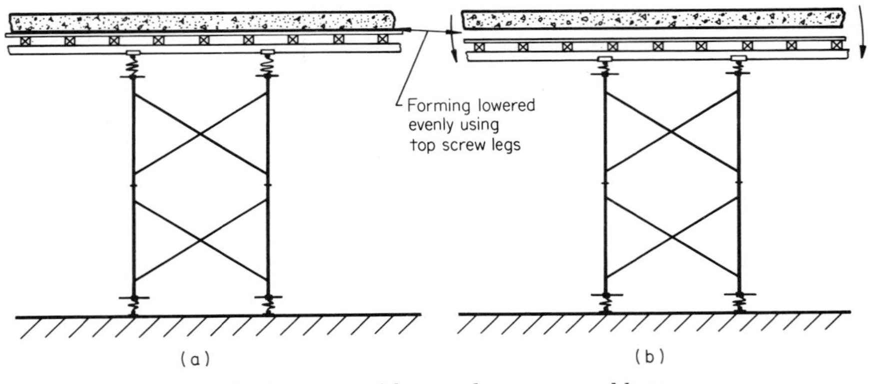

FIG. 13.9 Four-leg shoring tower with screw legs at top and bottom.

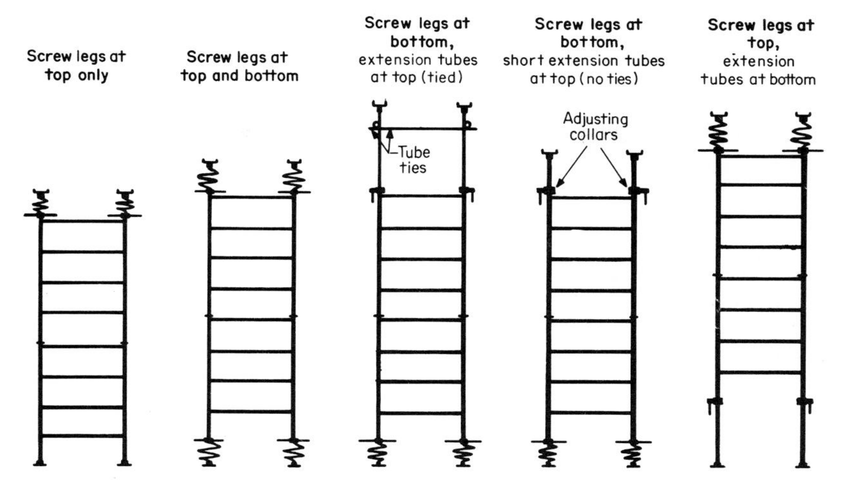

FIG. 13.10 Various frame shoring tower configurations. (Bracing at right angles to frames not shown.)

Shoring Towers

A shoring tower is a modular assembly of single or multiple-tiered pairs of frames connected by pivoted diagonal cross bracing.

The historical importance of the scaffold shoring tower lies in the fact that it provided the facility of at least four single-pole adjustable steel shores with extremely rapid erection. Starting to shore a corner of an installation with single-pole shores puts great reliance on the moment strength of the ledger–shore top connection in which the only connection is nails. The use of wood bracing and lacing members to plumb vertical steel shores and tie them together is a time-consuming and difficult process for usually three or more workers if the shores are over hand-reaching height. A scaffold shoring tower, on the other hand, is quickly and safely erected by two workers, and, in extreme cases, one-frame-high towers can be erected by only one worker. The four legs of a tower give the equivalent facility of at least four shores, as mentioned, but they are usually of a higher load capacity, have built-in devices for easy attachment of bracing, are quick and easy to plumb and level, and can be erected to great heights.

The use of two or more tiers of single-pole shores is prohibited by many states. However, Federal OSHA regulation [paragraph (d)(3) of Ref. 1] allows it but requires design and inspection of such installations by a structural engineer. Sectional frame shoring does not have such a requirement unless called for by a state's OSHA regulations or if the job is unusually complex.

Design of frame shoring can often be done by trained and experienced drafters or graduate engineers without needing the services of an experienced professional engineer, except for layout approval or difficult design applications. Sectional frame shoring must be installed and used in conformity with a shoring layout, generally designed and furnished by the concrete contractors, the shoring manufacturers, or their agents and distributors.

As the use of shoring frames expanded, the applications became more sophisticated and generally served to further reduce the use of single-pole shores.

The scaffolding industry makes cross braces available for spans as short as 2 ft (60 cm) and as long as 10 ft (3 m) between frames. When used with long-span horizontal shoring beams, bracing distances were expanded from 12 ft (3.6 m) up to 15 ft (4.5 m). Because of this, care must be taken to ensure that braces do not sag and pull the frames out of plumb. The combination of shoring frames and horizontal shoring beams is covered more fully later in this chapter.

Frame shoring towers are very flexible modules because they come in a wide range of bracing span lengths and frame widths. Although 4-ft (1.22-m) width is the standard—working nicely in conjunction with 4-ft (1.22-m)-wide plywood sheets—frames 2 and 3 ft (60 and 90 cm) wide are also popular because they are adaptable as shoring for beam soffits separately from slab shoring frames owing to soffit height differentials (Fig. 13.11).

Demand and supply developed enormously, peaking out in the early 1970s as other, proprietary shoring systems, such as column- and/or wall-supported flying deck systems, large "tables" with trusses and room forms, also started to become popular. However, large shoring projects can require as many as 25,000 frames or more!

Accessories

By being able to combine various accessories and brace sizes, contractors now have more flexibility. Tower heights, taking advantage of standard 12- and 24-in (30-and 60-cm) extra-length screw legs and piccolo extension tubes, can be easily assembled for any job height by using the relatively standard three frame heights of 4, 5, and 6 ft (1.22, 1.52, and 1.83 m) with which most shoring heights can be reached. Others, such as 3 ft 6 in (1.07 m) and 3 ft (0.91 m), are also available.

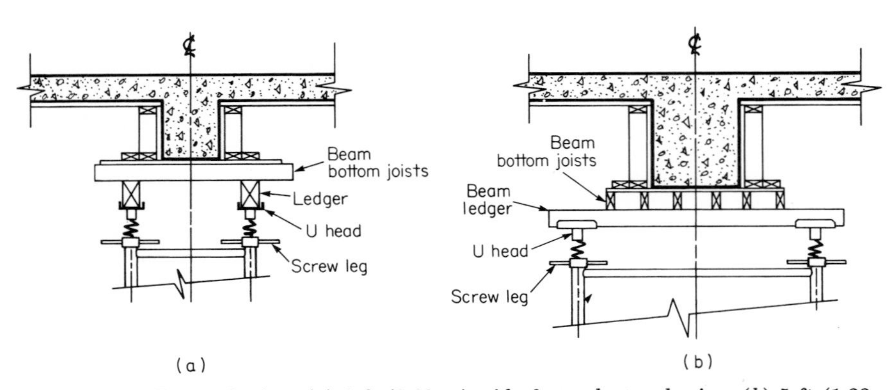

FIG. 13.11 Beam shoring: (*a*) 2-ft (0.60-m)-wide frame beam shoring; (*b*) 5-ft (1.22-m)-wide frame beam shoring.

Typical accessories, shown in Fig. 13.12, provide threaded adjustment on tops, bottoms, or both ends of the tower. U heads are available for almost limitless ranges of size and types of ledgers, whether wood or steel (Fig. 13.13). The vertical sides of the U heads give support to wood ledgers during installation; however, they are not designed or intended to resist overturning of the ledgers under shoring loads.

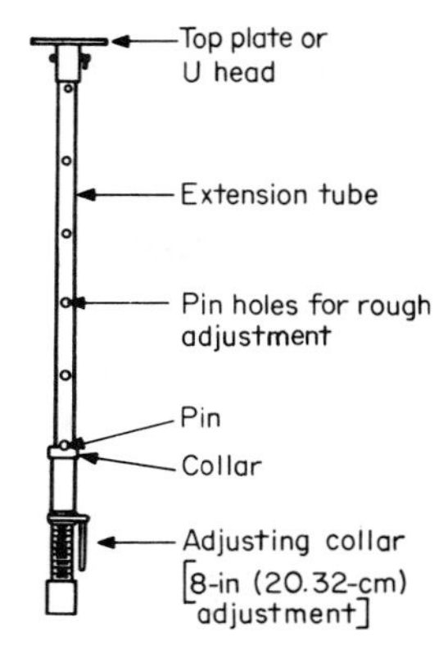

FIG. 13.12 Extension tube (piccolo) with screw adjustment.

A wide selection of steel I beams is popular and available for use as ledgers. These are generally used with "top plates" for attachment of the I beams to the frame legs. They are designed to permit lateral overlapping of ends of beams at frame leg junctions. Also available are rectangular top plates that permit attachment of other-sized I beams when the top plate is positioned at right angles to the normal. One of these is shown in Fig. 13.14.

STABILITY OF TOWERS

As a rule of thumb, any width-brace combination of tower size whose height is 4 or more times greater than its narrowest base dimension portends a stability hazard when not connected by ties to other towers; this hazard is substantially increased when people are working on the tower. Two-foot (0.61-m) wide frames and multiframe high towers having 2-, 3-, or 4-ft (0.61-, 0.91-, or 1.22-m) bracing are similarly unstable if not laterally tied. The commonly accepted practice is to connect rows of towers to each other with tube and coupler horizontal lacing

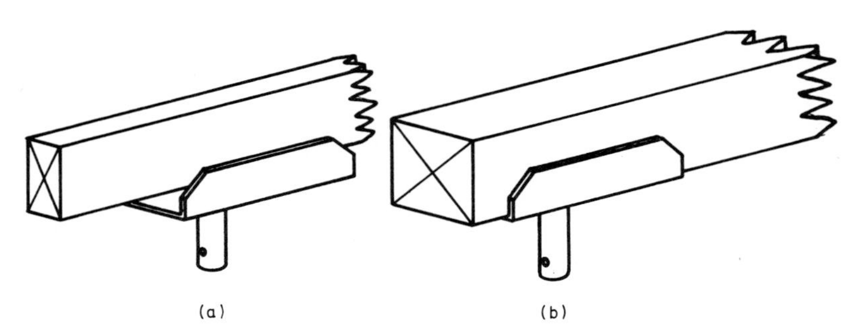

FIG. 13.13 U-head details: (*a*) 4 in (10 cm) wide × 8 in (20 cm) long and (*b*) 8 in (20 cm) wide × 14 in (36 cm) long.

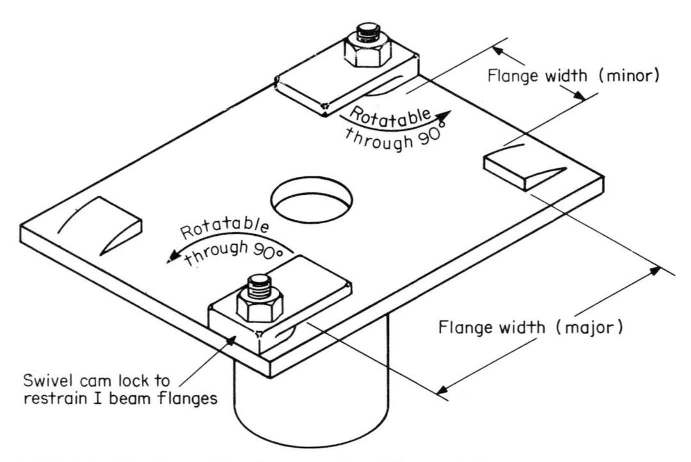

FIG. 13.14 Top plate for securing I-beam ledgers.

members, or where practical, with additional cross bracing so that rows of frames are continuously cross braced in one plane (Fig. 13.15).

With 4-ft (1.22-m)-wide frames having 4 ft (1.22 m) or longer brace lengths, towers should be tied (laced) to each other at a height of approximately 16 to 18 ft (4.9 to 5.5 m) from the base in line with the plane of the frames. If substantially high towers are involved, repeat the lacing-bracing at every third frame in height as work progresses. Lumber is also used to effect lacing-bracing, utilizing specially shaped nailer plates (Fig. 13.16).

Lacing and bracing with 2-in (5-cm) nominal tubing and 2⅜- × 2-in (6- × 5-cm) couplers affords a significant degree of moment connection. This lacing should be installed in both horizontal planes at each three-frame level in the

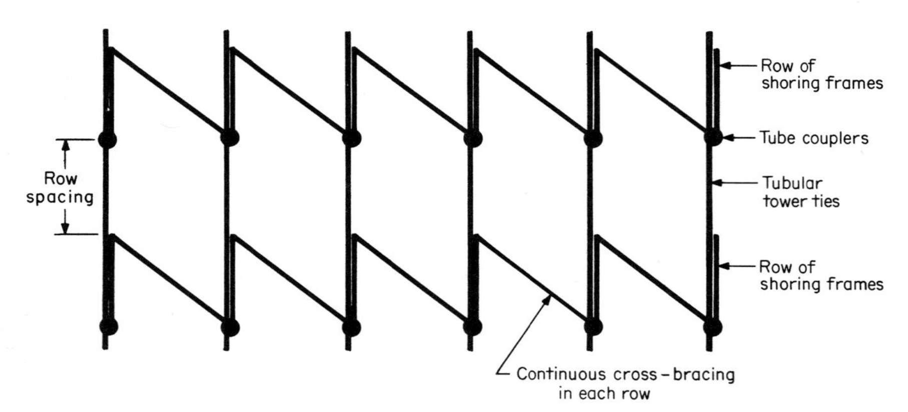

FIG. 13.15 Lacing of shoring tower rows in one direction, continuous cross bracing in other direction.

tower, unless continuous cross bracing can be used in one direction (plane at right angles to frames).

For relatively "clean" installations (i.e., having a flat, firm base), it is sufficient to attach these "lacing" members to the towers as continuous members with suitable scaffold couplers between the tubing and *one* leg of each tower, even though the tube will pass by a second leg of the tower without being connected. (See Fig. 13.17.) Installations involving multivariable base and shoring height conditions should have each lacing member coupled to *two* legs of each tower that it passes. When shoring to or from sloping surfaces, each tower leg in contact with the forming of the sloping area should be tied to its adjacent, similar member with ties attached approximately parallel to the sloping surface(s). Figure 13.18 illustrates this, in addition to showing some recommended ways of proper fastenings to avoid slippage of the members in contact.

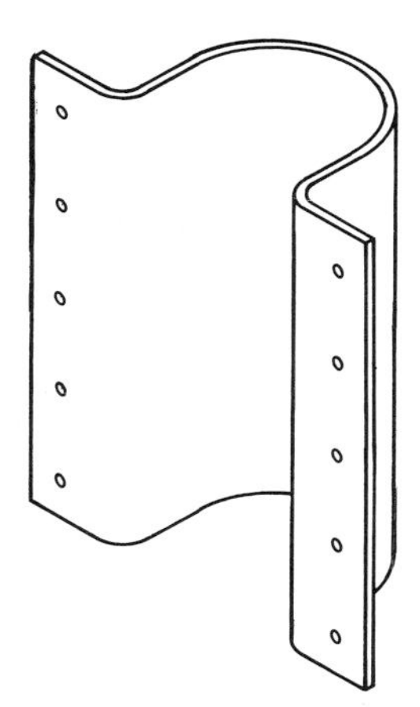

FIG. 13.16 Nailer plate fits around frame leg, nailed to lumber brace or tie.

Continuity of these lacing-bracing members is quite important to the stability

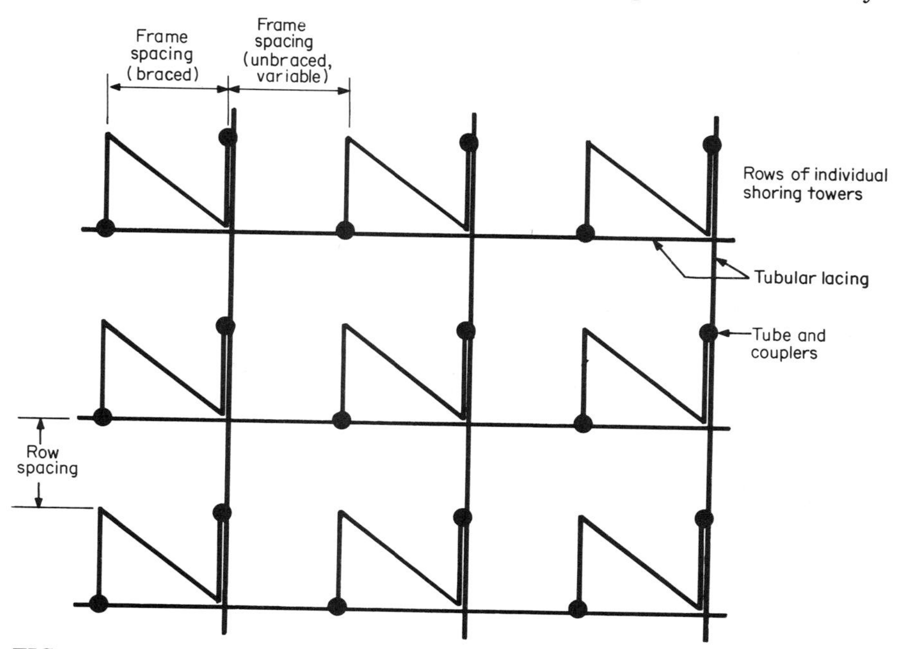

FIG. 13.17 Shoring tower laced in both directions.

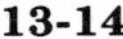

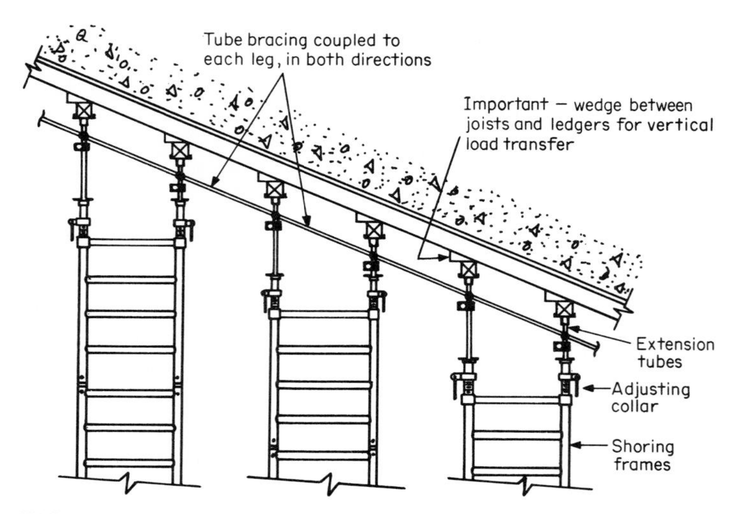

FIG. 13.18 Shoring sloping surfaces.

of the installation as a whole since each two or more connections develop a rudimentary moment connection which is a valuable stabilization feature. Also, screw legs can be tied together by these lacing-bracing members, thereby distributing the tendencies of one or more legs to deflect from the vertical owing to lateral forces.

In designing a shoring system it is very important that it be considered as a synergistic whole—possibly greater than the simple sum of its parts. The interface between ledgers-stringers and frame legs is an extremely important one and should take precedence over all considerations except overload of frame legs or ledgers. It is essential that ledgers be installed so as to: (1) have maximum surface contact with the U head, (2) be suitably placed so that ledger reaction loads are transferred *concentrically* to frame legs by the use of centering wedges and other suitable measures as shown in Fig. 13.8*b*, and (3) ensure that screw legs are not extended beyond the manufacturer's recommendations. Sloping ledgers must be wedged to provide a horizontal seating surface in the U head, as shown in Fig. 13.18.

SCREW-LEG EXTENSIONS

Frame-leg capacity is generally lowered when using long screw-leg extensions. Variations in manufacturer's designs preclude generalizations as to the degree of load reduction. When a screw leg, completely closed, is inserted in a frame leg, it adds a certain, nonreducible extension to the leg. This is known as the "dead-leg" dimension and must be added to the frame and sprocket stack-up height before adding the required screw-leg adjustment range for the situation. Several common screw-leg configurations are shown in Figs. 13.19 and 13.20.

As for type A and type B conditions shown in these figures, similar conditions exist at the tops of the towers between screw legs and U heads or top plates.

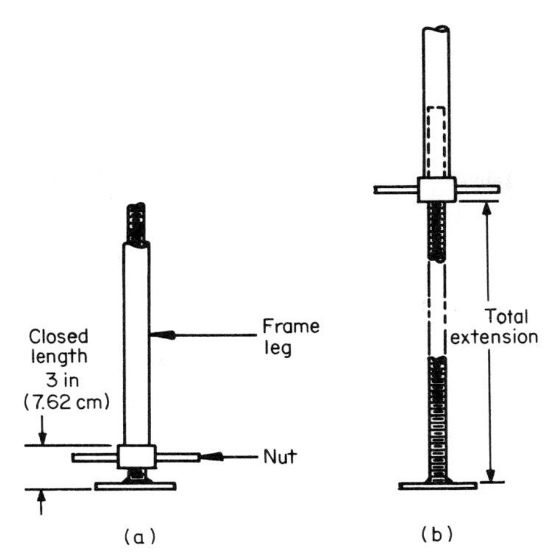

FIG. 13.19 Type A screw legs: (*a*) closed and (*b*) open.

Generally, total extensions up to 12 in (30 cm) do not significantly decrease the loading. Above 12 in (30 cm), the effect is variable and largely dependent on the unsupported column distance between the base of top plate and closest frame-leg structural connection, generally the cross-brace attachment. Based on this concept it is obvious that a type A screw leg will give a greater reduction of leg strength than a type B. Only the manufacturer will be able to give the precise allowable loading for various conditions of use for specific equipment.

Extra-Long Screw Legs

Some manufacturers have available 24-in (60-cm) and longer extension screw legs for use with heavy-duty frames. Since proprietary equipment varies greatly, the specific manufacturer of any special legs should be consulted for allowable loads.

Other Extension Devices

The previously mentioned piccolos, also known as extension tubes, shore staffs, etc., give great additional flexibility to the shoring tower. Piccolos are tubes, generally 2-in (5-cm) nominal OD and 5 to 6 ft (1.52 to 1.83 m) long, which telescope inside the frame legs. They give a larger range of adjustment than the threaded screw leg, having pinholes at frequent intervals [3, 4, or 6 in (7.5, 10, or 15 mm)] to accept a hardened steel pin usually of ½ or ⅝ in (12.7 or 15.9 mm) diameter, which affords rough adjustment of extension. Fine adjustment is achieved with a realtively short length of a screw-threaded adaptor collar serving as a transitory connection between the extension tube and the frame leg. (See Fig. 13.12.) Depending on total length and the size of frame used in conjunction, extension-tube shoring lengths vary from about 1 to 5 ft (30 to 150 cm).

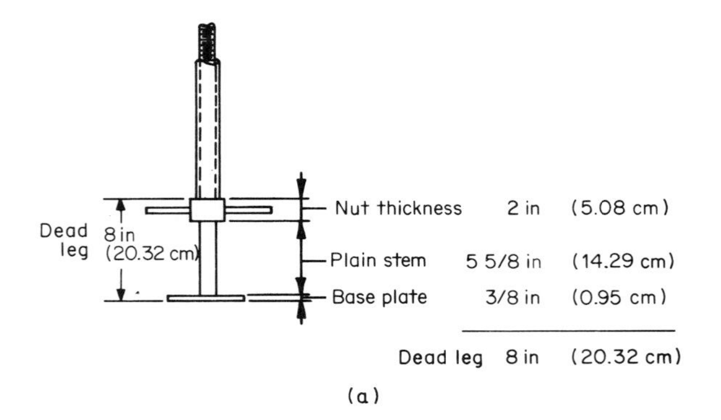

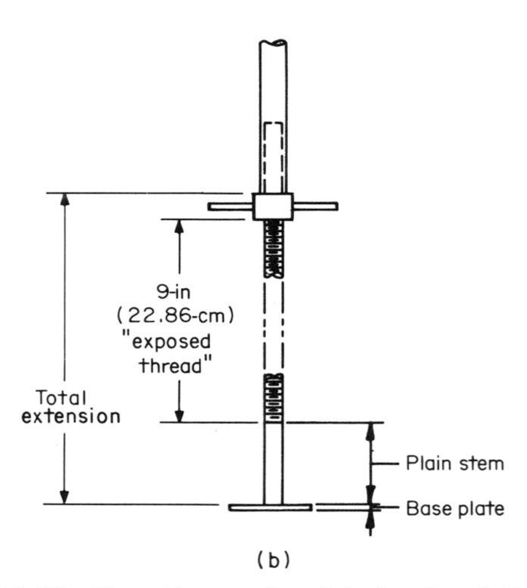

FIG. 13.20 Type B screw leg: (*a*) closed and (*b*) open. [*Important:* Most manufacturers call out the type of leg and *either* the fully extended leg (extension) from the frame leg, *or* give the amount of extension in terms of "exposed screw thread."]

Certain proprietary manufacturers utilize a pair of extension tubes welded into a "head" frame (extension frame) configuration which can achieve high stability properties owing to the facility to brace the welded members together with pivoted cross bracing which is readily adaptable to the extension frame.

Another means of bracing is by horizontal tube and coupler ties connecting each of four legs of a tower together in two planes. Where conditions allow, rows of towers can use long lengths of tubes in straight runs.

Nailer plates have also been developed to brace these legs with 2- × 4-in (5- × 10-cm) or 2- × 6-in (5- × 15-cm) wood lacing and bracing. The moment

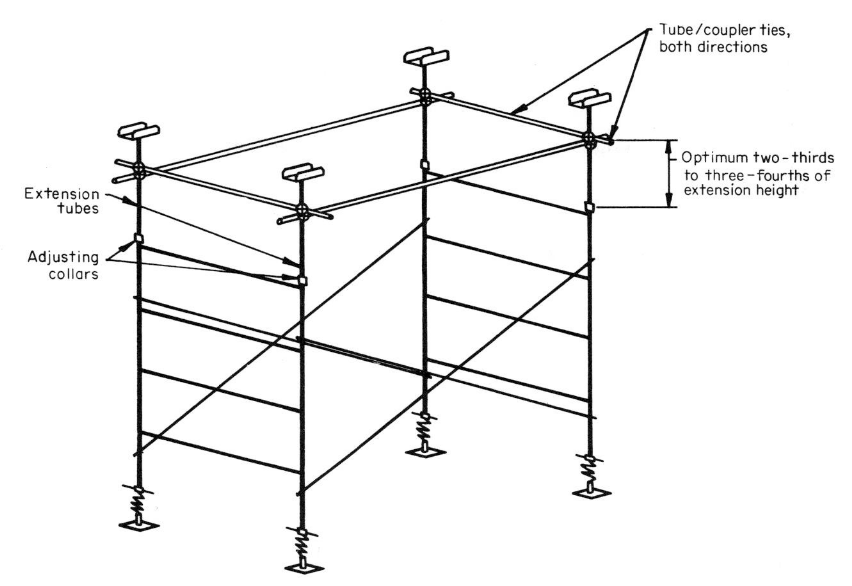

FIG. 13.21 Shoring tower with braced extension tubes. (Refer to Table 13.1.)

strength of nailer plates is not high, and generally stronger bracing is achieved using cross braces with special adaptors or tubes and couplers.

To illustrate the variables involved when using data published by manufacturers (Fig. 13.21), Table 13.1 shows a condensed comparison of two apparently similar types of extension tubes or staffs. To all intents and purposes the materials used are essentially the same. Manufacturer A does not publish recommendations on extension tube bracing, whereas manufacturer B does.

TABLE 13.1 Leg Load Derating When Using Extension Tubes (Piccolos)

Allowable loads in pounds per leg

	Manufacturer A			Manufacturer B	
	No. of frames high			Frame heights not specified	
Extension, ft	1	2	3	Legs unbraced	Legs braced
Up to 2	9100	8600	8100	10,000	10,000
3	8500	7500	6900	8,000	10,000
4	7300	6500	5700	7,000	9,000
5	6500	5550	4600	6,000	8,000
6	6000	4750	3800	N/A	N/A

SELF-WEIGHT OF SHORING

The self-weight of heavy-duty shoring affects leg load design. Weights of frames, braces, and other appurtenances vary greatly among manufacturers. Taking an arbitrary heavy-duty shoring system, one section of a tower would weigh:

2 frames at 65 lb (29.5 kg) each

4 coupling pins at 2 lb (0.9 kg) each

2 cross braces at 20 lb (9 kg) each

Totalling 178 lb (80.6 kg) per section

A tower 10 frames high would weight $10 \times 178 = 1780$ lb (806 kg)

Dividing by 4, the load per leg = 445 lb (1980 N)

It is a good rule of thumb that 5% or more should be deducted from the allowable leg load for self-weight. Since 445 lb (1980 N) is close enough to 5% of the 10,000 lb (44,480 N) allowable, it is recommended that design in this case be limited to 9500 lb (42,256 N) per leg. In general, for heavy-duty frames, deduct self-weight when towers are six or more frames high.

A similar but standard frame tower would be 10 to 20% lighter, say 140 lb (36.5 kg) per section or 35 lb (15.9 kg) per leg, per section. Since standard frames have a much wider range of allowable leg loads owing to size and configuration, the 5% rule is a good one to apply here too. For instance, a four-frame high tower with a nominal allowable leg load of 4000 lb (17,792 N) would have a self-weight per leg of $4 \times 35 = 140$ lb (63.6 kg) which is 3½% and which would be tolerable. A good rule of thumb here is to deduct self-weight when towers are four or more frames high.

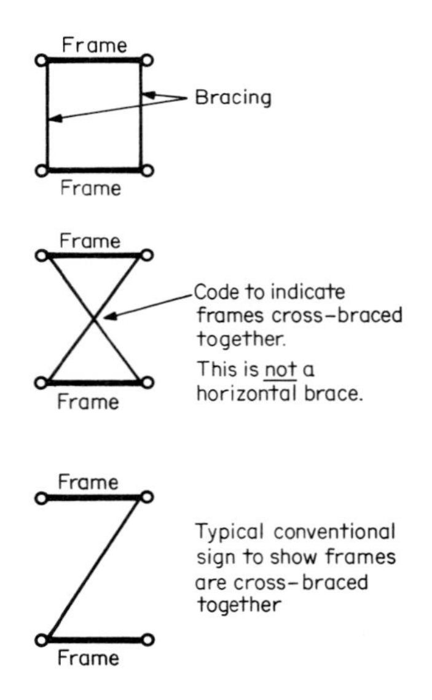

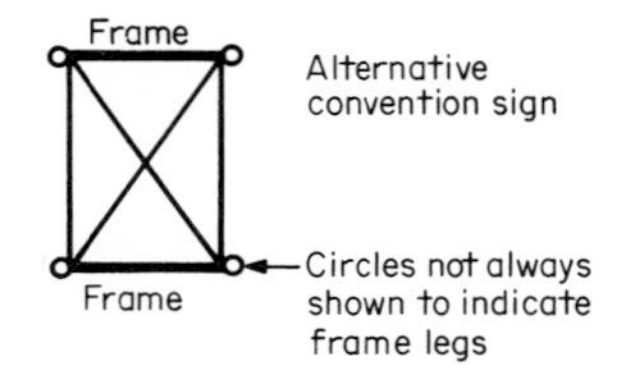

FIG. 13.22 Conventions for indicating shoring towers in plan view.

FRAME LAYOUTS

Frame towers on layouts may be shown with or without circles to indicate frame legs. Conventional ways of indicating shoring towers in plans are shown in Fig. 13.22.

It is difficult to design an efficient shoring layout unless the *total forming system* is considered as a whole. For any given concrete load, the items

TABLE 13.2 Interrelated Items

Item	Optimum size and spacing	Depends on
Plywood	Optimum	Joist spacing
Joist spacing	Optimum	Joist span between ledgers
Ledgers	Optimum	Span between frame legs
Frame legs	Spacing	Finite maximum load/leg
Sills	Size/length	Leg loads

listed in Table 13.2 are also closely interrelated. Any one of these items, if not optimized, could possibly have an adverse effect on the design of other components of the system. Typical reasons for not using optimum size and spacing of components and their consequences could be:

Design requirement	Reason for not using optimum size	Result
¾-in (19-mm) plywood	Only ⅝-in (16-mm) plywood is available	More joists at closer centers than need be*
4- × 4-in (10- × 10-cm) joists	Only 3- × 4-in (7.5- × 10-cm) joists available	More joists to supply and place at closer centers; inefficient use of plywood*
4- × 8-in (10- × 20-cm) ledgers	Only 4- × 6-in (10- × 15-cm) ledgers available	Frame legs at closer spacing than optimum and inefficiently loaded

*When these two results are combined, substantial excess handling occurs. This will be discussed later in detail.

SHORING DESIGN EXAMPLE

Various aspects of shoring design and comparisons of options will be explained through the use of an example.

Consider the forming and shoring of an 11-in concrete slab:

Plywood Available: ¾ in (19 mm)

Ledgers Available: 4 × 8 in (10 × 20 cm)

Joists Available: As required

Design to determine whether standard or heavy-duty frames have best utilization.

Design Load: 11-in (28-cm) slab = 140 lb/ft^2 (6703 N/m^2)
Live load = 30 lb/ft^2 (1436 N/m^2) (per SSI)
Total = 170 lb/ft^2 (8139 N/m^2)

For ¾-in (19-mm) plywood: With face grain at right angles to joists, maximum spacing is 21 in (53.34 cm) center to center [21 and 20 in (53.34 and

50.8 cm) are not evenly divisible into the 8-ft (2.44-m) length of a sheet of plywood].

Try: (*a*) 19.20-in (48.77-cm) spacing [one-fifth of an 8-ft (2.44-m) plywood sheet]:

Load per foot of joist = 170 lb/ft^2 × 19.2 in/12 in = 272 lb/ft (3.97 kN/m)

or (*b*) 16.0-in (40.64-cm) spacing:

Load per foot of joist = 170 lb/ft^2 × 16 in/12 in = 227 lb/ft (3.31 kN/m)

Limitations: Limit the wood design to:

Maximum bending stress = 1800 lb/in^2 (1241 N/cm^2)

Horizontal shear = 150 lb/in^2 (10.34 N/cm^2)

Modulus of elasticity = 1.76 × 10^6 lb/in^2 (1.21 × 10^6 N/cm^2)

Limiting deflection = span/360

These limiting stresses contain allowances of +25% for short-term loading for bending and shear and +10% for deflection.

Conclusions

1. Use 3- × 4-in (7.5- × 10-cm) joists at 19.2-in (48.77-cm) spacing for a 4-ft (1.22-m) span between ledgers if 4-ft (1.22-m)-wide heavy-duty frames are to be used, or
2. Use 4- × 4-in (10- × 10-cm) joists at 16-in (40-cm) spacing for a 5-ft (1.52-m) span between ledgers if 5-ft (1.52-m)-wide standard frames are to be used.

Ledgers—Standard Frames: Using 4- × 4-in (10- × 10-cm) joists at 16.0-in (40-cm) spacing, the *ledger span* will govern the bracing length between frames.

Concentrated point load applied by joists to ledgers at 16.0 in (40 cm) is 227 lb/ft × 5-ft joist span = 1135 lb (5048 N) per point. Calculations show that a 4- × 8-in (10- × 20-cm) ledger has a maximum allowable span of 5.5 ft (1.68 m) with this loading.

Conclusions

Use 5-ft (1.52-m) bracing to avoid design overload. The design for this example may now be finalized as:

¾-in (19-mm) plywood

4- × 4-in (10- × 10-cm) joists at 16 in (40 cm) center to center

4- × 8-in (10- × 20-cm) ledgers with a 5-ft (1.52-m) span leg to leg

5-ft (1.52-m)-wide standard frames braced, spaced 5 ft (1.52 m) apart

5-ft (1.52-m) spaces between rows of frames

Naturally, some dimensional concessions will require "closing up" some elements to allow for spandrels, columns, elevator shafts, floor openings, etc., where rows of elements cannot be practically reduced and require "expansion." There, single-pole steel shores can be used to good advantage. The layout will look as shown in Fig. 13.23. For a 100-ft (30-m) row of frames, there are twenty 5-ft (1.52-m) spaces, which equals nine 5- × 5-ft (1.52- × 1.52-m) single towers spaced 5 ft (1.52 m) apart, plus a double-braced 10-ft (3.04-m) tower at one end of each row. Each frame leg will support 170 lb/ft^2 × 5 ft × 5 ft = 4250 lb (18,900 N). All elements of this system would be efficiently loaded.

Although many standard frames can take 5000-lb (22,240-N) leg loads under optimum conditions, the type of frame available or long screw-leg thread extension may require a lower value and thus necessitate a load reduction by reducing one or more of the leading dimensions. The example uses a 4500-lb(20,016-N) limiting value.

Examination of Nonoptional Ledger Size

Presuming the contractor did not have 4- × 8-in (10- × 20-cm) ledgers and wished to use or buy 4 × 6s, (10 × 15 cm) let us examine what this would do to the efficiency of the system. With all other conditions the same 4- × 6-in (10- × 15-cm) ledgers could only span 3.5 ft (1.06 m) (a nonstandard brace length). There are few choices available to compensate for this change. One choice would be to use 3-ft (0.91-m) bracing for the towers with 3.5-ft (1.06-m) spaces, that is, 3-ft (0.91-m) braces alternating with 3.5-ft (1.06-m) spaces, between the towers. The 5-ft (1.52-m) spacing between rows would remain. The reduced leg load will now be

$$170 \text{ lb/ft}^2 \times \frac{5 + 5 \text{ ft}}{2} \times \frac{3 + 3.5 \text{ ft}}{2} = 2763 \text{ lb } (12{,}290 \text{ N})$$

$$\text{Efficiency} = \frac{2763}{4500} = 61\%$$

The frame quantity increase, with attendant waste of labor, is 100%/61% × 100% = 164%, that is, 64% more frames and necessary accessories, plus additional labor involved.

Since many contractors are more disposed to use 4 × 6s (10 × 15 cm) for ledgers than 4 × 8s (10 × 20 cm), it is interesting to follow through the economics involved in comparing the two ledger choices because of their interrelation with amount of shoring equipment required. Assume a size for the building

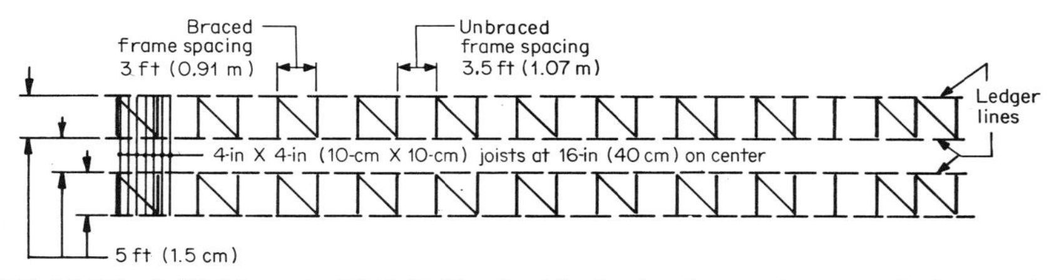

FIG. 13.23 Initial layout of 5-ft (1.52-m)-wide shoring frames for numerical example.

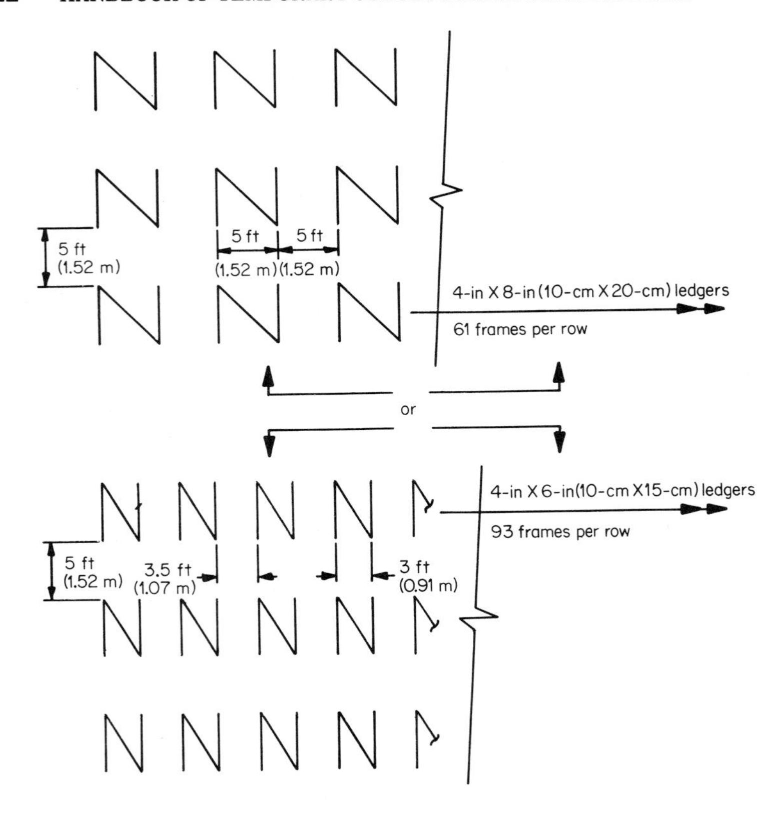

FIG. 13.24 Alternative frame layouts for 11-in (28-cm) slab in numerical example.

shored area to be three slabs, each 300 by 85 ft (91.2 × 95.8 m) and an average total *shored* period per slab of 21 days. Figure 13.24 shows the alternative frame layouts with the two ledger sizes. Figure 13.25 shows a view of an assumed shoring tower two frames high, with adjustable screw legs at top and bottom, using U heads. For cost comparison a rental period of 2 months is assumed. See Table 13.3*a*.

However, this example has not yet taken into account the *timing* of the job. Unless all shoring and formwork are relocated completely, the contractor will need equipment for *at least* 1½ floors to complete the work in the required time; many would use sufficient equipment to shore *two* slabs at one time. Obviously, this will increase rental costs; the labor cost would remain essentially the same except for additional material handling expenses. Therefore, increasing the rental costs by 50% for 1½ floors of equipment would increase the above shoring costs to those shown in Table 13.3*b*.

TABLE 13.3 Shoring Costs

	Layout with 4- × 8-in (10- × 20-cm) ledgers	Layout with 4- × 6-in (10- × 15-cm) ledgers
(*a*) Cost for one floor		
Frame positions per row	61	93
Number of rows	9	9
Total frames (2 high)	1098	1674
Total screw legs	2196	3348
Total U heads	1098	1674
Total bases	1098	1674
Estimated 2-month rental at 5% of list prices per month	$15,000	$22,400
Estimated shoring labor, place and remove	370 h	560 h
Labor cost for three slabs at $15.00/h + 20% overhead	$19,980	$30,240
Shoring cost: Rental plus labor	$34,980	$53,140
Average cost per square foot (square meter) 76,500 ft^2 (7107 m^2)	$0.46 ($4.92)	$0.69 ($7.48)
(b) Cost with rental overlap		
Rental plus labor	$42,480	$64,590
Cost per square foot	$0.56	$0.84
(Cost per square meter)	($5.98)	($9.09)

Offset against the cost difference of $64,590 − $42,480 = $22,110 would be the initial cost of the 4- × 8-in (10- × 20-cm) ledgers for 1½ floors which, assuming a cost of $500 per thousand board feet for Douglas fir select structural grade, would be

$$\frac{4 \times 8}{12} \times 300 \text{ ft} \times (9 \times 2 \times 1.5) \text{ lengths} \times 1.15 \text{ (waste)} \times \frac{\$500}{1000} = \$12{,}420$$

It would not be realistic to write off this lumber expense against one small job such as this. If amortized over four jobs—a very conservative number—a realistic comparison would be based on one quarter of $12,420 which is $3105, i.e.,

Extra cost of shoring and labor	\$22,110
One-quarter cost of 4- × 8-in (10- × 20-cm) ledgers	−3,105
Saving	\$19,005
Equivalent to cost difference of	\$0.25/ft² (\$2.67/m²)

This is a substantial saving over using 4- × 6-in (10- × 15-cm) lumber.

Similar economies can be applied to many shoring installations by optimizing the allowable strength relationships between the formwork and supporting shoring, especially when the shoring is to be totally disassembled before removal to other successive locations. The number of shoring uses involved will dictate the labor costs, and the potential labor and rental savings will be greater in a direct proportion to the number of uses.

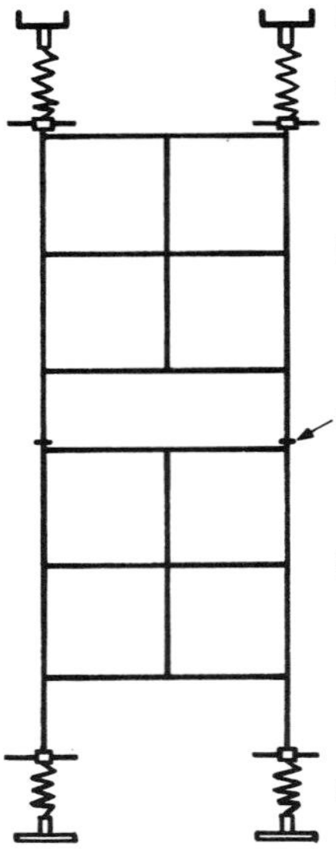

FIG. 13.25 Frame stack-up for numerical example.

Standard vs. Heavy-Duty Frames

Generally, slabs under 12-in (30-cm) thickness can be shored efficiently with standard frames. We previously concluded that with 4-ft (1.22-m)-wide heavy-duty frames we could use 3 × 4 or 4 × 4 in (7.5 × 10 or 10 × 10 cm) joists on 19.2-in (48.77-cm) centers at 4 ft (1.22 m) or more span. If 4-ft (1.22-m)-wide heavy-duty frames were to be used, the joists, load would be

$$170 \text{ lb/ft}^2 \times 19.2 \text{ in}/12 \text{ in} \times 4 \text{ ft} = 1088 \text{ lb (4839 N) per ledger point}$$

at 19.2-in (48.77-cm) centers. Calculations show that 4- × 8-in (10- × 20-cm) ledgers will span 6 ft (1.83 m), but this will not load the frame legs efficiently:

$$4 \times 6 \text{ ft} \times 170 \text{ lb/ft}^2 = 4080 \text{ lb/leg (18,148 N/leg)}$$

However, heavy-duty frames are available with 8-in-wide ledger U heads, enabling them to seat two 4- × 8-in (10- × 20-cm) ledgers side by side. Therefore, the ledger-point load from joists will be

$$\frac{1088 \text{ lb}}{2} = 544 \text{ lb per point (2420 N per point)}$$

With such a light load, 4 × 8s (10 × 20 cm) can span 8 ft (2.44 m), which is conveniently a standard brace length. If 8-ft (2.44-m) towers are spaced 8 ft (2.44 m) apart along the ledger lines, they allow for excellent utilization of 16-ft (4.88-

m)-long ledger lumber. However, the new leg load will be 170 lb/ft² × 4 × 8 ft = 5440 lb (8139 N/m² × 1.22 m × 2.44 m = 24,190 N), only 55% efficient.

What can be done to improve utilization? One could use 4- × 4-in (10- × 10-cm) joists at 12 in (30 cm) to increase their span, but not sufficiently to make much difference. The best alternative might be to use heavier 4- × 6-in (10- × 15-cm) ledgers as joists, with 4- × 8-in (10- × 20-cm) ledgers. Since 4- × 6-in (10- × 15-cm) joists at 16-in (40-cm) centers can span 8 ft (2.44 m), there will be 4-ft (1.22-m) frame rows spaced 8 ft (2.44 m) apart having 8-ft (2.44-m) bracing-spacing distances. Exploring this method, even though use of 4- × 6-in (10- × 15-cm) joists is not always popular, would give the layout shown in Fig. 13.26.

The 4- × 6-in (10- × 15-cm) joists would impart

$$\frac{8 + 4 \text{ ft}}{2} \times 16 \text{ in}/12 \text{ in} \times 170 \text{ lb/ft}^2 = 1360 \text{ lb per point (6107 N per point)}$$

at every 16 in (40 cm) to ledgers, that is, 680 lb per point (3053 N per point) to each of the two ledgers side by side which would be capable of spanning 8 ft (2.44 m) with this loading. They would not be loaded efficiently over the 4-ft (1.22-m) spans, of course. If the 4 × 8s (10 × 20 cm) are not in "good as new condition," a recommendation would be to close one of the spacings down depending on how the building layout will allow this. If 4- × 6-in (10- × 15-cm) joists *cannot* be used, then the use of heavy-duty frames for an 11-in (28-cm) slab such as this will be quite uneconomical. However, other wood sizes and materials, such as extruded aluminum sections for joists, as well as steel and aluminum sections for ledgers, can be used, thus enabling frame leg loads to be optimized.

Laying out the heavy-duty frames to the slab size, some adjustment is necessary for fit because the 8- and 4-ft (2.44- and 1.22-m) dimensions are not evenly divisible into the width of 85 ft (26 m), and then the balance is subdivided into equal modules of frame-plus-rowspace modules. Since the 4-ft (1.22-m) frame plus 8-ft (2.44-m) space module is a most convenient one for 12 ft (3.66 m) and longer joist lengths, use it as much as possible and finish with one odd-sized adjusting row; that is,

$$85 - 4 \text{ ft} = 81 \text{ ft (24.7 m)}$$

$$\frac{81 \text{ ft}}{4 + 8 \text{ ft}} = 6 + 9 \text{ ft (2.74 m) leftover}$$

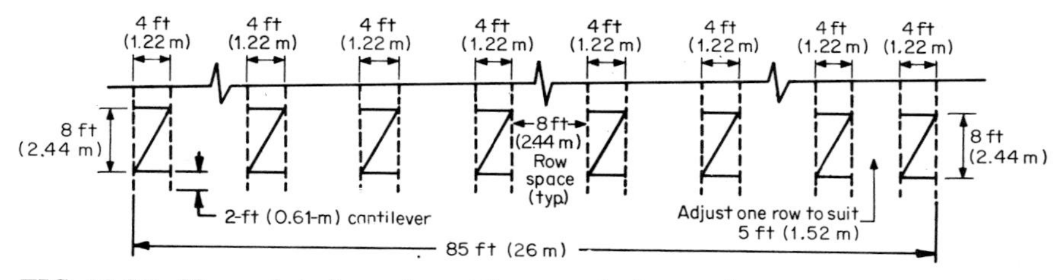

FIG. 13.26 Heavy-duty frame layout for numerical example.

Split the 9 ft (2.74 m) into a 4-ft (1.22-m) frame width and a 5-ft (1.52-m) space and finalize the design to be:

Eight 4-ft (1.22-m) rows of frames
Six 8-ft (2.44-m) spacing rows
One 5-ft (1.52-m) spacing row

as shown in Fig. 13.26.

Final adjustment of these dimensions would be to reduce one of the spaces if the shoring is within a confined area such as a basement or an area with spandrel beams. Generally, lay out the frame rows to get optimum economy and minimum quantities of joists. A careless choice of row spacing could result in a layout with many of the joists terminating in long, useless, cantilevered lengths which could result in excessive plywood deflection.

Economics of Heavy-Duty Frames

In the foregoing example, the 300-ft (91.4-m) building length is divisible into thirty-seven 8-ft (2.44-m) spaces between frames (38 frames per row) with 4 ft (1.22 m) left over. Divide the 4 ft (1.22 m) into two 2-ft (0.61-m) cantilevers at the ends of each row, as shown in Fig. 13.26. Should such cantilevered lengths give undesirable, excessive deflection, their ends can be "picked-up" or additionally supported with single-pole shores.

For this example, the bills of materials and their comparative costs for sufficient equipment for one floor are:

TABLE 13.3*c*

		Heavy-duty frames and accessories	Standard frames and accessories
Frames and braces		608	1,098
Screw legs		1,216	2,196
U heads		608	1,098
Bases		608	1,098
Dollar values	High:	$94,000	$150,000
	Low:	$75,200	$120,000
RENTAL at 5%/mo	High:	$ 4,700	$ 7,500
	Low:	$ 3,760	$ 6,000

The values shown are typical at the time of writing for the two types of equipment and necessary accessories.

It is obvious that heavy-duty frames do not cost twice as much as standard frames, even though they can carry double the load. Their price per item is about

24% higher; their weight is approximately 10% higher which does not involve any meaningful difference in erection or dismantling times. The reader is reminded that such prices will escalate approximately in accordance with manufacturing material and labor costs. Therefore, an escalation allowance of 10% per year is recommended. Actual selling prices are normally determined by the manufacturer or distributor at the time and point of sale, and the above figures should be used only for comparison. Rental rates also vary, and therefore contractors are best advised to seek specific cost information from suppliers.

Both the selling and rental costs are for only *one* floor of equipment; they must be factored for job timing requirements as previously mentioned, i.e., factor by 1.5 for enough equipment for one and one-half floors. The rental values are per month, not per job duration. This factor is henceforth referred to as the "overlap factor."

For the purpose of further comparisons, consider these quantities from the previous example:

Total area of three floors = 76,500 ft^2 (7100 m^2)

Three floors of concrete = 2600 yd^3 (1988 m^3)

Design load for three floors at 170 lb/ft^2 (8140 N/m^2) = 13,005 kips (57,846 kN)

With these quantities the purchase price range for *one* floor of shoring would be, using the given high and low values:

Purchase Cost Comparisons

	Heavy-duty frames, $	Standard frames, $
Cost per ft^2 of 11-in slab	0.98–1.23	1.57–1.96
(m^2 of 28 cm slab	10.60–13.20	16.90–21.10)
Cost per yd^3 of concrete	29–36	46–58
(m^3 of concrete	37.80–47.30	60.40–75.50)
Cost per kip design load	5.8–7.2	9.2–11.5
(kN design load	1.30–1.62	2.07–2.60)

The preceding figures must be multiplied by the overlap factor.

As with the 4- × 8-in (10- × 20-cm) ledgers, only a *portion* of the capital outlay would normally be applied against a short-term project such as this; this is an accounting and tax matter beyond the scope of specific recommendations here. However, "book" life expectancy of frame shoring is frequently taken as 5 for financial write-off, the physical life expectancy being generally longer and in proportion to the care taken in its use, handling, maintenance, and storage.

The rental total costs of the required shoring based on equipment for one and one-half floors and for 2 month's rental at 5% per month of the given purchase prices would be:

	Heavy-duty frames, $	Standard frames, $
Rental per ft^2	0.15–0.18	0.24–0.29
(m^2	1.59–1.99	2.54–3.17)
Rental per yd^3 of concrete	4.33–5.42	6.92–8.65
(m^3 of concrete	5.67–7.09	9.05–11.31)
Rental per kip design load	0.87–1.08	1.38–1.73
(kN design load	0.19–0.24	0.31–0.39)

NOTE: The total erection and dismantling labor for the three floors of heavy-duty shoring would be approx. 1100 hr versus 1680 hr for the standard frames (exclusive of the forming). The difference could be a $10,260 labor saving.

Estimating Rental Costs

It is most important to remember that rental rate per floor per month is usually a guide, and most equipment suppliers will quote lower monthly rates for long-term projects than for short-term ones. Also, longer-term jobs can often be completed in a shorter shoring time per floor by the use of more equipment. This serves the purpose of expanding the work to be done into more segments and enabling the various trade personnel to be more continuously employed with fewer layoffs or less downtime. Consequently, subject to *reshoring* requirements and procedures covered subsequently in this chapter, the use of two or even three floors of shoring equipment on multi-reuse work will increase the shoring cost per square foot by increasing the overlap factor to 2 and 3, respectively, but shortening the shored time per floor will give other economic advantages, such as earlier completion time and subsequently sooner building occupation.

These two variables, overlap factor and shored time per floor, must be carefully assessed when estimating shoring costs. A very common field error is to use the rental rate per month instead of the rental rate of equipment for the shoring duration per floor. The shoring cost per floor in the preceding example is, of course, derived from the total rental cost for a 2-month duration divided by the number on floors completed in that time, i.e.,

$$\text{Rental} = 2 \text{ months}/3 \text{ floors}$$

Obviously, the miscalculation of using the rental rate per month for 3 floors will cause the false assumption that the job will take 3 months instead of the actual 2 months resulting in 50% higher unit square foot and other costs.

The use of "flying shoring" techniques, utilizing steel-frame shoring, can further reduce labor costs for relocation of shoring and formwork from floor to floor with attendant time savings in job completion.

FRAME HEIGHT STACKUP

(The example is not metricated.) In the foregoing example we deliberately omitted any height calculations, preferring to do these separately.

Let us assume that the shoring is to rise from consolidated stone fill (rough grade) and that the height from grade to slab soffit is 14 ft 6 in.

The best way to determine frame combination is as follows:

1. Add up precise thickness of lumber. (¾ in for plywood + 5½ in for 4 × 6 joist + 7¼ in for 4 × 8 ledger + 2 in for sill = 15½ in.)
2. Add to this the amount of "dead leg" in the screw legs and stack up. (Assume 6 in at top + 6 in at bottom = 12 in.)
3. Subtract this total from shoring height. (14 ft 6 in − 2 ft 3½ in = 12 ft 2½ in.)

This is the height which must be filled by shoring frames plus variable adjustments on the screw legs. Let us assume that screw legs have a maximum of 10 in extension each.

At this height, two frames are certainly needed:

1. Try two 6-ft-high frames + 1-in coupling pins = 12 ft 1 in. Subtract 12 ft 1 in from 12 ft 4½ in = 3½-in adjustment called for. This will not be satisfactory unless one screw leg is dispensed with, not a practical solution since rough grade needs adjustment at the bottom for grade variations. At the top, at least 3- to 4-in adjustment must *always* be left to allow easy stripping of the form lumber before dismantling the scaffold.
2. Try two 5-ft frames + 1-in coupling pins = 10 ft 1 in. Subtract 10 ft 1 in from 12 ft 4½ in = 2-ft 3½-in adjustment needed. If screw legs are limited to 10-in thread extension (two of them 1 ft 8 in), we do not have sufficient height.
3. Try one 6-ft frame + one 5-ft frame + 1 in = 11 ft 1 in. Subtract 11 ft 1 in from 12 ft 4½ in = 1-ft 3½-in adjustment needed. This combination leaves 15½ in for screw leg adjustment, say 8 in for bottom legs and 7½ in for top legs. (Why should they be roughly equal? No reason except for symmetry.) A time-saving practice is to rough-set the screw legs before use (to approximately 8 in). Use a measuring stick or spacer and adjust legs until there is 8 in of screw leg extension showing between the underside of the handle nut and the base plate or head. It is faster to adjust the legs at ground level than while they are supporting the frames at the base and/or lumber at the top.

The average setting time is about 1 min each for clean, lubricated legs. Therefore, for 280 heavy-duty legs allow 4 to 5 man-hours; for 432 standard legs allow 7 to 8 man-hours.

There are two other items that should be considered: (1) grade settlement under load (should not exceed ½ in) and (2) compression of formwork lumber under load.

Grade settlement can be minimized using properly designed sills on compacted grade. Obtain from the soil consultant a ground settlement value for leg-base load intensity, or the per unit pressure that will be experienced. Single 2 × 10 plank sills are very local in their effect, so generally assume that their effective load distribution is only 1 ft^2 per leg. For a preliminary review of grade adequacy use the applied design load per leg as the pound per square foot intensity.

If a rough stone grade or compacted subgrade will withstand these pressures, all that remains is to place the sills so that they are in full contact with both base plates and grade. If the leg loads cause high settlement (i.e., more than a nominal ½ in), the sills must be designed for the actual conditions as described later in this chapter.

Consideration must also be given to local compression of the sills and form lumber. A good rule of thumb is to use 1/16 in for every lumber surface (see Fig. 13.27) in contact with another wood member or steel component loaded to more than 60% of capacity or allowable load; if loaded to less than 60% of capacity, use 1/32-in compression for each lumber surface in contact, except for plywood.

Shoring frame stackup charts are sometimes available from manufacturers to aid the contractor in selecting combinations of frame stacks. Table 13.4 is typical of such charts. It should be noted that although 6-ft- and 5-ft-high frames are well standardized in size, the smallest height frames vary with the manufacturer and may be 3 ft, 3 ft 6 in, or 4 ft high. The table gives heights for the median 3-ft 6-in-high sizes; therefore; adjustment by ±6 in to the stackup heights is necessary in accordance with the size of frame to be used.

Adjustments may be made to the charts for use with extension tubes, extension frames, and similar accessories. Adjustments should be made to the dead leg lengths used in the charts for those corresponding to the design and/or manufacture of the shoring equipment being used.

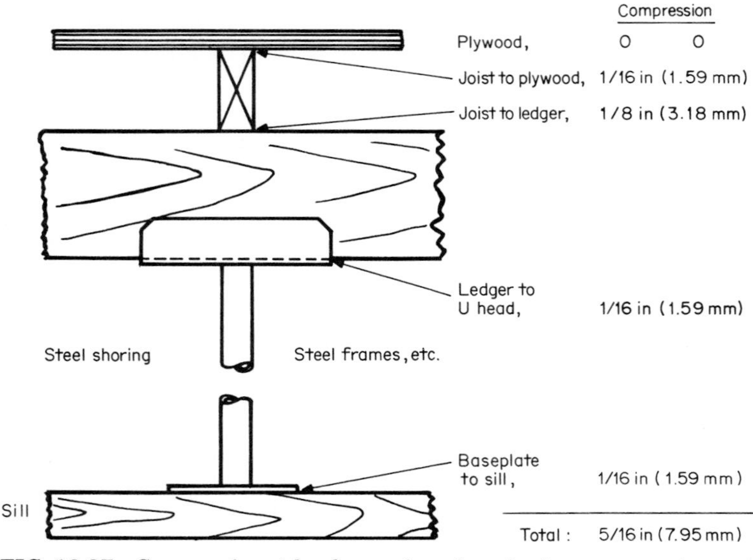

FIG. 13.27 Compression at lumber surfaces in a shoring tower stackup.

TABLE 13.4 Heavy-Duty Shoring Frame Stackup Chart (*Continued*)

Combination	Dead screw leg[6] length at		Frame heights			Frame stack[3]	Shoring heights with 12 in adjustable screw legs[1]	
	Bottom, in	Top, in	3 ft 6 in	5 ft	6 ft		Minimum[2]	Maximum
1	4 N/A[4]	6		1		5 ft 0 in	6 ft 2 in	6 ft 10 in
2	6	6		1		5 ft 0 in	6 ft 4 in	8 ft 0 in
3	4 N/A[4]	6			1	6 ft 0 in	7 ft 2 in	7 ft 10 in
4	6	6			1	6 ft 0 in	7 ft 4 in	9 ft 0 in
5	6	4 N/A[4]	1	1		8 ft 7 in	9 ft 9 in	10 ft 5 in
6	6	6	1	1		8 ft 7 in	9 ft 11 in	11 ft 7 in
7	6	4 N/A[4]	1		1	9 ft 7 in	10 ft 9 in	11 ft 5 in
8	6	6	1		1	9 ft 7 in	10 ft 11 in	12 ft 7 in
9	6	4 N/A[4]		2		10 ft 1 in	11 ft 3 in	11 ft 11 in
10	6	6		2		10 ft 1 in	11 ft 5 ft	13 ft 1 in
11	6	4 N/A[4]		1	1	11 ft 1 in	12 ft 3 in	12 ft 11 in
12	6	6		1	1	11 ft 1 in	12 ft 5 in	14 ft 1 in
13	6	4 N/A[4]			2	12 ft 1 in	13 ft 3 in	13 ft 11 in
14	6	6			2	12 ft 1 in	13 ft 5 in	15 ft 1 in
15	6	6	1	2		13 ft 8 in	15 ft 0 in	16 ft 8 in
16	6	6	1	1	1	14 ft 8 in	16 ft 0 in	17in 8 in
17	6	6	1		2	15 ft 8 in	17 ft 0 in	18 ft 8 in
18	6	6		3		15 ft 2 in	16 ft 6 in	18 ft 2 in
19	6	6		2	1	16 ft 2 in	17 ft 6 in	19 ft 2 in
20	6	6		1	2	17 ft 2 in	18 ft 6 in	20 ft 2 in
21	6	6			3	18 ft 2 in	19 ft 6 in	21 ft 2 in
22	6	6	1	3		18 ft 9 in	20 ft 1 in	21 ft 9 in
23	6	6	1	2	1	19 ft 9 in	21 ft 1 in	22 ft 9 in
24	6	6	1	1	2	20 ft 9 in	22 ft 1 in	23 ft 9 in
25	6	6		4		20 ft 3 in	21 ft 7 in	23 ft 3 in
26	6	6		3	1	21 ft 3 in	22 in 7 in	24 ft 3 in
27	6	6	1		3	21 ft 9 in	23 ft 1 in	25 ft 9 in
28	6	6		2	2	22 ft 3 in	23 ft 7 in	25 ft 3 in
29	6	6		1	3	23 ft 3 in	24 ft 7 in	26 ft 3 in
30	6	6	1	4		23 ft 10 in	25 ft 2 in	26 ft 10 in
31	6	6			4	24 ft 3 in	25 ft 7 in	27 ft 3 in
32	6	6	1	3	1	24 ft 10 in	26 ft 2 in	27 ft 10 in
33	6	6	1	2	2	25 ft 10 in	27 ft 2 in	28 ft 10 in
34	6	6		5		25 ft 4 in	26 ft 8 in	28 ft 4 in
35[5]	6	6		4	1	26 ft 4 in	27 ft 8 in	29 ft 4 in
36	6	6		3	2	27 ft 4 in	28 ft 8 in	30 ft 0 in
37	6	6		2	3	28 ft 4 in	29 ft 8 in	31 ft 0 in
38	6	6		1	4	29 ft 4 in	30 ft 8 in	32 ft 0 in
39	6	6			5	30 ft 4 in	31 ft 8 in	33 ft 0 in
40	6	6		6		30 ft 5 in	31 ft 9 in	33 ft 1 in
41	6	6		5	1	31 ft 5 in	32 ft 9 in	34 ft 1 in
42	6	6		4	2	32 ft 5 in	33 ft 9 in	35 ft 1 in
43	6	6		3	3	33 ft 5 in	34 ft 9 in	36 ft 1 in
44	6	6		2	4	34 ft 5 in	35 ft 9 in	37 ft 1 in

TABLE 13.4 Heavy-Duty Shoring Frame Stackup Chart (*Continued*)

Combination	Dead screw leg[6] length at		Frame heights			Frame stack	Shoring heights with 12 in adjustable screw legs[1]	
	Bottom, in	Top, in	3 ft 6 in	5 ft	6 ft		Minimum[2]	Maximum
45	6	6		1	5	35 ft 5 in	36 ft 9 in	38 ft 1 in
46	6	6			6	36 ft 5 in	37 ft 9 in	39 ft 1 in
47	6	6		7		35 ft 6 in	36 ft 10 in	38 ft 2 in
48	6	6		6	1	36 ft 6 in	37 ft 10 in	39 ft 2 in
49	6	6		5	2	37 ft 6 in	38 ft 10 in	40 ft 2 in

[1]To use 24 in adj. legs add 12 or 24 in to maximum heights for one or two legs, respectively.
[2]Minimum height *includes* 4 in of screw adjustment at top, necessary for stripping.
[3]Coupling pins between frames assumed at 1 in each. Adjust for other dimensions.
[4]N/A means "no adjustment." Dead accessory length of 4 in: adjust for manufacturer.
[5]Short frames not included over 26 ft 4 in stack height; all heights reachable with 6 × 5 ft frames.
[6]Screw legs are assumed to be type B, Fig. 13.20.
Note: 1 ft = 0.305 m.

For any required height, the combination having the least number of frames will be the most efficient. However, this must be balanced against sizes *available.* Generally, 3-ft- and 3-ft 6-in-high sizes are the least available; 4, 5, and 6 ft are readily available. It is noted that a 5-ft or 3-ft 6-in frame takes approximately the same time to erect as the 6-ft frame. It is also important to understand that when only one screw leg is required, the rule for using the screw legs at the top or at the bottom is of one's own choice.

Assume a building having soffit heights of 12 ft 6 in to the slab and 11 ft 8 in to the interior beams. The possibilities of frame combinations from Table 13.4 are the following:

	Slab	Beams
Soffit heights	12 ft 6 in	11 ft 8 in
Less lumber allowance	−1 ft 1⅛ in	−1 ft 1⅛ in
(assuming ¾-in plywood, 4-in joist, 8-in ledger, and 2-in sill)		
Shore height	11 ft 4⅞ in	10 ft 6⅞ in

First Choice: Combination #6 9 ft 11 in to 11 ft 7 in (slab)
Combination #6 9 ft 11 in to 11 ft 7 in (beam)

Second Choice: Combination #8 10 ft 11 in to 12 ft 7 in (slab)
Combination #6 9 ft 11 in to 11 ft 7 in (beam)

Third Choice: Combination #9 11 ft 3 in to 11 ft 11 in (slab)
Combination #6 9 ft 11 in to 11 ft 7 in (beam)

Some combinations utilize one screw leg adjustment positioned at top or bottom of the tower. If support is from rough grade, the adjustment should be at the bottom. If support is from a prior concrete slab, use the single screw leg at the top since it makes for easier stripping. The third choice utilizes screw leg adjustment at both top and bottom. At this height economy would dictate use of the first or second choice, since the time needed to adjust screw legs is substantial and should be held to a minimum.

However, towers of *more* than two frames in height should always have adjustments both at top and bottom. Any irregularity in the support slab manifests itself as an "out-of-plumb" condition which becomes more serious with an increasing number of frames in the height. A ¼-in floor deviation over the 4-ft width of frame becomes 1¼ out of plumb at 20-ft height. This is an insidious condition which can be remedied only by using hardwood shims or blocking and only if discovered early enough. Once the shoring is three or more frames in height, it is best to consider the lower screw legs as "spacers" merely for height attainment and do all the adjusting with the top legs. The stripping will also go faster if the form lumber is not "racked" by premature lowering of the bottom legs. Generally, *never change lower leg adjustment from beginning to end of a job,* unless different height conditions such as a penthouse or machine-room floor are experienced.

MUD SILLS AND SHORING BASES

This is an area in which problems of excessive soil settlement often arise because of lack of specific instructions to field erection personnel. Without guidelines, there are often unforeseeable results despite workers doing their best based on prior expertise.

When, after having layed out a sill system and having erected the shoring, it is decided that the sills need "beefing up," it is very expensive or even impossible to do so.

This section offers some practical means to predetermine the sill design for a given soil or foundation condition. It must be emphasized that the settlement of soil under loads is a very complicated and technical subject involving many variables and factors. The investigation and calculations should not be attempted by unqualified personnel and should be requested from the engineer having responsibility to the contractor.

The information needed by the sill designer is the allowable load per unit area of bearing surface that will give a specified amount of settlement over a specified time. Assume that a recommended specification for a concrete bridge structure is ½-in (12.7-mm) settlement over a 7-day period. The sill designer and the contractor can agree to permit, say, any reasonable settlement over the time period such as 1 in (25.4 mm). This figure must be used to set the vertical shoring adjustment high by this specified amount. In this section on frame shoring/formwork, mention was made of calculating estimated shoring lumber compression, which must be added to the calculated soil settlement. The "time" factor in soil settlement is extremely important, and should be equated (by the engineer having responsibility) to the estimated time after which the structure will be self-supporting and will cease to deflect under its self-weight. Seven days is a typical time for this to happen, but it may vary up to 20 or 30 days depending on the design and job specifications. If soil settlement is negligible, then the shoring will continue to carry the structure weight until it is "stripped," or removed. If soil settlement is of consequence, then the soil settlement should virtually cease after it has deflected sufficiently to take up the structure's self-weight settlement. It should be obvious that self-weight settlement time and stripping time are related but different and are controlled by different factors.

For heavy structures such as (but not restricted to) bridges (see Fig. 13.28), the sill designer can begin after receiving the allowable pounds per square foot (newtons per square meter) psf load given by the engineer based on the specific *time* and settlement parameters required. The leg load divided by the allowable psf intensity will give the sill bearing area required per leg. The sill acts as an upside-down beam, with the point leg loads as beam support reactions and the soil as the uniformly distributed load. Unless this is truly uniformly distributed, the sill cannot be calculated as a beam. Variations in soil surface, if not leveled, or built up with such as level compacted sand or a light concrete "skim" slab, can cause the "sill beam" to deflect in a very nonuniform manner.

For relatively "soft" soils, heavy timbers such as 10 × 10 or 12 × 12 in (25 × 25 or 30 × 30 cm) can provide good bearing qualities because of their beam stiffness and the assumption that the loads will be uniformly spread over the undersides of them. Highly compacted sandy soil surfaces are also very suitable for long timbers, not necessarily as large as 12 × 12s (30 × 30 cm).

Many sill conditions must be treated more carefully, on a leg-to-leg basis. A good pragmatic assumption for these conditions is that most lumber will extend a stress influence line of approximately 45° outward and downward from the edges of a leg-applied load. If base plates are rectangular, always place them with the long side in the direction of the sill length. The influence area is primarily determined by the depth of the sill, bounded by any limiting dimension. In Fig. 13.29*a*, an example is shown using continuous or padded 4- × 10-in (10- × 25-

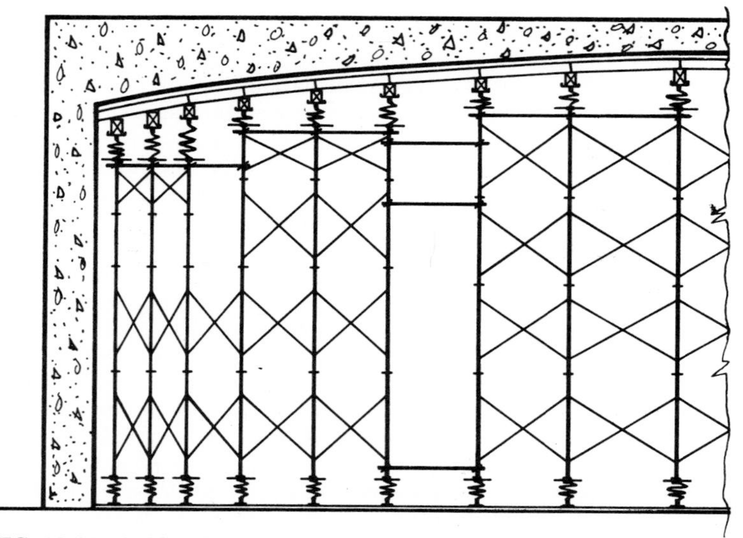

FIG. 13.28 Bridge shoring on a continuous sill.

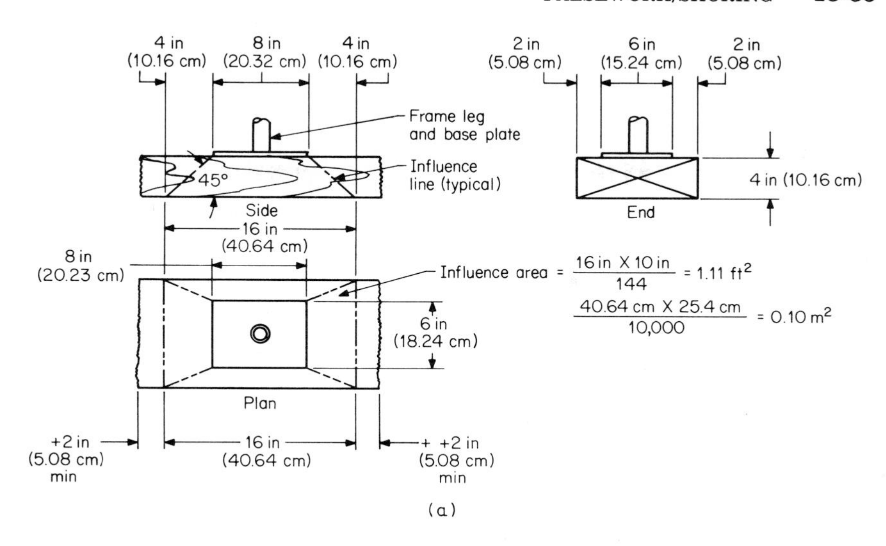

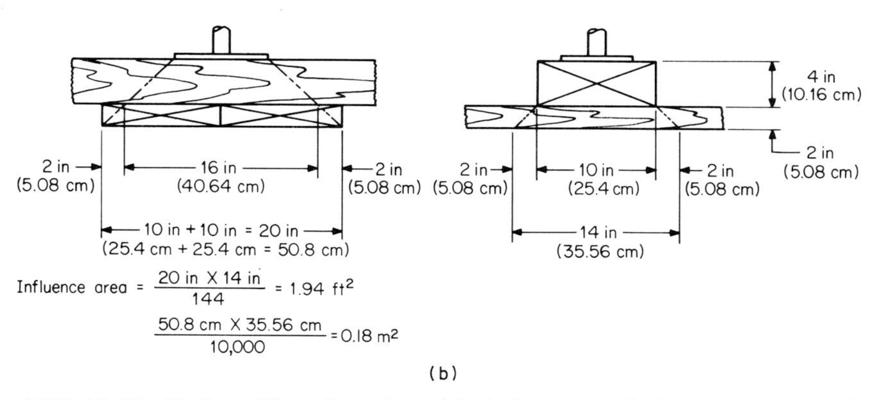

FIG. 13.29 Various sill configurations: (*a*) single 4- × 10-in (10- × 25-cm) wood sill (assumed full, rough); (*b*) 4- × 10-in (10- × 25-cm) sill plus two 2- × 10-in (5- × 25-cm) pads; (*c*) 4- × 10-in (10- × 25-cm) sill plus two levels of 2- × 12-in (5- × 50-cm) pads; (*d*) 4- × 8-in (10- × 20-cm) pads plus two 6- × 12-in (15- × 50-cm) sills (very soft condition).

cm) sills on the flat, which in conjunction with a 6- × 8-in (15- × 20-cm) base plate gives an influence area of 1.11 ft^2 (0.1 m^2). Fig. 13.29*b* shows that by adding two 2- × 10- × 24-in (5- × 25- × 60-cm) pads under each 4 × 10 in (10 × 25 cm) the influence area is increased by 75% to 1.94 ft^2 (0.18 m^2). If the 2 × 10s (5 × 25 cm) were changed to 4 × 12s (10 × 30 cm) (or a double layer of 2 × 12s), the area would increase to 3 ft^2 (0.18 m^2) per leg per pad as in Fig. 13.29*c*.

The depth of the pads should be added to the influence line length at each end, i.e., the 2- × 10-in (5- × 25-cm) pads should be 14 + 2 + 2 in = 18 in (46 cm) total length. For 4-in (10-cm) pads, the lengths would be 14 + 4 + 4 in =

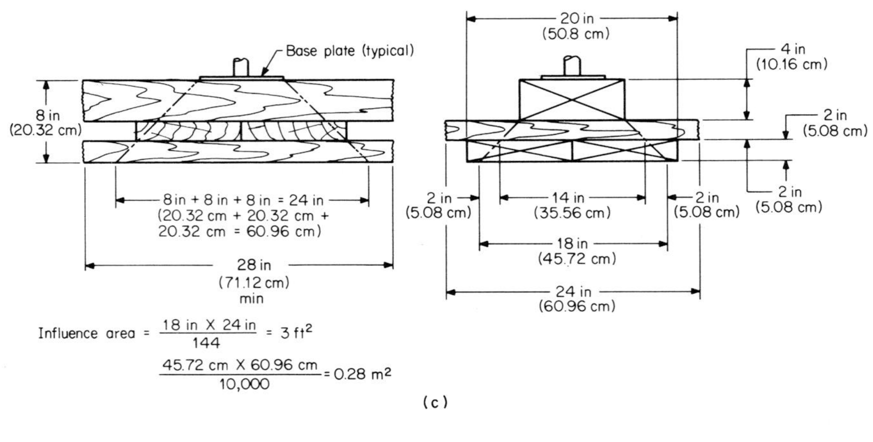

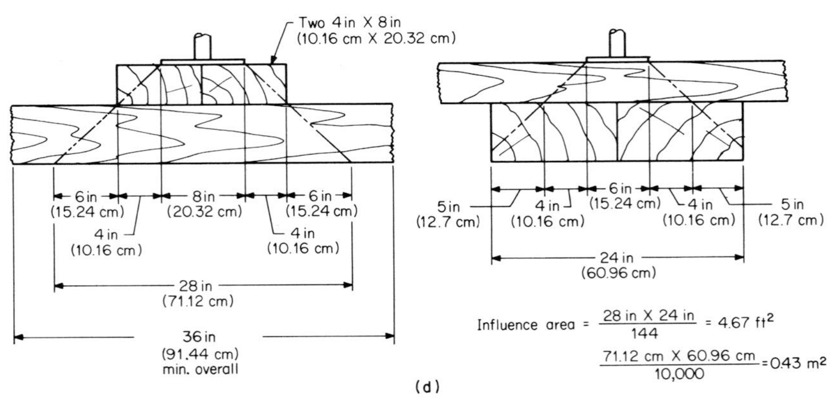

Fig. 13.29 ***(Continued)***

22 (50 cm). Figure 13.29*d* shows a typical sill design for very soft soil and gives a bearing area of 4.67 ft^2 (0.43 m^2).

When erecting shoring from concrete slabs, it is important to use sills rather than seat base plates directly on concrete. During erection it is difficult to keep rows of frames properly in line because steel-concrete friction is low and the shoring "walks." Also, use of sills helps avoid "punch-through" on thin slab sections such as concrete dome, joist, and voided slabs. Normally 2-in (5-cm) sills are quite adequate for this purpose.

For average concrete construction from grade, especially gravel and rock base or hard undisturbed excavation, 2-in (5-cm) or thicker sills should always be used. If the ground is only roughly leveled, a single or doubled-up 2-in (5-cm) sill will act like the upside-down beams previously referred to with decidedly nonuniform load distribution. The sills will settle more under the legs and less between the legs which will result in an incalculable amount of settlement. This, in turn, will affect leg-load reactions resulting in towers moving out of plumb.

Therefore unequal transmissions of loads throughout the complete shoring system could occur with obvious results of undesirable deflection and possibly overload and failure in some areas.

The maxim to be used here is to use 2-in (5-cm) sills very conservatively and, if in doubt, do not use them without approval of the engineer having responsibility.

RESHORING

When, in multistory buildings, a concrete floor is placed in forms supported by shoring, the total dead and live loads of concrete, falsework, workers, placing equipment, vibration, wind, and other forces must be supported and reshored from the lower floors of the structure until the poured concrete floor is self-supporting and sufficiently rigid to resist plastic deformation. The floors below the floor being poured are connected by various means of vertical shoring known as "reshores." New construction/pouring loads must be distributed among previously poured, reshored, lower slabs in some proportion to their relative flexural stiffnesses. After removal of reshores, the poured slabs become an integrated part of the total structure and thence perform their structural contribution as part of the total building design.

Various concrete behavioral phenomena, such as creep, excessive deflection, plastic deformation, temperature and concrete strength imbalances, varying compressions of reshores, and hidden inadequacies of errors in the actual design or construction, are known to exist, but to a large extent these phenomena are not controllable in a predeterminable or precise manner. There is a wealth of available documented analytical research which is largely based upon *assumptions* that the reshored poured lower floors will behave under construction pouring loads with precision. This, unfortunately, is not true. The aforementioned behavioral variables do, indeed, adversely affect theory with occasionally disastrous results.

Often an "instinctive" approach based upon field experience can be successful by means of economic overkill, using far more reshores than necessary, even to the extent of carrying them all the way down to the ground. This can result in hazardous overloading of the lower reshore members, even on three- or four-story buildings. At the other extreme, when using *too few* reshoring members and reshored floors, it has been said that

> If the (reshoring) support is insufficient, the best to be expected is a series of dished slabs, deflected beams and radial cracks around the columns. The worst is a local collapse of a previously placed slab which triggers a chain reaction of collapse and may carry down for the full height. Since these are (usually) shear failures and there is no warning, the path of failure will search out the areas of deficiency and bypass those areas which, though weak, still have the minimum necessary factor of safety. In modern designs of apartment and office buildings there is little surplus strength in a floor. Even with no Live Load there is no strength available to act as a bumper capable of resisting the impact of a similar mass falling one story. And as the accumulation of falling mass comes down, there is no stopping the chain reaction.

From available articles and analytical studies of the problem of reshoring it must be concluded that there is no one definitive source of guidance available. One of the main reasons for this lack of guidance is the often unpredictable behavior of poured concrete in structural form. We lack precise knowledge of creep, whether in the curing or aged state; we lack precise assurance of the quality of the concrete-steel bond, how much form-coating release agent got on the rebars, how many cups of coffee were poured into the forms (as was once discovered to be the cause of a "spongy" concrete column), how morning and night temperature shifts affect a monolithic pour, how seasonal variations affect checkerboard pours, etc.

Because of these factors, the builder is often unaware that crucial shear and flexural members are overloaded and their safety factors so diminished that a structure may be on the verge of a minor failure—or possibly a major one.

A contractor may take a job *including* responsibility for reshoring, without having a properly designed reshoring procedure to follow. A design could always be set out in the job specifications *by the only party having real knowledge of the parameters of the structural design:* the *designer.*

Perhaps, by examining what knowledge is available, we can propose some helpful suggestions.

To begin, there are three major approaches to the problem of reshoring: instinctive, analytical, and empirical. The empirical contains some of the best aspects of the instinctive and analytical approaches.

The Empirical Approach

A simplistic evaluation is available of the number of floor levels that must be connected with reshores to provide proper support. The following *assumptions* are made for this method:

1. Temporary load-carrying capacity of a slab is one-third above its design value.
2. Fresh wet concrete in the forms weighs 1.2 × dead load (allowing 20% for formwork and miscellaneous items).
3. Reshored floors all deflect equally; reshores do *not* shorten (or compress), and all deflect equally. The total load of system is divided into *shares* for respective floors in proportion to their stiffness.
4. Slabs are all equal in geometry and reinforcement; stiffness is in proportion to modulus of elasticity of the concrete. Since modulus increases more rapidly than does strength, stiffness is taken conservatively in proportion to concrete strength.
5. Concrete is up to its 28-day strength, and temperature conditions have been maintained at 50°F or above.
6. Concrete strengths (as percent of carrying strength) are listed in Table 13.5*a* with straight-line variation for intermediate ages.
7. Load-carrying capacity of floor slabs is shown in Table 13.5*b*.

The requirements shown in Table 13.5*b* are minimal. They may not be sufficient to prevent plastic yield and permanent deflection owing to empirical

TABLE 13.5*a* Concrete Strength Percent of 28-day Strength

Age, weeks	Type I cement	Type III cement
4	100	100
3	85	100
2	70	90
1	50	75

TABLE 13.5*b* Carrying Capacity of Floor Slabs in Terms of Slab Weight

Ratio of LL/DL		0.50	0.75	1.00	1.25	1.50	1.75	2.00
carrying capacity in terms of dead load		2.00	2.33	2.67	3.00	3.33	3.67	4.00
Pour rate, floors/week	Type of cement	No. of levels below floor being poured that must be interconnected*						
	I	3	3	3	2	2	2	2
	III	2	2	2	1	1	1	1
	I	5	4	4	4	3	3	3
	III	4	3	3	2	2	2	2
	I	7	6	5	5	4	4	3
	III	5	4	4	3	3	3	3

*Includes number of floors of formwork still in place *below* slab being poured. Number of levels of *reshores is one less* than interconnected floors shown.

assumptions. Wood reshores compress nonuniformly and allow additional, unpredictable slab deflections. Steel shores have negligible compression within the heights of typical building floors.

It is most inadvisable to remove reshores at any level within 2 days of the pouring of a new level. Reshore removal results in a *major readjustment of the loads in reshores* at all higher levels, and of course, possible changes in slab deflections, shear loads, and other interelated forces. Part of the load readjustment will be in the reshores supporting the slab upon which the currently poured slab is supported. Mistiming will result in a readjustment of loading in the shores, and undesired deformation in plastic concrete will inevitably be experienced later. One is further advised against removal of shores under *any slab two floors below the one just concreted,* unless the schedule is so slow that 28-day strength data are available for the floor supporting the formwork.

The foregoing is a suggested empirical guide and thus, of necessity, must be treated strictly as such—not as a specification or a rigid procedure. Its application to any project must be made by the design or consulting engineer using job-based design criteria and procedures. The reader is urged to examine more exhaustive studies, such as Ref. 2.

The Analytical Approach

A most erudite and thorough analytical study of this subject was made by Paul Grundy and A. Kabaila in Ref. 3. Their conclusions are used by many as an authoritative guide on the subject, and the authors take into account a variety of factors and criteria that are too extensive for repetition here.

Some Conclusions on Reshoring

1. Almost all reshoring should be done on a one-for-one basis for as many floors as necessary to safely withstand the service loads and construction loads from the floors above. Slab-floor markings can be of great assistance in assuring that vertical reshore continuity is maintained throughout *all* the shored and reshored levels.

2. Reducing reshoring on a 2:1 or 3:1 basis at lower reshored floors is a questionable practice. It should be realized that reduction is a false economy that may result in stress reversals in the slabs, punch-through, or redistribution of stresses that may be incompatible with the stress designs used by the designer. It can also result in buildup of excessive shear at critical points, increasing the potential for collapse. Many collapses are triggered by *shear* failure.

3. Before the top construction floor is poured, the shored slab(s) by which it is supported should be allowed to take up its own dead-load deflection. This is easily done by slightly releasing the adjustments of the shoring frames or shores so as to allow this. Shoring should then be retightened so as to snugly support the self-deflected slab. (Both over- or undertightening can be structurally harmful or even dangerous.)

4. Vertical shoring load failure is seldom experienced by steel-frame shoring because of high predictability of load design and builtin safety factors applied to allowable loads. Design error, bad sills, overloading, lateral instability, or lumber failure are generally to blame when steel shoring is involved. However, steel-frame shoring is not generally designed or intended for *reshoring* loads.

5. While discussing failure causes, it is important to keep in mind that wood is not homogenous but organic. It cannot be batch-tested with any real assurance of consistency. Each piece has its own individuality, and test comparisons of apparently identical pieces of wood will give diverse results. When new, with standardized safety factors built into allowable load data, wood is relatively reliable. After subjection to multiple reuse, its behavior is erratic in flexure. Therefore, when designing form lumber make sure to allow for ample margin for wood that is used or aged.

6. Wood posts used for reshoring compress significantly and thus allow flexure of reshored beams and slabs in excess of design calculations when carrying reshore loads from upper floors.

Monitoring Reshoring Loads

Recent collapses of multistory apartment buildings point to the necessity for standards to guide the industry. One way to assure adequate reshoring support for multilevel concrete loads is by monitoring the loads on shores and reshores with minihydraulic load cells. With experience and recorded data, concrete behavioral effects can thus be precisely measured even if not wholly and accurately predictable.

Obviously, the majority of multilevel concrete structures are completed without using monitoring gauges but with applying combinations of basic methods and common practices. However, because of unpredictable concrete behavior and the risk of construction deficiencies, such practices can give no real assurance of the efficacy of the shoring or reshoring system until the building is completed or suffers a failure. Monitoring is one way to be sure during construction, even though it is not the only possible procedure.

STANDARDIZATION OF SHORING FRAME TESTS AND DETERMINATION OF SAFE ALLOWABLE LOADS

In the infancy of steel-frame shoring, most manufacturers and users accepted load testing of individual frames to obtain allowable working loads. A commonly accepted value was 5000 lb (22,240 N) for each frame leg. The Scaffolding, Shoring, and Forming Institute, Inc. began, through the facilities of its members and those of certain professional laboratories, to research the behavior of scaffold frames in multitier frame configurations. Standard test procedures were developed and frames especially made for such tests. These frames were similar but not identical to those manufactured by member companies. Tests were made with four-leg towers, with the loading applied concentrically to each leg.

The results of this research had enormous impact on the use of scaffold frames for shoring. The frames were found to behave as four-legged columns when tested, with typical S-type buckling failure configurations similar to those of braced structural-steel columns. Unless the welded rigidity of a frame was very low, column buckling usually occurred in the plane of the cross bracing.

Of greatest importance was the finding that multitier shoring towers had significantly lower load capacities than one- or two-frame-high towers. Figure 13.30 illustrates these typical findings. The graph shows 40% greater capacity for one-frame-high towers than three-frame-high towers. There was a consensus that safest use would result from manufacturers' using their own individual allowable leg-load values based on three-frame-high tests. No known manufacturer authorizes the use of more than 100% of the three-frame-high load data.

Second, the amounts of extensions of screw legs and piccolos at tops and/or bottoms of the tested towers adversely affected frame strengths because they increased the unbraced column lengths of the frame legs from the highest or lowest cross-brace connection position. Third, the varying unbraced lengths between the positions of attachments of cross braces to the frame legs were found to result in important strength variations when comparing the relative perfor-

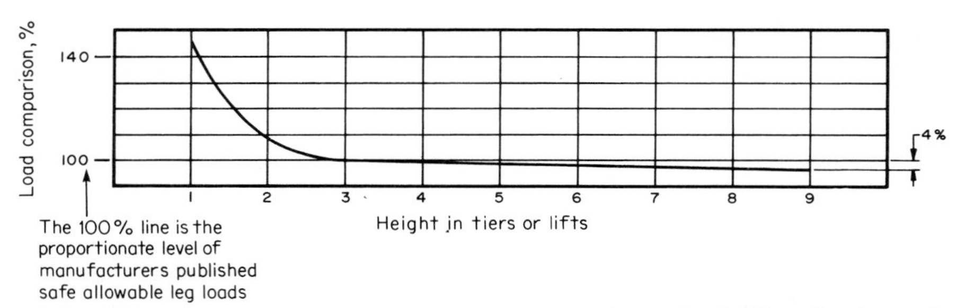

FIG. 13.30 Representative results of frame load tests by the Scaffolding, Shoring, and Forming Institute, Inc.

mances of frames of different configurations and heights. Fourth, arch, or walk-through-type frames had lower capacities than frames having horizontal members at the tops and near the bottoms of frames. Fifth, narrow frames such as those 2 ft (60 cm) wide were not always as strong as those 4 or 5 ft (1.22 or 1.52 m) wide because of their lower sectional moment of inertia when assembled into a four-legged braced column.

The final determination was a more realistic safety factor for frames used for shoring. It has been long-accepted standard practice to use a safety factor of 4 for scaffolding frames. Because of the greater control of conditions and use of frames in predetermined design layouts, it was concluded that for shoring a safety factor of 2.5 was adequate under all normal circumstances. Special applications would need special load considerations, of course.

Users are cautioned that the test data were obtained from nontypical frames to standardize industry practice by using identical test criteria. Some examples in this chapter used 5000 and 10,000 lb (22,240 and 44,480 N per leg) for convenience only. For each type of frame combination, load data for specific conditions must be obtained from the manufacturer of the equipment or the authorized representative.

SINGLE POST SHORES

There are many shoring applications where preengineered single steel post shores of adjustable height are used. They are normally constructed of a lower tube which contains a base plate, an adjusting nut to allow fine height adjustment, and a perforated telescoping staff member with a pin to allow large adjustments. (See Fig. 13.31.) The shore staff also includes a flat plate or a U head to which the formwork is attached.

Post shores should be used only where the loads imposed on the posts are pure compression loads. The load-carrying capacity of a post shore can be increased by reducing its effective height through the use of horizontal bracing in two directions. This bracing is used to increase the vertical load-carrying capacity, but does not take into consideration horizontal forces. The horizontal forces must be resisted by diagonal bracing. The load-carrying capacity of the post shore is defined by the manufacturer using a safety factor of 3.

Applications where lateral loads are imposed on formwork supported by post shores must be designed with extreme care and caution. In addition to horizontal post bracing, as a minimum, each post-shore system must be braced as shown in Fig. 13.32. If bracing consists of lumber, timber bracing clamps, or nailing plates, it is imperative that the load-carrying capacity of these devices be known and that the quantity and frequency of bracing be adjusted to conform to these loads. If guy wires are used for system stability, the vertical compression effect on the load-carrying capacity must be taken into consideration.

Exercise extreme care and be sure all types of loads imposed are understood and resolved to assure that the post shore is not overloaded.

In addition to the horizontal loads caused by wind, construction equipment movement, vibration of flowing concrete through pumping equipment, conveyors, or motorized concrete buggies, care must be given to lateral loads caused by deflection of horizontal shoring beams and truss members.

Double-tiered post shores should never be used. If the working height exceeds the maximum post-shore extension, an alternative shoring system, such as welded frame shoring, should be used.

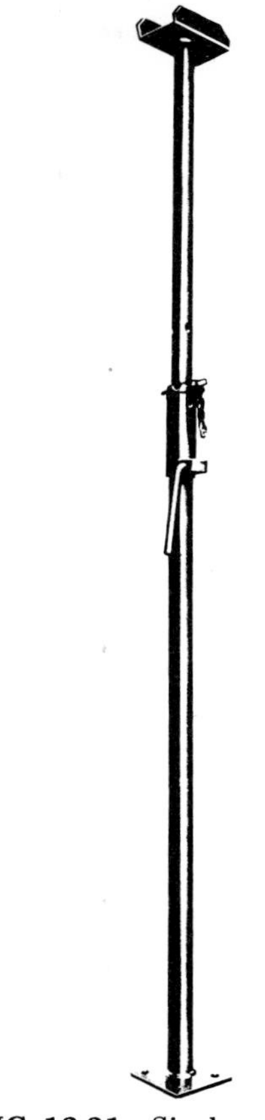

FIG. 13.31 Single post shore.

Ledger bracing

Formwork loads are transmitted directly to the post shore through a system of joists and ledger beams (also known as stringer beams). Since the ledger beam is in direct contact with the post shore, it must be considered when bracing for stability. Ledger beams should not exceed a nominal height-to-width ratio of 2.5:1. If this ratio is exceded, stabilization between ledgers is required.

Layout

Prior to developing a design layout, it is important to read, know, and understand all local, federal, and industry safety rules and regulations. In addition to the preceding design considerations, when making a shoring layout it is important that sufficient bearing under the post shore be available to safely support the design load. If ground conditions are not known at this time, they must be

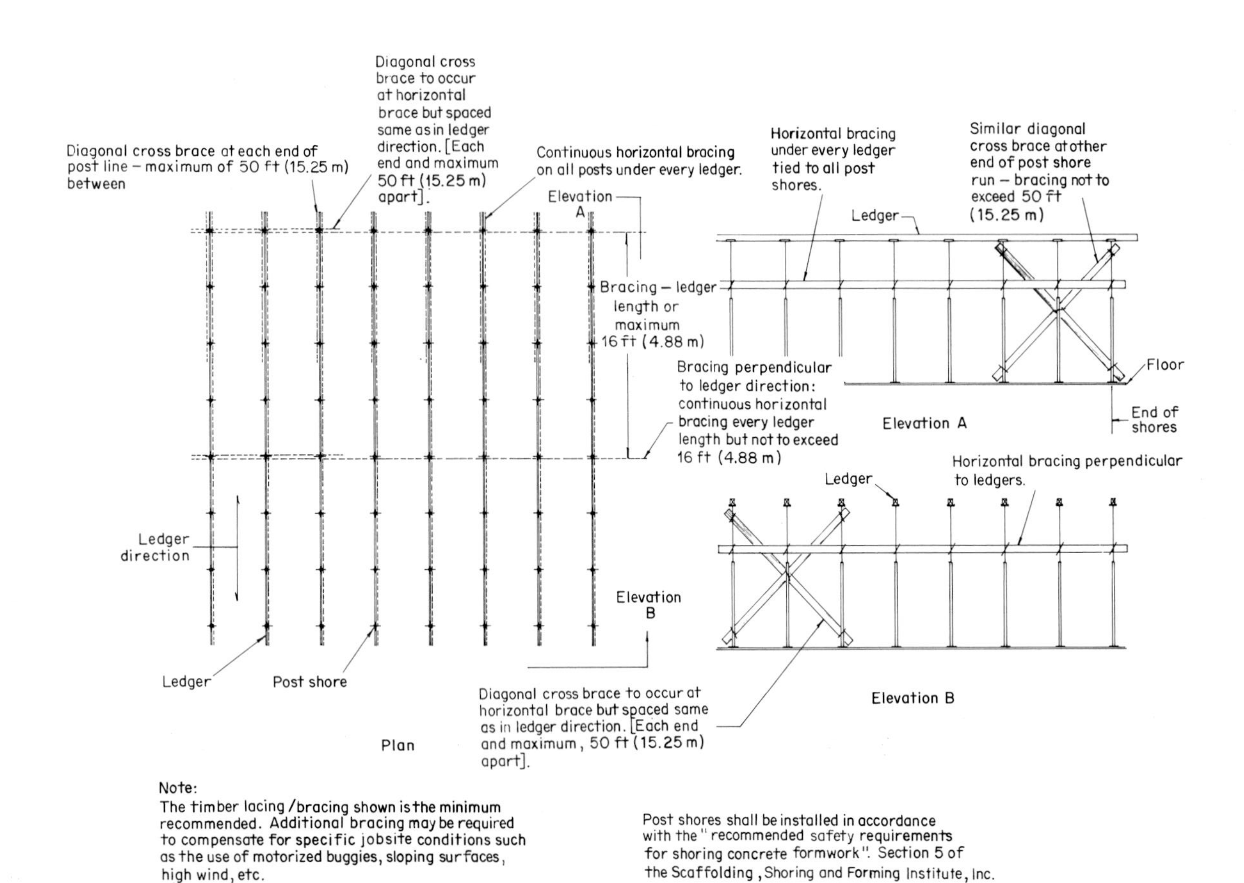

FIG. 13.32 Typical post-shore layout.

ascertained prior to post-shore installation. Avoid eccentric load placement on U heads or top plates by centering all ledgers.

If post shores are used in conjunction with welded frame shoring, the post shores should be tied to an adjacent shoring frame.

Stripping post-shore formwork

After the concrete has been cured or the engineer of record considers the concrete sufficiently cured to carry its own weight, post shores should be stripped by relieving the adjusting nuts in a planned sequence. Care must be exercised to avoid overloading individual posts. Once the entire load is relieved, the post shores and formwork can be removed safely. Post shores should not be struck at the base and allowed to topple.

FLYING SHORING SYSTEMS

The term "flying shoring" is generally used to describe the system of shoring and forming suspended reinforced concrete floors with large modular panels that are craned up from floor to floor as the building progresses (see Fig. 13.33).

FIG. 13.33 Flying shoring table. Shown "flying out" from its slab shoring position. The table is constructed of two cross-braced Interform aluminum trusses 55 ft (17 m) long supporting aluminum joists and plywood. The form deck area shored and moved in one operation is 1000 ft^2 (93 m^2). Tables of up to 2000 ft^2 (186 m^2) are possible.

The first flying shoring system came into being in about 1960 and consisted of steel scaffold-type welded frames, braced together and stabilized with wood joists, wood stringers, and plywood on top making the forming surface. The complete unit, after the concrete was placed and set, was lowered by screw jacks in the scaffold frame legs. Roller attachments were added to the unit and rolled to the edge of the building, and the wire rope slings were attached, moved out, and flown by crane exactly like the flying truss system. This type of frame system is still being used today very successfully. It is an improved version with steel stringers and aluminum or steel joists replacing the wood stringers and joists.

Sometime later, the rolling shore bracket which supports the flying deck form came into being. There are two types of flying deck forms: the flying flat table, supported by columns or walls, and the flying truss. The rolling shore bracket consists of a roller assembly in the shore head and an adjustment screw below the head for final adjustment and for lowering the flying deck form a short distance after the concrete is placed and set.

The rolling shore brackets are bolted to columns or walls. The flying deck form consists of formwork stringers, usually steel joists that span from stringer to stringer, plywood, and the entire flying flat table tied together. The completed unit is then placed on the rolling shore bracket. After the concrete is placed and set, the table is lowered a few inches, rolled out toward the edge of the building, and again flown the same way as the flying truss or flying frame system. An advantage of this system is that it leaves the entire floor area below the formwork entirely free and open and reduces the need for reshoring. This system also is used to a great extent today with horizontal telescoping shoring beams as the joists in many installations. It should be pointed out that, in using the rolling shore bracket system, the concrete in columns must have sufficient strength to support the loads imposed by the bracket, and high early strength cement is commonly used.

Flying Truss Systems

Flying truss systems consist of a table supported by high-strength steel or aluminum trusses. Trusses are of various depths and extend the full length of the form. Depending on table width, there are two or three trusses per table. Trusses are held in a rigid position with lateral cross bracing of adjustable or preset steel tubing which allows the setting of variable truss spacing. Trusses are spanned laterally with joists of wood, steel, or aluminum and secured to the top chord with clips, screws, or steel banding. Joist spacing typically is 2 ft (60 cm) on center; however, this can vary according to slab thickness. Wood nailing strips are secured to the top flanges of some joists.

Joists constitute the basic support on which the plywood forming surface is mounted. Soffit boards may be incorporated into the support system lateral to the joists when additional nailing surface is needed for dome or pan forms.

Most trusses are supported on telescoping legs with vertical holes with pins to lock at the approximate height desired and with screw jacks which allow fine adjustment to final grade. Some systems may use only screw jacks for adjustment. See Fig. 13.33 for a typical installation.

The Forming Cycle

Stripping

To describe the forming cycle, it is best to start with stripping as the introduction to flying. Supplementary jacks are placed under the lower chord of the trusses. Adjustable legs are telescoped back up into the trusses and pinned, and screw jacks are released and removed. The form is now quickly broken free.

Floor movement of forms

If the table is to be moved straight ahead for flying, it is lowered onto sets of rollers positioned on the floor directly under the trusses. Rollers are sized so that their flanges are far enough apart to easily accommodate the bottom chords of the trusses.

If the form table is to be moved laterally before going out of the building, corner and middle leg positions may be fitted with casters and the form is moved manually to wherever it is to be exited from the building.

Depending on their size and weight, flying truss forms can be moved over floors on casters by crews of from two to five workers. Many contractors, however, prefer to use power equipment for movement between stripping and flying.

Stripping a table, regardless of size, and readying it to fly can reasonably consume from 5 to 15 min. The form, with tag lines in place, is moved over the floor edge beyond the building line. Crane slings then are connected to the top chords of the trusses through small hatches opened in the form surface. The form is now ready to be flown.

Flying

The form is swung clear of the building and up to the level it had recently supported. Usually on high rise projects each table will occupy the same position on the new floor that it occupied below.

Flying forms generally are handled with medium capacity tower cranes. The operator often cannot see the form while booming it from the building side and will need direction to clear the floor edge, columns, or tables already set as the form is flown and brought in for a landing. Thus, usually the flying operation is accompanied by radio or hand communication between the rigger foreman and crane operator.

Grading

At the approach of the form for landing, four workers take it in hand and, together with the crane operator, guide the form to its next premarked position.

The four corner legs are extended first and locked into position at a predetermined setting, and the screw jacks are inserted. With the table in correct alignment, it is lowered into position on the slab.

The remaining legs are adjusted to the correct height at a later time as the crew now moves to position the next table.

SHORING DESIGN LOADS

Each structure must be thoroughly analyzed in terms of its physical configuration and content, relating these factors to the characteristics of the particular shoring and falsework to be used.

Vertical loads

It is customary to reduce the vertical load analysis to load per unit of projected area, such as pounds per square foot (newtons per square meter), so that the distribution of loads can be accurately determined when relating them to the supporting shores and falsework.

When determining this load, consideration must be given to all components including such items as rebar, aggregate, and additives. These additives can vary the weight from as little as 100 lb/ft^3 (1600 kg/m^3) with lightweight aggregate to more than 300 lb/ft^3 (4800 kg/m^3) with heavy aggregate, high concentrations of rebar and lead additives as used in nuclear plants.

Consideration must also be given to the work practices anticipated during construction with due provision made in design calculations for these practices. Some examples are: method of placement with attendant impact, motorized equipment such as buggies and vibrators, number of workers concentrated in one area, temporary storage of bundles of rebar, floor hoppers, and numerous other practices common during construction. Loads resulting from these conditions are customarily referred to as "live loads" in shoring calculations.

Scaffolding, Shoring, and Forming Institute, Inc. studies of actual field conditions provided data which were used to develop criteria for determining live and dead loads. These are as follows: A figure of 150 lb/ft^3 (2400 kg/m^3) should be used for calculating dead load where the concrete is commercial ready mix with normal rebar. An inverse relationship exists between dead and live loads when the dead load falls below 70 lb/ft^2 (3350 N/m^2). While a live load of 30 lb/ft^2 (1436 N/m^2) (20 lb/ft^2) (958 N/m^2) for live load and 10 lb/ft^2 (479 N/m^2) for formwork dead load can be used for slabs over 5½ in (14 cm) thick, the live load must be progressively increased as the dead load becomes smaller so that the combined live and dead loads are not less than 100 lb/ft^2 (4788 N/m^2). When motorized carts are used for the placement of concrete, the live loads cited should be increased by not less than 25 lb/ft^2 (1197 N/m^2). Table 13.6 shows design loads for various slab thicknesses and weights of concrete.

In calculating the total design load for a beam, consideration must be given to increased formwork weight. It has to be added to the dead load and treated as a line load acting along the beam. As an example:

Depth of beam	Additional load, lb/ft (N/m)
Up to 3 ft (90 cm)	0 (0)
Up to 5 ft (150 cm)	5 (73)
Up to 8 ft (240 cm)	6 (87)

TABLE 13.6 Design Loads of Concrete Slabs per Square Foot*
Allowance made for 10-lb/ft^2 forms and 20-lb/ft^2 live load

Wt. of concrete per ft^3	150 lb/ft^3 standard weight		120 lb/ft^3 lightweight		100 lb/ft^3 lightweight	
slab thickness, in	Non-motorized	Motorized	Non-motorized	Motorized	Non-motorized	Motorized
2 to 5.5	100	125	100	125	100	125
6.0	105	130	100	125	100	125
6.5	111	136	100	125	100	125
7.0	117	142	100	125	100	125
7.5	123	148	105	130	100	125
8.0	130	155	110	135	100	125
8.5	136	161	115	140	100	125
9.0	142	167	120	145	105	130
9.5	148	173	125	150	109	134
10.0	155	180	130	155	113	138
10.5	161	186	135	160	117	142
11.0	167	192	140	165	121	146
12.0	180	205	150	175	130	155
14.0	205	230	170	195	146	171
16.0	230	255	190	215	163	188
18.0	255	280	210	235	180	205
20.0	280	305	230	255	196	221
22.0	305	330	250	275	213	238
24.0	330	355	270	295	230	255

*For thicknesses of concrete greater than shown on the above chart, use dead weight of concrete plus 30 lb. If motorized, use dead weight of concrete plus 55 lb.

Note: 1 in = 2.54 cm; 1 lb/ft^3 = 16.01 kg/m^3; 1 lb/ft^2 = 47.9 N/m^2.

The figures cited in the table are guides to the most common methods used in shoring calculations and are minimum for that purpose. They are not intended to replace calculations based on actual conditions.

Once the basic design loads have been determined for those areas to be shored, it becomes necessary to support those loads with falsework, timber beams, horizontal shoring beams, steel frames or posts, singly or in combinations in such manner as to safely support those loads without failure or excessive deflection.

Horizontal loads

Shoring and falsework structures must resist all foreseeable lateral loads such as wind, dumping of concrete on inclined surfaces, cable tensions, and stopping and starting of motorized placing or finishing equipment on deck forms. These temporary structures must also withstand the sidesway effects caused by rapidly placed liquid concrete, vibration of concrete by power vibrator, and the effects which occur when concrete is unsymmetrically placed on slab or beam form.

In the absence of precise information on lateral loadings, it is suggested that the recommendations by the ACI in Ref. 4 for "Minimum Lateral Loads Acting in Any Direction" be used. This recommendation is as follows. Slab forms: 100 lb/lin ft (1459 N/m) of slab edge or 2% of total dead load on the form (distributed as a uniform load per lineal foot of slab edge) whichever is greater. "Consider only the area of slab formed in a single placement." These recommendations are only minimum requirements for slab form bracing, and a complete structural analysis of bracing requirements should be made when unusual construction conditions exist or are anticipated.

SHORING POSTTENSIONED CONSTRUCTION

Posttensioning is a complex and sophisticated procedure involving great accuracy of operation with high stresses and forces. There are occasions when these forces transmit additional loads to the shoring equipment over and above the dead loads and live loads for which they are normally designed. In almost all cases, the shoring remains in place until the posttensioning operation is completed.

One must therefore consider the effect of these loads and forces in all posttensioned jobs. The need for this consideration is clearly stated by the ACI in Ref. 4, Chap. 5, Sec. 5.5, "Forms for Pre-Stressed Concrete Construction," Article 5.5.2.1, Design: "Where the side forms cannot be conveniently removed from the bottom or soffit form after concrete has set, such forms should be designed for additional axial and/or bending loads which may be superimposed on them during the pre-stressing operation." Since the shoring equipment supports these forms, the shoring layout must be designed giving consideration to any additional loads which may affect the equipment.

Procedures

First, determine if the construction uses the posttensioning method. The fastest way is to ask the contractor; otherwise a thorough study of the structural drawings is necessary.

1. Look for tendons in the shape of a parabola, as opposed to horizontal reinforcing bars, shown in longitudinal sections of beams and slabs. Posttensioned slabs are usually of the following types:

   a. Deep, flat slab, generally with voids, such as used in bridge designs

   b. Pan-joist slabs, with tendons in the joists (generally long span)

   c. Grid-dome slabs with tendons in the ribs (generally in long-span direction)

2. Look for anchorage points; find out whether these are at piers, abutments, columns, and/or shear walls or at beams running at right angles to the tendons. This latter condition is of utmost importance *if you are shoring beams which carry anchorages from slabs or other beams.* The posttensioning forces applied at anchorage points are generally designated by "force vectors " at the anchorage point, or in tabular manner. They are generally shown in kips (thousands of pounds). It is important to know the values of these force vectors.

3. Obtain and check the job specifications, in addition to the drawings. Specifications generally devote several sections to describing the posttensioning system, end anchorages, and grouting and shoring requirements, in sections covering "Posttensioning," "Formwork," and "Shoring" or "Falsework."
4. Determine whether the shoring equipment will be left "in place" during the posttensioning operation (usually yes).
5. Determine the sequence of posttensioning, i.e., what sections or areas will be posttensioned first.
6. Determine if the posttensioning method and sequence will superimpose additional loads on the equipment. Figures 13.34 and 13.35 illustrate two common conditions in which the tensioning will tend to "lift" the slab, ribs, or beams off the shoring support and transfer all or part of the "dead weight" of section A to section B. Section B is prevented from deflecting by the shoring.

   The persons to give a logical answer as to whether there will be superimposed loads, and if so in what amount, will generally be the structural design engineer, the contractor's engineer, or possibly the posttensioning subcontractor.

   Failing to obtain satisfactory answers, the quotation *and* estimate layout should include the following notes: "This shoring is not designed to carry additional loads that may be superimposed upon it as a result of posttensioning operations. All reshoring and the reshoring of the superimposed posttensioning loads are the responsibility of others."

   If information is obtained (preferably in writing or on a drawing) on the

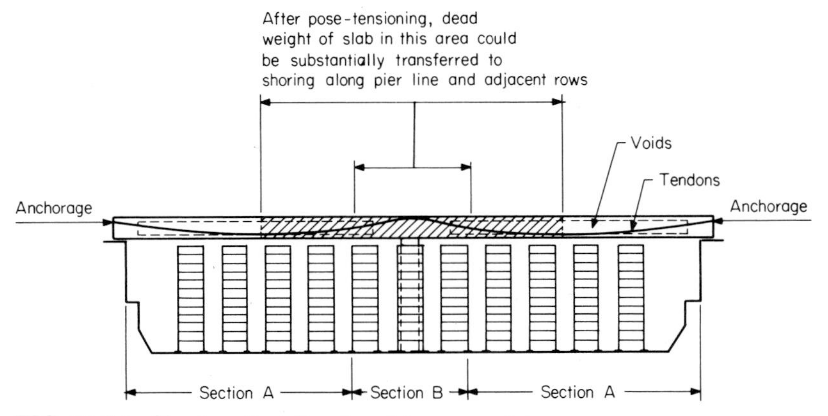

FIG. 13.34 Shoring of posttensioned two-span slab.

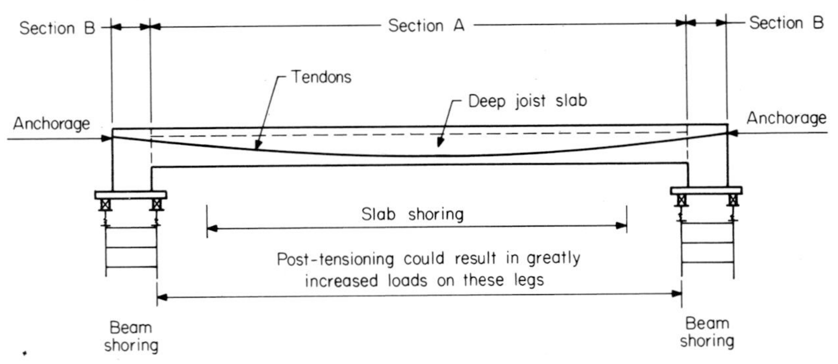

FIG. 13.35 Shoring of posttentioned joist slab.

positions and amounts of the additional superimposed loads, the following notice should be used instead: "This shoring is designed on the consideration, as directed by the structural designer (note: or other appropriate party—substitute as necessary), that the posttensioning operation will induce additional loads on the shoring equipment in the amounts and locations indicated on this layout."

7. Obtain the contractor's consent to contact the parties named in step 6.
8. Have the contractor determine, or obtain the contractor's permission to determine, whether the engineer has a special requirement concerning the procedure to follow for the removal of shoring, if this is not clearly spelled out in the drawings or specifications.
9. Similarly, determine if the engineer wants to review or approve the shoring layout. Although the design and safety of formwork is generally the responsibility of the contractor, the engineer or architect may wish to approve the formwork and shoring designs. If possible, it would be desirable in all cases to have a written or stamped approval as a confirmation of acceptance of any design assumptions made in the layout.

SHORING IN ALTERATION WORK

This subject is complex; it is therefore imperative to obtain complete information on loads and other pertinent factors. Based on this information, the shoring layout should be designed and shoring carried out under the supervision of a licensed structural engineer.

Shoring in alteration work differs from shoring a new building because each situation has to be studied individually to determine the expected loads. Factors to be considered are:

1. Occupancy, because whether the building to be altered will be occupied or not changes the live load to be considered on slabs and floors.

2. Structural characteristics:
   a. Slab thicknesses and strength, which will affect the size of bearing area for reshoring (similar to plywood thickness which affects joist spacing in placing fresh concrete).
   b. Direction of reinforcing steel rods.
   c. Whether the altered sections are load bearing members or not (columns, main walls, etc.).
3. Single or multiple floors.
4. When alteration work is to be done inside the existing structure, some regular shoring considerations such as wind load and mud sills will not apply.

ERECTION TOLERANCES

All vertical shoring equipment must be erected and kept plumb in both directions. The maximum allowable deviation from the vertical is ⅛ in (3 mm) in 3 ft (90 cm) and it should never exceed 1 in (25 mm) in 40 ft (12 m). If this tolerance is exceeded, the shoring equipment should not be used until readjusted within this limit.

Eccentric loads on shore heads and similar members should be avoided. The capacity of a shoring leg or adjustable base is decreased by a large percentage when an eccentric load acts on it. [As an example, a jack that would support 33,000 lb (146 kN) when loaded concentrically will bend at 11,000 lb (49 kN) if the load is placed 2 in (5 cm) off center.]

FALSEWORK DRAWINGS

All jobs requiring vertical shoring should have a drawing prepared or approved by a qualified person. This drawing should have a complete plan of the area to be shored along with elevations showing the makeup of the shoring equipment. Unusual conditions such as heavy beams, ramps, and cantilever slabs should be covered in detail to assure a safe and proper installation. A copy of the drawing should be available and used on the jobsite at all times.

Shoring layout drawings should be available at the jobsite and should be strictly adhered to. All shoring layouts include general comments which provide additional information. Examples of these notes follow.

General Notes

This drawing is provided as a service to illustrate the assembly of the manufacturer's products only. It is not intended to be fully directive nor cover engineering details of such products or equipment or materials not furnished by the manufacturer nor the interconnection therewith. Inasmuch as the manufacturer does not control jobsite assembly or procedures, grade, or quality of materials or equipment supplied by others, it is the responsibility of the contractor to integrate this drawing into a composite drawing suitably complete for construction purposes consistent with safe practice and overall project objectives. The following points should be covered:

1. All dimensions and details shown on this layout must be checked and verified by the contractor before proceeding with the work.
2. The concrete supported by the shoring shown on the layout is assumed to weigh ___ lb/ft^3 (kg/m^3).
3. The design layout includes a live load (including forms) of ___ lb/ft^2 (N/m^2) which does not include provisions for motorized concrete equipment. If motorized equipment is used, add 25 lb/ft^2 (1200 N/m^2) to the above figure.
4. Approximate amounts of screw jack extensions have been noted. These extensions may require adjustment due to field conditions. However, the maximum screw jack extensions for this layout are limited to ___ in (cm) top and ___ in (cm) bottom.
5. The contractor will design and provide suitable sills to properly distribute imposed shoring loads.
6. All stringers, ledgers, or other members resting on the manufacturer's equipment must be centered directly over the shoring legs.
7. Splicing of ledgers must occur at the centerline of jack screws.
8. Ledgers where supported by U heads will be equally packed or wedged.
9. In setting elevations, allow for lumber and soil compression.
10. **Timber Notes**
    a. Timber calculation and design criteria are based on dressed sizes conforming to the American Softwood Lumber Standard, Voluntary Product Standard 20-70, U.S. Department of Commerce, National Bureau of Standards, and to the *National Design Specifications for Stress-Grade Lumber and Its Fastenings,* 1968 edition, revised November 1970.
    b. The timber falsework details shown are manufacturer's suggestions and are based on the following minimum design values. Properties are for: Douglas fir-larch No. 1 stress graded 1500 (per National Forest Product Association) size classified 2 to 4 in (5 to 10 cm) thick by 6 in (15 cm) and wider with a 25% increase for short duration loading 10% for *E*.

Fiber stress in bending 1500 lb/in^2 (1034 N/cm^2) (Engineered Use)	1800 lb/in^2 (1290 N/cm^2)
Horizontal shear 95 psi (65.5 N/cm^2) (if checks are not in excess of width of piece, allow 33% increase to 126 lb/in^2 (86.9 N/cm^2)	150 lb/in^2 (103 N/cm^2)
Compression perpendicular to grain	480 lb/in^2 (331 N/cm^2)
Modulus of elasticity E	1.76×10^6 lb/in^2 (1.21×10^6 N/cm^2)

    c. Plywood design is based on American Plywood Association technical data (Concrete Formwork Brochure No. S71-90, January 1971). Grains of face plies are perpendicular to the supports. Plywood continuous over two or more spans. Values are for B-B plyform class I (exterior type), deflection limited to L/270. If B-B plyform class II is used, increase thickness of plywood shown on layout by ⅛ in.

11. The final responsibility for the formwork design and placement remains with the concrete contractor.

12. The shoring system, as shown, is designed on the assumption that formwork will be restrained from lateral movement by the contractor. Sufficient lateral support must be provided where necessary to prevent the imposition of lateral loads on the shoring system.

13. Shoring equipment should be erected and used per published safety rules and regulations of the Scaffolding, Shoring, and Forming Institute, Inc.

14. The print is the property of the manufacturer and is furnished for the exclusive use of the customer on the condition that it is not to be copied or used by others without written consent.

15. All steel beams must be secured to U heads with beam clamps or by another approved manner, as they are installed.

16. All lower frames of a tower must be plumb and level before erecting the remainder of the tower. Check again for plumbness prior to placement of concrete.

17. Reshoring design and procedures are the responsibility of others and should be thoroughly checked by the architect and/or engineer to determine proper placement and that sufficient capacity exists to support areas being reshored.

18. Tower leg loading should be as uniformly distributed as possible. Never load only one leg of a frame or one ledger of a tower.

19. Horizontal Shore Notes

- **a.** Job architect or engineer is to determine that the structural steel beams supporting horizontal shoring beams are capable of carrying construction loads during placement of concrete.
- **b.** An intermediate shore must never be placed under the lattice section of the horizontal shoring beam.
- **c.** An allowance of up to ⅝ in (16 mm) must be made for the depth of the horizontal shoring beam bearing prongs, in order to determine the correct elevation of the ledger.
- **d.** Do not nail horizontal shoring beam prongs to ledger.
- **e.** Care must be taken to follow the designed allowable clear span of the horizontal shoring beam.
- **f.** Beam hangers must be designed and loaded in accordance with manufacturer's specifications and designed to fit the shape of the horizontal shoring beam bearing prongs.
- **g.** All beam sides supporting horizontal shoring beams must have a minimum of a 2- × 4-in (5.08- × 10.16-cm) vertical stud directly under each

horizontal shoring beam or a design which takes horizontal shoring beam concentrated loads into consideration.

h. The manufacturer assumes no responsibility for placement of beam ties or kickers for beamside forming.

20. Post shores, if shown, are to be the manufacturer's post shores. Any substitutions of other post shores are to be approved by the supplier.

21. The manufacturer does not provide all items illustrated on these drawings. Those items not supplied should be erected according to drawings furnished by their manufacturer or supplier. The final responsibility for design and placement of these items remains with the contractor.

22. No deviation from these drawings is permitted without written consent of the manufacturer.

23. The customer or lessee bears the sole responsibility for ensuring that any erection or use of the goods shown on these drawings conforms to all laws, ordinances, and local codes and for checking the accuracy of field details and dimensions.

FIRE PROTECTION

The best fire protection is a good, clean, orderly jobsite with everything in its place and a place for everything. Flammable liquid storage and proper fireproof storage containers a proper distance from buildings are a must. The proper fire control equipment in proper locations as prescribed in Federal OSHA Sections 1910 and 1926 and in National Fire Protection Association code rules must be followed. A thorough and proper survey of the jobsite, buildings, and adjacent areas by the job superintendent with the aid of a local fire protection agent can point out additional fire and safety precautionary measures and controls which must be taken. A fire drill must be run through to alert all to their responsibilities and assignments, as well as to the location of exits and fire control equipment. When space heaters are used, precautions must be taken in view of combustible materials on the jobsite. Any regulations covering the use of this type of equipment must be complied with.

Material shipped and stored on the jobsite must be analyzed for its flammable or inflammable qualities and stored and protected accordingly. Certification of any lumber or material as to its flame-retardant protection or combustible nature should be reviewed, noted, and tagged.

NEW PRODUCT TRENDS

For many years aluminum was thought to be unsuitable for construction equipment because it was considered susceptible to trucking and job-handling damage. In the early 1970s escalating labor costs and the demand for greater jobsite productivity acted catalytically in the reexamination of the use of aluminum in construction.

The scaffolding and shoring equipment industries had, by the end of the 1970s, developed numerous new items to enhance equipment productivity potential and had redesigned steel shoring to be manufactured from strong and relatively lightweight aluminum extruded sections. The construction industry welcomed these new products, placing the equipment industry in an up cycle of building up availability of aluminum shoring products. A short list of the major equipment types follows.

Aluminum shoring trusses

These are generally used on high rise and other repetitive shoring work. They tend to eliminate the use of steel shoring frames for repetitive-use flying shoring.

Aluminum shoring joists

Developed first for use with flying trusses, these joists are now in popular use for all shoring applications. They are cost-effective; their light weight [average 4 lb/ft (58.4 N/m)] and high strength make them desirable to replace lumber joists on many jobs. Their ability to support concrete slabs over long spans enables shoring frames to be spaced farther apart and thus obtain higher leg loading closer to the shoring frames' safe allowable leg loads.

Aluminum shoring ledgers (stringers)

These are stronger, heavier versions of aluminum joists and can replace steel I-beam ledgers with weight and handling advantages.

Aluminum shoring frames

These are relatively new, and their design is such that shoring loads of up to 15,000 lb (66,720 N) per leg can be accommodated safely, having a weight saving of about one-third over heavy-duty steel shoring frames which have average capacities of only 10,000 lb (44,480 N) per leg.

Single post aluminum shores

While not yet available in large quantities, these are expected to become popular in the 1980s.

The design advantages of aluminum are numerous; it enables almost unlimited flexibility of design shapes to obtain the most efficient strength-to-weight ratio. Aluminum extrusion dies are cheaper and simpler to build, whereas special shapes formed from steel are very expensive resulting in the almost exclusive use of standard available tube, pipe, I-beam, and other rolled steel shapes.

A typical section of shoring using steel frames, I-beam ledgers, aluminum joists, and plywood is shown in Fig. 13.1. In general, fears that aluminum would not stand up to construction-site use have been unfounded.

THE SCAFFOLDING, SHORING, AND FORMING INSTITUTE, INC.

In 1959 a group of scaffolding manufacturers got together and created an association with a mandate to promote safety and the safe use of their products by construction and other industries.

The Scaffolding, Shoring, and Forming Institute, Inc. publishes safety rules for the majority of uses of almost all types of scaffolding and shoring products. It also has available a series of booklets of safety requirements for major types of scaffolding. Another series of hip-pocket-sized booklets deals with how to erect steel frames (scaffolding and shoring) safely. Also available are a number of wall-poster sized illustrations of various hazards and their prevention. Three groups of color slides are available at nominal cost; they come with descriptive booklets exhibiting dos and don'ts for scaffolding, shoring, and suspended powered scaffolds. The address of the Institute is 1230 Keith Building, Cleveland, Ohio 44115.

REFERENCES

1. Federal OSHA Regulation 29 CFR, part 1926. 701 Subpart Q.
2. *Lessons from Failures of Concrete Structures,* ACI monograph No. 1, American Concrete Institute.
3. Grundy, P., and A. Kabaila: "Construction Loads on Slabs with Shored Formwork in Multistory Buildings," *ACI Journal,* December 1963, pp. 1729–1738.
4. *Recommended Practice for Concrete Formwork* ACI Standard 347-68, American Concrete Institute, Detroit. 1968.

14

Concrete Formwork

by R. Kirk Gregory

Formwork is a classic temporary structure in the sense that it is erected quickly, highly loaded during the concrete pour, and within a few days disassembled for future reuse. Also classic in their temporary nature are the connections, braces, tie anchorages, and alignment and adjustment devices which forms need.

The term "temporary structures" may not fully connote, however, that some forms, tie-hardware, and accessories are used hundreds of times, which necessitates high durability and maintainability characteristics and a design that minimizes labor.

Unlike conventional structures, the formwork disassembly characteristics are severely restricted by the concrete bond, rigidity, and shrinkage, which not only restricts access to the formwork structure, but causes residual loads that have to be released to initiate stripping.

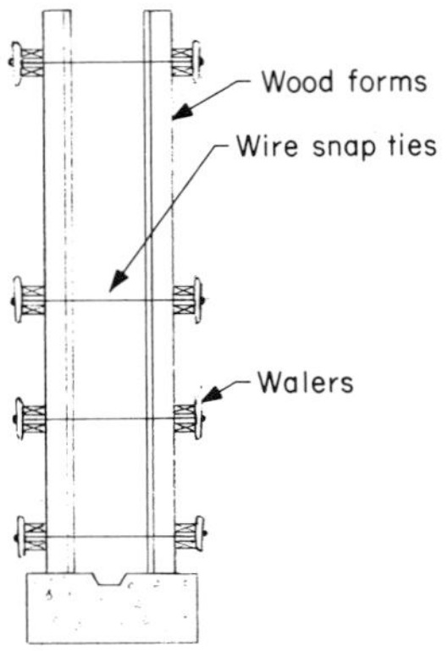

FIG. 14.1 Wood forms and wire snap ties. *(American Concrete Institute.)*

The general practice of using plywood and lumber with snap ties and brackets (Fig. 14.1) is well covered in the American Concrete Institute (ACI) special publication SP-4[1] and in ACI standard 347–78.[2] These set factor-of-safety criteria on form ties, anchors, and lifting inserts (Table 14.1) which were included by reference in the American National Standards Institute's ANSI-A10.9[3] and thence by the U.S. Department of Labor, Occupational Safety and Health Administration's OSHA 1518.701.[4] The American Plywood Association[5] (APA) also publishes wood-formwork guides.

There is not enough space here to adequately cover the design of wood formwork, so this chapter will only summarize the design criteria, types of forms, and

TABLE 14.1 Design Capacities of Formwork Accessories*

Accessory	Safety factor	Type of construction
Form tie	1.5	Light formwork; or ordinary single lifts at grade and 16 ft (4.9 m) or less above grade.
	2.0	Heavy formwork; all formwork more than 16 ft (4.9 m) above grade or unusually hazardous.
Form anchor	1.5	Light form panel anchorage only; no hazard to life involved in failure.
	2.0	Heavy forms—failure would endanger life-supporting form weight and concrete pressures only.
	3.0	Falsework supporting weight of forms, concrete, working loads, and impact.
Form hangers	1.5	Light formwork. Design load including total weight of forms and concrete, with 50 lb/ft^2 (2390 N/m^2) minimum live load, is less than 150 lb/ft^2 (7180 N/m^2).
	2.0	Heavy formwork; form plus concrete weight 100 lb/ft^2 (4790 N/m^2) or more; unusually hazardous work.
Lifting inserts	2.0	Tilt-up panels.
	3.0	Precast panels.
Expendable strand deflection devices†	2.0	Pretensioned concrete members.
Reusable strand deflection devices†	3.0	Pretensioned concrete members.

*Design capacities guaranteed by manufacturers may be used in lieu of tests for ultimate strength.
†These safety factors also apply to pieces of prestressing strand which are used as part of the deflection device.

jobsite conditions that influence form selection and will describe special applications that present structural challenges.

DESIGN LOADS

Fresh concrete exerts pressure on formwork, similar to fluids or soils, that depends on the density of the concrete and the height of the pour. Unlike fluids, concrete consolidates due to vibration and the cement starts setting so that if poured slowly at a steady rate of rise, the maximum pressure can be controlled. ACI 347–78[2] contains charts and formulas that give pressure as a function of rate of pour and temperature (Fig. 14.2). This maximum pressure is distributed as shown on the typical pressure diagram in Fig. 14.3. However, often overlooked are the conditions under which these charts and formulas are valid. Full liquid-

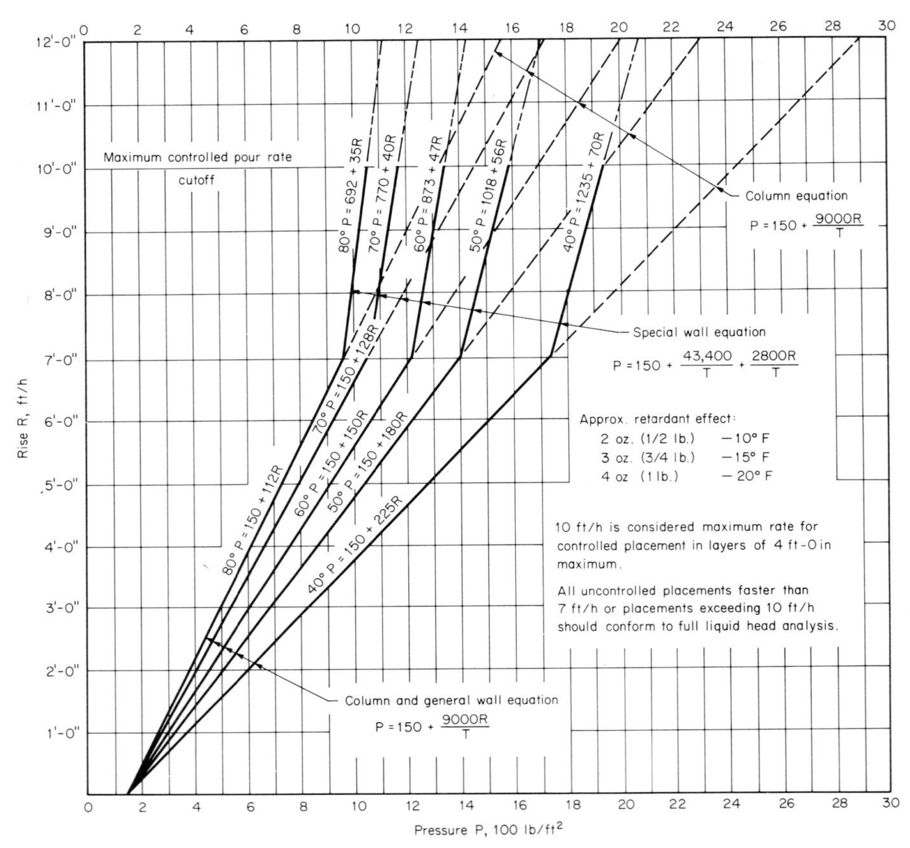

FIG. 14.2 Concrete pressure against forms. *(American Concrete Institute.)*

head pressure, 150 lb/ft^3 (2400 kg/m^3) x height for normal concrete, must be used when: (a) slump exceeds 4 in (10 cm); (b) external vibrators are used; (c) retarder or workability admixtures are used; and (d) layers in excess of 4 ft (1.2 m) are poured.

When concrete is pumped into the bottom of the forms, the form pressure is the full liquid-head plus additional pump pressure to overcome frictional flow resistance. This extra pressure depends on the concrete mix, the pump, and flow restrictions. Excessive pump speed can overpressure the forms. Even with optimum conditions, 50% extra pressure at the pump inlet should be included in the design loads.

If there are significant restrictions to the flow of the concrete being pumped from the bottom, such as prepacked aggregate, precast elements inside the forms, large anchorage/embedments, or box-outs, the extra pressure should be 100% more than the full liquid head.

Pressure acts perpendicular to the form surfaces. This causes uplift on battered wall forms or uphill thrust on sloping slabs which therefore require positive anchors, braces, or deadloads to hold them down. Some of the pressure effects not readily apparent are:

1. Force imbalance when pouring against embankments, previous pours, and sheet piling.
2. Upward reaction from inclined wall braces when one-sided forms are poured.
3. In-plane forces due to pressure on bulkheads, which become severe on thick pours, such as bull-nose piers.
4. In sloping tunnel roofs, the intensity of the pressure depends on the height of liquid head, which can be much greater than the slab thickness.
5. On thick circular walls, hoop-tension forces require positive anchorage.

Secondary load criteria relate to wind load on braces, motorized buggy turning and braking forces, walkway loads, lifting forces during tilt-up of gang forms as well as moving them from pour to pour, and means of supporting the dead weight of gangs during erection and stripping.

While the concrete pressure is considered to be acting on the full height of the pour for design purposes (Fig. 14.3), it must be recognized that concrete has to be poured onto the bottom first, which tends to pivot the forms on the lower ties and kick the forms together at the top. If such partial loadings govern the form design, they must be considered.

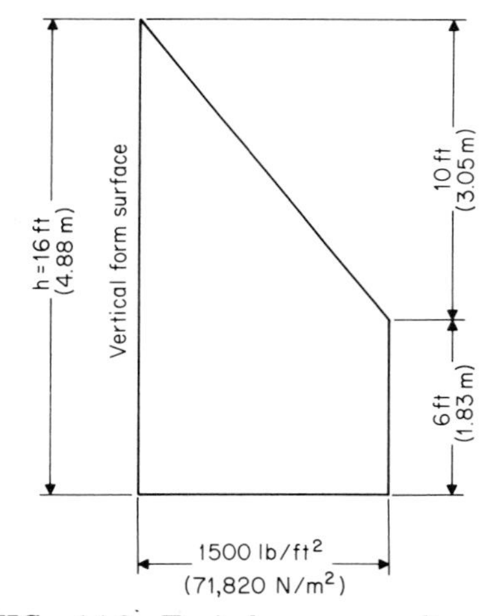

FIG. 14.3 Typical pressure diagram. [Maximum pressure under controlled rate of pour, such as 6 ft (1.83 m)/h at 40° F (4.4°C).] *(American Concrete Institute.)*

If at any one time during the pour the fresh concrete is higher along one side than along the opposite side within the formed area, a net horizontal thrust is created which can cause dislocation and sliding of the entire unit. This could also occur on a box culvert with inside bracing in lieu of ties between opposite walls.

SOME DESIGN CONSIDERATIONS

The formwork structure must be integrated with the tie system that restrains the concrete pressure in a manner that allows quick erection, form-tie unloading, easy disassembly, and tie "removal." Since the ties positively anchor opposing forms, the proper locations of the ties at corners and pilaster, in relation to the formwork structure, are just as important as the strength and the modes of anchorage and removal of the ties.

Modular dimensions that easily combine into any size form assembly with minimum waste, good stackability, and adequate alignments make forming practical.

Formwork weight is always a factor whether hand-setting or crane-handling the gangs.

Multiple lift forms require anchors to carry dead weight, walkways for access, braces for wind loads, and plumbing.

Coreforms, pilaster forms, etc. get severely locked in by the concrete pressures and its shrinkage. Some positive means to relieve this pressure greatly facilitates stripping, as discussed later in this chapter.

JOBSITE CONDITIONS

The nature of the job (including local conditions) is one of the primary factors in formwork selection. In addition to job specifications, variations in the shape of the concrete, reuse potential, form cycle time, gang-forming feasibility, proximity of form inventory and/or supplies, crew experience, crane capability and cost, inserts, penetrations, and concrete pour rate are only some of the factors that influence the type and design of form-tie systems.

If the concrete is to be exposed to the public, then most of the hand-set modular systems will need liners due to their grout leakage and the many "triple-joint" patterns. The larger modular systems have higher pressure capacity and fewer, tighter joints to grind and finish, but the crane capacity, reach, and cost must be considered. However, if walls have many jogs, pilasters, or counterforts of varying sizes, the hand-set modular system will fit better with fewer job-built fillers. This trade-off can be solved with a new integrated system where the large [8-ft x 12-ft, 8-ft x 16-ft, 8-ft x 20-ft, 10-ft x 20-ft (2.44-m x 3.66-m, 2.44-m x 4.88-m, 2.44-m x 6.1-m, 3.0-m x 6.1-m)] rentable modules directly connect and fit hand-set systems to easily form and fit pilasters, jogs, chamfers, and other irregular shapes.

Construction sequence may be another factor. Buildings typically have high floor heights in the first few floors, so that a rentable formwork system will handle the first taller walls and columns, then reduce the rental equipment for the typical floor heights.

The architect may specify cone-ties on 2-ft (60-cm) centers which fit hand-set modular forms and job-built forms equally, but not the 4-ft x 4-ft (1.22-m x 1.22-m) or 4-ft x 8-ft (1.22-m x 2.44-m) tie patterns of the large systems. Dummy cones can create the 2-ft (60-cm) cone patterns, but at extra cost. On the other hand, odd cone spacings, such as 16 in, 19.2 in, or 32 in (40.6 cm, 48.8 cm, or 81.3 cm), can be worked out with job-built forms.

Buildings and power plants usually have extensive electrical and mechanical requirements for conduit or plumbing form penetrations and/or box-outs for window and door duct openings. This requires nailing, drilling, and patching that is fast and easy with plywood form faces, but not with steel faces.

Most of the jobsite conditions have to be evaluated on their own merits, such as access, staging areas, contractor-owned forms, crew experience, weather, budget, and time schedules.

FORMWORK MATERIALS

Although lumber and plywood have been the usual materials because of their economy, availability, and jobsite workability, more durable materials are required for modern, large projects and for contractors having continuing repetitive formwork needs of a similar nature, such as large tracts of home basements. Table 14.2 is a list of form materials and their principal uses (it was reproduced in part from ACI 347–78[2]).

Lumber and plywood formwork design has been well covered by ACI SP-4[1]

TABLE 14.2 Form Materials and Uses

Material	Principal use
Lumber	Form framing, sheathing, and shoring
Plywood	Form sheathing and panels
Steel	Panel framing and bracing Heavy forms and falsework Column and joist forms Stay-in-place forms
Aluminum	Lightweight panels and framing; bracing and horizontal shoring
Hardboard, particle board	Form liner and sheathing; pan forms for joist construction
Insulating board, wood, or glass fiber	Stay-in-place liners or sheathing
Fiber or laminated paper pressed tubes or forms	Column and beam forms; void forms for slabs, beams, girders, and precast piles
Corrugated cardboard	Internal and under-slab voids; voids in beams and girders (normally used with internal "egg-crate" stiffeners)
Concrete	Footings, stay-in-place forms, molds for precast units
Fiberglass-reinforced plastic	Ready-made column and dome pan forms; custom-made forms for special architectural effects
Cellular plastics	Form lining and insulation
Other plastics: Polystyrene, Polyethylene, Polyvinyl chloride	Form liners for decorative concrete
Rubber	Form lining and void forms
Form ties, anchors, and hangers	For securing formwork against placing loads and pressures
Plaster	Waste molds for architectural concrete
Coatings	Facilitate form removal
Steel joists	Formwork support
Steel frame shoring	Formwork support
Form insulation	Cold-weather protection of concrete

and APA,[5] including the load-span characteristics and the highly durable overlaid plywoods.

Steel has high strength, stiffness, and durability (except light-gauge sheet metals), but as a form surface it loses nailability, light weight, and jobsite workability (other than by torch). However, for columns, dam faces, pier caps, and other surfaces where conduit or plumbing penetrations and box-outs are limited and where weight is manageable, all-steel formwork is practical if about 15 reuses are needed.

The most highly successful materials are steel frames with replaceable plywood faces. This combination affords the jobsite workability to minimize labor and yet gain large tie spacing as well as long-life structural frames, walers, and accessories. Overlaid plywood of high grade further extends the form-face wear and grout degradation of the plywood, yet retains nailability and workability.

The most successful of these systems embodies proprietary designs of high-carbon steels (to cut weight) that protect the edges of the plywood and absorb tie loads and stripping, wracking, and lifting stresses. Since ties fit between panel joints (instead of through the plywood), the steel frame absorbs the tie loads and the wear. Only a few bolts or rivets attach the plywood to the frame; thus there are only a few overlay penetrations for grout to degrade the plywood. By patching the gouges and nail and conduit holes, and by faithful use of release agents, these panels are so durable that they are rented by form suppliers. Some suppliers maintain large inventories of various-size panels, fillers, corners, pilaster forms, chamfers, etc.; they also offer assistance in planning and engineering layouts and provide special equipment for lifting, aligning, and bracing as well as cleaning and refacing.

Fiberglass-reinforced-plastic form materials produce such excellent as-cast concrete surfaces that they have found success as column forms, void forms for waffle slabs, and one-way concrete joists as well as for seamless gang forms, beam forms, and column capitals. Since fiberglass forms are made by a molding process, custom shapes are ideally formable with fiberglass materials.

Fiberglass underwater pier-repair sleeves offer excellent resistance to corrosion, erosion, chemical degradation, and marine borers.

Aluminum and magnesium also offer lightweight modular forms, but suffer in durability due to chemical attack of the alkaline concrete and due to large deflections and damage susceptability indigenous to these materials. Brick-textured aluminum form-face systems are successful in the basement and housing formwork applications. Aluminum extrusions enable design flexibility to provide bolt slots, nailer pockets, and other special features. Beams, channels, tees, and tubes assemble into very strong but lightweight flying trusses. These nailable beams and double-channel walers provide large gang-wall forms which are exceptionally lightweight and straight due to the extrusions.

Plastics have been used as form liners, water stops, rustication strips, and tie cones with varying success. "Structural-foam" plastic forms and domes have emerged overseas, but have not found a domestic market.

Irrespective of the form material, ties and hardware are generally steel to resist the high loads and localized bearing stresses at the anchorages. Fiberglass ties have been tried, but low strength with high cost could not offset the nonrust advantage. Sizes, designs, and special features of ties are described elsewhere in this chapter.

FIG. 14.4 Form boards of various sizes. *(Symons Corporation.)*

MODULAR FORMS

While a 4-ft x 8-ft (1.22-m x 2.44-m) sheet of plywood was the early module for forms, the term "modular form" has grown to refer to prefabricated all-metal or metal-supported-plywood systems of panels, fillers, corners, and ties whose integrated design of tie and connecting hardware is engineered to assure dimensional control, speed of erection, and ease of stripping, as well as structural integrity and reliability.

Domestic sizes commonly available are 2 ft x 8 ft (61 cm x 2.44 m) and 4 ft x 8 ft (1.22 m x 2.44 m) to fit plywood with a minimum of scrap, but other sizes also exist (Fig. 14.4). The 2-ft x 8-ft (61-cm x 2.44-m) system has been the most popular since it can be handled by one worker and because of the many sizes

FIG. 14.5 4-ft × 8-ft plywood modules with steel frames and walers. *(Symons Corporation.)*

available to fit any job. Thus the name "hand-set" system evolved. Aluminum modules of 3 ft x 7 ft 6 in (91 cm x 2.28 m) are also hand-set systems for house basement walls.

The availability of cranes and forklifts allows the use of larger modules such as the 4-ft x 8-ft (1.22-m x 2.44-m) plywood with steel frames and walers (Fig. 14.5) or larger "waler-less" panels up to 10 ft x 20 ft (3 m x 6 m). These modular systems have their own integrated design of corners, ties, chamfers, braces, and other parts.

GANG FORMS AND TIES

While this refers to large crane-handled formwork units, it also applies to hand-set forms that can be assembled into large units (Fig. 14.6) and still provide for insertion and anchorage of the ties in such a manner that they can be removed, released, and broken to allow gang stripping and reuse. Lumber and plywood gang forms are feasible with a pass-through type cone-tie system with wedge or screw connections so that the tie load can be readily released. The durability and life expectancy is a cost trade-off with first cost, tie cost, wood repair, and replacement cost.

Building shear walls and bridge-pier caps and stems are well suited for 8 ft x 12 ft (2.44 m x 3.66 m) up to 8 ft x 20 ft (2.44 m x 6 m) steel-faced "waler-less" forms that are structured so that the ties are spaced 4 ft (1.22 m) along the long edges. Thus the tie holes on shear walls are hidden behind the baseboards and

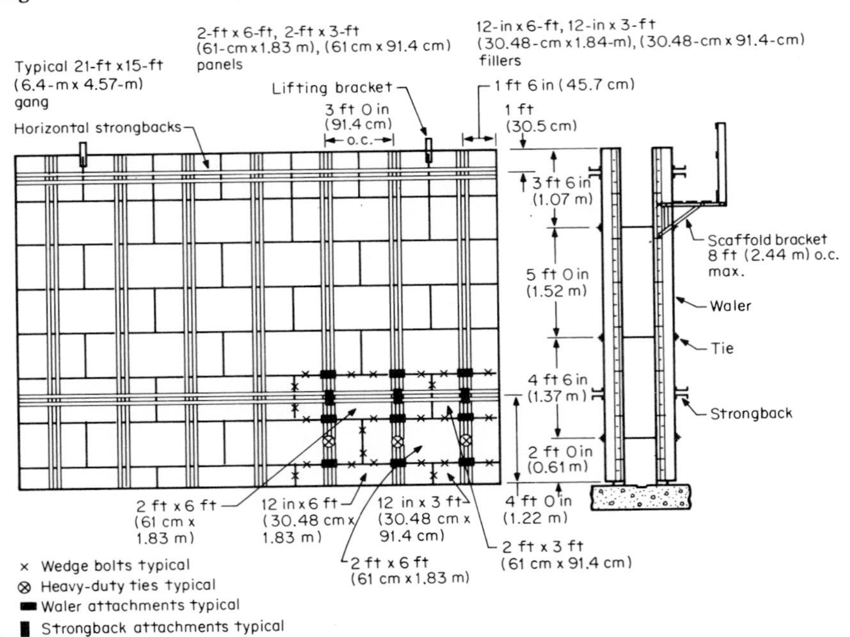

FIG. 14.6 Gang form made up of hand-set modular panels. *(Symons Corporation.)*

there are very few joints to seal, align, and finish. Just the large 4-ft x 8-ft (1.22-m x 2.44-m) tie spacing and high-pressure capacity of 1,200 to 1,500 lb/ft^2 (57,450 to 71,820 N/m^2) alone found tie-cost advantages and usefulness in spite of the high form cost and weight. One of the main features of these panels is the "girder action" which made pier-cap forming practical *without* all the high shoring and falsework. These big steel-faced forms have been called "girder forms" because they can be assembled into a boxlike structure for pier-cap construction wherein the panels function as girders to carry the dead weight to the piers without the normal shoring. To carry huge tensile forces across panel joints and to prevent buckling of the compression flanges requires an integrated system of accessories (girder bolts, braces, yokes, soffitt angles, cover plates, and support brackets).

Initially gang-wall forms were assemblies of hand-set panels with longer panel ties and special hardware that anchored the tie behind (rather than in between panels) the forms so that the tie could be removed and the entire assembly moved at once. Separate walers also were used to support the gang and employed larger ties and tie spacings. As crane and forklift capacity and availability grew, gang forming grew to offset the rising cost of labor. Steel walers and still bigger ties emerged along with faster concrete-placement capability. This trend combined with architectural concrete growth to span new 8-ft x 4-ft (2.44-m x 1.22-m) prefabricated modular systems that use separate steel walers to gain flexibility in tie locations. While the frame size matched common plywood sheets, steel faces were also used.

One of the shortcomings of all these bigger panel systems is fitting the odd dimensions—especially in between corners, pilasters, jogs, and counterforts. Therefore it is wise to integrate the designs so that the hand-set system structurally connects to the big panels, thus enabling the big panels to fit all dimensions. Also, comingling the wood-faced panels where a cluster of penetrations or box-outs occurs is a benefit.

Culvert and room forms are gang forms that either roll or hoist to the next pour. Sometimes the slab and walls are poured monolithic, which tends to lock-in the forms like elevator core forms. This requires a means to unload the forms and get clearance for movement. These applications are similar to the collapsible core forms (discussed elsewhere in this chapter) using the triple-hinge corner that eliminates slip joints, starter walls, and the cost of invert forms.

TIE SYSTEMS

Recent changes to the ACI standard 347-78[2] require a tie factor of safety of 2.0 on pours which are "over 8 ft (2.44 m) above grade or unusually hazardous." Previously this requirement was "over 16 ft (4.8 m)." This is intended to reduce the progressive failures that have occurred when only one tie is unlatched or fails.

Very high-strength steel wire and flat straps minimize the cost per tie; however, the tie anchorage and release must be quick yet reliable to minimize labor cost. Also, the tie design must provide for easy removal of protruding stubs that would rust, stain, be ugly, or would be safety hazards. Figures 14.7 through 14.10 depict some common ties. Note that many form and tie systems do not require separate load-gathering wales. The cone ties function as form spreaders and assure a neat, adequate break-back.

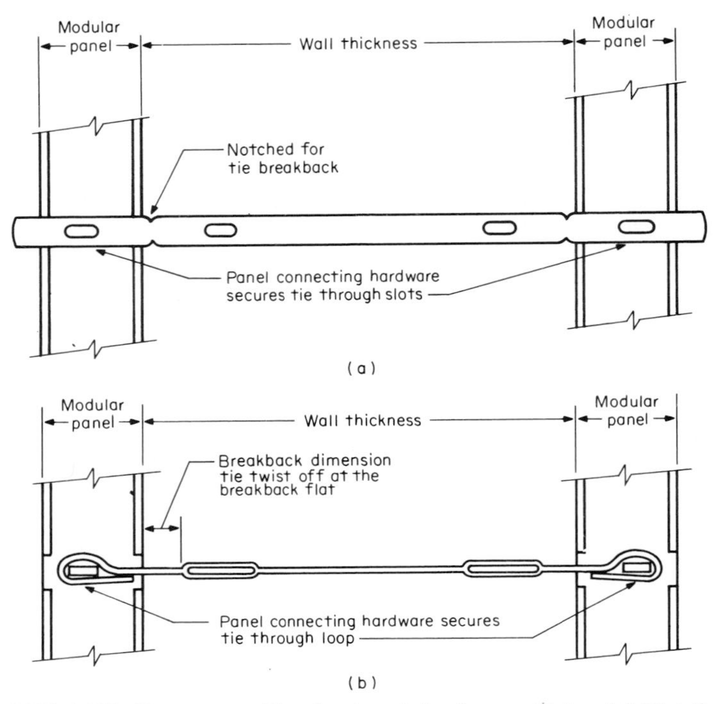

FIG. 14.7 Two types of hand-set modular form wall ties: (*a*) Flat tie and (*b*) wire panel tie. *(Symons Corporation.)*

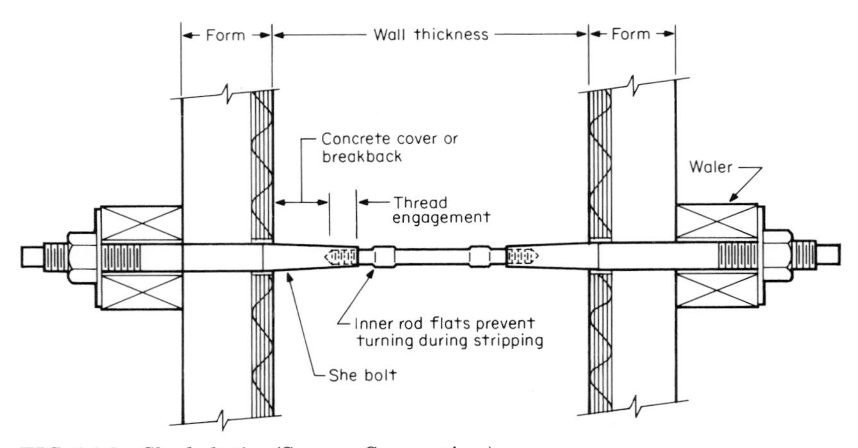

FIG. 14.8 She-bolt tie. *(Symons Corporation.)*

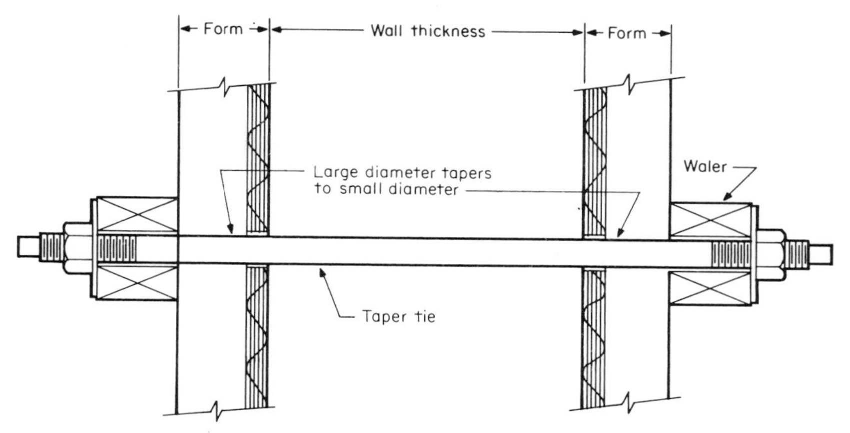

FIG. 14.9 Coil tie. *(Symons Corporation.)*

A hingeable plastic cone has evolved for field installation on wire ties; however, it does not spread the forms. Its cost is justifiable only in an emergency to avoid crew delays. While it may appear reusable, the labor economics and reuse life are yet to be demonstrated.

Hand-set and gang-form "snap ties" are readily available for up to 4,000-lb (17,800-N) tie loads (see ACI SP-4[1]), which limit tie spacings to about 2 ft (60 cm), or about 4 ft^2 (0.37 m^2) of form area on each side. Larger ties have screw-type anchorages to carry the loads yet be readily releasable to strip the forms.

Snap ties, she-bolts, and coil ties leave a steel inner piece that must be recessed within the wall to prevent rust staining of the wall. She-bolts (Fig. 14.8) and coil ties (Fig. 14.9) have threadable inner units that are left in the concrete but can be reengaged for additional usage as anchors for subsequent pours.

She-bolts and taper ties (Fig. 14.10) will pass through both gang forms after setting. Figure 14.11 shows the tapered nose and female thread to engage the recessed inner unit yet allow easy stripping. The reusable she-bolts, taper ties, bolts, and bearing plates are often rentable from suppliers who generally offer

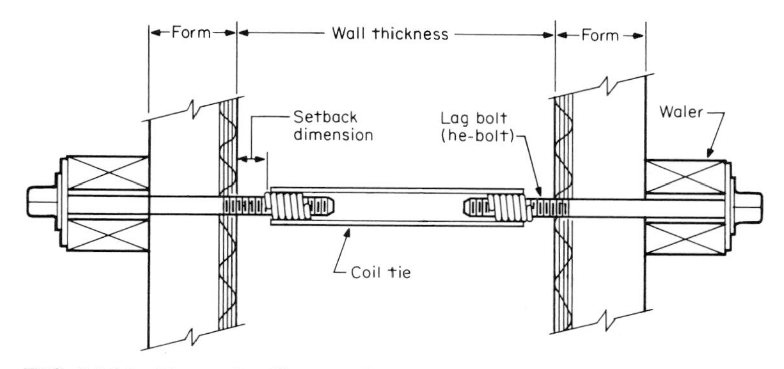

FIG. 14.10 Taper tie. *(Symons Corporation.)*

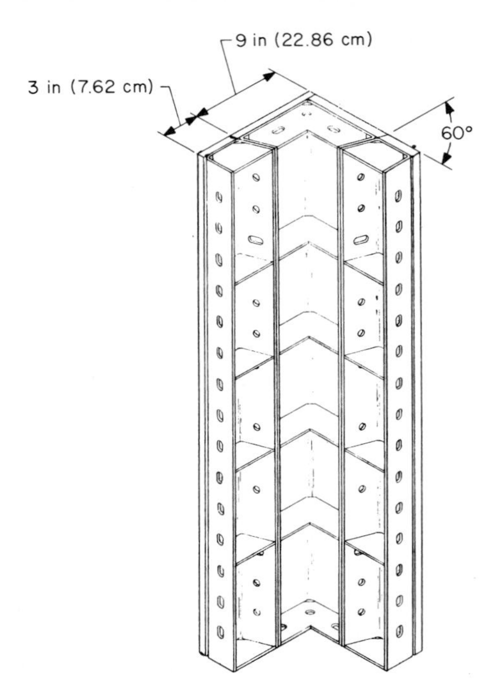

FIG. 14.11 Stripping corner. *(Symons Corporation.)*

assistance in form designs, tie selection, layouts, and logistics. Various sizes and strengths of she-bolts are available; however, the threads from various suppliers need to be checked for interchangeability.

Removable ties provide good long-term material and labor economics, but leave holes all the way through the walls or columns. While these holes can be filled, the watertightness is dependent on the filler or membrane materials.

Threaded taper ties are the most common removable type and they come in various diameters, tapers, lengths, and strengths up to 96,000 lb (427,000 N) ultimate strength. A pullable flat tie is successful on thin [8-in (20-cm)] walls if pulled soon enough (about 2 days).

"Slim Jim" is a patented rubber-encased tie-rod that removes when pulled because the outside diameter of the rubber shrinks.

Top-ties are external ties that attach to the forms or vertical walers extending above the forms.

WALL-FORM COMPONENTS

Inside corners on wood forms are usually wrecked during stripping due to the concrete pressure and shrinkage that creates residual loads on the forms even after the concrete is set. With most steel forms, these residual loads and stripping stresses are carried by the steel frames, not the plywood. There are also stripping corners that are specifically designed to relieve these residual loads (Fig. 14.11). Many walls have pilasters, but since their dimensions vary widely, the use of adjustable pilaster forms is advantageous. They require minimal building and stripping labor and are of low cost because they are reused repeatedly without damage.

Outside corners are simply steel angles with holes that fit the connection bolt pattern.

Culvert forms, inside and outside bay corners, hinged corners, filler angles, bulkhead-forms, stoop forms, brick ledges, step fillers, waler clamps, and haunch brackets are the many rental components that make a system complete enough to be practical. Versatility such as this is essential for today's complex forming to minimize high labor costs.

Waling and Aligning

Double 2-in x 4-in (5-cm x 10-cm) wales with snap ties passing between the 2-by-4s is the conventional formwork structure that is just as straight as the 2-by-4s available. However, alignment perpendicular to the waler requires additional "strong-backs," braces, and kickers to plumb and straighten the forms. Modular forms also require aligning in both directions; however, the clamping brackets for wood waler attachment to steel forms need to be positive and reliable, yet be quick to minimize labor costs. Steel alignment walers are also available in longer, straighter lengths and are much more durable than lumber.

Scaffolds

Forms over 8 ft (2.4 m) high require a work scaffold and ladder access. The scaffolds are usually attached directly to the forms; however, if walers or strong-backs are well attached to the forms, they may support the scaffolds. Modular systems have scaffold brackets and attachments engineered for quick yet safe installation.

Lifting

Lifting of gang forms requires proper lift brackets whose attachment points on the gang forms must withstand inclined loads from the slings during lift-up maneuvers. When the gangs are set down on multilift walls, a wall bracket anchored to the previous pour is required to carry the weight of the outside gang. Since the previous pour is not fully cured concrete and the loads are eccentric, the inserts, anchors, and brackets must be carefully designed and tested.

COLUMN AND PIER FORMS

Column forms are exposed to severe pressure because, while the amount of concrete required to fill them up is relatively small, the great height of fresh concrete creates large hydrostatic pressures at the base. This gave birth to adjustable steel column clamps (Fig. 14.12) for wood forms about 65 years ago. Their high strength, reusability, speed of installation and removal, and better accuracy made these clamps the primary forming method for square and rectangular columns, until wood and labor costs became so high that crane-handled steel-framed column forms became more economical. At first they were handled as two separate L-shaped assemblies, but hingeable forms with quick latches were developed subsequently which require less labor and less highly skilled manpower to maintain square corners and dimensional accuracy. Currently there is a trend toward 2000 lb/ft^2 (100 kN) or more column forms for faster pours of tall columns.

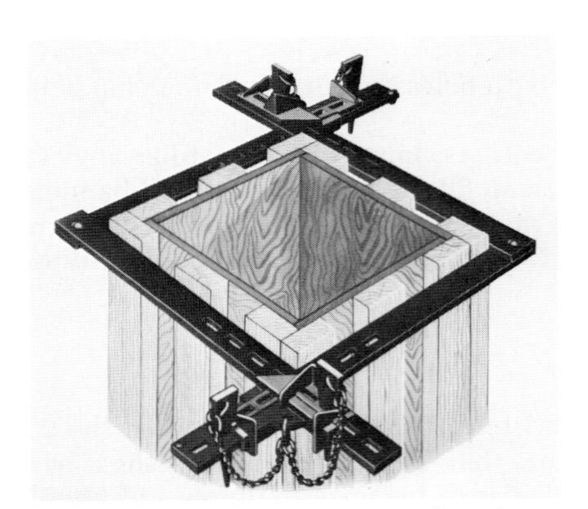

FIG. 14.12 Adjustable steel column form clamp. *(Symons Corporation.)*

Cylindrical column forms have been popular on bridge piers and parking garages where no wall framing or weather sealing was required. Waxed paper tubes spiral wound are available in many diameters and lengths; however, they have to be kept dry until used and cause a pronounced spiral-wound joint pattern and discoloration on the concrete. All-steel two-piece column forms are predominant in highway bridge-pier construction because many states have adopted standard diameters and because their long life provides good economics. They also function well with steel wall forms on bull-nose piers.

Fiberglass round forms evolved due to the costs and stripping and handling problems of heavy steel-column forms on high-rise buildings. These are flexible enough to wrap around the column reinforcement to set and spring open to strip, leaving only one vertical joint to finish. These are now rentable in many diameters up to 36 in (90 cm) and lengths from 6 ft (1.8 m) up to any length, such as 40 ft (12 m), that can be shipped and handled on the job.

Custom-shaped columns, capitals, and column-spandrel beams are often part of the architectural design of buildings. These are available in steel or fiberglass, depending on the design. Generally radius corners or other nonflat shapes are economical in fiberglass if reused more than 10 times.

Underwater pile repair sleeves of fiberglass act as the form during pouring of concrete or epoxy filler, then stay in place to provide a smooth, chemically resistant, errosion-resistant shield.

Custom-automated 22-in (56-cm)-diameter cylindrical column forms designed to be pumped from the bottom to 42 ft (12.8 m) high with heavy piano hinges, ram-actuated latching wedges, and ram opening and closing achieve very high productivity and very low labor costs.

BEAMFORMS

Beamforms with form ties utilize conventional form design with studs and wales; however, when architectural finish or other factors preclude ties, then outside means of support are required to withstand the lateral pressure.

The practice of transferring these lateral loads into the adjacent decking and/or deck joists takes special precaution in design to provide stripping relief of the decking. With flying deckforms, this practice is not recommended unless the beamform is part of the flying table. The edge spandrel beamforms will still require kicker braces unless the beamform itself has been designed to withstand the lateral pressure, which is feasible with steel but more difficult with wood and fiberglass.

The outside of a spandrel beam receives the most severe brace load because the pressure is applied to the beam side which is also the slab edge form. This lateral load will cause overturning moments on the shoring that must be withstood. Whenever the spandrel beam is supported with a tall shoring tower, analysis of tower stability is critical to provide the following:

1. Wider towers so that the dead-load "righting" moment exceeds the overturning moment.
2. If the green columns are strong enough, the lateral load can be transferred with beams, trusses, and/or guy wires to the exterior columns.
3. Diagonal guys down to the previous deck, provided the shoring tower will take the additional load from the vertical component of the guy load.
4. Horizontal guys back to the top of interior concrete columns.

On sloping-deck garages the beams tend to be deep and narrow with large column spacings. Beam side pressure and lateral pressure on the sloping deck can accumulate from bay to bay, and the forms will move "uphill" on the slope unless the lateral loads are carried by kicker braces.

CORE FORMS

Elevator shaft and stairwell inside forms became known as core forms since concrete is usually poured on all four sides, one floor height at a time. Architects cluster utilities, elevators, and stairwells into a building "core" to enhance the architecture and broaden the views.

The concrete pressure and shrinkage locks in a core form on all four sides and creates severe stripping problems even with gang ties. The box-outs for door openings, inserts for elevator hardware, and conduit penetrations all add to the stripping problem. To avoid the high cost of wrecking and rebuilding wood core forms, stripping corners and inside-hinged corners evolved to achieve reusability. Crane handling an L-shaped gang or a collapsible coreform was an improvement but still awkward.

This gave birth to the triple-hinged corner (Fig. 14.13) that provides coreform stripping relief by foreshortening within these corners so that the wall forms move straight back into a collapsed rigid box. This is easy to rig, lift, and set on top of core walls for cleaning, oiling, and for insert and box-out installation without taking crane time and storage space to lower to the slabs or in scarce staging areas. Yet the collapsed box is quickly reset and expanded back to core size for reuse.

Wall-mounted brackets below the core form support its weight and, due to the adjustable bracket design, provide the means of fine-grading the core form. Due to its rigid box configuration, this also plumbs the form-face surfaces. A safe falsework platform attached to these same wall brackets allows stripping and collapsing operations without the crane. The crane is needed only for lifting the core form and setting atop the core walls (for cleaning and oiling). Valuable crane time is minimized—during both stripping and resetting. Haunch-corner core forms are now available and function in a manner very similar to the more common square-cornered cores.

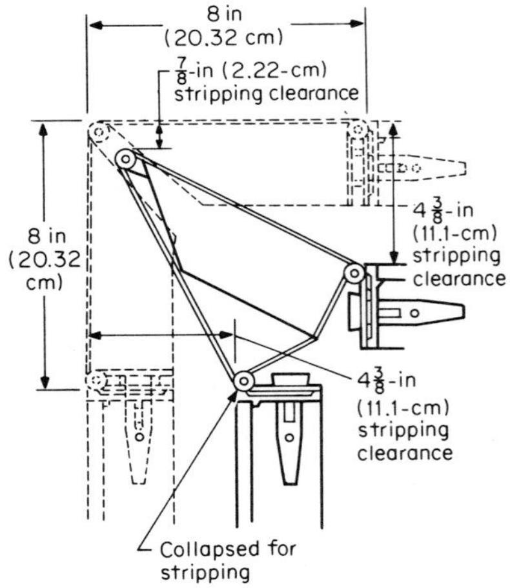

8 in (20.32 cm)
$\frac{7}{8}$-in (2.22-cm) stripping clearance
8 in (20.32 cm)
$4\frac{3}{8}$-in (11.1-cm) stripping clearance
$4\frac{3}{8}$-in (11.1-cm) stripping clearance
Collapsed for stripping

FIG. 14.13 Triple-hinged corner. *(Symons Corporation.)*

CURVED-WALL FORMS

Treatment plants include large circular tanks with high walls. To form curved walls with 2-ft (60-cm) modular forms, there is only a slight departure from a truly circular surface and this is much less expensive (Fig. 14.14) because most of the equipment can be rented and suppliers have developed refined accessories and techniques.

There are several waling methods to hold the forms in the desired curve. Using plywood templates precut to the radius is the jobsite method, but this requires a lot of labor and special layout skills. Custom rolled-steel walers are available but expensive. Straight walers can be used with adjustable stand-off clamps. Curved pipe walers are satisfactory and can be rerolled to a different radius if there is a sufficient quantity, say about 1000 ft (300 m); however, special attachment clamps are required.

FIG. 14.14 Curved wall formed with flat modular forms. *(Symons Corporation.)*

Circular tank forming with 4-ft × 8-ft (1.22-m × 2.44-m) modular flat forms and curved horizontal walers has found limited use due to the higher chord offset.

BLIND AND ONE-SIDE FORMS

In excavations when the concrete placement can be made against the sheet pile, slurry wall, or soldier-beam embankment, only the inside forms are required. This is one-sided forming whether the forms are tied with anchors in rock, stud-welded to the steel, or just braced.

Blind forming is a special situation where adjacent buildings or building-line restrictions prevent stripping access to the outside forms. The outside tie anchorage has to be removable from the inside so that the outside gangs can be stripped and lifted straight up. Threaded ties through the wall into captive nuts, such as taper ties, are required.

On one-sided forms where the ties anchor to the soldier beams, a flexible waler system is required because the soldier driving causes highly variable spacing. Another important detail is the capacity of the tie anchorage with good welds or rock anchors. If only kicker bracing is used, the angle should not be too steep because the upward component of the brace reaction may lift the form. Also, the kickers have to be adequately supported with mud sills or deadmen and anchor bolts.

SLOPING FORMS

The large 45° haunches common in power-plant turbine pedestals and the girders on stadium bents experience severe lateral forces due to high concrete pressures and the steep angles. It must be recognized that concrete pressure acts perpendicular to the form surface. (This causes the well-known uplift on battered walls that requires tie-downs). On haunches the vertical component is usually carried by the form ties on haunch forms; however, long-span plate-girder-form stadium bents may need anchorage or special support brackets for this lateral (uphill) force.

BULKHEADS WITH WATERSTOPS

While bulkheads are not considered part of the formwork, on treatment-plant long walls and large tanks that use the "skip" pouring sequence to avoid later shrinkage cracking, the labor to form the keys, place and hold water stops and cut around the rebar penetrations, then to wreck it all out, is very expensive.

New adjustable, reusable, labor-saving, steel split keyforms that clamp and hold the water stop have proven to cut labor by a factor of 4.

CANTILEVER FORMS

On mass concrete pours, such as dam faces or thick foundation mats, the forms may cantilever above the previous pour by extending the walers below the forms to preset anchors. This requires very strong walers and anchors; however, cantilever forming is amenable to automation with hydraulic and mechanical lifters and latches.

ARCHITECTURAL-CONCRETE FORMS

Exposed architectural concrete has become popular with architects because it is a more creative material and less costly than brick and marble facings; however, it requires a wide range of special formwork details. Overlay plywood, tight form joints, well-sealed ties, gasketed joints on steel forms, foam-rubber-sealed laps to previous pours, and rustication strips are some of the steps necessary to minimize or mask unsightly form joints, tie holes, and construction joints. Overly deflected forms become readily apparent on flat architectural concrete, so special attention to form pressure and stiffness is required.

Form liners of various materials impart fins, grooves, brick, rock, and other textures to the concrete: hence the indigenous flaws in concrete (bugholes, discolorations, form bulges, and joints) are not so blatantly apparent on sculptured as on flat surfaces.

Fiberglass column and beam forms give very attractive architectural concrete; however, adequate aligning and bracing is required. Fiberglass column connectors and capitals, domes, and pans are also attractive due to radius corners, seamless surfaces, and uniform concrete color.

Attractive concrete also requires carefully pretested mix design, proper selection of release agent, and thorough vibration and revibration to minimize honeycomb, bugholes, pour lines, and discolorations. A test wall and/or tests in unexposed areas (basement or garage walls) is the most reliable method of developing all the counter-opposing factors into a harmonious construction plan. The test pour is also a training ground and visual aid to the architect.

UNDERWATER FORMING

Without caisson sinking and dewatering, the repair of underwater piles has become practical with flexible fiberglass sleeves. They snap around the old pile and interlock so they can be filled with epoxy or concrete and stay in place as barriers against erosion, borers, and chemical attack.

The sleeves require circumferential bands spaced along the sleeve to withstand the concrete pressure. Steel straps, "load-binder" belts, or fiberglass bands are only temporarily required during concrete placement and curing. The extent of the banding depends on the diameter, pressure, and type of interlock. New interlocks have been developed with a hook-type engagement that will carry much of the hoop tension created by the concrete pressure. This interlock reduces the number of bands required threefold, which thus reduces the diver labor to install and remove the bands. Pumping into the bottom of any form creates pressures much higher than the liquid head, which must be considered in the strength and spacing of the bands. Furthermore, these hook interlocks prevent future joint loosening due to wave action, temperature cycles, and mechanical impact.

Reusable fiberglass heavy-duty column forms have also been used underwater to form concrete encasement for deteriorated piles. Techniques have been developed to embed the permanent sleeves in the splash zone where they are most needed and, in the same pour, to form the concrete encasement below the sleeve to the mud.

BATTERED WALL FORMS

If one or both wall forms have a slope to construct tapered walls, the inside concrete pressure has an uplift component. Tie-down anchors or dead-weight ballast are required to prevent the tendency of "floating" of the forms. Special spherical castings and nuts take tie angles, but the various tie lengths need preplanning.

CUSTOM FORMS

Custom steel tunnel forms and travelers, self-rising automated dam-face cantilever forms, integrated column-spandrel forms, hammerhead pier-cap and stem forms, and special column or wall forms, as well as custom fiberglass shapes too numerous to list are only a few customs that have been manufactured whenever the shape or the reuse potential justifies the cost. The dimensions, shapes, design pressure, stripping motions, traveling maneuvers, etc., are customized through normal proposal, drawing, and shop-drawing procedures.

FORMS FOR PRESTRESSED CONCRETE

Due to the high-strength steel and concrete employed, the degree of precision in this formwork is higher in order to achieve correct dimensions of the structural member and accurate positioning of the tendons. This is critical because unbalanced cross sections or mislocated tendons induce eccentricities which are magnified by the high tendon stresses that may result in weakened or even unsafe structures. This accuracy is also required to meet the higher fit-up tolerances during assembly, especially where precambered and prestressed members fit side by side to build a level floor or roof or a load-bearing wall.

Precasting plants normally use permanent forms or casting beds with an anchorage abutment for pretensioning type of manufacturing, which is beyond the scope of this handbook. However, there are limits to the size or weight that can be handled, hauled, or lifted into place, which has led to on-site and cast-in-place methods of posttensioning.

Circular poststressed tanks, silos, and chimneys employ very specialized patented methods, and are outside the scope of this handbook.

Cast-in-place posttensioned members require shoring during pouring and curing until the tendons are pulled; however, this avoids the severe tolerance stack-up, lifting requirements, and welding fit-up of on-site precast. On the other hand, when cast-in-place tendons are pulled, the dead load shifts away from midspan shoring to the girder shoring. If the girder tendons can be partly stressed, then additional shoring may be avoided.

The formwork considerations for posttensioned members are very similar to those for conventional concrete, except for the following factors:

1. On-site precast posttensioned members require tighter tolerances so that the parts properly fit together when assembled and without excessive accumulations of multipart tolerances.
2. Form materials must be tougher and more durable to achieve the necessary reuses without creeping tolerances that cause fit-up problems later, which are almost impossible to correct.
3. Architectural surfaces require stiffer forms, tighter joints, smoother finishes, and more care to implement chamfers, rustications, and deep textures in such a manner as to prevent spalling during stripping and handling.
4. For some members such as long tees or piling, concrete shrinkage, foreshortening, or deflections during tendon stressing may induce additional loads on the formwork to be considered during design. This is in addition to the camber requirements.
5. Steam or other elevated-temperature cure methods may deteriorate some materials or induce differential expansion distortions to be considered. For example, unsealed wood rustication strips may swell and spall the edges unless kerfed snd sealed. Void forms may swell or undergo thermal expansion to cause difficult stripping or damage to the concrete finish.
6. Interaction between chemicals, such as retarders, with plastic chamfers, formliners, spacers, and styrofoam void forms may be aggrevated by higher-temperature cure methods.
7. Tendon drape tie-downs and/or flotation forces from embedded void forms may need to be considered in the form design.
8. Rebar or tendon congestion may prevent internal vibration, plus external vibration may be more productive, which induce severe localized forces and form-work resonances that require tough rigid-form materials and designs.

FORM-RELEASE AGENTS

A proper release agent not only makes stripping easier, but it prolongs the life of the plywood and reduces cleanup labor by 50%. Some agents chemically react with the concrete, leaving a soapy residue that may be objectionable due to discoloration, dusting, or preventing the bonding of curing compounds, paint, or paneling. All release agents are *not* alike; pH balance, uniformity, and active ingredients should be checked before use.

FORM REMOVAL

Formwork for columns, walls, and piers is generally removed in 12 to 48 h after placing of the concrete, depending on construction and wind loads, maturity for anchor bolts, concrete mix design, etc. Where textured architectural concrete is involved, and in elevated slabs and pier caps, where gravity loads are significant, a longer (3 to 7 days) cure of the concrete is required before form removal.

The following recommendations for form-stripping time are reproduced from ACI 347-78.[2]

When field operations are not controlled by the specifications, under ordinary conditions formwork and supports should remain in place for not less than the following periods of time. . . . If high-early-strength concrete is used, these periods may be reduced as approved by the engineer/architect. Conversely, if ambient temperatures remain below 50°F (10°C), or if retarding agents are used, then these periods should be increased at the discretion of the engineer/architect.

Walls	12 hours
Columns	12 hours
Sides of beams and girders	12 hours
Pan joist forms	
30 in (75 cm) wide or less	3 days
Over 30 in (75 cm) wide	4 days

	Where design live load is: less than dead load	greater than dead load
Arch centers	14 days	7 days
Joist, beam, or girder soffits		
Under 10 ft (3 m) clear span between structural supports	7 days	4 days
10 to 20 ft (3 to 6 m) clear span between structural supports	14 days	7 days
Over 20 ft (6 m) clear span between structural supports	21 days	14 days
One-way floor slabs		
Under 10 ft (3 m) clear span between structural supports	4 days	3 days
10 to 20 ft (3 to 6 m) between structural supports	7 days	4 days
Over 20 ft (6 m) clear span between structural supports	10 days	7 days.

TOLERANCES

ACI 117-81[6] covers concrete tolerances of all kinds and ACI 347-78[2] covers formwork aspects that affect deviations from print dimensions, cross sections, and grades, such as column alignment, plumbness or slab flatness, or wall straightness. Furthermore, ACI 347-78 establishes four classes of permitted irregularities in formed surfaces as checked with a 5-ft (1.5-m) template which is a straight edge, radius edge, radius arc, or other shape as specified:

Irregularity	Class A	Class B	Class C	Class D
Gradual	⅛ in (3 mm)	¼ in (6 mm)	½ in (12 mm)	1 in (25 mm)
Abrupt	⅛ in (3 mm)	¼ in (6 mm)	¼ in (6 mm)	1 in (25 mm)

Class A is a surface exposed to public view where appearance is of special importance.

Class B is intended for coarse textured concrete intended to receive plaster.

Class C is a general standard for permanently exposed surface where other finishes are *not* specified.

Class D is a minimum quality for a surface where roughness is not objectionable.

The ACI classes provide a basis of quantifying tolerances for surface variations due to forming quality. Form offsets and fins are "abrupt," whereas warping, unplaneness, and deflection are "gradual" flaws.

DRAWINGS AND SPECIFICATIONS

Architect-engineer (A-E) structural drawings normally do not cover formwork or accessories, but they detail the end result required, such as the shapes and dimensions plus tie-cone size and spacing, rustication strips at construction joints, and the other specifics of the project. Their specifications also cover the concrete strength needed prior to stripping forms and decentering shores, the special tolerances and finishes.

Where architectural textures, treatments, or finishes are specified, a jobsite preconstruction mock-up should be constructed large enough to illustrate procedures and materials such as sealing form joints and ties, inside corners and core-form-stripping clearances, vibration technique, allowed tie-hole and honeycomb patching, sand-blast or bush-hammer treatments, and release agents and curing compounds. This mock-up must be specified and detailed by the A-E so that it is included in all the bids and preconstruction preparations so that a full-scale acceptance standard is established.

Formwork drawings are prepared by contractors along with their suppliers to instruct builders and erectors so as to optimize equipment, reuse, manpower, materials, pour sequence, and other jobsite conditions. Such drawings show formwork elevation, plan, and section views, tie size and location, formwork pressure limits, tie loads, bracing, work platforms, falsework, supports, pour sequence if required, as well as materials, dimensions, connections, and standard details. These are generally submitted to the A-E for review and approval prior to construction.

Formwork-drawing details for fabrication include member sizes and spacings as well as overall form dimensions, tie type and locations, wale size and location; on elevated forms, the means of formwork support, work platforms, and wind load bracing or guying; and other information to instruct field personnel how to erect, pour, strip, and cycle forms safely and efficiently.

For some projects, other information that may be valuable is:

Concrete pour sequence and timing of subsequent lifts, if required by the formwork design

Preattached inserts, anchors, block-outs, and rustications

Vibrator access and inspection ports

Bulkhead details for keyways and waterstops

Pour pockets and clean-out access

Formwork camber, if required

Special bracing for one-sided (pit) forms, including anchors and mudsills

Stripping relief between pilasters and inside coreforms, such as crush plates or wrecking bars

Form coatings and when applied

Sequence of tie or form removal for worker safety as well as concrete strength

Standard panels or details are shown schematically or included by reference. Where field job building, such as around pipe penetrations and where large door openings are necessitated by nonrepetitive or field fit conditions, these areas should be so indicated.

Formwork drawings should be rechecked against structural drawings to verify:

Critical dimensions

All construction joint locations

Special tolerances on camber and deflection

Architectural concrete details and textures

Inserts for other trades

Precast and posttension coordination

SAFETY

The Scaffolding, Shoring, and Forming Institute publishes both field erection safety brochures and an 80-slide program of safety "do's and don'ts" that should be used in the planning stages and the crew training. These publications embody the latest ANSI and OSHA code provisions plus practical safety pointers.

Design safety factors on form ties, anchors, hangers, and inserts are recommended by ACI 347-78;[2] however, they depend on the type of construction (see Table 14.1). The responsibility for safety is the contractors', because they control the crews, the material quality, the schedule, and all of the factors that determine tie and anchorage safety.

Design safety on formwork panels and walers is not specifically covered by OSHA other than by reference to ANSI.

Allowable stresses of the materials are those used in standard structural design unless actual test data support higher values for a product design. Lumber and plywood stresses for short-term loading are permitted to be introduced due to the brief duration of concrete pressure loading.

INSPECTION

It is essential that formwork be erected as designed so that formwork members are not overloaded inadvertently. If qualified supervisors are not available, the form designer should inspect the erection.

No field changes in the design should be permitted without approval of the form designer.

Inspection must assure that all the form ties have been properly installed and latched, that threads are fully engaged and uniformly tightened, uplift anchors are provided, bracing is sound, connections are tight, and that work platforms, guard rails, access ladders, and safety belts are provided. ACI suggests a three-stage inspection:

1. Preliminary—After forms are built, but prior to oiling or rebar placement.
2. Semifinal—Just prior to final cleanup.
3. Final—Immediately before concreting; check forms, spreaders, inserts, and fixtures for dislocation, and whether surfaces are clean, oiled, and, if specified, wetted.

FORMING ECONOMICS

Since formwork costs are the major portion of concrete costs that are controllable by the contractor (Fig. 14.15), success depends on thorough, objective analysis of the prevailing project and jobsite conditions and the resulting form design and selection.

The Key to Material Cost: Reuse Life

Job-built wood forms have a low first cost, but they can be reused only 10 to 15 times. The labor cost to repair and reface job-built wood forms is so high that prefabricated steel modular forms have proven to be more economical because they last 10 to 20 times longer, repaying the higher initial cost many times.

Steel-faced forms have a proven high-reuse life, if the face thickness is at least 10 gauge [⅛ in (3 mm)], preferably $\frac{3}{16}$ in (5 mm), except for small items such as inside corners. However, nailable form faces save labor required for attaching

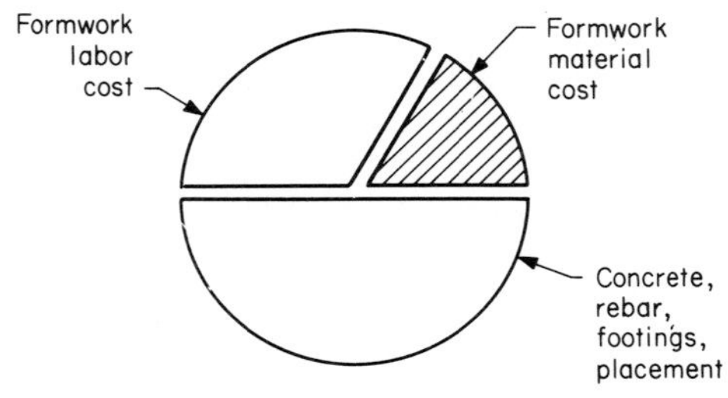

FIG. 14.15 Concrete formwork labor and material costs. *(Symons Corporation.)*

inserts, for box-outs for doors, windows, and ducts, and for penetrating the face with conduits, pipes, and other items. Therefore, specially developed coated plywood form faces, positively attached to steel frames, yet replaceable when needed, have proven to be the best compromise between the durability of steel and the practicality of wood.

Modular formwork systems, where the steel frame protects the edge of plywood from the inevitable wear and tear during stripping, jobsite handling, and shipping, have proven records of 100 to 200 reuses before plywood replacement. Since these frames last over 10 years, they have investment tax credit and depreciation tax incentives. These systems are so durable that even the plywood and all the connecting hardware (except snap ties) are rentable.

Since these systems can be rented, they may be charged to the job under way; and after follow-up jobs are secured, the rental-purchase option can be exercised to convert the rental agreement to purchases. This reduces the up-front capital requirements.

Productivity: The Key to Labor Costs

The rising cost of labor necessitates formwork designed to minimize crew size, delays, and wasted motion. The labor cost comes from actual experience with particular form systems and will vary by job type.

To determine true labor costs, time-and-motion studies have been conducted of actual jobs with time-lapse cameras which recorded every event on the job. From this film the productive manpower to accomplish each phase of the full formwork cycle was measured along with the square footage of formwork.

Formwork operations that were measured included five phases (including penetrations, insert or box-out attachment, bulkheads, and pilasters): (1) setting first side of formwork; (2) placing ties; (3) setting second side of formwork; (4) wailing, aligning, and bracing; and (5) stripping, cleaning, and moving.

The cameras also identified nonproductive delays caused by material shortage, crane scheduling, improper sequence, and other inefficiencies which would have otherwise been *mis*interpreted as "productive effort." Some of the delays were unavoidable results of weather or of equipment breakdown. But 60% of the delays were avoidable with better planning, training, standardization, and supervision. For example, it was found that 30% of the total effort was in plumbing, aligning, waling, and bracing, because scrap-lumber braces had been planned. Looking for such scrap and cutting and trying to make it work was twice as slow as using steel adjustable turnbuckles anchored to steel stakes.

Two crews with the same contractor had 100% different productivity because one foreman planned ahead better and had developed more effective techniques in eliminating delays.

Studying different types of jobs established practical (optimum could be 35% higher) productivity standards in square feet of contact area per man-hour as follows:

On multilift industrial jobs, 37 ft^2/man-hour (3.5 m^2/man-hour)

On residential basement jobs, 72 ft^2/man-hour (6.7 m^2/man-hour)

On hand-set columns, 53 ft^2/man-hour (4.9 m^2/man-hour)

On gang-forming with steel walers, 37 ft^2/man-hour (3.5 m^2/man-hour)

On gang-forming with wood walers, 55 ft^2/man-hour (5.1 m^2/man-hour)

On hand-set curved walls, 34 ft^2/man-hour (3.2 m^2/man-hour)

Comparing this productivity with three well-known construction-estimating guides on job-built forming showed that modular-forms productivity was 67 to 190% higher than job-built. This translates into 30 to 50 cents/ft^2 ($3 to $5.50/$m^2$) savings on each pour depending on carpenter rates and the type of forming.

Rental + Productivity = Small Job Success

Combining formwork material and labor costs indicates that above 15 reuses, it is more economical to purchase modular forms than to job-build forms. For small jobs, rental of modular forms at about 50 cents/ft^2 ($5.50/$m^2$) per month is more economical than $1.10/$ft^2$ ($11.80/$m^2$) to job-build, especially if there are four reuses in that month [12½ cents/ft^2 ($1.35/$m^2$) per use]. Rental payments on modular forms can be applied to the purchase price as follows: After the first month's rent, apply 100% of rental to purchase; after the second month's rent, apply 90%; after the third, 80%; after the fourth, 70%; after the fifth, 60%.

Due to re-plying, cleaning, and refurbishing costs, there are advantages to rental other than capital conservation; however, in general, for a job with 40 or more reuses, purchase of modular forms is more economical than rental (neglecting resale value).

Plywood and lumber prices have historically been more volatile than modular forms. The assurance of ready availability of modular forms combines with such price stability and with the above productivity standards to minimize the risk in bidding and schedule commitments.

Reuse the Ties to Control Tie Costs

Bigger and stronger ties cost more per unit, but are more reusable (thus rentable) and carry proportionately more area of forms, and they allow faster pours, which saves placement labor, crane time, and often avoids overtime finish labor. The use of big ties is particularly advantageous on large gang forms because cranes are plentiful and are needed anyway to handle rebar bundles, concrete buckets, and other heavy items.

Gang-form snap ties spaced at 2 ft (60 cm) on centers support 4 ft^2 (0.37 m^2) on each side; thus they cost 4 cents/ft^2 (43 cents/m^2) per use because the tie is not recovered; steel walers at 4 ft (1.22 m) on centers with taper ties at 4 ft (1.22 m) on centers cost more initially, but they last indefinitely (zero cost per use) and are rentable. A typical rental rate for such a waler and taper tie is about 30 cents/ft^2 per month. If the job allows the cycling of the forms four times per month, the cost *per use* is 7½ cents/ft^2 (81 cents/m^2). However, this taper tie leaves a 1¼-in (32-mm)-diameter hole all the way through the wall, which may be objectionable on some jobs such as treatment plants. In this case, she-bolts can be used that leave a ¾-in (190-mm)-diameter inside rod but no through-hole. Their rental cost is 6½ cents/ft^2 (70 cents/m^2) per use for the recoverable parts, plus 2 cents/ft^2 (22 cents/m^2) per use for the inside rod for an 8-in (20-cm) wall.

Furthermore, fewer ties to place, attach, and strip, and fewer tie holes to patch save labor cost. Assuming tie labor cost at $10 per tie, this amounts to $1.25/ft^2 ($13.50/m^2) for snap ties, only 32 cents/ft^2 ($3.50/m^2) for taper ties, and 50 cents/ft^2 ($5.50/m^2) for she-bolts. This comparison is oversimplified, since faster pour rates and gang-form movement offer additional labor savings with the big pass-through type of ties. The ultimate goal is faster cycle time (and maximum reuses), which cuts cost of rental equipment, overhead, and interim financing.

FUTURE FORMWORK DEVELOPMENT

In the foreseeable future there will be more durable coatings for plywood to provide longer reuse life and perhaps even nonwood form faces that are nailable, patchable, and replaceable. New high-strength steels for higher-performance forms will meet the demand for faster pour rates and the resulting higher concrete pressures. Bigger and stronger ties will support higher-pressure forms and wider tie spacings. Self-rising shear-wall and core forms will evolve to reduce labor and crane time. Underwater seawall and dam-face repair techniques will become more feasible.

REFERENCES

1. M. K. Hurd, *Formwork for Concrete,* 3d ed., ACI Special Publication no. 4, 1977.
2. *Recommended Practice for Concrete Formwork,* ACI Standard 347-78, American Concrete Institute, Detroit, Mich., 1978.
3. *Concrete Construction and Masonry Work,* ANSI A 10.9, American National Standards Institute, New York, 1970.
4. *Forms and Shoring,* OSHA 1926, 701, U.S. Dept. of Labor, Occupational Safety and Health Administration, Washington, D.C., 1979.
5. *Plywood for Concrete Forming,* Brochure V345, American Plywood Association, Tacoma, Wash., 1971.
6. *Standard Tolerances for Concrete Construction and Materials,* ACI 117-81, American Concrete Institute, Detroit, Mich., 1981.
7. *Steel-Ply System vs. Job-Built Forming Productivity and Cost,* Publication 77-17, Symons Corporation, Des Plaines, Ill., 1977.
8. R. L. Peurifoy, *Formwork for Concrete Structures,* McGraw-Hill, New York, 1976,
9. *Manual of Concrete Inspection* ACI Special Publication no. 2, American Concrete Institute, Detroit, Mich., 1981.

15

Slipforms

by J. F. Camellerie

Slipform technology has gone a long way since its emergence at the beginning of the twentieth century. As late as 1950 forms were being raised using hand screw jacks and job-built wooden yokes. The deck of a slipform operation reminded one of an ancient slave galley. Every time the deck superintendent blew the whistle, the men pulled heavily on their bars in some sort of unison until they had turned each of five jacks to raise the form 1 in (2.5 cm). The whistle blew 6 to 12 times an hour and more, and when workers finished their shift they were pretty well beat. In those days this technology was seldom used for any type of structure other than bins and silos, although the first apartment house was slipformed in Tennessee as early as 1956. The silo work was relatively simple, taller buildings were built of steel, labor was eager and not too demanding, and generally speaking, there was not too much pressure for innovation in the United States during the first half of the century.

In Europe, however, where the taller buildings were being built of concrete, there was an incentive to develop better jacking equipment and more sophisticated forming systems. Slipforming, which first came into being in the United States, reached maturity in Europe with the development of hydraulic jacks and a broad application to commercial building, housing, bridge piers, etc. In the second half of this century the United States turned seriously to high-rise concrete construction and the Swedes and other Europeans who had developed the hydraulic jacking systems found a good market in this country. By 1960 slipforming as we now know it was well established in the United States.

During the first half of the century and into the 1950s, contracting firms had developed that specialized in slipform construction. During this period almost all slipform construction was bulk materials storage for grain and heavy industry and the work was performed exclusively by these specialized firms. They designed, built, and owned their own slipforming equipment.

During the 1950s, however, the situation changed. The European firms, with Heede International at the forefront, made sophisticated hydraulic jacking systems as well as steel yokes and other specialized accessories available on a competitive rental basis. They backed up the rented equipment with highly skilled technicians who brought with them the advanced technology which had been developed in Europe. This made it possible for contracting firms that had not specialized in slipforming and had not accumulated equipment or trained crews to undertake difficult slipform projects with considerable success. Although most of the slipform construction is done by contractors who are specialists in slipforming, a good part of the construction is performed by firms on a one-job basis with the help of technicians provided by equipment-rental companies and consultants.

CONSTRUCTION PRACTICES IN THE UNITED STATES

Commercial and institutional work is often performed by firms that do not specialize in slipforming, but the industrial work—comprised mainly of silos and to some extent chimneys—is always contracted to specialists. Not only the construction is let to the specialists, but the design as well, so that most silos and chimneys are let on a design-and-construct basis. It is interesting to note that at the present time only 20% of U.S. firms doing slipforming own their own slipforming equipment. Seventy-five percent of slipforming equipment presently in use is owned by less than half a dozen rental firms, one of which is predominant.

A general contractor or owner-builder who undertakes a slipform project has several options:

1. To design, build, and operate the slipform complex, and in doing this, to enlist the assistance of expert consultants who will help at every step along the way.
2. To subcontract the work to a contractor who specializes in this field. (There are approximately 50 in the United States.)
3. To buy the forming system completely designed and fabricated ready to be erected at the site and perform the slipforming with the contractor's own forces.

The rental cost of slipform equipment in the United States is usually based on the number of lineal feet of slide, on the capacity and quality of the equipment and jack rods involved and whether the jack rods will be left in place or recovered. To obtain the lineal feet of slide for pricing, multiply the number of jacks required by the total height of the slipform operation. This rental fee usually includes the jacks (with spares), the pump or pumps, hydraulic lines, leveling devices, and yokes. Erection is by the contractor. At an additional fee, form design and field technicians may also be obtained. Forms are not normally available on a rental basis except in the case of slipforms for tall chimneys. Special hoists and lifting equipment are often available for rent from some of the rental firms.

It was at one time axiomatic that slipform operations were executed in a continuous round-the-clock manner, resulting in a monolithic structure without horizontal joints. In silo work, around-the-clock operations are still the rule; in fact most industries specify that silos are to be slipformed in this way, but they almost always allow a stop in the operation over weekends. In commercial building work, daily stops are the rule and continuous slipforming the exception. The amount and complexity of reinforcing steel, embedments, openings, conduits, windows, doors, beam pockets, slab keys, dowels, etc. militate strongly against slipforming most commercial structures around the clock. It is often desirable, if not absolutely necessary, to allow time between slides to accumulate on deck the necessary reinforcing bars and embedments for the next "jump" or "slide" as well as to lay out and set in place a good portion of the embedments and vertical steel. Daily stops, usually at one floor height, often work out well with the floor cycles.

Obviously, labor practices have a tremendous impact on construction practices in general. This influence is even more evident in slipform operations which are so labor-intensive and so subject to premium pay. In any particular locality the union attitude toward shifting, overtime pay, lunch breaks, etc. influences the contractors' decision to slide around the clock or in daily jumps. The availability of labor and the ability of the union hall to supply shifts is also critical; it may be necessary to arrange the schedule in such a way as to encourage craftsmen and laborers to participate and stay with the job. An unfriendly union attitude tends to discourage continuous slip operations because contractors will be less vulnerable to walkouts if they choose a daily slide.

FIG. 15.1 USF&G headquarters building, Baltimore, Md. (© Tadder/ Baltimore, 1981.)

There are, of course, other considerations involved in deciding whether to go around the clock or in daily jumps: aesthetics could be one; time requirements could be another; mechanical or elevator installation could militate for a continuous slide; horizontal joints may be undesirable, as in the case of nuclear containment shells.

Slipforming, like all construction work and concreting operations in particular, is very sensitive to the weather; in fact, more so than other techniques for placing concrete. The adverse consequences of problems can be more serious in hot weather when the concrete has a tendency to set too rapidly. Cold weather reduces concreting problems and could result in a very good quality of product, but there is always the danger of frozen concrete, the inefficiency of working under harsh conditions, and almost certainly a reduction in slide rate and a corresponding increase in the cost of placing the concrete. Fall or spring are the ideal times and slipform operations are often planned for these seasons of the year. Since round-the-clock concreting greatly compresses the construction time required, such an operation offers better possibility of planning the slip operation for an ideal time of the year without affecting the overall completion schedule.

The commercial structures which are most likely to be slipformed under current construction practices in the United States fall into three categories: the concrete-core and structural-steel-frame hybrid structure; the bearing-wall type of structure in which all walls are slipformed; and the combination concrete-core and precast-concrete structure. Shear walls may be slipformed in combination with the cores or possibly in lieu of the cores.

There are some excellent examples of slipformed concrete cores to be found around the country. The USF&G Building in Baltimore is the most outstanding example with a 535-ft (160-m)-tall core. The Manufacturers and Traders Trust Building and One Main Place are two excellent examples in Buffalo, and the IBM Building and Penn Center are good examples in Philadelphia. The Knights of Columbus building in New Haven is an interesting variation. Strangely enough, Chicago, the city of tall buildings, has no outstanding commercial slipforming.

The slipforming of apartment houses and hotels has been most prevalent on the west coast, in Hawaii, in Florida, and in Puerto Rico. The Torre del Mar Apartments in San Juan and the Sheraton Waikiki Hotel in Honolulu are outstanding examples of these bearing-wall type of structures. Another outstanding example is Bay View Terrace in Milwaukee, a 25-story frame completed in 35 days.

FIG. 15.2 Palacio del Rio, San Antonio, Tex. *(Linden-Alimak, Inc.)*

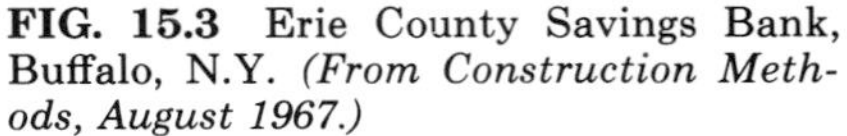
FIG. 15.3 Erie County Savings Bank, Buffalo, N.Y. *(From Construction Methods, August 1967.)*

FIG. 15.4 Cleveland Trust Company, Cleveland, Ohio. *(Howard Photographers.)*

The American Bible Headquarters Building in New York City and the Palacio Del Rio Hotel in San Antonio are excellent examples of combining slipformed cores and shear walls with precast units.

Figures 15.1 through 15.6 show examples of slipforming tall structures.

DESIGN CONSIDERATIONS

A sliding form differs radically from a "fixed" form in that a fixed form is basically a mold into which the plastic concrete is cast and allowed to harden before the form is removed or "stripped," whereas the slipform is in reality a die through which the plastic concrete is extruded. This means that the concrete emerging from the die must be at just the right state of the setting process as it passes through the form. Since the "die" is often fairly extensive in area, it is obvious that a reasonably predictable uniform rate of concrete set is a necessity to the operation. Herein lies the challenge. Fortunately, this requirement has sufficient tolerance to allow slipforming to be executed under conditions which are normal in the construction industry.

FIG. 15.5 TVA nuclear containment shell, Sequoyah Nuclear Plant. *(Linden-Alimak, Inc.)*

Another basic difference between fixed and sliding forms lies in the integrity of the sliding form. Fixed forms are made up of coordinated elements or groupings of elements which must interface exactly but which are separate components singly stripped. The slipform, on the other hand, is one single "structure." Whereas the fixed form will remain stationary during the entire period that it is forming the concrete, the sliding form must be constantly moving during this period. This means that the slipform must be designed to travel and act as one single component.

Another very important function (and necessity) of the slipform is that it provides support for the necessary working platforms, storage facilities, and vertical transport equipment. Actually, when one designs a slipform, the design procedure has little in common with conventional form design. A sophisticated structure has to be designed which is capable of mobility and flexibility and yet must be able to carry without undue distortion substantial vertical, lateral, and overturning loadings (see Figs. 15.7, 15.8, and 15.9).

Rigidity

There are two basic philosophies in the design of slipform structures. Some designers emphasize a flexible structure that will take considerable racking and other strains by deflecting to accommodate such movements. This type of structure can more easily get out of alignment, level, or plumb, but it will also react more readily to corrective measures. In the days of the hand jack, this flexibility

FIG. 15.6 IBM building, Philadelphia, Pa. *(Lawrence S. Williams, Inc.)*

was a practical necessity to allow the jackers to raise the first jacks against heavy pressure. The alternative is to design a very rigid forming structure which resists distortion and tends to "bridge" problem areas. Once this type of form has developed problems in alignment or plumbness, it is more difficult to bring it back "on track" than it is for the more flexible design. It also has a somewhat greater tendency for the forms to pinch or bind.

Dimensional Stability

In almost all cases, the deck area of a slipforming structure is composed of a group of "islands" separated by the empty spaces that will receive the concrete to form the walls. These individual "islands" have a tendency to shift relative to each other and to rotate about their own axes, resulting in variations of wall thickness and possibly of the overall dimensions of the building. The forming structure must be designed to keep the movements of these "islands" within tolerance and to provide adequate means of adjustment. The design must also assure the dimensional stability of the "island" shapes themselves as well as the dimensional stability of the perimeter forms, and at the same time provide means for making adjustments if necessary. The design should also address the tendency of the structure as a whole to rotate about its vertical axis and to drift laterally in any direction. Prevailing winds could be a factor in this regard.

Flexibility

In designing the forms, careful attention must be given to deflections. The most common slipform system used is based on two horizontal wales spanning between jacking yokes and placed vertically 2 ft (60 cm) apart or less with the form sheathing and/or battens spanning vertically between the wales and overhanging below the bottom wale. The forms usually project above the top wale to bring them to the level of the working deck. The forms will deflect outward between the wales under the hydrostatic pressure of the plastic concrete, and since it is in this area that the concrete is losing plasticity, excessive deflection will in effect result in a horizontal mechanical key which will prevent the form work at the level of the lower wale from sliding past this "key." The allowable deflection between wales will be a function of the flexibility of the wale system. The deflection of the portion of the form that overhangs the lower wale is not as critical. Since the forms are battered and the concrete wall thickness is set a foot or two above the bottom of the form, the lower edge of the form will normally not be in contact with the concrete. Nevertheless, there will be occasions when the form will rub against the concrete at the bottom and when this happens, flexibility is desirable to reduce excessive drag forces against the hard concrete.

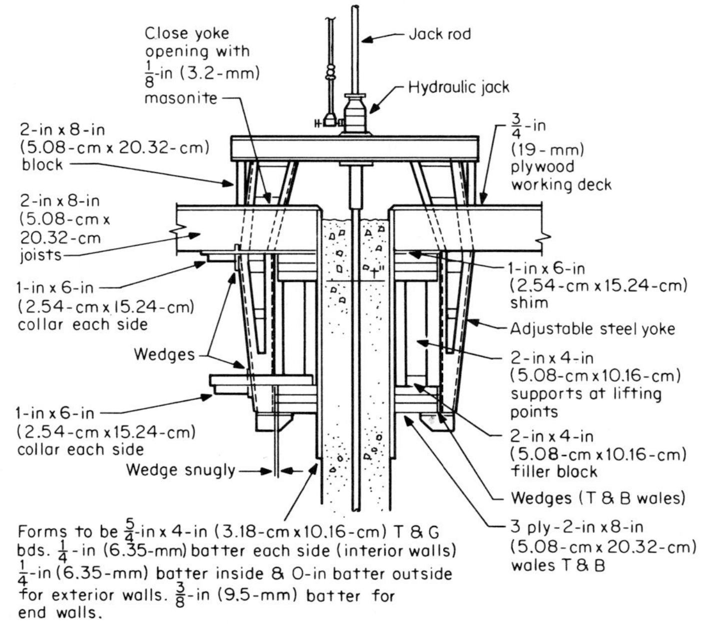

FIG. 15.7 Section through slipform. *(From J. Havers and F. Stubbs, Handbook of Heavy Construction, 2d ed., McGraw-Hill, 1982.)*

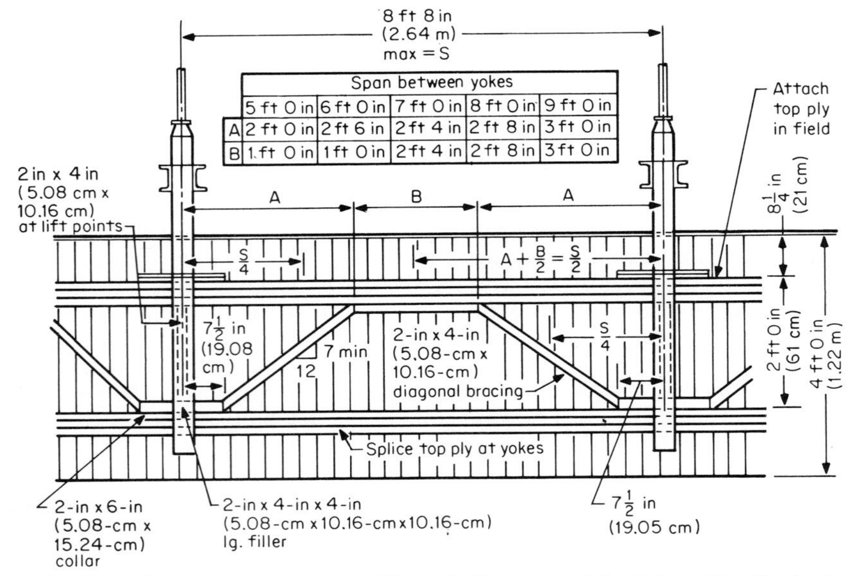

	Span between yokes				
	5 ft 0 in	6 ft 0 in	7 ft 0 in	8 ft 0 in	9 ft 0 in
A	2 ft 0 in	2 ft 6 in	2 ft 4 in	2 ft 8 in	3 ft 0 in
B	1 ft 0 in	1 ft 0 in	2 ft 4 in	2 ft 8 in	3 ft 0 in

FIG. 15.8 Side view of slipform. *(From J. Havers and F. Stubbs, Handbook of Heavy Construction, 2d ed., McGraw-Hill, 1982.)*

On the other hand, excessive deflection will increase the danger of concrete "fall-outs" if the plastic concrete is allowed too low in the forms. In addition, excessive deflection of the lower section of the form will increase the effects of overpours and shingles and can have a bad effect on the appearance of the concrete, especially in the case of architectural work.

Leveling

Since the slipform structure travels along an axis perpendicular to the plane of the deck, it is critical that this deck be maintained level at all times so that the concrete structure being built will rise plumb. The slipform design must provide the proper number, location, distribution, and capacity of jacks to achieve uniform raising of the forms as well as provide devices to monitor simultaneously the level of each and every portion of the deck. Control systems for making correction to keep the deck level must also be developed. Conversely, the deck can be purposely tilted out of level in such a direction as to correct tendencies of the construction to go out of plumb.

Speed of Operation

The average speed at which the slipform will be operated must be decided and minimum and maximum speed limits set as a basis for designing the concrete mix, the forms themselves, the storage capacities, and the equipment for raising workers and materials. This consideration will often include decisions as to the use of concrete additives, heating, use of ice, form insulation, enclosures, etc.

Labor practices in the geographical area will also have significant impact on these decisions.

Concrete Placing

The method of raising and distributing concrete into the forms must receive early attention as this will have significant impact on the concrete mix as well as on the form design. For tall slides, concrete is commonly raised by means of deck-mounted hoists, deck-supported cranes, or pumping. Cranes at grade are often used when the height allows. In all cases deck-receiving hoppers are required on the deck. Obviously, the slipform structure must be designed to carry the hoists, cranes, hoppers, and other required equipment. Distribution from the receiving hopper to the forms is most often affected by means of hand buggies, often "georgia" buggies, 30-in (75-cm)-wide buggies which are more easily maneuvered through the vertical steel and other obstructions. Traffic patterns must be worked out for concrete distribution, including their effect on placing of rebar, embedments, openings, etc. Other means of distribution that are used when the specific situation permits are pumping, conveyors, crane and bucket, elephant trunks, and power buggies. Naturally, the decks and hydraulic systems have to be designed to accommodate the loadings imposed by the equipment.

Working Platforms

All slipform jobs require as a minimum a working platform at the level of the top of the forms; indeed the ability of the forming system to carry this working platform automatically and at the proper height to do the work is a major advantage of the technique. Based on the amount, complexity, and distribution of the rebars, embedments, and box-outs, design decisions must be made regarding the need for perimeter platforms outside the building in addition to an internal

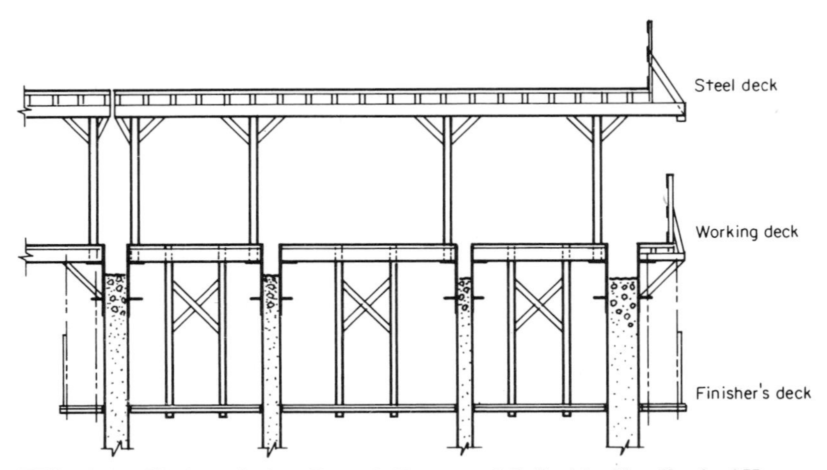

FIG. 15.9 Slipform decks. *(From J. Havers and F. Stubbs, Handbook of Heavy Construction, 2d ed., McGraw-Hill, 1982.)*

working platform. It may be decided or even required that the interior decks be partial decks or that they have large opening areas to reduce the weight of the slipform structure or to allow lowering of materials. A second level of working platform or scaffold will probably be required to provide access for inspection of the concrete, finishing, and curing. On some projects a third level may be required, or at least desirable, to facilitate the receiving and storage of rebars and embedments, the placing of embedments, rebars, and/or concrete; and to provide shelter or alleviate congestion due to intensive activity. The slipform structure must be designed to handle all these loads, including the effect of wind, in such a manner as not to affect the speed or trueness of the slip operation. Enclosures, heated or otherwise, wind screens, sun shading, and safety and control devices are often supported on the working platforms, and careful attention must be given to the detail design of this equipment and its attachment to and effect on the platforms.

Redundancy

Redundancy is a serious consideration in the design of slipform systems. These operations are highly labor-intensive and compress work activities into time spans that are only a fraction of what they would otherwise be. For that reason, work stoppages and slowdowns have very heavy impact on the construction cost. In addition, such stoppages or slowdowns can seriously affect the quality of the finished product and can even result in a disastrous binding of the forms, requiring removal of placed concrete and restarting in midair. All equipment must be selected and backed up so that failure of any piece will be quickly remedied and so that provision is made to keep the operation going even if it is at a reduced rate. This redundancy must also be considered in designing the forms as well as in arranging concrete, rebar, and embedment delivery, power supply, road access, and communications systems.

Demountability

Once the job is completed, this more or less complicated structure with all the attendant equipment and facilities must be disassembled and safely lowered to the ground, sometimes under difficult conditions. It therefore behooves the designer to plan this operation before finishing the design of the slipform structure and to include in the design those features which will facilitate the "form-stripping" operation.

Economy

Economy is mentioned here almost as an afterthought and perhaps appropriately so. Although the cost of a slipform "structure" is very high, as forming and structural costs go, the cost per square foot of concrete formed should be very small when compared to the cost of other forming techniques. The reuse factor in slipforming often reaches 100. In addition, scaffolding, shoring, stripping, and cleaning costs are greatly reduced. Although every effort must be made to keep the cost (and weight) of the slipform to a minimum, the major economy lies in an efficient placing operation minimizing construction time and labor costs.

Economies which endanger the smoothness of the actual slip operation must be summarily rejected with the same forcefulness as rejecting unnecessary costs and equipment.

JACKING SYSTEMS

Slipform jacks are available in 3-, 6-, and 22-ton (2.7-, 5.4-, and 20-t) capacities. Most of those used in the United States are hydraulic jacks, but electric and pneumatic jacks are also in common use. All the available jacking systems are synchronized to operate all the jacks simultaneously from a central control station. All the jacking systems have controls (manual, automatic, or semiautomatic) to keep the jacks level or at least to indicate when they are out of level. There is also a range of rental steel yokes to choose from, varying as to strength, degree of deflection under load, and clear distance under the yoke beam. Yokes can be designed to meet any desired characteristics and can be fabricated in any good fabrication shop.

In most cases the capacity of jacks required will be resolved very simply by the configuration of the structure to be constructed and by the area and loading of decks and scaffolds to be supported. The loading on each jack is equal to the length of forms supported by that jack, times the estimated drag forces per lineal foot of form, plus the square feet of decks and scaffolds supported, times the anticipated loading per square foot for each of the decks or scaffolds involved. To this must be added all dead loads and the weight as well as impact of any equipment to be used. If design is based on working loads, then it is suggested that the jacks be selected with a factor of safety of 2 against their ultimate capacity. This allowable capacity will usually coincide with the capacity stated by the manufacturer, but it should be checked. In other cases the allowable jack capacity will be derived in accordance with the appropriate code or specification. Every effort must be made to equalize the loading on the jacks, as overloaded jacks will tend to drag behind the others and get out of level. Although it is desirable to use jacks of the same capacity in any one job, it is often better or even necessary to use more than one type.

Generally speaking, lower-capacity jacks have the advantage of spreading the load over more supports, thereby reducing the strength requirements for the forms and the yokes as well as increasing the redundancy in the case of malfunction. On the other hand, the smaller interval between jacks makes it more difficult to place rebars, inserts, and openings. Depending on jack size and deck loadings, the interval between jacks will normally be between 4 and 9 ft (1.2 and 2.7 m) when standard equipment is used. Using high-capacity jacks and specially designed yokes, the interval between jacks can be increased substantially.

Hydraulic jacks are relatively light and compact, easy to install, and quite reliable under almost all conditions. Breaking hydraulic oil lines and spilling oil into the concrete occurs only infrequently. These jacks come with mechanical leveling devices which must be set by hand at each jack and moved in approximately 2-ft (60-cm) increments. If properly set, they will automatically level all jacks at these intervals. In addition, an independent water-level system is usually used with gauges at every jack as a check to insure that the jacks do not get badly out of level. In most cases, keeping the jacks level within ½ in (12 mm) is satisfactory. These jacks are cylindrical in shape and climb up a jack rod by hydraulically activating a spring-loaded piston having a 1-in (2.5-cm) travel and then releasing the pressure to allow the piston to reset for another move. Jaws alternatively grip and release the jack rod as the jack climbs upward. These jacks can be raised individually by a hand pump for adjustment or in an emergency.

The jack rods are usually ¾-in (19-mm) pipe, 1¼-in (32-mm) solid rods, or 2½-in (64-mm) pipe, depending on capacity. These rods are placed inside the forms and are prevented from buckling by the concrete as it hardens. Since the jack rods depend on the concrete for stability, it is necessary to support the jack rods against buckling any time they are out of the concrete, such as in the case of a jack rod going through a door opening. Whenever possible, jack rods are therefore laid out to miss major openings in walls. An office core or wall having many openings militates the use of higher-capacity jacks having maximum spacing to allow flexibility to miss the openings. At times, the distance between the underside of the yoke beam and hard concrete exceeds the allowable ratio of unsupported length to radius of gyration of the jack rod, thus requiring a larger rod or intermediate support.

When economy justifies the cost of recovering the jack rods rather than leaving them in place, a recovery tube is attached to the yoke and suspended into the concrete. This tube, slightly larger in diameter than the jack rod, leaves a chase in the concrete which supports the rod without bonding to it. At the end of the slide, the rods are jacked out from the top and, if required, the chases are grouted full.

The jack rods are commonly 10 ft (3 m) long with threaded and machined ends to allow them to be screwed together to a good load-bearing contact. These jack rods may double as reinforcing steel if the connection is capable of transferring the induced stresses. Channel-type jack rails that stay entirely outside the walls are available in Europe, but these require precast block inserts for attachment of the jack rails and are quite awkward and expensive to use.

Pneumatic jacks operate very much like the hydraulic jacks except that they use air instead of oil. They are considerably larger in diameter and have a tendency to malfunction in cold weather.

Electric jacks have the advantage of being automatically self-leveling, as they are activated by a water-level system running through all the jacks with the water level controlled from a central control point. Each jack has its own electric motor which operates an arm attached to the jaw system. The solenoid, which starts or stops the motor, is in turn activated when the water level is raised to a new level and stops when the jack reaches the new level. These jacks are larger than the hydraulic jacks and require electric and water connections as well as individual oil reservoirs.

All the above jacks utilize the same type of jack rods.

The hydraulic jacks are operated from pumps; the pneumatic jacks are operated from air compressors which normally have a capacity to handle as many as 100 jacks. If more jacks are required, additional or larger pumps are used and separate circuits are developed and synchronized. The layout of the hydraulic piping to the various jacks is important to obtain as smooth a lift as possible, but most of all to ensure that the system has adequate capacity to lift the form at the rate required. After each lift of 1 in (2.5 cm), a recovery time is required for the hydraulic pressure in all the jacks to be released and for the oil to return to the reservoir, leaving all the jacks recocked and ready for another lift. Improper piping will increase this recovery time and decrease the lift-rate capacity of the system. Most jacking systems are capable of operating at a speed of 20 in/h (50 cm/h), which is adequate for almost all slipform operations. In the rare instance that higher speeds are required, the jacking speed can be increased by adjusting the stroke or by other means.

CONSTRUCTION LOADS

It is very difficult to estimate the drag on the forms, as the drag will vary significantly with different concrete mixes and weather conditions. It is also affected by the configuration, angles of corners, bevels, surface, and cleanliness of the forms. The use of 100 to 200 16/lin ft (1460 to 2920 N/m) of two-sided, 4-ft (1.2-m)-deep forms supported has proved to be satisfactory in the past.

Deck loads will depend on the purpose for which the decks will be used. If they are to be used for storage of materials, the maximum loading expected must be calculated and used. In most instances, 75 to 100 lb/ft^2 (3600 to 4800 N/m^2) will prove adequate. The usual live load used, if no storage or equipment is planned in the area, is 75 lb/ft^2 (3600 N/m^2) on sheathing and joists and 50 lb/ft^2 (2400 N/m^2) on beams, girders, and trusses. In addition, the decks must be checked for concentrated loads such as buggy wheel loads. Finishers, scaffolds on which no storage is allowed, are usually designed for 25 lb/ft^2 (1200 N/m^2). Lateral loads due to power buggies, hoists, wind, etc. must be reduced to a minimum and then designed for. In the absence of specific code requirements, a force of 10 lb/ft^2 (480 N/m^2) of projected area is usually used for wind. The effect of power buggies may be taken as 50% of the vertical loading, and the lateral forces due to hoists and other equipment are as published by the equipment manufacturer.

The allowable stresses in wood, steel, etc. are as specified by local codes or by nationally recognized societies having jurisdiction. When codes such as OSHA specify that platforms, scaffolds, or equipment are to be designed for a certain multiple of the load, the ultimate strength of the material will be used to satisfy this requirement. Except in rare instances, seismic loads are not considered in designing slipforms.

DIMENSIONAL STABILITY

Positive steps must be taken to ensure the dimensional stability of the slipform structure so that the concrete product will be within specified tolerance. The wall thicknesses are preserved by a system of blocking and collars connecting the forms to the yokes. This system of connection must be positive enough to hold the forms securely in position, but at the same time allow for adjustment if this should become necessary. In addition to the yokes at the jacking points, "dummy yokes" are sometimes used at intermediate points as spacers for the forms. The wales are so designed as to prevent excessive deflection, corners going out of square, or hinges developing at splices. This is accomplished by means of staggering splices between wale plies for long wales and at corners, adequate nailing at connections, and stitching to develop horizontal shear between plies, bolting, steel connectors, braces, and cross-ties. Blocking between the forms is at times possible and desirable. The use of spacer beams and/or tie-rods is common to preserve proper distance between parts of the slipform. It is most important that this system to maintain dimensional stability be designed as a system with each device coordinated with the others so that every possible distortion is limited to acceptable levels without overlap in function. Lateral thrusts from moving equipment must receive special attention.

There can develop a tendency for the structure to shift as a whole. This must be guarded against by keeping the deck as level as possible and by eliminating or reducing lateral loads to a minimum. Control of such shifting and of rotational tendencies will be discussed under tolerances and alignment. The "island" concept as described under design consideration must be kept very much in mind.

FORMING MATERIALS AND TECHNIQUES

Just about any material that has a smooth, hard surface capable of resisting abrasion and that is reasonably easy to clean without damage can be used as a slipform if properly supported and backed up.

The most common material used is wood in one of several forms: plywood; 1-in (2.5-cm) boards, plain edged or tongue-and-grooved; overlaid plywood; and finn form, a many-plied variation of plywood. The plywood is sometimes used as is or is backed up with 1- or 2-in (2.5- or 5-cm)-thick battens. Curves are obtained by layering thin plywood over a wood frame.

Steel is also used when reuse and special conditions warrant the additional cost of steel forms. The major use of steel forms are for chimney construction, which requires that the forms lap to accommodate the changing diameter. Steel slipforms as much as 8 or 10 ft (2.5 or 3 m) high were in common use during the period of missile silo construction and are still in use in the slipforming of vertical underground shafts.

Fiberglass-reinforced-plastic forms offer considerable promise, but to the author's knowledge have been used only once to do a slipformed project. In addition, a test panel has been used successfully on a project in Puerto Rico. On the Bay View Terrace Apartment Building in Milwaukee, ornate plastic panels were used as stationary inserts supported against sliding battens.

The use of stationary inserts (or form liners) of all sorts supported by a sliding batten framework allows a whole range of architectural treatments including panels of brick and glass blocks, concrete sculptured panels, and ceramics. Sliding inserts supported from tie-rods attached directly or indirectly to the slipform yokes may be used to reduce wall thicknesses or to form chases or weight-reducing lightening cells. In addition, liners or cleats may be firmly attached to the slipform to produce a corrugated, ribbed, or striated finish. The use of varied width and thickness boards is another means of achieving a pleasing architectural finish that honestly reflects the nature of concrete and the slipform method.

OPERATIONAL TECHNIQUES

Having properly designed the slipform complex, it is equally important that this extrusion machine be properly operated. On important projects such as nuclear containment shells, operational manuals are produced which give in detail the plan for placing concrete, reinforcing steel, and embedments as well as for organization, quality control, survey methods, schedules, and standby provisions.

Concrete

Since concrete is the major ingredient in a slipform structure, considerable effort must be spent in arriving at the optimum concrete mix, giving proper consideration to materials and additives as well as to temperatures and conditions under which the concrete must be placed. The use of type I normal portland cement is recommended in almost all instances. Type II modified cement is also used for subgrade work, for mass concrete placements, etc., but not for cold-weather placements of thin or medium-thickness walls. Type III high-early-strength cement may be used only with the greatest caution, if ever, and certainly not in any but the very coldest weather. Care must be taken that the type I cement selected for the job does not, in fact, have type III characteristics. This can be ascertained by reviewing fineness of grind, initial set time, and early-strength tests. Cements stored at temperatures in excess of 100°F (38°C) must be avoided or handled and mixed with special care. The cement content must be selected not only on the basis of strength requirements, but also while keeping in mind that a minimum amount of cement, say 6 bags per cubic yard (7.8 bags per cubic meter), is desirable to produce a "fatty" mix which will slide well and produce good finishes. In addition, the amount of cement used per cubic yard for higher-strength concretes must be kept to a minimum, as concretes with higher cement content tend to set more rapidly, possibly too rapidly for the particular slipforming operation under consideration.

Fine and course aggregates must be carefully selected as to maximum size, good gradation, and particle shape. Light-weight aggregates can be used, but care must be exercised in controlling water absorption to ensure a fatty slippable concrete in the forms.

Basically, the mixes and material requirements for slipformed concrete are more exacting, but not different from requirements for conventionally placed concrete: both are usually pumpable.

Temperature control of the concrete is extremely important. Realistic but stringent maximum and minimum temperatures should be specified to suit the ambient conditions, placement methods, and exposure conditions. These temperature controls should be imposed at the mixer, in the truck, and in the forms. Control of concrete temperature can be accomplished by proper, prudent use of admixtures carefully measured and evenly dispersed to ensure uniformity throughout the concrete placement. In lieu of, or in addition to, admixtures, the use of ice in the summer and an additional bag of cement in the winter are highly recommended. The substitution of ice for water in the mix is highly effective in keeping down concrete temperatures and provides safety against uneven setting times in different sections of the form. Using ice as a full or partial substitute for water results in more uniform and predictable concrete set and definitely limits the length of time the concrete set will be retarded.

The sources for concrete and concrete materials should be inspected to ensure that adequate quantities are on hand to supply the job and that proper storage, handling, mixing, and transporting techniques and equipment are available and in good condition to ensure the concrete desired. Provisions should include sufficient redundancy in mixing, transporting, and placing capacities to ensure a continued operation in the face of breakdowns, traffic problems, rejected concrete, and other contingencies.

Concrete Placement

The concrete-handling system must be carefully designed to receive the concrete, raise it, and distribute it into the forms in the shortest possible time and without excessive loss of slump in the summer or heat in the winter. Particular care must also be taken to prevent segregation of the materials or irresponsible addition of water. In hot weather, shade may be required for trucks and equipment as well as painting the trucks and equipment white or a light color to reduce absorption of radiant heat.

The concrete is placed in uniform, level layers of 2 to 12 in (5 to 30 cm) thick, the 12-in (30-cm) layer being the most desirable, the 2- or 3-in (5- or 7.5-cm) layer being used only when the slip is seriously delayed and concrete must be placed frequently enough to prevent a cold joint. Since the elapsed time between placements should not be greater than 1 h and may be even less under certain circumstances, 2-in (5-cm) placements could accommodate a minimum slide rate of 2 in/h (5 cm/h). By placing 12-in (30-cm) layers at 3-min intervals, and assuming a 1-in (2.5-cm) jack stroke, a slide rate of 20 in/h (50 cm/h) can be assured. Higher speeds may be attained under certain circumstances by increasing the placement thickness, reducing the time interval between jack movements, or increasing the jack stroke. It is highly desirable to keep the forms as full of concrete as possible and every effort should be made to do this. If 6-in (15-cm) placements are being made, for instance, a placement should be made as soon as possible every time there is 6 in (15 cm) of empty form. If the forms are allowed to become excessively empty (as in the case of a delay in obtaining concrete), shingles or overpours will result and the danger of uneven placements and setting times will be increased.

Vibration of the concrete is, of course, very important. One-inch vibrators are used except in the case of very thick walls, in which case larger vibrators up to 3 in (7.5 cm) in diameter are in order. The vibrators should penetrate to the layer immediately below the one being placed. If a vibrator is allowed to penetrate under its own weight and is pulled out as soon as it stops, this will result in just about the right amount of vibration. Excessive deep vibration could cause a fall-out of concrete from under the form. Every effort should be made to vibrate the concrete well into the cover area between the wall surface and the steel curtains, into corners, at openings, and around embedments and heavy concentrations of reinforcing steel.

There is a tendency to "dump" too much concrete at one spot along the walls and then to move the concrete laterally using the vibrators. Such excessive dumping of concrete by the buggy, bucket, hose, etc. should be discouraged, but when it does happen, the concrete should be distributed using shovels and leaving the deck clean of any concrete. Concrete left on deck tends to harden and then be pushed into the form with a subsequent placement. A broom-clean deck should be insisted upon, and clean-out openings must be provided for the disposal of any concrete which was accidently spilled and hardened on deck.

In a 4-ft (1.2-m)-deep form, the amount of hardened concrete should not be allowed to be less than 12 or 15 in (30 or 38 cm). If the height of hardened concrete falls below this minimum, the rate of slide must be decreased to give the concrete a chance to set. By the same token, if the height of set concrete exceeds 30 in (75 cm), every effort must be made to increase the rate of slide; if this is not possible, the placement layer thickness must be decreased. The thickness of hardened concrete is usually measured by forcefully ramming a ⅝-in (16-mm) rebar by hand into the concrete and measuring the distance from the top of the form to the end of the bar.

If for any reason—including rain or a "bleeding" of the concrete—water is accumulated in the forms, this water should be immediately ladled out. Holes are usually drilled in the main slipform decks to provide drainage and prevent creating a "watershed" that will tend to collect rainwater in the forms.

It is possible under certain circumstances for a repetitive placing sequence to cause rotation or shifting of the form. If such a condition is anticipated or encountered, the sequence must be alternated or otherwise modified to eliminate this effect.

Concrete Curing and Finishing

Concrete normally comes out from the underside of the slipform in an ideal condition for finishing. A float-and-brush finish is most often applied; however, other finishes are sometimes specified or possibly no finish at all. Excellent repairs to eliminate honeycomb, bugholes, etc. may be made at this time. Particular care must be taken to obtain clean lines around openings, at corners, at embedments, etc., but it is just about impossible to remove shingles if these have occurred.

Embedments such as weld plates have a tendency to get covered by a thin film of concrete that may be as much as ½ in (12 mm) thick and will be hard to locate and "dig out" later when the slipform operation is finished. The finishers should be alert to spotting these inserts and cleaning them up while the concrete is still relatively plastic.

The finishing is done from scaffolds attached to and hanging from the forms at appropriate heights to facilitate the work. They are provided on the inside or outside of the forms, or both. Many slipform designers provide finisher scaffolding whether finishing is required or not, since these provide a means for inspection, a working platform in case of problems, a support for enclosures, and a working platform for applying curing compounds.

Wall curing is now seldom accomplished by means of water. In the past, perforated pipe was strung out along the bottom of the forms to provide a continuous spray, but this method has many problems including gouging the concrete, misses, discontinuity, and creating messy water conditions at the base of the structure. The use of curing membranes applied from the finisher's scaffold is presently more popular with fugitive dyes often used to guard against misses.

In specifying membrane curing compounds, consideration must be given to subsequent finishes to ensure that the compound used will be compatible with the finish contemplated. Care must also be exercised to ensure that the fugitive dye, if used, will actually disappear completely within the required time period. Very often the enclosed inside areas of slipformed structures retain considerable warmth and humidity during the slide operation, making the use of curing compounds in these areas redundant. When required on inside faces of walls, the curing compound selected must not emit toxic fumes harmful to the construction laborers.

Reinforcing Steel

Since the placement and inspection of reinforcing steel for slipform operations must be done under conditions peculiar to the technique, the detailing for cutting and bending of the steel must meet certain requirements. The beams of the jacking yokes cross the walls at some height above the deck. Since these yokes move upward as the work progresses and since the reinforcing is stationary, it follows that horizontal steel running along the walls cannot be placed above the yoke beams. It is therefore necessary to thread the horizontal steel through the yoke beams, jack rods, and vertical steel within the area between the deck and the yoke beam. The length and configuration of the horizontal steel must therefore be so detailed as to facilitate placement under these conditions. Rebar placing is further complicated by the fact that concrete is placed at frequent intervals, thus covering the reinforcing steel shortly after it is placed. To facilitate control and placing, the reinforcing is detailed in layers so that all steel throughout the plan of the structure is at the same spacing and elevation. This allows well-organized crews with very specific areas of responsibility to place one complete layer at a time to suit the rate of jacking. It also allows inspectors to check the steel layout to ensure that all reinforcing steel, especially corner bars, tie bars, bars around openings, etc. are in place, that proper splices are maintained, and that there is adequate concrete cover between the steel and the face of the wall. In addition, the spacing of the horizontal steel must be such as to allow time for placing each layer within the time allowed by the planned jacking rate. Since minimum steel areas are required by the design, trade-offs are required between spacing and size of bars to obtain as great a spacing as possible consistent with avoiding heavy unmanageable steel and concrete-placing difficulties.

Vertical reinforcement does not present the problems connected with horizontal steel, but it must be carefully detailed, placed, and held in place to miss jacking yokes, openings, and embedments, to provide support and facilitate placing of the horizontal steel and to allow "buggy paths" for concrete placement. It may also have to be detailed around hoisting and other equipment to be used during the slide. The length to which the vertical steel will be detailed will depend on the slipform sequence (daily or continuous slide), on the size of bars (weight and length for handling and tendency to "wave in the breeze"), and on the type of template provided. If daily slides are anticipated, with concreting and rebar placing on alternate days, then the vertical bars are either detailed in floor heights or in two-floor heights with laps alternated between floors. This procedure may also be possible under certain circumstances with concrete placement on a daily basis. If a continuous around-the-clock operation is contemplated or if additional time is not available for placing reinforcing steel between daily slides (overtime is a possibility), then the vertical steel is spliced in several "runs" with splices staggered at 2- or 3-ft (60- or 90-cm) intervals to allow a fairly continuous placing of vertical steel throughout the operation. The vertical steel is held near the top or at some distance below the top and is tied at the bottom to the top of the previously placed bar with the correct splice length. In other words, the vertical bar is dropped through the template hole and tied to the bar to which it is being spliced. The height of the template above the deck will depend on whether an upper deck is provided and on the size and length of vertical bars used.

Reinforcing steel should be on the job well ahead of the start of the slide and properly "shook out" in a convenient lay-down area to serve the slide. The reinforcing steel should be received and organized on the order that it will be used and a steel storage should be established and continuously replenished on deck. Reinforcing steel may be raised by crane or hoist.

Embedments

Embedded items present a special problem in slipforming because there is no stationary form against which to position them. Actually, the inserts must be at least ½ in (12 mm) less in dimension than the walls to provide clearance tolerance to the moving forms. The vertical reinforcing steel provides a ready stationary base to which inserts and embedments may be attached, but as there is a strong tendency for the forms to drag embedments upward and as wire ties will not be sufficient to hold them down, welded attachments to the steel will be necessary. Since many specifications prohibit welding to the reinforcing steel, additional reinforcing bars may have to be added for the purpose of supporting embedments.

Small embedments such as threaded inserts are very difficult, if not impossible, to position and keep in place unless they are ganged together by means of steel rods or bars, keyways, or positioning boards. Weld plates should be designed oversize to allow for inaccurate spacing. Whenever possible, groups of small weld plates should be combined into one large plate, but not so large as to make handling excessively difficult. Care must be taken also to ensure that anchors are so detailed as to prevent interferences with reinforcing steel and other embedments in the concrete. Prestressing cables, electrical conduit, and piping can be placed in a manner similar to the reinforcing steel while the prestressing anchors, electrical boxes, etc. are set as embedments.

Very large wall penetrations have been placed during slipform operations using a crane, but this required carefully planned clearances for lowering the inserts and previously prepared adjustable seats to receive them. Such wall penetrations are not uncommon in nuclear containment shells. Whenever possible, it is very important that all inserts, large and small, be placed well ahead of time so as to prevent hectic activity when the form rises to the particular elevation. There is a tendency for several inserts to be at one elevation, say a floor elevation, causing a dangerously prolonged delay in the slip operation while they are being placed.

Manpower

A minimum-size crew is required almost without regard to the size of the job. This includes an operator for the jacking system, an electrician, one or two carpenters to act as form watchers and make emergency repairs, a top-and-bottom worker for concrete hoisting, an operating engineer, an oiler if a crane is used, and a layout worker for placing inserts, openings, etc. and checking plumb. Concrete-placing crews vary with the amount of concrete to be placed per hour, wall thickness, length of run from deck hopper, and impediments to the placing. These crews often consist of one buggy worker for every 2 to 5 yd^3 (1.8 to 4.5 m^3) of concrete to be placed per hour and anywhere from one to three placing crews consisting of one or two shovelers and a vibrator operator. Reinforcing crews will usually consist of several two- or three-worker units placing horizontal steel, at least one unit placing vertical steel and jack rods, and a unit bringing up the steel to the deck. The rebar crews required for slipform projects are very similar in size to rebar crews required for conventional construction. Additional carpenters, iron workers, electricians, pipe fitters, and "engineers" may be required to place embedments in their respective crafts. Finishers and finishers' helpers will also have to be provided as necessary and probably a laborer or two to apply membrane curing compound. To this add the deck superintendent and appropriate foremen for all trades.

TOLERANCES AND ALIGNMENT

Slipformed structures have been known to go out of plumb by 6 or 8 in (15 or 20 cm) and at times while under the supervision of experienced but careless slipform superintendents. This statement is not made to indicate that slipform work is inclined to be excessively out of tolerance, but rather to indicate that careful planning and execution are necessary to prevent this happening. Actually most slipformed building cores that are known to the author have stayed substantially within ½ in (12 mm) of the planned dimension. This is not to say that specifications should call for ½-in (12-mm) or even 1-in (25-mm) tolerance. ACI 347, *Recommended Practice for Concrete Formwork,* specifies a tolerance from plumb of 1 in/50 ft (2.5 cm/15 m) of height with a maximum of 3 in (7.5 cm), and stricter requirements for building cores and shafts. A maximum limit of between 2 and 3 in (5 and 7.5 cm) in the total height of the building for those over 100 and 200 ft (30 and 60 m) in height, respectively, is reasonable and acceptable to most contractors for commercial buildings. Stricter requirements will be very difficult to attain. Specifications for silos and other structures not requiring such precision should specify a greater tolerance. Wall-thickness variation is limited by ACI 347 to − ⅜ in (1 cm) and +1 in (2.5 cm).

As previously indicated, it is imperative that forms be properly designed and that the slipform operation be competently executed. Beyond this, a system of "tracking" the slipform must be initiated to ensure early warning of variances from plumb so that corrections may be made. Targets must be provided on the forms and at the base of the structure to provide readings which will show variations in north-south as well as east-west directions. These instrument shots should be taken at 12-h intervals as a minimum and more often for structures which have a strong tendency to go out of plumb. Plumb bobs (in oil cans if necessary), optical plumbs, laser beams, and transits are used for this purpose. Corrections are normally made by tilting the deck to counteract the "drift." Eccentric loads on outriggers, cable drags on the ground, turnbuckles to previously cast concrete, and other remedies are used, but extreme caution is recommended.

Twist or rotation of the structure is very difficult to correct. If a problem is anticipated or encountered, structural-steel anchors can be set in the concrete at strategic points and come-alongs installed between the anchors and the yoke beams to counter this rotation. Come-alongs or turnbuckles to previously placed concrete, drags on the ground, tilting of yokes, and other corrective actions may be used, but again with utmost caution.

The best means of militating against going out of plumb is by assuring a level deck at all times and by equalizing loads on the jacks. In addition to having automatic or semiautomatic self-leveling jacks, independent water-level systems should be installed with graduated tubes at every jacking point. The level of the jacks should be checked regularly and frequently and corrections made wherever jacks are from ½ to 1 in (12 to 25 mm) out of level.

There is considerable concern about placing tolerances for openings and embedments. Although tolerances should never be more stringent than necessary, close tolerances can be achieved if proper placing and tying methods are used.

DRAWINGS

Three types of construction or "shop" drawings are required for a properly organized slipform operation; detailed form-design drawings for construction of the forming complex; complete plans and elevations showing detailed sizes, locations, elevations, and identification numbers for all embedments, openings, keys, etc.; and rebar-placing drawings and bar listings. These drawings are normally prepared by the contractor and approved by the design architect or engineer.

The *form drawings* consist of the following:

Dimensional plan showing all dimensions

Deck plan showing deck design, layout of all jacks, yokes, beams, and equipment, and general notes indicating materials, equipment, design loadings, design slip rates, etc.

A plan at the finisher's level showing all scaffolding and access facilities

A plan of the upper deck, if one is required

Plans for the hydraulic, pneumatic, or electric lines, control centers, lighting fixtures, etc.

Drawings showing crane and other equipment support

Drawings showing complete details of all structural components

Drawings showing form panels

Drawings showing details of all openings, keys, and box-outs

Drawings showing detailed miscellaneous iron and structural steel for fabrication

The *elevation drawings* show elevations of every wall, column, and element of the structure as it will look at the time of completion of the slide, providing for—but omitting—floors and all other structural, mechanical, and architectural components not to be placed during the slide. Plans are used as required in support of the elevations to properly illustrate the configuration and interrelationship of the various openings, box-outs, and embedments. The types of items that will be shown on these drawings are:

Keyways, with or without dowel inserts

Beam pockets

Door and window openings

Weld plates

Penetrations

Utility pipes, conduits, boxes, etc.

Top and bottom wall cutoffs

Openings

Anchor slots

A slide elevation is usually established, setting the start of the slide as 0.00, and all elevations are given to this datum. Elevations used will be at the top of the item (as for weld plates) or at the bottom of the item (as in door openings) in accordance with the method by which the item will be placed in the field. On these drawings interferences are worked out with the jack rods and, if possible, eliminated by moving either the opening or the jack rod. Obviously, it is not always possible to avoid jack rods (and the entire yoke) passing through openings; so in these instances, the box-out must be designed to accommodate the yoke beams sliding through and to support the jack rods in this area. Wherever possible, floor keys are made discontinuous at the jacking points to allow early and better placing of the keyways. Each item must have an identifying number and all identical items will have the same number. In addition, a "slide schedule" is prepared which lists all items to be placed in the order in which they will be placed.

Rebar-placing drawings will usually be divided into three groups covering horizontal, vertical, and miscellaneous steel. The vertical steel is usually organized in "runs." A "run" maybe defined as a vertical set of spliced steel starting at a certain elevation, ending at another elevation, possibly having certain discontinuation at particular elevations, and having bars of a certain length and size at any particular elevation. Any one run will consist of the same number of identical bar mark numbers. All vertical steel is then shown in plan and dimensioned together with the jack rods and labeled by runs (run A, B, C, etc.). The composition of each run is shown on the vertical run elevations. This system is very helpful in simplifying placing and inspection in the field and has grown out of many years of experience.

The horizontal steel is usually shown in plan, with the verticals shown for relationship only. Coursing of the horizontal steel is usually established by number, starting with course number 1 at the bottom and continuing in numerical sequence to the highest course at the top of the slide. A vertical cross section of the wall is shown and all course spacing, elevations, and numbering are indicated thereon. Horizontal steel is then indicated in plan by reference to the course numbers. For example, the labeling for an individual bar may be "30 courses Mk W801 no. 8 bars @ 8 in o/c courses 28 thru 57." Another typical notation for a course could be "80 courses no. 8 @ 8 in o/c—10 Mk W801 and 1 Mk W805 per course." The mark numbers are usually by size, numbered in order for bent bars and by size and length only for straight bars. Subscripts may also be used to indicate location or function of the bars. In making up the bar list, every effort must be made to orient delivery of the steel to the slipform schedule. Color tagging to indicate delivery schedules could be very helpful.

Miscellaneous steel, such as additional rebars around openings and embedments, beam steel, hanger rods, etc., are shown in plans or elevations in such a manner as to most clearly indicate what is required and to avoid their being missed during the slide. The elevation drawings are often used for this purpose, or separate details are developed as necessary.

It may be mentioned at this time that diagonal bars are extremely difficult to place, are often improperly placed or omitted in the field, and should therefore be avoided. A combination of vertical and horizontal steel can be detailed to perform the same function as diagonal bars.

SPECIFICATIONS

Specifications for ordinary slipformed concrete need not be much different from specifications for other cast-in-place concrete. They should require that the contractor demonstrate competence to perform the work and that the contractor submit the concrete-mix design, complete form drawings, rebar-placing drawings, and slipform procedures to the architect or engineer for approval in a timely manner. If there is a requirement for a continuous slip operation, seasonal limitations, slide-rate limitations, redundancy, construction loads, jacking, or control systems, this must be clearly stated. Similarly, any specific tolerance or finish required must be clearly specified. For bins or silo work involving design as well as construction, compliance with ACI 313, *"Recommended Practice for the Design and Construction of Bins and Silos,"* should be required. In reviewing shop drawings, form design, equipment, procedures, and concrete mix, the architect and/or engineer should be guided by the various considerations discussed in preceding sections.

Architectural slipformed concrete is quite another matter. These specifications must be detailed and explicit and require a high degree of slipform expertise on the part of the architect who writes them. Almost all specifications written for slipformed architectural concrete work to date have been inadequate, resulting in some surprises and disappointments to the architect. Whether it is to be cast or slipformed, the achievement of good architectural concrete is really in the realm of sculpture and cannot be approached with a routine attitude. The good work done by contractors has been almost in spite of the specifications and drawings furnished by the designers. It makes as much sense to award an architectural concrete job to the lowest bidder as it would to award an art commission on that basis.

A detailed discussion of specifications for architectural slipformed concrete is in order at this point, but one really cannot discuss the specifications without simultaneously discussing the design to which these specifications are related.

Concrete

Whenever possible, slipformed concrete structures should be designed to strengths not exceeding 4000 to 4500 lb/in^2 (2750 to 3100 N/cm^2). Using the lower strengths will result in thicker walls and lower steel ratios, thereby reducing rebar congestion and allowing insertion of tremie chutes and vibrators for proper placing of the concrete. In addition, the resulting lower cement ratio will decrease the speed and heat of the hydration process, thereby reducing the danger of premature set, cold joints, and cracking, especially in the summer.

Apart from strength requirements, there are certain placing requirements affecting the cement ratio in the mix. Below a certain ratio, harshness, honeycomb, and cold joints will be encouraged because of the stiffness of the mix. A high cement factor may generate excessive heat, making the concrete hard to place, entrapping air, and increasing temperature stress in the green concrete. It is therefore recommended that the cement factor be no less than 6 sacks per cubic yard (7.8 sacks per cubic meter) and no more than 7½ sacks per cubic yard (10.2 sacks per cubic meter) for summer construction; no less than 7 sacks per cubic yard (9.1 sacks per cubic meter) and no more than 8 sacks per cubic yard (10.5 sacks per cubic meter) for winter construction. Excessive safety factors often applied by laboratories to ensure strength gain should be rejected if they result in too high a cement ratio. The use of facility records indicating strength-deviation history should be taken advantage of to reduce cement content in higher-strength concretes. Concrete trial batches should be mixed and tested for slump and air content at the anticipated placing temperatures.

Type I cement is in general the most economical, available, and satisfactory cement to use. Type II cement may be used in the summer, but used with caution in the winter. Type II (light-buff-colored cements) may be used as required. All cements to be used must meet the following specification requirements:

Maximum cement temperature—100°F (38°C)

Maximum fineness (Wagner)—2000

Minimum initial set time (Gillmore needle)—2 h

If a high early-strength cement is required, the above requirements must be deleted from the specifications. However, the use of high early-strength cements (type III or accelerator additives) is to be prohibited unless clearly required in specific instances.

Water acts as the lubricant in sliding form work. Therefore, excess water is highly desirable. In this type of construction, water does not tend to decrease the density of concrete to the extent that it does in concrete sections placed in higher lifts. A minimum amount of water is required for good placing. To go below this minimum will result in less dense rather than more dense concrete. It is therefore recommended that specifications call for the allowable slump that corresponds to a 4-in (10-cm) slump before air-entrainment and/or water-reducing agents are added. If air-entrainment or water-reducing agents are used, neither the water nor the cement should be reduced.

Aggregates must meet the requirements normally specified. Crushed stone, gravel, lightweight aggregate, or slag may be used. Crushed stone has no tendency to roll on a moving form, and is therefore slightly preferable to gravel. Good control is required to ensure sufficient free water in the forms when lightweight concrete is being used; in any case, a harsh concrete and finish is likely to result with lightweight concrete. The grading of all types of aggregates is important for handling ease and finish; the grading curves must, therefore, be smooth and reasonably straight.

Temperature is probably the most important factor in good concrete placing and perhaps the least emphasized. Temperature control of the concrete is imperative to obtaining good architectural concrete. The additional cost involved in controlling this temperature is well spent and the effect is of such magnitude as to indicate that money so spent will probably have the highest-quality yield per dollar spent. The following temperature specifications are therefore recommended:

Winter placing	Summer placing	Spring and fall placing
Max. 80°F (26°C), min. 70°F (21°C) in truck	Max. 55°F (12°C) in trucks	Min. 55°F (12°C) in truck or form
		Max. 65°F (18°C) in truck
Min. 60°F (15°C) in forms	Max. 70°F (21°C) in forms	or form

Admixtures may be used only when called for in the specifications, or when requested or approved by the architect or the field representative. Admixtures should not be used unless there is a clear need for their use. Nor should the use of admixtures be categorically disapproved. The method and precision of controlling the dosage must be clearly described before an admixture is approved. The use of heavy dosages is very dangerous and must be limited, as control becomes more difficult, errors become more serious, and side effects become more significant as the dosage increases. Admixtures should not be used in lieu of temperature control. Heating or cooling the materials is effective and involves none of the dangers associated with chemical control. Substituting ice for water in the mix is the best method of cooling concrete and is highly recommended.

Trial mixes must be developed and then checked by use of a slipformed mock-up at least 8 ft (2.4 m) wide and 12 ft (3.6 m) high and involving at least one stop and start, a corner detail, and a box-out if such features are involved in the prototype. This will also help in making the decision as to finish and will provide an acceptable standard for the contractor's viewing.

Concrete Placing

Uniform placing of concrete is essential to producing a high-quality concrete. Concrete must be placed in increments that result in uniform horizontal bands that are not so deep as to prevent proper vibration of the concrete and rising of trapped air to the surface. It is recommended that the specification state that the concrete be placed in equal horizontal bands no more than 12 in (30 cm) and no less than 4 in (10 cm) in depth.

An erratic rate of placing will result in variations in surface texture, wall thickness, and "shingles." A rate of placing concrete increments should be specified, taking into account heights of increments placed, temperature and composition of concrete, ambient temperatures, and box-outs, plates, etc. to be placed. The interval between placing increments must be limited to a maximum of 40 min in the summer and 60 min in all other seasons. In addition, it must be specified that the nonplastic concrete in the form be limited to a minimum of 12 in (30 cm) and to a maximum of 30 in (75 cm). Last, a uniform jacking rate of 6 to 15 in/h (15 to 40 cm/h) plus or minus 2 or 3 in/h (5 or 7.5 cm/h) should be specified. These requirements for uniform placing rate, uniform placing increments, and minimum and maximum heights of "set" concrete will be given the following precedence in importance:

1. Amount of "set" concrete in form

2. Minimum placing interval

3. Uniform jacking rate

4. Uniform placing increments

When all four of these requirements cannot be satisfied, this conflict must be removed by acceleration or retardation through control of concrete temperatures.

Construction site control of concrete placing and concrete-placing temperatures require proper planning of mixing and/or delivery, considering temperatures at which concrete will be placed, distance of plant from site, traffic problems, type of equipment, method of charging and discharging, waiting time, and temperature variations during handling of concrete. The specifications should require that the contractor purchase or mix the concrete in such a manner as to meet the control requirements of the specifications.

The length of time of mixing concrete after cement has been introduced into the truck should be limited to 40 min for concrete hotter than 60°F (15°C), 60 min for cooler concrete.

In spite of careful mix design and tight control of concrete delivery, placing, and temperature, there is a possibility that the proper control of placing requires a variation in the water content in the field. The specifications must allow the addition of some water in the field if this necessity is evident. This decision must be limited to the architect's field representative who in turn is free to consult with superiors. The addition of water is only to be used as a last resort if temperature-control activities have been exhausted.

A basic decision must be made as to whether daily lifts or a monolithic slide are required, and this requirement must be stated in the specifications. If weekend stops are allowed, this must also be stated. If it is required that the top of the form be raised a certain distance above the top of the concrete, this too should be clearly stated. Generally speaking, intermediate stops require the form be raised 12 to 18 in (30 to 45 cm) above the top of concrete. At the final stop, it may be specified to raise the form 24 to 36 (60 to 90 cm) above the top of the concrete.

Plan the construction schedule to secure the best slipform conditions. Early spring and late fall are preferred. Summer is least desirable. Generally speaking, ambient temperatures between 50 and 70°F (10 and 20°C), light rains, and uniformity of weather are desirable. Torrential rain, high winds, extreme temperatures, and rapid temperature changes are not desirable.

Concrete Finish

The specifications must state whether the concrete is to be finished, finished in certain areas only, or left absolutely untouched as it comes out of the forms.

The specifications must indicate the type of formwork surface that is required (finn form, board form, etc.).

The specifications must indicate the type of finish to be applied (wood float, rubber float, steel trowel, scraped, scrubbed with wire brush, etc.).

If the finish cannot be determined at the time of obtaining bids, the alternates under consideration should be stated. Most alternates will not result in a significant cost differential. Contractors must make exceptions as they see fit.

Formwork

Formwork shall be designed, erected, and used in accordance with *Recommended Practice for Concrete Formwork* (ACI 347-78), section 6.3.2, except as otherwise noted in the specifications.

Construction Tolerance

Tolerances should be the maximum that the design and/or aesthetic requirements allow. Tight tolerance is expensive and should be specified only when necessary. Tolerances specified may normally not be less than the following:

Variation from plumb in any direction:
- 1 in (2.5 mm) in any 50 ft (15 m)
- 2 to 3 in (50 to 75 mm) in total building height

Variation from grade:
- 1 in (25 mm)

Variation in wall thickness:
- −¼ in (−6 mm)
- +½ in (+12 mm)

Relation of critical surfaces to each other:
- 1 in (25 mm)

ECONOMICS

The cost of slipformed concrete is very sensitive to building cross section, height, steel density, embedments, quantity of concrete per foot of height, labor policies, weather, etc., and therefore varies over a very wide range. Obviously, the cost of formwork will remain the same regardless of whether the slipformed structure is 50 or 500 ft (15 or 150 m) high or whether the walls are 6 or 26 in (15 or 65 cm) thick. Openings in the wall not only cost money to form, but they slow down the slide, thereby increasing the cost per volume of concrete proportionally. Similarly, a basic crew of approximately 10 is required whether 10 or 20 yd^3 (7.5 or 15 m^3) of concrete are being placed per hour. Heavy steel, in addition to the increased cost of placing the steel, results in a reduced rate of slide and, therefore, greater concrete cost. If the weather is cold, the concrete will set more slowly, the rate of slide will decrease, and the cost per cubic yard of concrete will increase. The only way that a realistic cost may be obtained is by making a detailed estimate for the particular structure at the specific place.

A decision as to the economy of slipforming a structure should be based not only on the cost per cubic yard of the concrete, but on the cost savings that may accrue as a result of a decrease in construction time and easier construction of other elements of the structure.

BIBLIOGRAPHY

Recommended Practice for Concrete Formwork, ACI 347-78, American Concrete Institute.

Recommended Practice for Design & Construction of Concrete Bins, Silos and Bunkers for Storing Granular Material, ACI 313–77.

Building Code Requirements for Reinforced Concrete, ACI 318–71.

Hurd, M. K.: *Formwork for Concrete,* Special Publication no. 4, American Concrete Institute.

Havers, John and Frank Stubbs (eds.): *Handbook of Heavy Construction,* McGraw-Hill, New York, 1971.

Peurifoy, R. L.: *Formwork for Concrete Structures,* 2d ed., McGraw-Hill, New York, 1976.

"Slip-Form Details and Techniques," ACI title no. 55–67, *Journal of the American Concrete Institute,* vol. 62, no. 10, October 1965.

"Novel Structural Frame Combined with Slip-Form Construction Results in Record Breaking Construction Time," ACI Title no. 62–66, *Journal of the American Concrete Institute,* vol. 62, no. 10, October 1965.

"Continuous Concreting," *Construction Methods and Equipment,* vol. 55, no. 5, May 1973.

"Vertical Slipforming as a Construction Tool," *Concrete Construction,* vol. 23 no. 5, May 1978.

"Sophisticated Slipforming Produces a Floor a Day," *Engineering News-Record,* July 2, 1964.

"Slipform Casts all the Walls of a 25-Story Building," *Construction Methods and Equipment,* July 1964.

"K of C Building: A Super Blend of Form and Technique," *Building Design and Construction,* August 1970.

"Design Change sets Slipforming Record," *Engineering News-Record,* November 26, 1970.

"Slipforms Accelerate Erection of Concrete Towers," *Engineering News-Record,* June 25, 1964.

"Slipform Shapes Four-Unit Tower Core," *Construction Methods and Equipment,* April 1964.

"Slipform Adjusts for Alignment as it Forms Fluted Tower," *Construction Methods and Equipment,* January 1967.

"Tower Crane Rides Slipform Up 29-Story Building," *Construction Methods and Equipment,* August 1967.

"Up, Up and Away as Slipform Rises Rapidly to Shape West's Tallest Chimney in Calgary," *Heavy Construction News.*

"Application of Slipform Techniques to Dam Construction," in *Economical Construction of Concrete Dams,* published by the American Society of Civil Engineers, May 1972.

16

Access Scaffolding

by Colin P. Bennett

Scaffolding has been used for 5000 years to provide access areas for building and decorating structures taller than the people who worked on them. The word "scaffolding" refers to any raised platform or ramp used for ingress and egress for pedestrian movement and/or the passage of building materials. The word "staging" is often used synonymously with scaffolding since the Shakespearean era when plays were staged on a raised platform. A scaffold also means a raised platform used for executions. In broad context, the scaffolding in this case is a "work place" for those who do the hanging or beheading.

It can be surmised that the first crude forms of scaffolding were developed when humans moved from single story huts at ground level to higher structures; most certainly they used scaffolds made of tree trunks and limbs lashed together with fibrous vines.

By the middle ages, people became more adept at devising and using more complex mechanical devices: raw trees became smoothed poles held together with longer-lasting ropes, twisted or plaited from hemp fibers. In the far east, light hollow bamboo poles are still used extensively. Joints are made with bamboo bark strips that are wet and then tied. They dry out to become a considerably rigid joint binding.

It was not until the mid 1920s that the concept of using steel pipes fastened together with metal-formed or -cast clamps (couplers) instead of poles and ropes was introduced. This advance lowered erection costs and provided more predictability for the strength of the finished scaffold. Aluminum alloy pipes and couplers were developed for their lighter weight and speedier construction. Aluminum alloy is only two-thirds as strong as steel, but it is only one-third to one-half its weight. Because of higher initial cost, aluminum is restricted mostly to uses in the chemical processing industries where it is preferable because of its corrosion-resistance qualities.

Although pipe was the original basic structural component, thinner-wall structural tubing has become popular in the United States because it has similar strength with lower weight. Although the words "pipe" and "tube" are often used synonymously (especially in Europe), they differ substantially in standard sizes, strengths, construction, and nomenclature. Pipe has fairly thick walls to enable thread cutting at the ends with minimum strength loss. Standard rolled pipes, pressure- or butt-welded, are now generally of ASTM designation A36 steel, approximately comparable to the former A-7 designation. The strength properties of tubes used for scaffolding vary considerably with proprietary manufacturers as covered later.

In the late 1930s and 1940s the concept of welding pieces of pipe and tube into prefabricated "frames" or "panels" enabled simplification of tube and coupler scaffolds by sectionalization. Each frame replaced at least three pieces of tube and at least four couplers, thereby speeding erection times. Some early frames incorporated coupler devices built into them so that they could be joined longitudinally with standard scaffolding pipe runners. In Europe, frames using this principle are still in use. (See Fig. 16.1.) However, in the United States, the construction industry quickly took advantage of the convenience of frames, and by the end of the 1950s frames had become the most common type of scaffolding in use. They evolved to have quick longitudinal attachment of the frames with pivoted diagonal cross bracing made in various lengths to give *fixed* spacings between frames, the most popular being between 5 and 10 ft (1.5 and 3 m). Quick-acting mechanical devices were developed to attach the braces to the frames. Threaded studs with wing nuts progressed to sliding or hinging devices, as shown in Fig. 16.2. Today the original wing-nut type of fastening has been

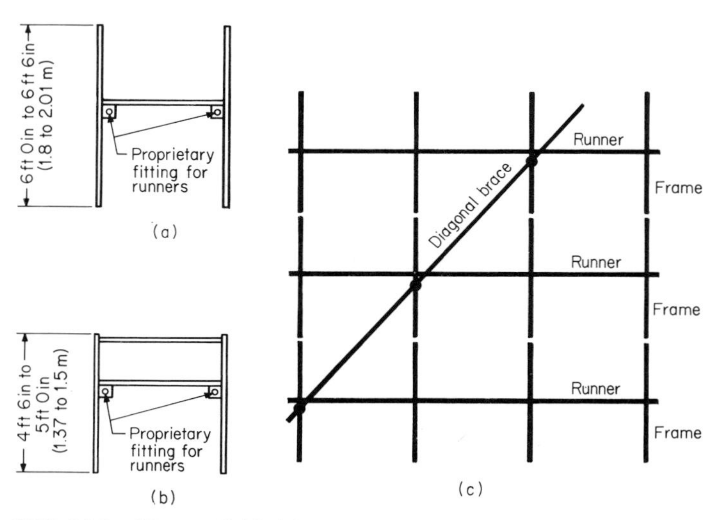

FIG. 16.1 Pipe scaffold: (*a*) section; (*b*) section; and (*c*) side view.

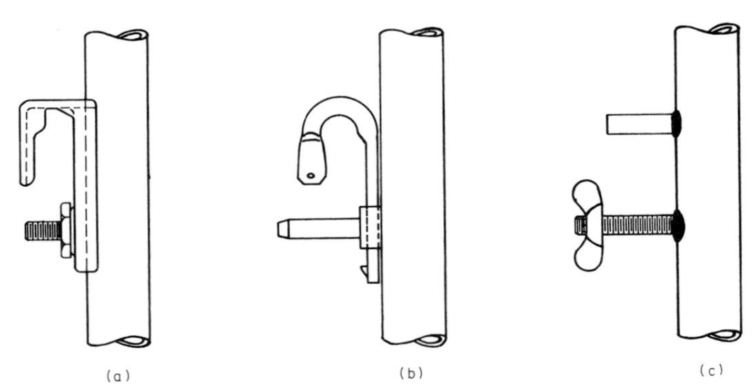

FIG. 16.2 Bracing attachments on sectional frame scaffolds: (*a*) slide lock A; (*b*) slide lock B; and (*c*) stud and wing nut C.

almost completely outmoded by the quicker-acting devices, all resulting in shorter erection and dismantling times and requiring no special wrenches or other tools as are needed for tube and coupler scaffolds.

Continuing sophistication in design of scaffolding components is commonplace, and most manufacturers carry a large line of special-purpose accessories. The two basic types of scaffolding, sectional frames and tube and coupler scaffolds, including small sections equipped with wheels and known as rolling towers, are commonly referred to as "built-up" scaffolds. This is to differentiate between these and other specialized types of scaffolding, such as those consisting of scaffold platforms suspended by cables from overhead structural components and known as "hanging" or "suspended" scaffolds of which there are a number of types. Technical advances in both built-up and hanging scaffold designs now include power-operated components: either air-powered, electromechanical, or electrohydraulic. These "motorized" scaffolds generally allow for fast relocation of the scaffold work platforms and are of two main types:

1. For vertical movement of a suspended scaffold platform
2. For vertical and/or lateral movement of a platform supported from a wheeled base

GENERAL DESIGN CONSIDERATIONS

Commonly, all types of scaffold have incorporated in their designs a minimum safety factor of 4. This criterion is generally stated as: "Scaffolds and their components shall be capable of supporting without failure at least 4 times the maximum intended load." An exception to this is suspension wire rope which, when used to support scaffolds, must have a safety factor of 6.

There are three ways of complying with this requirement:

1. Multiply the design load by 4 and derive the limiting strength of the component from the yield stress of the material in accordance with accepted engineering criteria and practices.

2. Construct samples or prototypes of the proposed design and subject them to test loads equivalent to 4 times the design load applied to the specimens in a manner that simulates actual field usage of assembled components and method of loading.
3. Base a design for a specific purpose on data from applicable results of a similar scaffold configuration obtained from method 2.

By nature almost all scaffolds are relatively indeterminate structures not capable of being analyzed precisely. Approximated results can be mathematically obtained, but to do so many design condition assumptions must be made. There is little assurance that such conditions will exist in the erected scaffold structure because of imprecise erection practices and physical limitations of the equipment used.

Therefore, design procedures generally adopted in this country are based on a conservative and judicious combination of approximate mathematical analysis followed by testing of design prototypes in a realistic manner. It is emphasized that testing of individual scaffolding components can be misleading and has little value unless also incorporated into a scaffold structure which duplicates the conditions and loading under which it is intended to be used.

The safety factor of 4 has been standard practice for decades and is now enforced by OSHA. It is not certain why this particular value was originally chosen, or by whom; it can be rationalized in a variety of ways, three of which are:

1. To provide a "cushion" against imprecise and improper erection and usage. This concept assumes components having known finite strength limitations.
2. To provide the designer with a similar margin to allow for the variations between theoretically designed components and behavioral differences resulting from their deviation from the theoretical due to the actual configuration of the components and methods of erection and use.
3. Maximum intended load (MIL) — 100%

   Plus 100% MIL overload — 100%

   Sudden application of the previous 200% MIL — 200%

   Total — 400%

The third rationalization is more specific than the other two, which are vague. The origin of the value of 4 is not as important as adherence to it, based on testing (as described previously). Theoretical analysis alone may result in a scaffold which can be inadvertently under- or overdesigned. The first is dangerous; the second, uneconomical. Finally the "proof" load of a scaffold if tested properly does not give the full story unless it is also tested to destruction. The gap between the proof-load test and a higher destructive test may be inordinately large. This may indicate overdesign and suggests that more economy in design may be prudent—provided the designer has taken every effort to ensure that the test method is realistic.

It is obviously impractical to physically test a 150- or 200-ft (45- or 60-m)-high scaffold. However, the total behavior of a high scaffold is dependent on the strength and adequacy of its ties to the structure, among other considerations. Therefore, to simulate a high scaffold, the test scaffold should incorporate at least two levels of wall ties vertically and three horizontally, and the MIL × 4 must additionally incorporate load allowance due to the self-weight of the scaffold plus effects of nonstandard loading, wind forces, guy wires, and other bracing forces applied to the scaffold. These latter *forces* can be relatively easily calculated and their resultants applied to the critical scaffold components *or* their calculated effects can be applied to the difference between the proof load and destruction load. Empirically this can be stated as:

Destruction test load − (4 × MIL proof load) = margin for extraneous forces

Figure 16.3 shows a concept of a test configuration based upon the above criteria. Although it is possible to physically load a scaffold sufficiently to produce failure or destruction, it is a very unsafe physical procedure to have to do. Therefore, a compromise is generally used whereby the test loads are applied by a system of equalized hydraulic cylinders and steel wire ropes as shown. "Failure" in

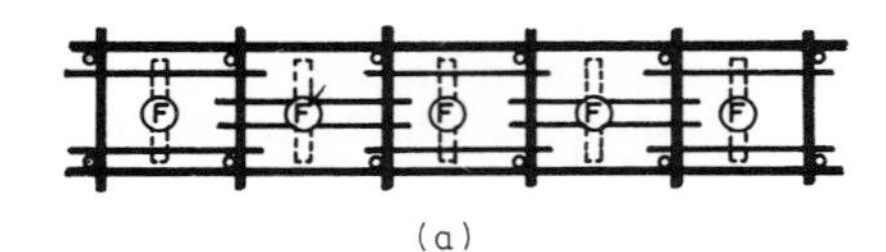

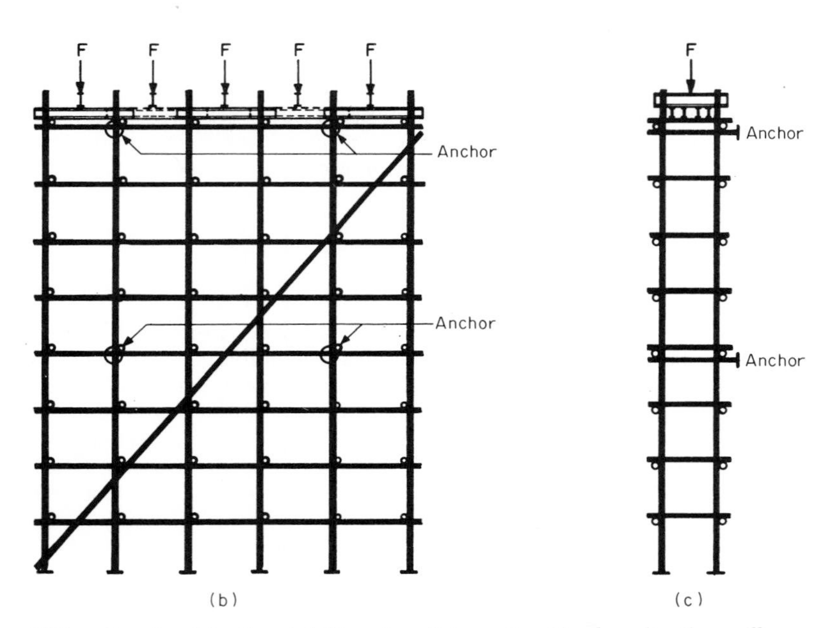

FIG. 16.3 Load testing. (*a*) Plan view. Schematic of load application grillage to simulate uniformly distributed loading. (*b*) Elevation of test scaffold (F = central test load applied in each bay. (*c*) End view.

this process is generally assumed to take place at the load intensity immediately prior to the inability of the tested scaffold to accept any additional loading. It can take place at some point *between* the limit of proportionality (or elasticity) and the physical rupture of one or more of the tested components.

Although use of wood as a structural scaffolding material is very limited, some areas still use this method mostly for low-rise housing work. Prefabricated horses, bricklayers' squares, and other special-use wood components are still used frequently by many builders. Testing of such components should follow the same general principles as for any other type of scaffolding.

CODES OF STANDARD PRACTICE

It is law that all scaffolding be designed, constructed, and used in compliance with the pertinent regulations and standards of the Department of Labor Occupational Safety and Health Administration in the *Federal Register* as follows:

Part 1926, Construction Safety and Health Regulations CFR29, vol. 39, no. 122, part II, October 1, 1979.

Subpart L—Ladders and Scaffolding (1926.450, 451 and 452)

Subpart N—Cranes, Derricks, Hoists, Elevators and Conveyors (1926.552,.553)

Part 1910, Occupational Safety and Health Standards CFR29, vol. 39, no. 125, part II, June 27, 1974 (General Industry Standards).

Subpart D—Walking-working Surfaces (1910.21, .28, .29)

Subpart F—Powered Platforms, Manlifts and Vehicle-Mounted Work Platforms (1910.66, .67)

Both regulations and standards are required to be applied in conjunction with all other, applicable provisions of their other accompanying subparts containing various general provisions such as those for personal protective equipment and fire protection. Some general standards are also applied to construction work (1910.29).

A number of states have their own federal OSHA approved regulations such as CAL/OSHA in California and WA/OSHA in Washington. In the event of conflict between federal and state OSHA regulations, the more (apparently) stringent and/or more specific provisions prevail. Some states have dual-jurisdiction status with both federal and state inspection machinery in effect.

One major item of nonuniformity is the height of guardrailing: federal OSHA construction regulations [subpart 1926.451 (a)(4)] call for guardrails on scaffolds more than 10 ft (3 m) high to be approximately 42 in (107 cm) high. [If the scaffold platform is less than 45 in (114 cm) wide in any direction, guardrails are required at 4-ft (120-cm) platform height.] Federal OSHA standards [subpart 1910.28 (a)(3)] require guardrails to be installed for platform heights of more than 10 ft (3 m)—not less than 36 in (91 cm) or not more than 42 in (107 cm) high—unless any open side of the scaffold is within a wholly enclosed area and "not having any side exposed to . . . openings." California requires guardrails to be 42 to 45 in (107 to 114 cm) high. There are no finite guidelines available; what is acceptable will depend upon local interpretation or common-sense. Obviously, a high guardrail is relatively safer for a tall person, whereas 36 in (91 cm) is probably quite safe for those of short to average height. It is logical, therefore, that a reasonably safe guardrail height above platform level would be between 36 and 45 in (91 and 114 cm).

OSHA parts 1910 *(General Industry)* and 1926 *(Construction)* were both reprinted and reissued February 9, 1979, with minor changes (vol. 44, no. 29). The following statement was added:

> The following requirements from 29CFR Part 1910 have been identified as applicable to construction (29CFR 1926.451 Scaffolding), in accordance with their respective scope and definitions.

These additions are as follows:

> 1910.28(a)(15) Materials being hoisted onto a scaffold shall have a tagline.
>
> (a)(18) Employees shall not work on scaffolds during storms or high winds.
>
> (a)(20) Tools, materials and debris shall not be allowed to accumulate in quantities to cause a hazard.

It also adds that the whole of "Subpart 1910.29, Manually propelled mobile ladder stands and scaffolds," now be incorporated into the 1926 construction regulations. This section deals with special-purpose rolling ladders and scaffolds used primarily in industry but sometimes also used in the finishing construction stages. Some comments on the regulations follow:

1. **Wire Mesh:** These are required to be installed between toeboards and guardrails "where persons are required to work or pass under the scaffold." An easy solution is to provide a ground barrier to prevent passage of persons close to the scaffold. Although ½-in (12-mm) wire mesh is specified, tarpaulins, plywood, or other suitable materials are generally acceptable.
2. **Planking:** Nominal planking [(i.e., dressed, 1½ in (3.8 cm) thick] is not recommended for heavy-duty [75 lb/ft^2 (3590 N/m^2)] use. This is an anachronistic holdout from the original ANSI A10.8-1969 which was published when spruce was the most popular wood used, and was then often rated with a fiber stress in bending of 1500 lb/in^2 (1034 N/cm^2). Shortage of scaffold plank grade spruce in the last 15 to 25 years has prompted most of the construction industry to switch to Douglas fir and southern pine which are stronger, denser, and more expensive. As discussed in the wood planking section, dressed Douglas fir and southern pine are quite capable of safely handling heavy-duty loading of 75 lb/ft^2 (3590 N/m^2), uniformly distributed over spans up to 10 ft (3 m).
3. **Wire Ropes:** These are to have a minimum safety factor of 6 when used for suspending scaffolds, instead of the normal 4 for all other materials.
4. **Open Sides and Ends:** (These are required to have guardrails, midrails, and toeboards at outsides of all scaffolding platforms.) A definition for this

term is not given (although self-explanatory applied to the outside perimeters of a scaffold platform). An editorial recommendation of such a definition is:

An open side of a scaffold platform shall be considered to exist on an internal side when one or more of the following situations exist:

- **a.** There is a distance of more than 12 in (30 cm) from the inside platform edge to the working surface of the wall or structure.
- **b.** The working surface of the wall or structure has multiple or continuous openings into which a man could fall to an inside building floor (such as exist when a building is only partially closed in).
- **c.** When a scaffold extends past the end of a structure and is not joined or connected to another working platform at the same level.

5. **Controversial Provision:** Under the construction regulations, subpart 1926.451(e), persons are allowed to ride on moving, manually propelled mobile scaffolds (rolling towers) under certain listed conditions even though industry groups and manufacturers are almost unanimous that this is an unsafe practice under any conditions regardless of the precautions enumerated in 1926.451(e). There is no provision for, nor warning against, this practice in subpart 1910.29 now applicable to construction operations.

It is important to emphasize that compliance with OSHA regulations is necessary by law. It is impossible to provide specific circumstances which lead to scaffold-related accidents in the field; only the broadest caution—using only safe procedures and being alert at all times—can prevent accidents. Therefore, it is recommended that appropriate ANSI standards for type and use of equipment always be followed whether or not their current versions are promulgated into OSHA regulations. Further, almost all scaffolding manufacturers publish and make freely available safety rules and instructions for their various products. Following these publications can further promote safety and reduce hazards. An industry association which publishes safety rules and other free safety-enhancement information is Scaffolding, Shoring, and Forming Institute, Inc., 1230 Keith Building, Cleveland, Ohio 44115. The National Safety Association in Chicago also has valuable information available. There are other sources in many states from which safety data are available, and there are a growing number of safety-concerned industrial corporations, such as E. I. duPont de Nemours and Consolidated Edison, that have important safety material available at nominal charges.

A list of scaffold-related ANSI standard publications follows:

A10.4 —1975 *Personnel Hoists*
A10.5 —1975 *Material Hoists*
A10.8 —1977 *Scaffolding*
A10.11—1971 *Safety Nets*
A10.13—1978 *Steel Erection*
A10.14—1975 *Safety Belts and Lifelines*
A10.18—1977 *Temporary Floor and Wall Openings, etc.*
A10.22—1977 *Workmens' Hoists, Rope-Guided and Non-Guided*

A120.1—1970	*Powered Platforms*
A92.1—1977	*Rolling Scaffolds, Ladder Stands*
A92.2—1979	*Vehicle Mounted Elevating and Rotating Scaffolds*
A14.1—1982	*Portable Wood Ladders*
A14.2—1982	*Portable Metal Ladders*
A14.4—1982	*Job Made Ladders*
A14.5—1982	*Portable Reinforced Plastic Ladders*

A number of judgments in product liability lawsuits have established that over and above regulatory compliance, *voluntary* compliance with national concensus codes such as ANSI is a positive defense and indicative of maintaining pace with current state-of-the-art manufacturing practices.

DESIGN LOADS

In accordance with OSHA and ANSI criteria and common practice for many years, arbitrary design load ratings for *scaffold platforms* are as follows:

Light-Duty Loading: 25 lb/ft^2 (1200 N/m^2) maximum working load for support of people and tools (no equipment or material storage on the platform)

Medium-Duty Loading: 50 lb/ft^2 (2400 N/m^2) maximum working load for people and material restricted not to exceed this rating, often described as applying to bricklayers' and plasterers' work, but not confined thereto

Heavy-Duty Loading: 75 lb/ft^2 (3600 N/m^2) maximum working load for people and stored material often described as applying to stone masonry work, but not restricted thereto

These ratings assume uniform load distribution. They are not well defined and are difficult for field personnel to interpret into practical applications. With the exception of the weight of stored materials (which in themselves are transitory because they either decrease due to dispersion along the scaffold and use in the building construction or occasionally increase as in scaffolds used to store materials from a building being demolished) scaffold loads most often consist of personnel, both stationary and transitory. Also, with few exceptions, scaffold loads are relatively concentrated loads: a strapped bundle or a pallet of bricks weighing up to 2000 lb (900 kg) is distributed between only one and three adjacent planks and then only for the width of the bundle or pallet. Heavier-density materials, such as stone or precast sections, are distributed over even smaller concentrated areas.

It is important to remember that the OSHA and ANSI load-rating system is intended for guidance of field personnel in the construction and use of nonspecifically engineered scaffolding applications. Exceptions are a few types of specialized and suspended scaffolds which *are* specifically required to be designed to carry stated loads or load ratings. The regulations and standards do not prohibit design and use of any scaffolding to carry any desired maximum load whatsoever, and the scaffold designer is not constrained to follow these load ratings for applications where there is control over the design, configuration and, most importantly, use of the scaffold.

Scaffolding layout drawings are necessary, of course, to establish the applicable load-use-spacing parameters. For tube and coupler scaffolds which are designed and built to other than these load ratings, OSHA requires that they be designed by a "qualified engineer" competent in this field. Sectional and tube and coupler built-up scaffolds over 125 ft (38 m) in height must be designed by a registered professional engineer.

There are three limiting factors which determine the frequency of the scaffold support numbers under a platform:

1. The allowable load-span relationship of the planking
2. The safe carrying capacity of the platform supports (bearers)
3. The type of platform required for the work—fixed relocatable or multilevel

A single fixed, high-level platform scaffold (as required for such alteration work as cornice removal and limited by plank span) could be built with the planking supported at more frequent intervals than the spacing of the vertical members, or a double level of planking could be installed. To some extent it is not logical to construct a closely spaced scaffold structure merely to give frequent support to the weakest member (planking) when the spacing of the members can be increased to more closely approach the safe allowable load limitations of the scaffold. As an example, the components for a 100-ft (30-m)-high scaffold spaced longitudinally at 7 ft (2.1 m) because of plank strength limitations could be reduced by 25% if they are spaced at 10 ft (3 m) and the planking is suitably reinforced. The equipment cost savings may not justify a special design, but the labor cost savings involved in erecting and dismantling the equipment certainly should merit serious thought. Figure 16.4*a* to *d* illustrates the respective configurations of this example. It must be remembered that scaffold labor becomes very slow at high heights because of slower material handling, and all scaffolds must be dismantled as well as erected. While scaffolds disassemble faster than they assemble, the material lifting and lowering times are approximately the same. For a scaffold 400 ft (122 m) long, labor cost savings of thousands of dollars could be achieved.

TUBE AND COUPLER SCAFFOLDS

Tube and coupler scaffolds are assembled from three basic structural elements: the uprights, or posts, which rise from the ground or other solid support; the bearers, which support the working platforms and/or provide transverse horizontal connections between the posts; and the runners, which attach to the posts directly below the bearers and provide longitudinal connections along the length of the scaffold. These three elements are usually connected with standard or fixed couplers which provide a 90° connection in two places. (The word "standard" derives from the European terminology of naming the upright posts as

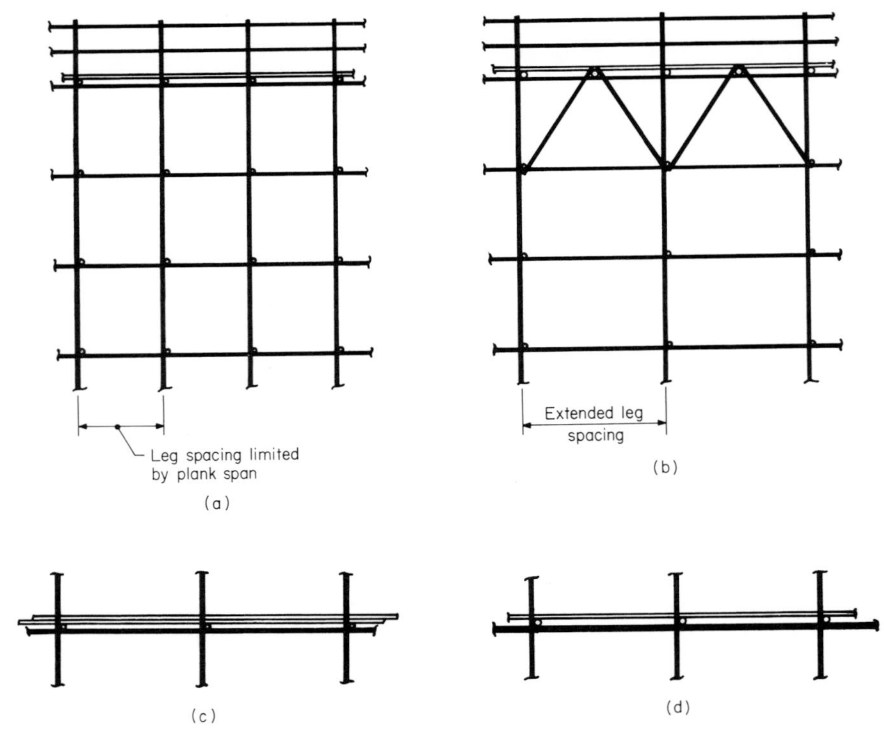

FIG. 16.4 Scaffold leg spacings. (*a*) Spacing limited by allowable span of planks. (*b*) Additional bearer in center of bay, runner reinforced with diagonal bracing. (*c*) Use of double level of planking. (*d*) Use of 2½-in (6.45-cm)-OD bearers instead of 2-in (5.08-cm)-OD bearers.

standards.) These three elements form the basic structure shown in Fig. 16.5*a* and are repeated in the horizontal and vertical planes to build the scaffold to its desired size and egg-crate type configuration with the components and fittings shown in Fig. 16.5*b* to *g*.

Diagonal bracing is used to stiffen the structure as necessary—most importantly in the longitudinal direction. Bracing is generally connected to the posts with "adjustable" or "swivel" couplers which have the facility of adjusting a full 360° so that two members in parallel planes are connectable at any angle. Diagonal bracing should always be attached to the posts as closely as practical to the "node" points formed by the runner-bearer connections. The use of lateral diagonal bracing in the transverse direction (i.e., in the plane of the bearers) is used extensively in Europe but to a much lesser extent in the United States, as discussed later.

Another important structural element is the building tie which connects the scaffold to the wall or structure and is needed to provide rigidity and anchorage of the scaffold in the transverse direction. Fig. 16.6 shows a number of methods

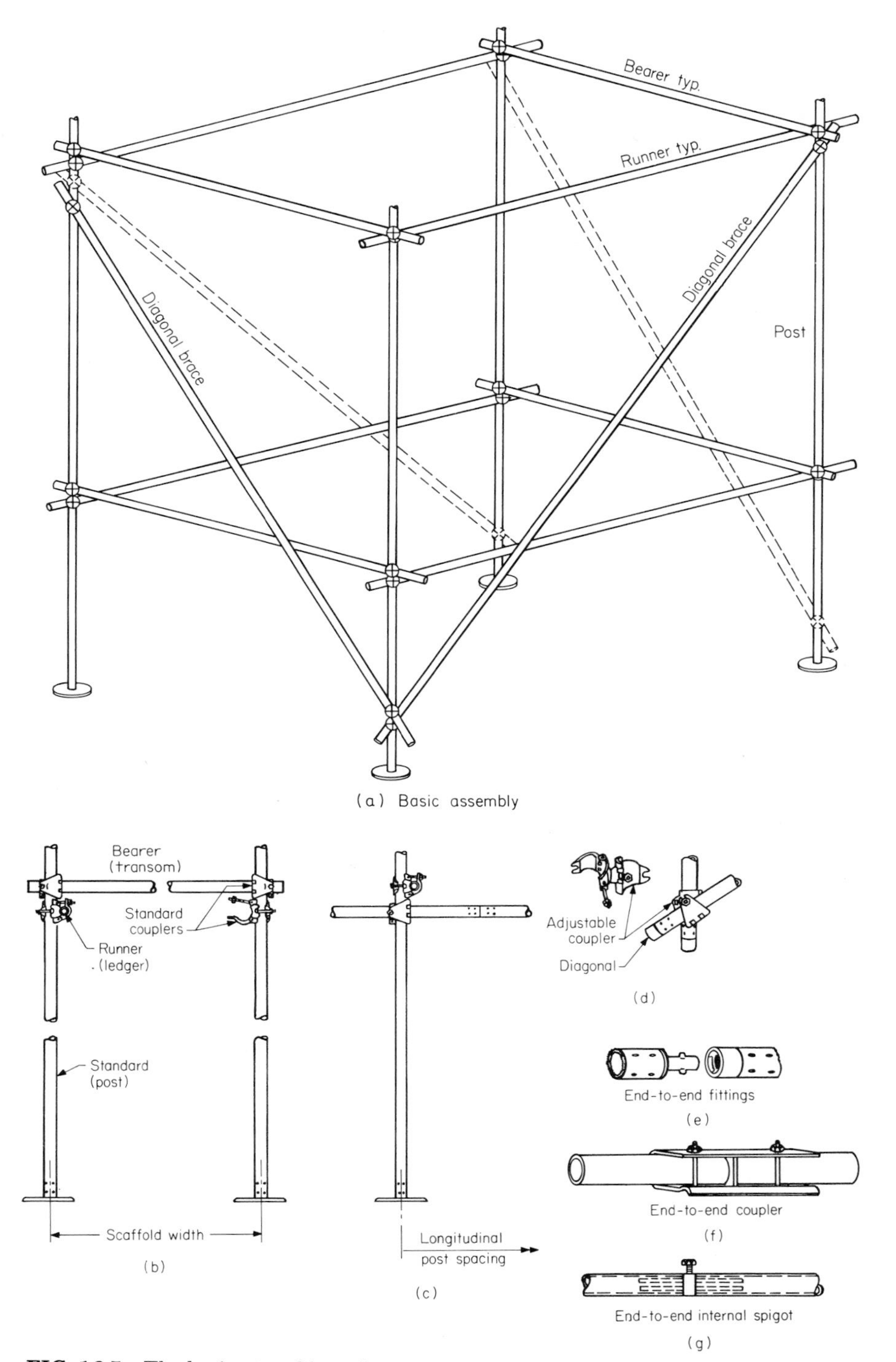

FIG. 16.5 The basic assembly and components of tube and coupler scaffolds.

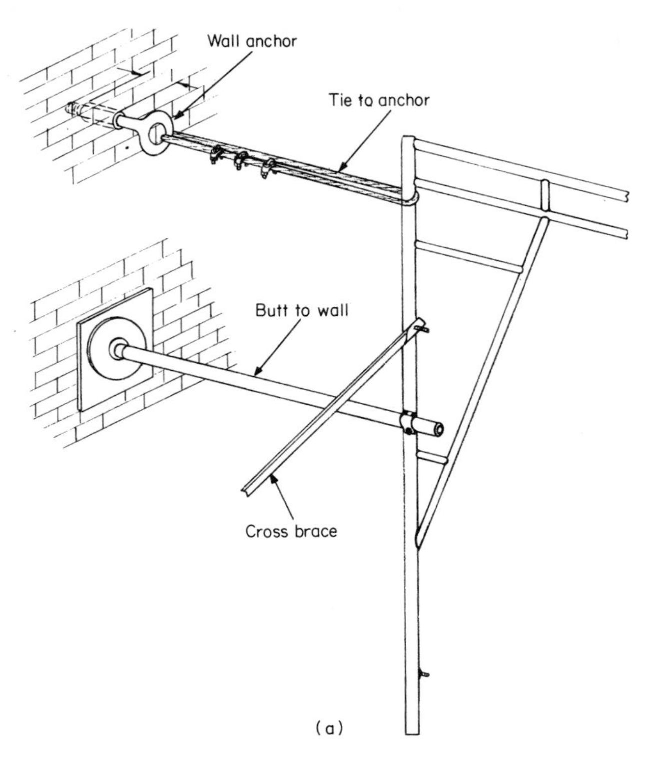
Wall anchor
Tie to anchor
Butt to wall
Cross brace

(a)

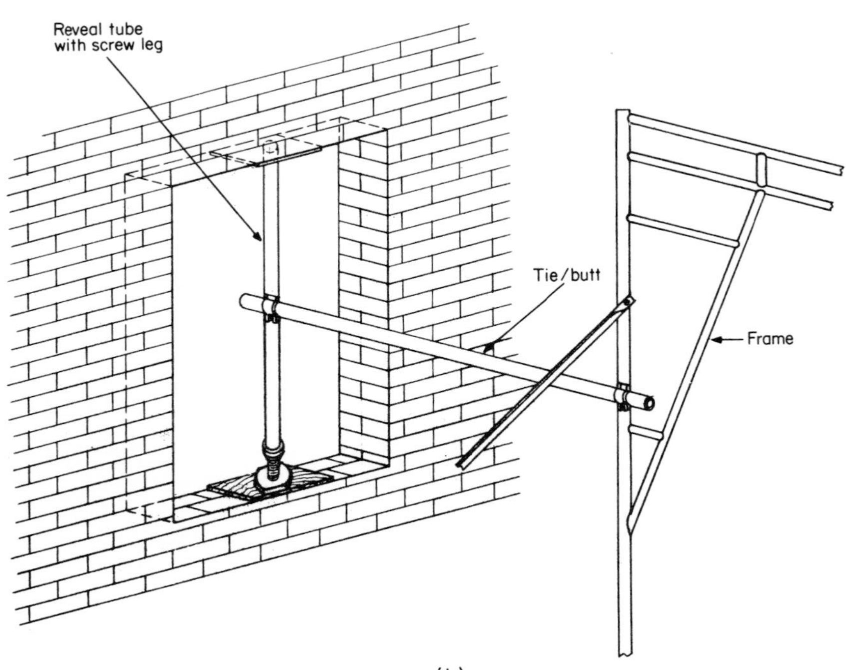
Reveal tube
with screw leg
Tie/butt
Frame

(b)

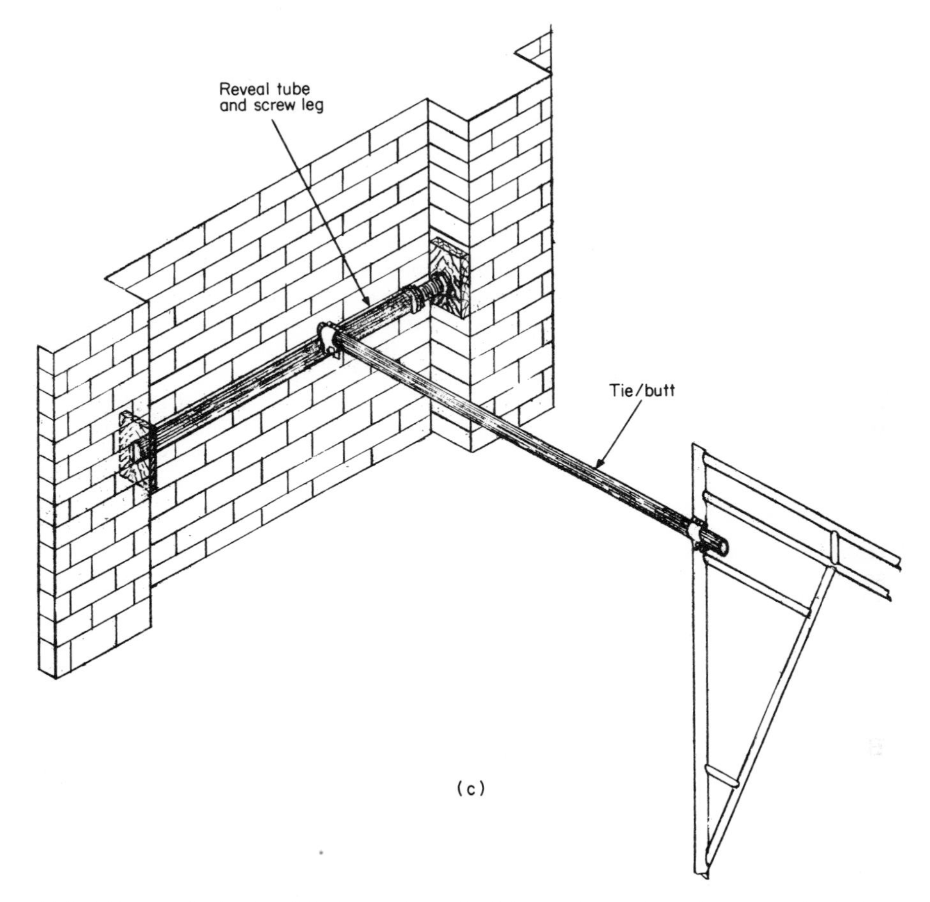

FIG. 16.6 Methods of stabilizing against a building: (*a*) wall tie anchorages; (*b*) window reveal tube; and (*c*) reveal between pilasters.

of accomplishing this. The scaffold in the transverse (narrow) direction forms a repeated series of braced, columnlike frames connected by runners along the length of the scaffold; these column frames or bents need to be laterally supported; otherwise they are unstable because of their height-to-width ratio and have low strength to resist wind and other lateral forces.

Application

Tube and coupler scaffolds can be assembled in numerous ways because of the flexibility of their assembly dimensions in the horizontal and vertical planes (within their structural limitations). Unlike sectional frame scaffolds they are not restricted by frame width in the transverse direction, by brace length in the longitudinal direction, or by frame height in the vertical direction. Consequently, they are preferred for access to work places having irregular dimensions and contours, e.g., churches, old auditoriums, and chemical processing structures.

Additionally, the chemical and petrochemical industries often have structures or vessels with multiple protrusions, such as walking platforms, curved stairs, or piping, which tend to make construction of scaffolds of sectional frames difficult and hazardous. Such protrusions are often added *after* the scaffold is in place and often pass *through* the scaffold; frequently, cutting sectional frames is required for removal. This is a wasteful process obviated by using tubes and couplers. Church interiors and exteriors and other old buildings with flying buttresses, multiple decorative corbeils, and cornices are similarly more safely and easily scaffolded with tubes and couplers. Scaffolding with tubes and couplers is the work of experienced specialists since it is not always possible to pre-plan a scaffold design for a church exterior (for instance) since accurate drawings of the structure are not always available and dimensions of embellishments, spires, and other obstacles frequently cannot be ascertained prior to the erection and must be done "on the job" as these various obstacles are reached and become measurable.

Basic Configurations

The basic configurations are as follows:

1. **Double Pole:** Also called "independent" wall scaffolds, these are used for access to vertical surfaces for construction, alteration, or surface finishing and repair. They consist of repetitive *pairs* of posts along the length, connected by bearers and runners.
2. **Single Pole:** Also called "putlog" wall scaffolds, these are used for construction of masonry walls. They consist of *single* posts 3 to 5 ft (90 to 150 cm) away from the wall surface spaced at regular or varying intervals along the wall. The different feature of this type of scaffold is that the inside ends of the bearers are supported at joints or courses in the wall being built instead of by inside posts. Flat plates are often attached to the bearers to seat in the mortar joints; when used this way, the bearers are called "putlogs." The putlogs are removed from the mortar courses and the holes thus left are grouted as the scaffold is dismantled. This method has little use nowadays because it is slow and cumbersome compared with other available methods, but is mentioned because it is included in most regulations and codes. Originally, it was devised as an inexpensive means of providing scaffold access for a very few workers who constantly relocate around the perimeter of the building.
3. **Tower Scaffolds:** These consist of one or a few bays in either horizontal plane, constructed to a required height for access to ceilings or for specialized load support requirements not conveniently achievable with sectional frames. They may be mounted on casters and become mobile scaffolds or *rolling towers* as shown in Fig. 16.7. A specialized use of tower scaffolds is to provide stair access to unusual structures such as cooling towers. The posts can be angled and offset so as to conform reasonably closely to the curved outer surface. Such applications as this exist when it is not practical or feasible to use a mechanized personnel elevator.

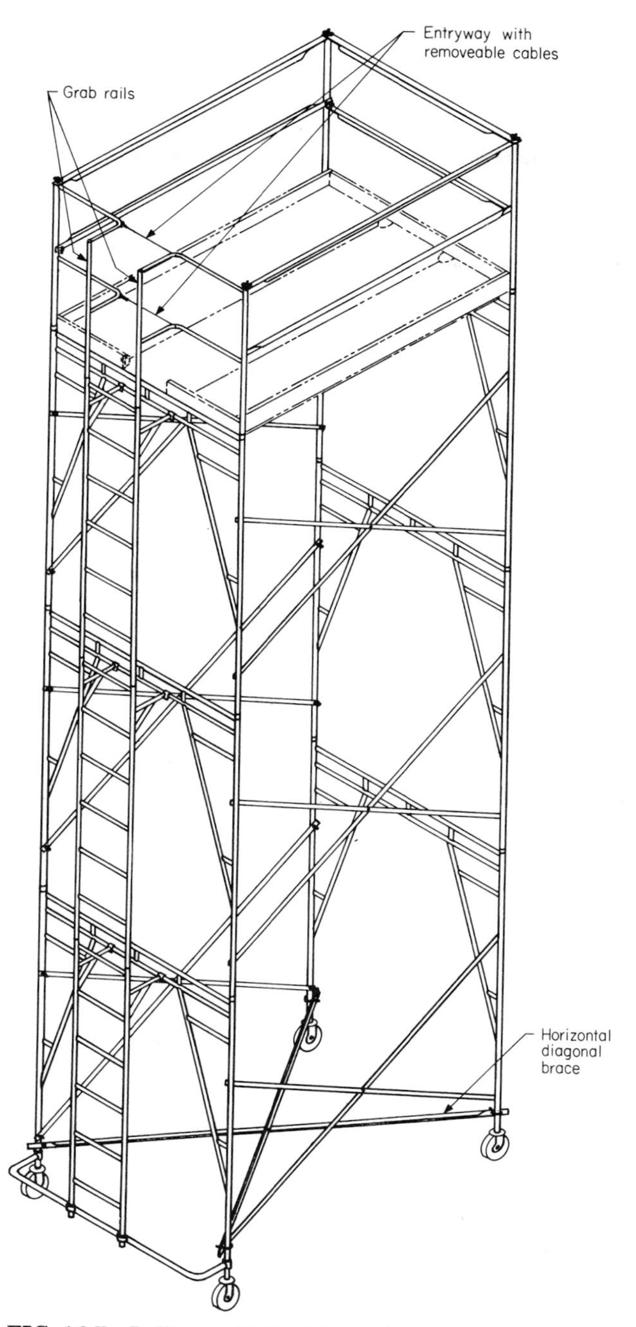

FIG. 16.7 Rolling scaffold with stand-off ladder.

4. **Material Platforms:** These are used to wheel building materials from a material hoist to set-back elevations of a building. Tube and coupler scaffolds are often more suitable for this purpose than sectional frames because of width and height restrictions of the frames. These are not in common use any longer since building designs are now made so as to lessen the incidence of space-wasting setbacks; however, renovations to existing set-back buildings still require such scaffolds.
5. **Wide Area Scaffolds:** These are used for building interiors for construction and renovation of walls and ceilings. Because of their flexibility in dimensional spacings, tube and coupler scaffolds are easily adaptable to irregular wall and column shapes, domed or sloping ceilings, sloping floors, positioning between seat rows, etc. While modern church and auditorium interiors tend to be simple and plain-surfaced and easily scaffolded with sectional frames, those with special acoustic-dependent surfaces and older buildings needing renovation are usually scaffolded better with tubes and couplers.

Tube and Coupler Materials

The most popular size pipe or tube used throughout the world is approximately 2 in (50 mm) OD. Standard sized pipe and its structural tubing equivalent are compared in Table 16.1. *Pipe* is generally classified by its *internal* diameter: ID in the United States and NB (nominal bore) in Europe. Both measure approximately 1²⁹⁄₃₂ in (48 mm) OD.

Tubing is generally classified by its *outside* diameter, which for scaffolding purposes is manufactured to 1²⁹⁄₃₂ in (48 mm) OD and often designated as "pipe-size" tubing. Although there is also exact 2-in (50.8-mm)-OD tubing available, the pipe-size tubing is often referred to as 2-in (50-mm) nominal OD pipe or tube.

European pipe of 1½ in (38 mm) NB conveniently equals 48 mm OD in metric measurement. It is still common European practice to refer to pipe as tubing.

The minor differences in outside diameter, whatever the pipe or tubing used, are of no importance to external couplers. However, *internal* couplers or joint pins do require that they be manufactured with proper sizing for joining pipes or tubes end to end.

In view of the generic use of sizing and name terminology, it is suggested that any specification qualify the given size by stating ID, OD, or nominal as well as the type of construction.

Example

1.9 in standard pipe

1½ in standard pipe

2 in nominal OD pipe

generally mean the same item

Their proper description in the United States is: 1.5-in nominal ID standard, black (or galvanized) steel construction pipe conforming to ASTM designation A-36 and to ASTM specification A501-71a and with A53-7. For convenience purposes in the construction industry this is generally referred to as 2-in nominal

TABLE 16.1 Comparisons between U.S. Structural Tubing, Pipe, and British Pipe Commonly Used for Tube and Coupler Scaffolding

	Welded and seamless steel tubes*	Cold-rolled close jointed steel tubes*	U.S. 1½ nominal standard pipe†	U.S. 1.9-in-OD structural steel tube†
Specification	B.S.1139	B.S. 1139	ASTM A-120 A-53 A-36	High carbon structural tubing
Ultimate tensile, kips/in²	49–67	49–67	58–80	75 min.
Minimum yield, kips/in²	30	30	36	50 min.
Minimum elongation, %	23–32	23–32	23 min.	23 min.
Nominal OD, in	2	2	2	2
Actual OD, in	1.906	1.906	1.90	1.90
Wall thickness, in	0.160	0.176	0.145	0.097
Moment of inertia, in⁴	0.337	0.362	0.309	0.225
Section modulus, in³	0.354	0.380	0.326	0.237
Area, in²	0.878	0.957	0.799	0.552
Radius of gyration, in	0.620	0.615	0.622	0.638
Weight per foot, lbs	2.98	3.25	2.72	1.95

*The word "tubes" here is British nomenclature.

†The U.S. manufactured pipe and tube specified here are the most common ones. There are others used, especially tubing which can vary from ASTM A-1020 to 1050; some scaffolding manufacturers use proprietary grades of steel available only from certain rolling mills. As such, they are not necessarily referenced by any ASTM specification or other standards.

NOTE: It is of utmost importance that the scaffold designer be in possession of all the above necessary data for the pipe or tube intended to be used.

1 kip/in² = 6.89 MN/m²; 1 in = 2.54 cm; 1 in⁴ = 41.6 cm⁴; 1 in³ = 0.000016 m³.

OD pipe; under the old ASA Schedule 40, which is still used, it has a standard wall thickness of 0.145 in.

Description of tubing should always be accompanied by the wall thickness, i.e., 1.9-in OD × 0.097-in wall or 2-in nominal OD × 0.097-in wall. Table 16.1 shows the popular pipe and tube sizes in general use. Larger sizes of pipe and tube are also used for special scaffolding purposes such as 2½-in (63.5-mm) nominal OD (2-in ID standard pipe, or 2⅜-in or 2½-in OD tube) and 3½-in (89-mm) nominal OD (3-in ID standard, extra-heavy or double-extra-heavy pipe) for fully enclosed hoist towers and 4½-in (114-mm) nominal OD (4-in ID pipe) for sidewalk bridging columns. Wall thicknesses vary with manufacturer and application.

European scaffolding practices vary tremendously from those used in the United States; only U.S. manufactured material, U.S. standard practice, ANSI Standard, and OSHA regulations are considered here (except in comparisons). The scaffold user should be aware that British-manufactured scaffolding, plus materials from many other European countries such as France, Germany, Spain,

Italy, Holland, Lebanon, and the Scandinavian countries, are currently sold in the United States and Canada. It is obviously impractical to cover every type.

Couplers (Fittings)

There is a substantial difference between coupler design and behavior. Most couplers used by the U.S. construction industry are of a rigid cast or have a forged massive coupler body, whereas most European couplers are more flexible and made of pressed plate, forged bodies of plain or spring steel (bolt-actuated and wedge-actuated designs). With typically European load testing methods, European couplers permit distortion of up to 10°. The most popular U.S. couplers have typical distortion figures of less than 1° under the same test methods.

Therefore, there is a profound difference in scaffolding design philosophy and standard practices between Europe and the United States. The far more rigid U.S. coupler provides such a higher degree of moment rigidity between each vertical and/or horizontal joint in a completed scaffold structure that, provided the necessary wall ties and butts are installed, the scaffold is solid and rigid with a *minimum* of diagonal bracing. The lengths of some coupler bodies along their connected tubes are as much as 4 in (10 cm). On the other hand, the more flexible European couplers typically have small bodies, between 1¼ and 2 in (3 and 5 cm); consequently, their scaffold *structures* require a much higher amount of diagonal bracing to stabilize and impart lateral rigidity to the scaffold.

Figures 16.8 and 16.9 show typical U.S. and British constructions of wall scaffolds. A comparison of nomenclature of scaffold members is shown in Table 16.2.

Table 16.3 shows the OSHA and ANSI criteria developed out to final leg loads at the base using weights of a typical proprietary manufacturer of tube and coupler equipment. Since the maximum load on a structural upright at the base of the scaffold is dependent upon applied uniform loading per square foot of platform area(s) supported, scaffolds narrower than these stated maxima will have lesser leg loads for any given duty rating of scaffolds. Narrow scaffolds—2, 3, and

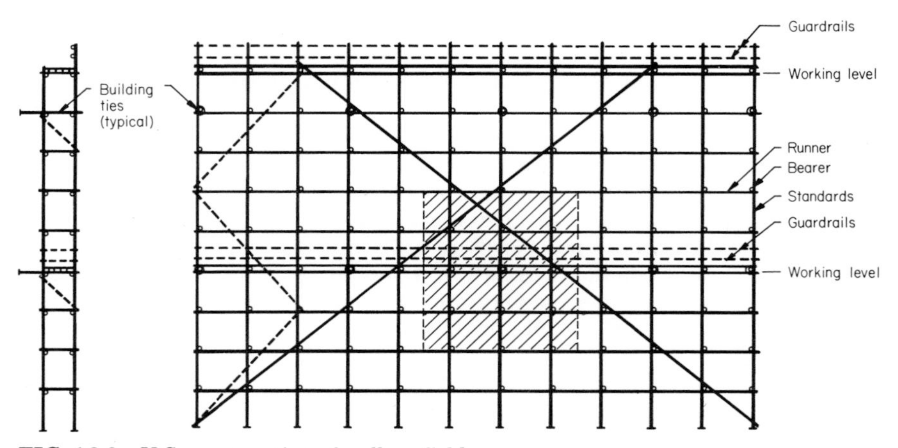

FIG. 16.8 U.S. construction of wall scaffold.

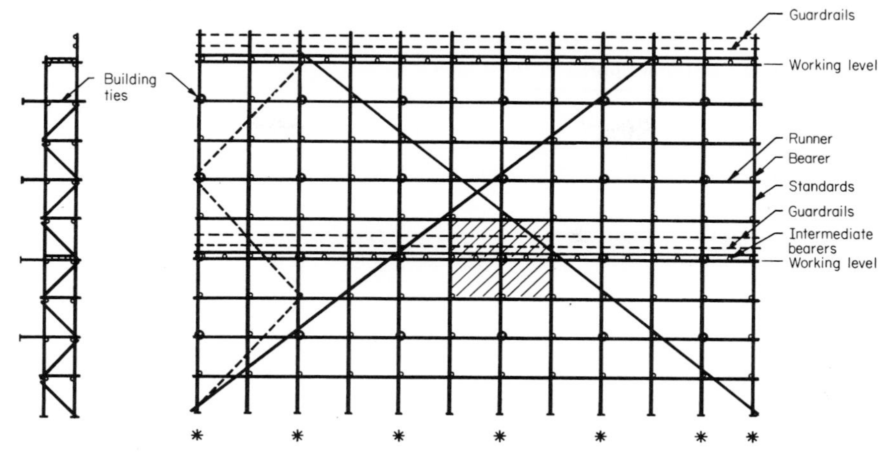

FIG. 16.9 British construction of wall scaffold.

4 ft (60, 90, and 120 cm) wide—are particularly common in the petrochemical and other plant processing industries, especially in the western states, for access to wall repairs and finishing, such as for stuccoing, painting and insulating.

It is interesting to note that in Table 16.3 the derived post loading at the bottom of the scaffold is illogical. The developed loads show the heavy-duty scaffold to be the most lightly loaded, the medium-duty sizes more heavily loaded, and the light-duty sizes the most highly loaded. Obviously this is the direct result of specifying the bearer spacings in accordance with plank loading limitations. As stated earlier in this chapter in the Design Load section, the user of a high scaffold with only one platform at the top to remove a stone cornice, which is normally assumed to be heavy-duty work, must erect a very inefficiently loaded

TABLE 16.2 Tube and Coupler Scaffolding Terms

Item	U.S. nomenclature	U.K. nomenclature
Tube connection	Coupler	Fitting
Structural members	Tube or pipe	Tube
Vertical supports	Posts (uprights, poles, legs)	Standards
Lateral pairs of posts	Bent	Frame
Spacing between bents	Bay	Bay
Longitudinal members	Runners	Ledgers
Plank support members	Bearers	Transoms
Vertical structural intervals	Lifts	Lifts
Diagonal bracing	Same	Same
Platform members	Planks	Boards

TABLE 16.3 Development of OSHA and ANSI Design Data for U.S. Tube and Coupler Scaffolds

Scaffold size 180 ft approx. × 70 ft high

Live load and post spacing	ANSI A 10.8—1977				OSHA 1926.451 (c)			
	WL[a]	PL[b]	Ht,[c] ft	Post load,[d] lb	WL[a]	PL[b]	Ht,[c] ft	Post load,[d] lb
25 lb/ft² (light duty) 6 ft × 10 ft	1	16	125	5187	1	8	125	3563
	2	11	125	5125	2	4	125	3704
	3	6	125	5063	3	—	91	3586
50 lb/ft² (medium duty)	1	11	125	4223	1	6	125	3402
6 ft × 8 ft[e]	2	1	125	3945	2	—	78	3393
5 ft × 8 ft	1	11	125	3633	1	6	125	2912
	2	1	125	3335	2	—	78	2858
75 lb/ft² (heavy duty) 6 ft × 6 ft[e]	1	6	125	3197	1	6	125	3398

[a]WL = number of working levels.
[b]PL = number of additional planked levels.
[c]HT = maximum height of scaffold unless designed by a qualified engineer compenent in this field.
[d]The post loads shown are developed from the dimensions and loadings specified. These values are *now allowable loads* for any make of equipment. Allowable loads must always be obtained from equipment manufacturers and the scaffold erected in accordance with their instructions.
[e]For 6-ft widths of medium- and heavy-duty scaffolds, bearers are 2½ in nom. OD. All other components of all scaffolds are 2 in nom. OD.
NOTE: 1 ft = 0.305 m; 1 lb = 4.45 N; 1 lb/ft² = 47.9 N/m²; 1 in = 2.54 cm.

scaffold to comply with the OSHA and ANSI regulations, that is, a 6-ft (1.8-m) longitudinal spacing of the bearers and posts. In the Wood Planking section of this chapter it is shown that there are a number of species of wood planking capable of safely supporting a 75-lb/ft² (3590-N/m²) live load over as much as a 10-ft (3-m) span. For heavier loads, a *double* level of planking at the working level would allow spanning of 10 ft (3 m) with ease. Another practical expedient would be to use a center-support bearer at the mid-span of each bay (as used in the United Kingdom), but it is believed that the point-load reactions at the center span of the *runners* may require them to be reinforced. To avoid overloading the 2½-in (63-mm) nominal OD bearers, there is probably no need for a 6-ft (1.8-m)-wide scaffold and, if only one working level is required, there is also no need for 2½-in (63-mm) nominal bearers except to support the work platform; all other bearers below the work platform could be 2 in (50 mm) nominal OD.

Figure 16.4*a* to *c* shows a number of methods by which such a special-purpose scaffold could be designed utilizing some of the foregoing alternatives and still erecting a safe, but more economically efficient, structure.

The major point to be observed from this example is that to an inspector it would not comply with OSHA regulations and, therefore, the design should be properly engineered with an engineer's stamped layout available. Possibly a local "approval" may need to be obtained depending on locale. It is an economy-based choice as to whether it is more advantageous to go to the expense of such a special design (with on-the-job savings) or to follow the layout dimensions per the regulations.

In recent years a number of couplers with pressed steel bodies or welded, shaped bars activated by pressed or stamped wedges instead of bolts have been introduced into the United States and Canada from Europe. This type of coupler precludes meaningful time-based comparison with bolt-activated couplers and their tightening efficiency, and continued holding power is questionable. However, they are quite cheap in comparison with most bolt-activated couplers, and there are always markets for such products. The hazard of accidental dislodgement of the wedges from a blow or vibration is a real one, but if the scaffolds are constructed using European assembly criteria by personnel experienced in their use, they probably are satisfactory. It is doubtful, however, that they would make as rigid a scaffold as the U.S. type of equipment when erected in accordance with OSHA and ANSI criteria and general practice in the United States.

Design Loads for Tube and Coupler Members

Because of the wide range of steel grades and manufacture available, finite loading figures are not given here. Specific information should be obtained from the manufacturer. The manufacturer's information should be based upon test results and not on theoretical calculations alone.

Figure 16.10 shows four duty-rated tube and coupler scaffolds. It is seen in the illustrations that the bottom post is usually capable of both translation *and* rotation in a light-duty scaffold, whereas the bottom level of bearers and runners can counteract both rotation and translation *provided* the couplers have high rigidity. The effects of loads in the transverse and longitudinal diagonal bracing are transferred to the posts *eccentrically;* distortion of a bearer under load will induce some bending forces in the post *in addition* to the effect of eccentric load application. If it were mathematically feasible to take all these engineering conditions into account as applying simultaneously to a post in a given set of circumstances, i.e., a totally "fail-safe" design, it would generally be concluded that the safe allowable load would be so small as to be uneconomical—which of course is not the fact.

In reviewing all the contributing factors, some generalities can be stated. The most heavily loaded post, or leg, will be at the base of a scaffold or that section of a post immediately below a *loaded* working platform within a few tiers above the base. In the former, the loads can be considered to be applied *concentrically* through the posts and contain working loads plus plank loads from nonworked platforms plus the self-weight of the scaffolding materials for the full height of the scaffold. In the latter, the tier immediately *below* the loaded working platform is subject to eccentric application of that platform load plus concentric application of other higher platform loads and scaffold self-weight.

In the final analysis, calculation of a pair of posts as a braced column system or as a type of "truss" supported at building tie positions will be unrealistic; the weakest element will probably be local buckling of a single tier or lift of post or leg, usually 6 ft 6 in (2.0 m) high, assuming eccentric loading of a platform immediately above the tier and assuming it capable of rotation at both ends. Also, it

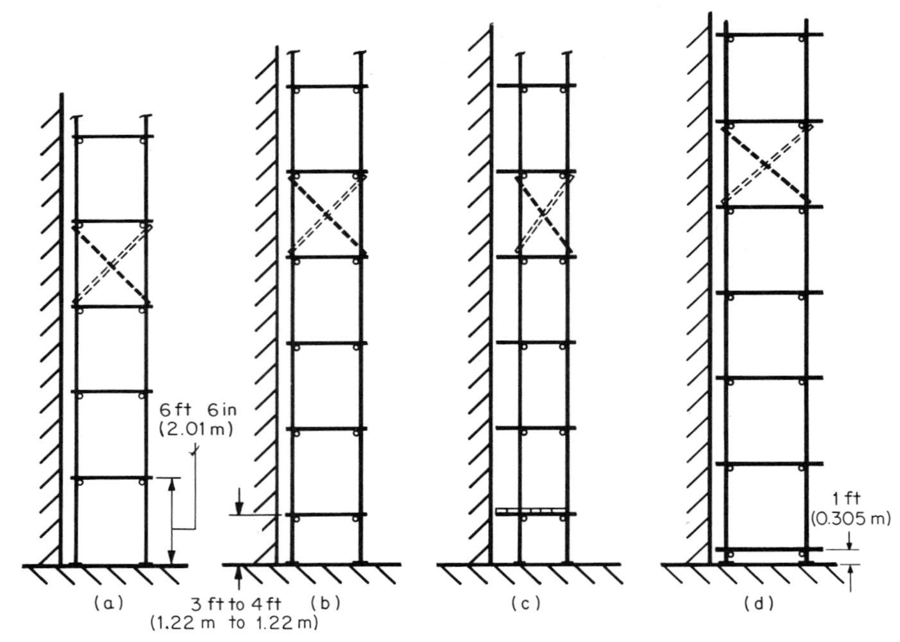

FIG. 16.10 Suggested construction methods for duty-rated tube and coupler scaffolds: (*a*) light-duty scaffold; (*b*) medium-duty scaffold; (*c*) alternative medium-duty scaffold; and (*d*) heavy-duty scaffold.

is not valid to consider it capable of translation since, in order to do so, a whole series of interconnected members must have previously deflected—or be capable of doing so simultaneously.

It should be noted that for the "nonengineered" height of 125 ft (38 m) the calculated maximum loading at the lowest scaffold post is over 4800 lb (21,500 N) for a fully loaded light-duty scaffold. It is recommended that this loading be regarded with considerable caution in the field and that, in any instance that the design load approaches 4000 lb (17,920 N) per post or more, the scaffold be subjected to checking by a competent engineer.

The allowable strength of bearers can be calculated as beams with simply supported ends when using couplers having high distortion and as having fixed ends when using couplers having high movement rigidity. A commonly used approximation is to calculate the bending moment on a bearer with *rigid* couplers by $M = wl^2/10$.

New Designs

In petrochemical and similar processing industries most vessels need internal and/or external maintenance regularly once per year. Thé development of tube and coupler scaffolds has tremendously increased labor costs because of the work required in tightening the coupler bolts and/or wedges.

To enhance productivity, a new type of scaffold—sometimes called "modular" or "systems" scaffolding—has been developed. There are a number of manufactured types available, and all have the same essential working principles listed in the following paragraphs.

Posts (standards)

Posts have locking rings, or cups, welded to them in height increments of $19\frac{1}{2}$ to 21 in (49.5 to 53.3 cm). These rings are used for the attachment of horizontal bearers, runners, and diagonal bracing.

Horizontals

Bearers, runners, and guardrails constitute the range of fixed-length members that have a fitting built into each end; these fittings are usually wedge-type, although other types exist. They attach to the posts at the locking ring or cup levels and their end-connecting devices obviate the labor-consuming necessity of constant bolt and nut tightening and loosening.

Diagonals

These are also fixed-length members of varying sizes for the appropriate horizontal members of the scaffold. With these components, flexibility of application approaching that of tube and coupler scaffolds is achieved, but it is accomplished with the simplicity of connection of sectional-frame scaffolds. It is believed that labor savings of up to 50% are achieved over the labor costs of installing similar tube and coupler scaffolds.

In Figs. 16.11 and 16.12 an example of this type of scaffold is shown; it is the QES (QUICK ERECT SCAFFOLD) manufactured by Patent Scaffolding Company, Division of Harsco Corporation. As many as 10 competitive and similar designs are available in the United States.

SECTIONAL SCAFFOLDING

The construction principle of sectional scaffolding is shown in Fig. 16.13. The most common materials used in the fabrication of steel frames is $1\frac{5}{8}$-in (41.3-mm)-OD tubing with a wall thickness between 0.086 and 0.105 in (2.18 and 2.67 mm). The most common grade of steel used for this purpose is AISI designation A-1050, a high-carbon alloy having a minimum yield stress of 50,000 lb/in^2 (34,475 N/cm^2) with a corresponding ultimate stress of over 75,000 lb/in^2 (51,712 N/cm^2). Some manufacturers use lower-carbon steels of A-1020 to A-1025. The higher-carbon steel is generally preferred because its lower ductility and greater rigidity make it more resistant to damaging and bending of the members and because it has greater strength. Many frame designs utilize smaller tubes for *internal* frame members, generally 1 or $1\frac{1}{16}$ in (25 or 27 mm) OD, of various wall thicknesses. These smaller secondary members generally perform bracing and rigidity functions. The welded joints are of two types: coped to conform to the

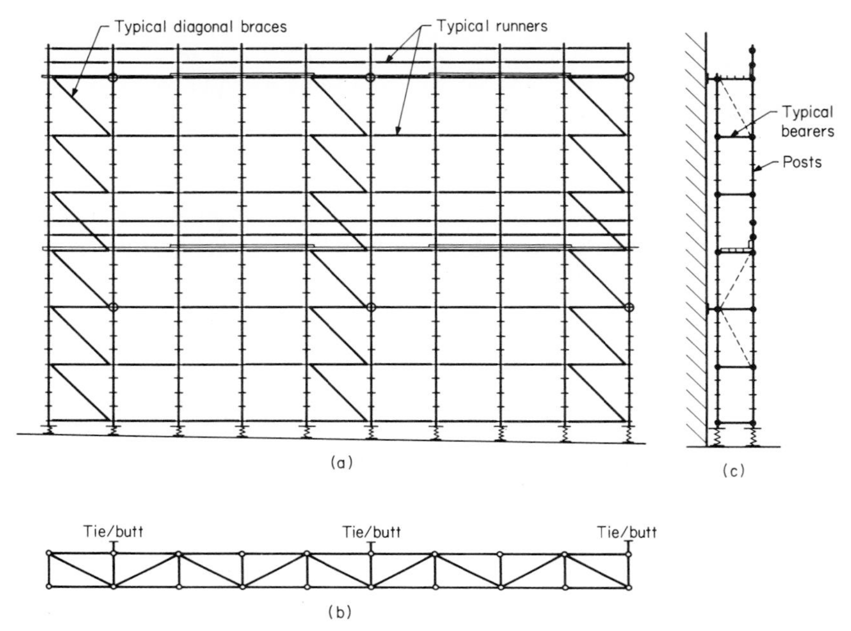

FIG. 16.11 "QUICK ERECT" modular scaffold system: (*a*) elevation; (*b*) plan view at anchorage level; and (*c*) vertical section.

contour of the tubes to which they abut or flattened at the ends as shown in Fig. 16.14. Generally, the coped joint is stronger.

The availability of frame configurations is almost limitless, depending on designs for specific uses and user preferences. Most frames are available in widths of 2, 3, and 5 ft (60, 90, and 150 cm); some special-purpose frames are available in 4-ft (120-cm) and 6-ft (183-cm) widths, the latter generally used in sidewalk canopies where the greater width is more convenient for the passage of pedestrians.

Standard frame heights are 3, 4, 5, and 6 ft, and 6 ft 6 in (90, 120, 150, 180, and 200 cm) high, with some areas using 4-ft 6-in (140-cm) high frames for masonry requirements. Special-purpose frames for sidewalk canopies are also available in heights of 7 ft 6 in, 8 ft, and 10 ft (230, 240, and 300 cm). Some typical representative frame designs are shown in Fig. 16.15, although each manufacturer has its own variants for internal dimensions and configurations.

Pivoted diagonal cross bracing, or X bracing, is generally made of approximately 1-in (25-mm)-diam tubing or 1¼-in (32-mm) angle, these being flattened at the ends which have holes punched in them for attachment to fastening devices on the frames. The X configuration is achieved by riveting or bolting two pieces together at their geometric centers; they therefore have the facility of triangulation when in use and the ability to be folded flat when stored or transported.

Cross braces are dimensioned so as to give whole-foot incremental spacing

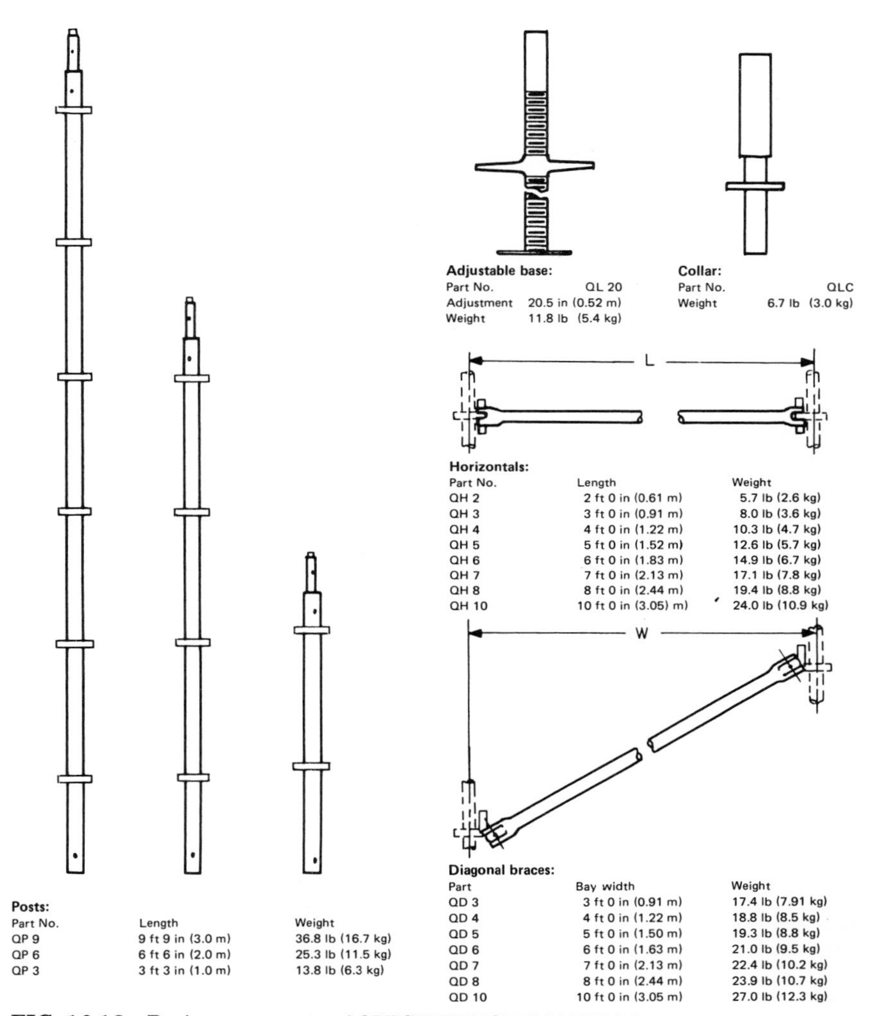

Adjustable base:

Part No.	QL 20
Adjustment	20.5 in (0.52 m)
Weight	11.8 lb (5.4 kg)

Collar:

Part No.	QLC
Weight	6.7 lb (3.0 kg)

Horizontals:

Part No.	Length	Weight
QH 2	2 ft 0 in (0.61 m)	5.7 lb (2.6 kg)
QH 3	3 ft 0 in (0.91 m)	8.0 lb (3.6 kg)
QH 4	4 ft 0 in (1.22 m)	10.3 lb (4.7 kg)
QH 5	5 ft 0 in (1.52 m)	12.6 lb (5.7 kg)
QH 6	6 ft 0 in (1.83 m)	14.9 lb (6.7 kg)
QH 7	7 ft 0 in (2.13 m)	17.1 lb (7.8 kg)
QH 8	8 ft 0 in (2.44 m)	19.4 lb (8.8 kg)
QH 10	10 ft 0 in (3.05) m)	24.0 lb (10.9 kg)

Posts:

Part No.	Length	Weight
QP 9	9 ft 9 in (3.0 m)	36.8 lb (16.7 kg)
QP 6	6 ft 6 in (2.0 m)	25.3 lb (11.5 kg)
QP 3	3 ft 3 in (1.0 m)	13.8 lb (6.3 kg)

Diagonal braces:

Part	Bay width	Weight
QD 3	3 ft 0 in (0.91 m)	17.4 lb (7.91 kg)
QD 4	4 ft 0 in (1.22 m)	18.8 lb (8.5 kg)
QD 5	5 ft 0 in (1.50 m)	19.3 lb (8.8 kg)
QD 6	6 ft 0 in (1.63 m)	21.0 lb (9.5 kg)
QD 7	7 ft 0 in (2.13 m)	22.4 lb (10.2 kg)
QD 8	8 ft 0 in (2.44 m)	23.9 lb (10.7 kg)
QD 10	10 ft 0 in (3.05 m)	27.0 lb (12.3 kg)

FIG. 16.12 Basic components of QUICK ERECT SCAFFOLD.

between the frames they were designed to be used with. The most popular spacings are 5, 6, 7, 8, and 10 ft (150, 180, 210, 240, and 300 cm). The distance between brace connections on frames varies somewhat with the manufacturer. A specific brace size when connected to frames of different size having different brace stud spacing dimensions will result in the frames being spaced at some fractional-foot distance which is sometimes convenient in enabling a scaffold to fit more closely to physical limitations of the building. The odd distance can be easily calculated, once the brace connection dimensions are known, by "squaring the triangles." For example, a brace designed to be used with 48-in (122-cm) brace connections

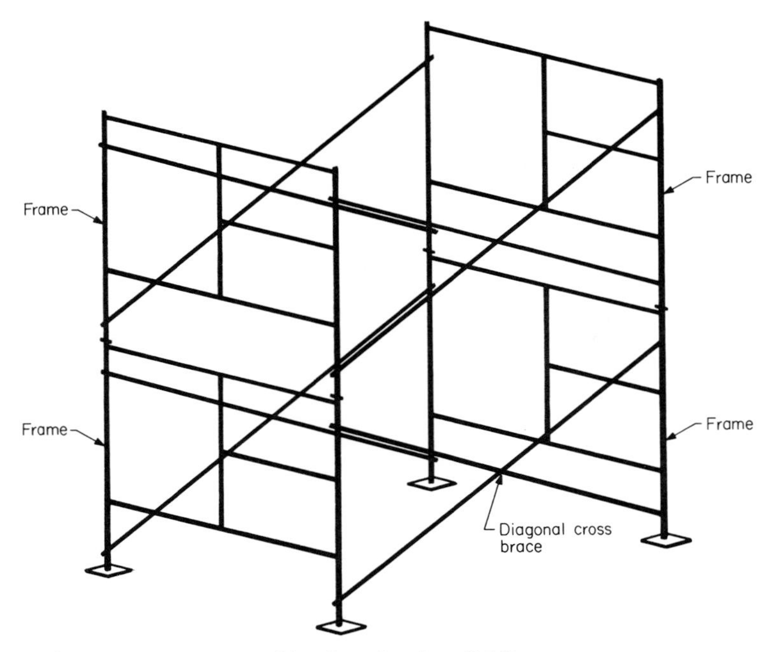

FIG. 16.13 Basic assembly of sectional scaffolding.

at 84-in (213-cm) spacing is required to be connected to a frame having a 24-in (61-cm) connection spacing. To find the new span, or frame space, proceed as follows:

$$\text{New spacing} = \sqrt{48^2 + 84^2 - 24^2} = 93.72 \text{ in (238 cm)}$$

Other basic components necessary for the modular construction of sectional frame scaffolds are the brace-attaching devices built into the frames. The design of these is as varied as the imagination of manufacturers, and it is impossible to show them all. Three types are shown in Fig. 16.2 as being representative of the choices available. With some minor exceptions, the braces were originally attached by means of threaded studs welded to the frame legs, securing being achieved by wing nuts, which necessitated their removal and rethreading every time a brace was attached or removed. This was undesirable because it was so time-consuming, spurring the design of the various fast-acting, non-thread-reliant devices now available.

The last basic component, (the coupling pin Fig. 16.16) is used to provide vertical alignment (and frequently positive connection) between frames when vertically stacked. These devices are also known as sprockets, connectors, joint pins, etc. Most are available with holes drilled in the upper and lower portions of the coupling pin which correspond to alignment holes in the frame legs through which solid or formed connecting pins can be installed to prevent the coupling pins from falling out and being lost—which at times can be a major expense. These connecting pins are available in various configurations such as hinge pins and those retained with cotters. Connecting pins also provide a means of preventing uplift of a frame from the frame below it, and in most types of manufacture they are utilized extensively to positively attach various accessories used in conjunction with the basic frames and braces. Some manufacturers weld coupling pins in the frames. This has both advantages and disadvantages.

The frame and brace module gives the flexibility of unlimited horizontal and vertical extension of a scaffold within the limitations of (1) available modules of frame size and brace spacing and (2) allowable height of the scaffold (see design section for limitations).

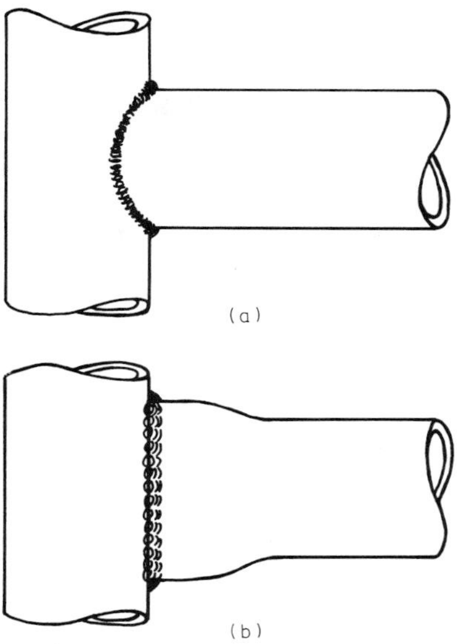

(a)

(b)

FIG. 16.14 Welded frame joints: (*a*) coped and circumferentially welded joint; (*b*) flattened and welded joint.

Standard Accessories

There is a major difference in the design concept of sectional scaffolding from that of tube and coupler scaffolding. With the latter, any element, such as inset cantilevers for plank support as used by masons and finishers, guardrailing, and other formed configurations, are achieved by suitable assembly of the three basic components. Sectional scaffold-

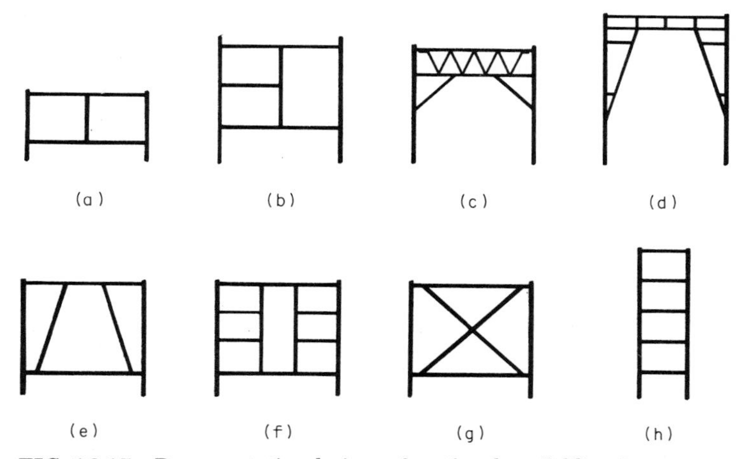

FIG. 16.15 Representative designs of sectional scaffolding frames.

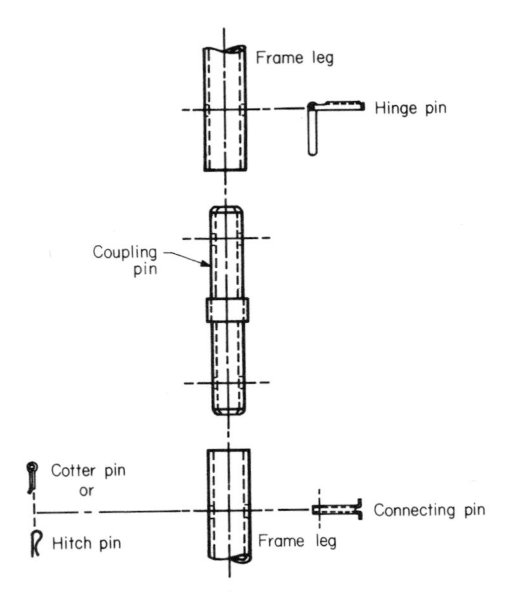

FIG. 16.16 Typical coupling pin or coupling sprocket.

ing requires that special-design accessories must be used for each individual purpose and, more importantly, *available* at the site; this entails substantially greater preplanning and scheduling. There are two types of accessories: (1) standard, such as base plates, casters, adjustable screw legs, vertical guardrail supports, and horizontal guardrails (and midrails) and (2) special purpose, such as sidewall brackets for inset planks of various types, trusses for spanning over openings or obstacles, attachment accessories, and accessories for vertical irregularities. Most manufacturers have special accessories to overcome the majority of problem conditions experienced in the field.

Codes of Standard Practice for Sectional Scaffolding

There are some special OSHA and ANSI provisions applicable to sectional scaffolding. Frames must be erected plumb and square with each other. Since frame legs are not adjustable for minor ground irregularities as are tube and coupler posts, the use of adjustable screw legs at the base is required except on known level flat surfaces, where plain base plates should be used. The use of connecting pins or other locking devices in conjunction with coupling pins is required "where uplift may occur." Drawings and specifications for all frame scaffolds over 125 ft (38 m) in height above the base plates are to be designed by a registered professional engineer. Most manufacturers have suitable printed drawings available for this purpose for standard construction; for special construction and loadings, individual design should be made for the application. In the OSHA "General Requirements" section, construction regulations subpart 1926.451(a), it is stated that "unstable objects such as barrels, boxes, loose brick or concrete blocks shall not be used to support scaffolds or planks." This requirement is generally more directed against sectional scaffolding, which does not use adjustable screw legs, than against tube and coupler scaffolding, which, within commonsense safe practices, can be self-compensating, as mentioned earlier. The use of such blocking is most prevalent with subtrades and is a most dangerous practice.

Requirements for guardrailing, as for all scaffolds, are stated in terms of installation of wood 2 × 4s (5 × 10 cm) for guardrails and 1 × 6s (2.5 × 15 cm) for midrails, but room is left for flexibility in the words "or other material providing equivalent protection." No specific strength of guardrailing is given, and the commonly used criterion is that contained in the "Floor and Wall Openings" section, subpart 1926.500(f), which states that the guardrail construction "shall be capable of withstanding a load of at least 200 lb (900 N) applied in any direction at any point on the top rail, with a minimum of deflection."

Requirements for access ladders are vague and over the years have resulted in substantial controversy since, lacking specifics, field inspectors have frequently applied construction provisions for fixed industrial ladders and portable wood and metal ladders to scaffolding frames containing a series of horizontal members used for climbing and intermediate planking levels. Needless to say, the frame designs do not generally comply with such inapplicable requirements. There are no definitions of climbing or access ladders contained in the regulations. The "Manually Propelled Mobile Scaffolds" section, subpart 1926.451(e)(5), calls for the provision of ladders or stairways to be affixed to or "built into" the scaffold, being so located that their use will not have a tendency to tip the scaffold; also it specifies that a landing platform must be provided at intervals not to exceed 30 ft (9 m). This requirement is similarly not cross-referenced to any other type of scaffold. Since the majority of mobile scaffolds are constructed from sectional frame scaffolding, this lack of specifics is important. The newly adopted sections from the general industry regulations, again dealing *only* with mobile ladder stands and scaffolds (towers), define a "climbing ladder" as "a separate ladder with equally spaced rungs usually attached to the scaffold structure for climbing and descending"; this definition was *not* adopted into the construction regulations so that the hiatus continues. The nonadopted revised ANSI Standard A10.8-1977 contains adequate language and provisions to clarify this controversial and important requirement.

Essentially, manufacturers of sectional scaffolding have three available means of climbing scaffolds: (1) climbing the frames, provided they are of the "mason" type having multiple and relatively regularly spaced rungs and having been designed for climbing. (2) Climbing hook-on ladder accessories which attach to the frame. Older types have the rungs immediately adjacent to frame legs and newer ones have a "stand-off" provision so that *clearance* is provided between the rungs and the frame legs. (3) Climbing internal stair sections positioned within the confines of the 5-ft (150-cm) frame widths and generally the 7-ft (210-cm) bracing length. Alternatively, properly installed job-made "cleated" ladders and portable wood or metal ladders can be used for scaffold access. Figure 16.17 shows the above-described methods.

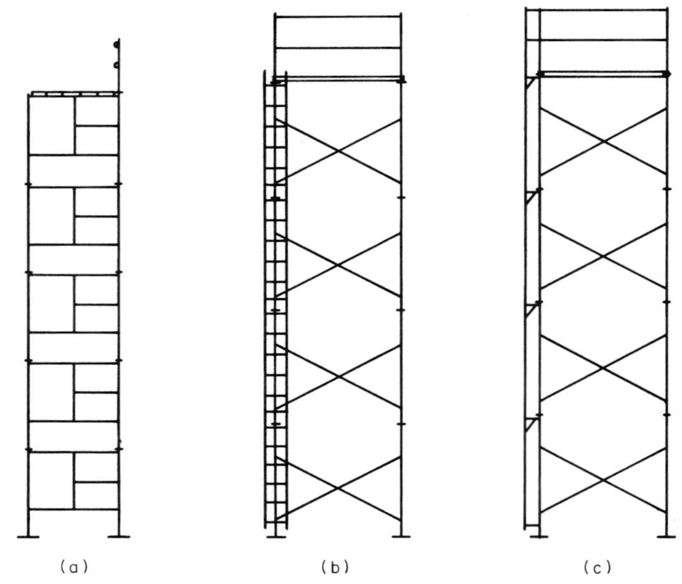

FIG. 16.17 Climbing scaffolds: (*a*) frame rungs; (*b*) hook-on ladders; and (*c*) stand-off attachable ladders.

Design Loads for Sectional Scaffolding

The design load statements made earlier in this chapter apply to sectional scaffolding. In general and with the exception of some cities' regulations the pound per square foot (newton per square meter) load ratings are not aligned to any particular type, configuration, or allowable spacing of scaffolding frames for use by specific building trades. Allowable plank loading vs. span is usually the determinant of the frame spacing. Since the majority of scaffolding work involves work platforms which are progressively relocated, the economies achieved by larger spacings between the scaffolding members are not applicable. It is *common practice* in most areas to use 7-ft (210-cm) frame spacing for masonry work and 8- or 10-ft (240- or 300-cm) frame spacing for painting and similar renovation work.

The stated safety factor of 4 is applicable with a caution. Platform loads at levels of application are transmitted to frame legs by means of bearers, generally called "head bars," and consequently apply a bending movement to the frame legs which can result in their having lower capacity than when concentrically loaded through the legs from upper sections of the scaffold. There are three determinants of frame capacity: (1) allowable load on the frame head bar or other load transfer member, which is usually less than the allowable concentric load of *one* frame leg [depending on design and method of loading 5-ft (150-cm)-wide frame head bars can typically support loads between 2000 and 3000 lb (9000 and 13,500 N), uniformly distributed; point loads applied close to the frame legs plus other intermediate point loads can give higher values, such a typical application being the use of 19-in (48-cm) or so prefabricated planking which applies loads through hooks or siderails at the edges of the planks (Fig. 16.18)]; (2) allowable leg loads with loading transferred through the head bars; (3) allowable leg loads with loading applied concentrically. Typically, the maximum leg load condition in a given scaffold will be a combination of (2) and (3) when a loaded platform is located close to the bottom of a high scaffold.

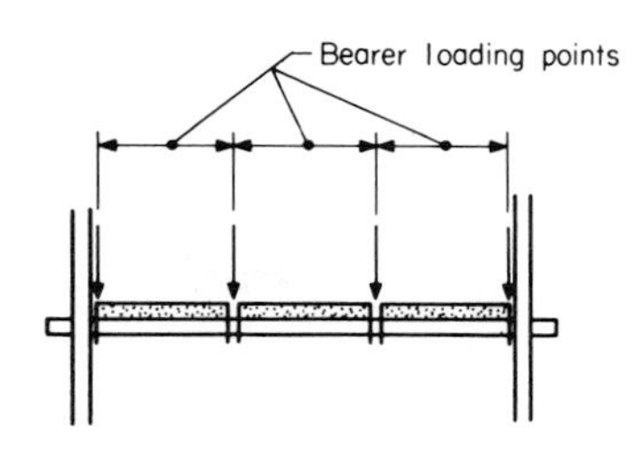

FIG. 16.18 Prefabricated planking.

A caution is advisable in the use of leg loads. The industry in general tends wrongly to think of sectional scaffolding in terms of 5000 lb (22,400 N) per leg safe capacity, a value which is applicable to only a few frames of the strongest configuration and *only* as applied to concrete shoring. *Shoring* loads are determined using a safety factor of 2.5, whereas *scaffolding* loads are determined using 4. Therefore, a frame rated for 5000 lb (22,400 N) per leg for shoring has to be inverse-proportionately adjusted in accordance with these differing safety factors by multiplying the allowable shoring load by 2.5 and dividing it by 4. Table 16.4 gives a series of shoring loads adjusted accordingly for scaffolding use.

Another frequently overlooked loading application is that used for caster-wheel attachments. Casters should be allowable load-rated for two conditions: static loaded and mobile loaded. Static load is a result of the scaffold weight plus working load; mobile load is the self-weight applied to the casters while the scaffold is being moved. Allowable mobile load is usually a small fraction of the static load, varying with manufacture and design, from approximately 20 to 35%. The scaffolding self-weight on each caster must be checked to ensure that it is not greater than the allowable mobile load for the caster, even though when stationary and in use with live loads applied, the load per caster can be well within the *static* allowable limit. This is particularly important when using high, trussed rolling towers and counterweighted rolling towers.

TABLE 16.4 Sectional Frame Allowable Shoring Leg Loads Adjusted for Scaffolding Use

Factor used is 2.5/4 = 0.625

Shoring load, lb per leg	Scaffolding load, lb per leg
5000	3125
4500	2813
4000	2500
3500	2188
3000	1875
2500	1563

NOTE: 1 lb = 4.45 N.

Mobile scaffolds with cantilevered loads, such as a rolling benchwork for a suspended scaffold (Fig. 16.19), must be restrained from overturning by counterweighting or physical restraints. Safe practice requires that the counterweighting or restraining forces required be calculated using 4 times the maximum applied load to determine the overturning moment. From this can be deducted the restraint moment of the inboard self-weight of the scaffolding equipment *before* determining the amount of counterweight or restraint force. A common miscalculation is to multiply the *total out of balance* overturning moment by 4, having included the self-weight restraint moment in this calculation. This method is wrong because, by so doing, the self-weight moment restraint has been multiplied by 4 which understates the external restraint force required.

Safe allowable loads on all structural scaffolding components should always be determined from the equipment manufacturer(s) or their authorized agents. Because of the great variance in materials, design, and allowable loads, it is most hazardous to connect intermingled components of different manufacturers to each other, although many are apparently physically compatible; even then, differences in tolerances and fit can cause incalculable strains between them, resulting in unsafe loading conditions.

Figure 16.20 shows how the loads applied to posts from platform-supporting bearers are eccentric in nature.

WOOD PLANKING

The strength of wood used for scaffolding planks is possibly the least comprehensible of all scaffolding elements. Wood is organic and affected by many factors, such as region of growth, rate of growth, moisture condition at time of saw-

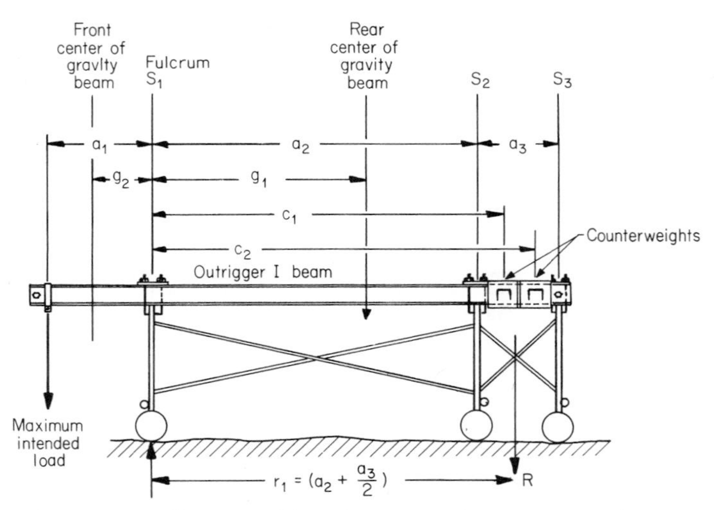

FIG. 16.19 Rolling scaffold used to suspend a two-point swinging scaffold (swing stage).

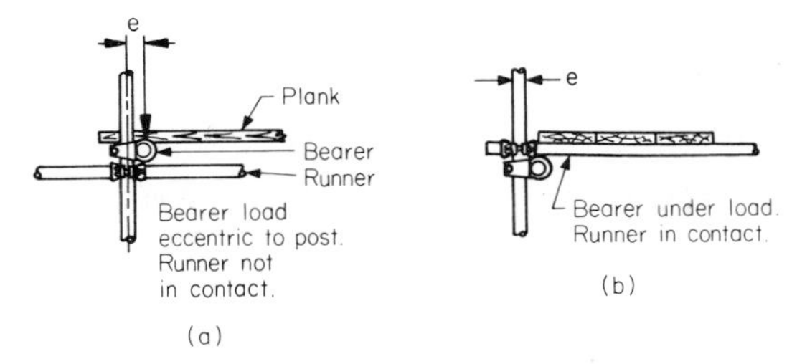

FIG. 16.20 Eccentricity *e* of loading on post.

ing and grading, and, naturally, conditions of use. These factors do not permit use of exact stress design data similar to those used for relatively homogeneous metals. The complexity is compounded further by the standard commercial practice to cut a batch of trees from a "stand," i.e., a certain small area which may contain a number of trees of similar but different species from the major one. These different species generally will have varying strength qualities.

Essentially, published wood strengths are derived from procedures published in ASTM Designation D2555, *Standard Methods for Establishing Clear Wood Strength Values.* Various statistical "weighting" factors are applied to the results of testing small "clear" samples of woods from a given stand or area using complicated statistical procedures. A clear sample is a piece of visually perfect wood.

The cut wood is then further graded by an official grading association authority using visual grading criteria. Subsequently, in accordance with ASTM Designation D245, additional downgrading factors are applied to the wood-strength values to arrive at published "working" stresses for various grades of that wood. Such factors as species preference or preponderance, species "grouping," commercial marketing practices, design techniques, safety, and wood graded for specific-product end uses all contribute to the complexity of determining end-strength values which can only be averages.

The final visual grading of the woods is relatively arbitrary, and therefore published working stresses should be used as a guide rather than finite values. In general, they have a ratio of approximately 1:4 below the clear strength values for the species, but not always.

Finally, these published values need to be factored by the end purchaser or user in accordance with conditions of use such as moisture content *during* use and use with the wider dimension on the "flat" such as for planking. Unless otherwise specified, strength values are for woods used as beams with the largest dimension vertical. Wood in use is considerably weaker when green or wet than when it is dryer since the wood fibers are more easily able to slide along one another and this reduces bending strengths and moduli of elasticity (stiffness). The most expensive woods are those which have been slowly kiln-dried to a very low moisture content and which exhibit the highest and most consistent strengths for a certain wood species and grade. Next are woods that have been gradually air-dried or seasoned. Both are generally too expensive for use as scaffolding planks and are mostly used for furniture and other ornamental purposes.

The one exception is lumber which has been mechanically (or machine) stress rated (MSR). This is a process in which each piece of lumber has its moisture content measured and is nondestructively tested and stamped by the machine with its allowable fiber stress F_b and modulus of elasticity E. Typically, the grading stamp will include the letters MSR followed by values for these two stresses such as "1200f-1.3E." The 1200f value of F_b given is that for "single member" loading, explained later. Each grade authority book of rules will give tables showing other design stress values assigned to the grade mark. This MSR lumber also has to pass some additional visual grading inspection and is quite expensive in comparison with visually graded lumber.

Most scaffold plank material is cut green, or unseasoned, and, if stacked properly during shipping and in yards before use, may season somewhat. The degree of moisture content is not easily assessed or determined by the user. Even though its sawn density may include less than 19% moisture content, or even 15%, it can absorb additional moisture from humidity of the air and during its placement on scaffolds when subjected to rain and snow. A recent change in grading practices now permits wood having a moisture content of less than 15% to be classified as "dry." Green or unseasoned wood is assumed to have not more than 19% moisture content by density.

The importance of yard stacking under cover, with good airflow between the planks after receipt from the mill source in order for the wood to season slowly, cannot be overemphasized. This allows the wood to dry out slowly with the least shape distortion such as crooking, warping, bowing, and end splitting. Consequent exposure to climatic moisture will subsequently have the least effect on its strength in use. Putting green wood immediately into use, especially in hot sunlight, will usually distort much of it too severely for further use and at best may drastically shorten its service lifetime because of end-cracking and "cupping."

Selection of wood for scaffolding plank is made from select structural or dense select structural grades by an appropriate grading authority for the species in accordance with individual visual criteria which vary somewhat in knot sizes allowed, slope of grain, and other factors; they are not given here but are available from the various grading authorities listed. Administrative and marketing considerations often make it expedient to combine different species having relatively similar physical properties into a single marketing combination. Published data of allowable stresses generally give two values for maximum fiber stress in bending (F_b): single and repetitive. Single values assume that any single piece of wood may contain maximum strength-*reducing* characteristics allowed in the grade, on the basis of each individual member carrying its *full* design load permanently. Repetitive values assume that members are used in a system with common supports or joining surfaces (such as plywood) so that applied loads are distributed over a number of members. Scaffolding plank can be considered to fall into the repetitive category, since large, single point loads are virtually never applied to only one plank. An exception to this might be the point load from a wheel of a wheelbarrow or buggy, but even then the loaded total weight of the vehicle will be distributed over one or more wheels or between a single wheel and the driver's foot when the foot is supporting the carrying handles. Also, such loads are seldom static and are usually transitory in nature. Pallets of bricks or blocks generally apply their loads over two to three planks, and the duration of full loads is short, maybe 2 to 4 days.

Wood has the unique ability to withstand high overloads for short periods of time in accordance with Table 16.5. All published wood strength data give work-

TABLE 16.5 Working Strength Increase for Duration of Loading

Maximum load duration	Type of loading	Allowable increase of working stresses, %
1 s	Impact	100
1 min		72 (Interpolated)
1 h		47 (Interpolated)
1 day	Wind/earthquake	33
7 days	Concrete formwork	25
1 month	Snow	17
2 months	Snow	15
1 y		7 (Interpolated)
10 y	Normal	0
50 y	Permanent	−10

ing stresses for permanent (normal) duration during loading, a category in which scaffold planking does not belong. Therefore, the criteria assumed hereafter are quite conservative since no increase is assumed in the limiting stresses resulting from short-term loading. Thereby, an additional safety factor, which can compensate for planks in less than perfect condition, is built in.

Table 16.6 gives strengths for scaffold plank grades obtained from the various grading associations as published in 1978. It is not always possible to correlate these values to those assigned to the generic species values published.

TABLE 16.6 Published Stress Values for Visually Graded Scaffold Planks

Grading authority	Wood species	Scaffolding plank grades	F_b, lb/in^2	$E \times 10^{-6}$, lb/in^2
WWPA	Douglas fir	Scaffold no. 1*	2200	1.6
	and larch	Scaffold no. 2*	2000	1.6
WCLB	Douglas fir	Dense premium	2300	1.7
	Douglas fir	Premium	2100	1.6
Para. 171aa	Douglas fir	Dense select structural	2050	1.7
Para. 171a	Douglas fir	Select structural	1900	1.6
Para. 171bb	Sitka spruce	Premium	1450	1.3
Para. 171b	Sitka spruce	Select structural	1350	1.3
SPIB†	Southern pine	Dense industrial 72	1750	1.6
	Southern pine	Dense industrial 65	1600	1.6

*These are not cross-referenced to select structural or other grade values, and the method of derivations of these values is not disclosed.

†Values given apply to lumber used under "exposed-to-weather" conditions and are quite realistic; however, the F_b stresses printed have *not* been increased by 15% for "repetitive" use as allowed in the grading rules; the increase would result in the values of 2010 and 1840 lb/in^2, respectively.

NOTE: 1 lb/in^2 = 6.89 kN/m^2.

It is the author's belief that use of these optional allowable stresses to predict precise behavior is not realistic for many reasons. Some reasons have already been enumerated: they assume perfect visual grading; they ignore subspecies inclusion possibilities and other statistical procedures which result in some percentage of any purchased "lot" having lower values than these. Such lots may contain small amounts of subspecies woods and woods from different areas in different stages of seasoning and having similar appearance but possibly less strength. Certain percentages of such are allowed in the grading rules. Douglas fir "north" is appreciably stronger than Douglas fir "south" for instance.

Table 16.7 gives a cross section of values obtained from the most complete single source of species stress data: the booklet *Design Values for Wood Construction,* latest issue, June 1978, published as a supplement to the 1977 edition of *National Design Specification* by the National Forest Products Association of Washington, D.C. The supplement gives stress values for all major species in one reference source; it gives the grading authority from which the stresses are obtained and lists the various modification factors applied by the grading authorities. It does not give stresses for "special" end-use grades such as scaffolding plank.

Without having the assurance afforded by buying (expensive) machine stress rated lumber, the lowest possible evaluation of published strengths must be made taking into account all the potential strength-affecting factors.

After reviewing the generic groupings in Table 16.7, the following four groupings have been chosen objectively for general reference as a practical guideline to use after processing. In all cases, in Table 16.7, stress values have been factored by 0.92 for "wet" use rather than the 0.97 value per footnote *, Table 16.7, except for southern pine. The final choices are shown in Table 16.8, together with the appropriate modifying factors and end-use stresses after modification. The limiting stress values used have considered the likelihood of such species groupings and intermingling and are extremely conservative as a measure of commonsense practicality. Caveat emptor is the guiding principle which should be applied since the final purchaser usually has little knowledge of the real origins of the wood purchased.

Using factored values, allowable simple spans have been calculated for whole-foot spans from 7 to 10 ft (210 to 300 cm) and the nationally prescribed loadings of 25, 50, and 75 lb/ft^2 (1200, 2400, and 3600 N/m^2), using maximum allowable deflection at any span to be equal to the plank thickness [R = 1.875 in (4.75 cm) rough, D = 1.5 in (3.81 cm) dressed] as shown in Tables 16.9 and 16.10.

PREFABRICATED PLANKING

Also available are a number of types of manufactured planking, including the following major types:

1. Laminated wood (plywood-type plies on flat)
2. Steel fabricated
3. Aluminum frame with plywood deck
4. Aluminum frame with aluminum deck

TABLE 16.7 Listing of Published Working Stresses for Woods Commonly Used for Scaffold Planking

Species groups	Grade	F_b lb/in^2 repetitive	$E \times 10^{-6}$ lb/in^2	Grading authority	Factoring needed		
					F_b flat	F_b wet	E wet*
Group I—Douglas fir							
Surfaced green *or* dry Douglas fir–larch (including DF–L north)	DSS	2400	1.9	WCLIB	1.22	0.86	0.97
	SS	2050	1.8	WWPA	1.22	0.86	0.97
Douglas fir (south)	SS	1950	1.4	WWPA	1.22	0.86	0.97
Hem–fir	SS	1650	1.5	NLGA	1.22	0.86	0.97
Mountain hem–fir	SS	1650	1.3	WWPA	1.22	0.86	0.97
Group II—southern pine							
Surfaced green, used *any* condition	DSS	1850	1.6	SPIB	1.22	Incl.	Incl.
	SS	1600	1.5	SPIB	1.22	Incl.	Incl.
Dense industrial #72†	DS	2050	1.6	SPIB	1.22	Incl.	Incl.
Dense industrial #65 (2½ in thick +, wet use)	DS	1800	1.6	SPIB	1.22	Incl.	Incl.
Group III—western spruces							
Englemann spruce‡	SS	1350	1.3	WWPA	1.22	0.86	0.97
Sitka spruce	SS	1550	1.5	WCLIB	1.22	0.86	0.97
Spruce–pine–fir	SS	1450	1.5	NLGA	1.22	0.86	0.97
Group IV‡							
Western white pine	SS	1300	1.4	NLGA	1.22	0.86	0.97
Other west white woods	SS	1300	1.1	WWPA	1.22	0.86	0.97
Lodgepole pine	SS	1500	1.3	WWPA	1.22	0.86	0.97
Idaho white pine	SS	1300	1.4	WWPA	1.22	0.86	0.97
Ponderosa pine	SS	1400	1.2	NLGA	1.22	0.86	0.97
Ponderosa, sugar, lodgepole pines	SS	1400	1.2	WWPA	1.22	0.86	0.97

Grading Authorities

WWPA	Western Wood Products Association, Portland, Oregon 97204
WCLIB	West Coast Lumber Inspection Bureau, Portland, Oregon 97223
SPIB	Southern Pine Inspection Bureau, Pensacola, Florida 32594
NLGA	National Lumber Grades Authority (Canada), Vancouver B.C., Canada V6E 2H1

*All grading associations give a "wet" factor for E of 0.97. However, one specifies whether for "flat" or "end" use, and the latter is assumed since it is normal. WWPA uses lower factors for "decking" which is the way plank is used. The flat wide side exposes much more of the wood surface to moisture penetration. It seems logical to use this 0.92 factor for *all* woods, instead of the given 0.97.

†Dense industrial 72 and 65 are the grades from which southern pine scaffolding planks are selected with stricter knot and growth characteristics limitations. Southern pine (also known as yellow pine) consists of four principal subspecies: longleaf, shortleaf, slash, and loblolly. It is grown only along the eastern seaboard, from southeastern New York and New Jersey to Florida, and some in the southern parts of South Carolina, Georgia, Alabama, and Mississippi and into eastern Texas and Oklahoma. Beware of woods sold in the northwest as "southern pine" since unless specially shipped from the south on order, they are most likely to be of the western pine grouping. See note‡.

‡Sitka spruce is the strongest now available. Canadian spruce stands under the NLGA ratings may be grouped as "spruce–pine–fir." The western pines listed are mainly white in color except Ponderosa which is botanically a yellow pine. This latter is sometimes sold mistakenly for southern pine which is also yellowish-white. The difference can be determined by closer spacing of growth rings on southern pine which is also up to 40% heavier. Southern pine is yellow, but yellow pine is *not* necessarily southern pine. Western pines are about equal in strength to western spruce, although many will not pass scaffold plank visual grading criteria for knots, slope of grain, etc. In the following tabulation, they will be grouped with spruce however.

NOTE: 1 lb/in^2 = 6.89 kN/m^2; 1 in = 2.54 cm.

TABLE 16.8 Recommended Design Stresses for Scaffold Planking

	Basic design stresses			Factored values for use	
	F_b Repetitive, lb/in^2	$E \times 10^{-6}$ lb/in^2	F_b factors	F_b Repetitive, lb/in^2	$E \times 10^{-6}$ lb/in^2
Douglas fir (dense premium)	2300	1.7	$\times 1.22 \times 0.86$	2415	1.6
Douglas fir (others)	2000	1.5	$\times 1.22 \times 0.86$	2100	1.4
Southern pine	1800	1.6	$\times 1.22$	2200	1.6
Sitka spruce	1500	1.5	$\times 1.22 \times 0.86$	1575	1.4
Other spruces and western pines	1300	1.2	$\times 1.22 \times 0.86$	1365	1.1

NOTE: 1 lb/in^2 = 6.89 kN/m^2.

The most readily available laminated planks are marketed under the name Micro-Lam. Their structure is akin to 1½-in (38-mm)-thick plywood, available in desired widths and lengths. High bending strength and longer life are claimed for this material although initial cost is close to double that for solid Douglas fir or southern pine.

The fabricated metal or metal-plywood combination planking is available from many proprietary manufacturers in fixed, 1-ft (30-cm) incremental lengths, the most popular being 7, 8, and 10 ft (210, 240, and 300 cm). They are equipped with hooks at each end to seat on scaffold supporting members. Depending on manufacturer, they are rated at various load capacities up to and including 75 lb/ft^2 (3590 N/m^2). The ones with metal walking surfaces generally incorporate some slip or skid resisting provisions, and the aluminum-plywood deck types are available or treatable with skid-resistant paint. The planks with aluminum frames are, of course, susceptible to damage from rough handling and are gen-

TABLE 16.9 Allowable Standard Load Ratings for Various Spans

Span	Condition*	7 ft	8 ft	9 ft	10 ft
Douglas fir, dense premium	R	75	75	75	75
	D	75	75	75	50
Douglas fir, others	R	75	75	75	75
	D	75	75	50	25
Southern pine	R	75	75	75	75
	D	75	75	75	50
Sitka spruce	R	75	75	75	75
	D	75	50	50	25
Other spruces, pines	R	75	75	75	50
	D	75	50	25	25

*R = rough; D = dressed.
NOTE: 1 ft = 0.305 m.

TABLE 16.10 Allowable Spans for Standard Load Ratings, in Feet*

Species	Condition†	Loading lb/ft²		
		25	50	75
Douglas fir, dense premium	R	10	10	10
	D	10	10	9
Douglas fir, others	R	10	10	10
	D	10	9	8
Southern pine	R	10	10	10
	D	10	10	9
Sitka spruce	R	10	10	10
	D	10	8	7
Other spruces, pines	R	10	10	9
	D	10	8	7

*It is recommended that because of lower ring growth and density with consequently more rapid moisture absorption spruces and western pines be used in the rough condition only, irrespective of type of duty *unless* they are never used for greater than 7-ft spans, i.e., cut to lengths not less than 8 ft and not more than 9 ft. All single-span short planks should be cleated.

†R = rough; D = dressed.

NOTE: 1 ft = 0.305 m; 1 lb/ft^2 = 47.9 N/m^2.

erally more favored by subtrades for rolling scaffold applications than for general-duty construction scaffolding, where the solid or laminated wood planking is preferred for more arduous usage and flexibility.

PLANK TESTING AND CARE

There is some controversy as to the desirability of testing new, solid wood planking before use. Advantages are that testing may reveal hidden flaws in strength which are not visually apparent. A method of testing is described in Rossnagel's *Handbook of Rigging,* 3d edition (McGraw-Hill, 1964), and is used by many. The method will often reveal a "defective" plank but may *worsen* a flaw in a plank which would otherwise give satisfactory performance. The method relies on proof load testing one side of a plank and then the other, under controlled load and deflection conditions.

Opposed to this practice of two-sided proof loading is the Forest Products Laboratory, of Madison, Wisc., a department of the U.S. Department of Agriculture, which in its Technical Note No. 264, November 1962, considers that it is beneficial for all solid planking to be used always *with the same side up.* When in use, the upper surface fibers of the plank are in compression, or squeezed together lengthwise; the lower side fibers are in tension and tend to pull apart lengthwise. Shock loads, temporary overloads, and rough handling can create damage to the upper fibers, called "compression failures," which show up as "wrinkles" across the grain and are sometimes so faint as to be unnoticeable. When the plank is turned over and loaded, these compression failures are now put in tension and can very easily pull apart and cause the plank to fail. It is admittedly difficult to control the one-sided use of long planking used over multiple scaffold bay spans. Short planks used for single spans can be easily restricted to single-sided use only by placing cleats at either end of the plank on the underside in accordance with the fixed span(s) with which it is to be used, as shown in Fig. 16.21. If noticeable compression failures are found on one side, that side should be used as the upper surface. Multiple-span planks (Fig. 16.22) are not practical to cleat since they are generally overlapped at joints; they could be marked to denote the working (top) side, however.

For longevity, the ends of planks should always be protected from propagation of splits by means of a number of end-protective devices available, such as spiked, formed steel strips which are hammered into the top and/or bottom surfaces and annular ring nails. (See Fig. 16.23.)

Stacked planks should be intermediately separated by dunnage or firring strips in accordance with handling methods or machinery, ensuring that all such strips are placed vertically over each other—not offset so as to cause local bending. (See Fig. 16.24.) Where possible, keep planks dry, protect them from rain, or arrange them so that they will drain rapidly. The top level of planks in a pile will receive most rain exposure so be extra cautious in their use.

WOOD DETERIORATION

Aging of wood alone does not cause it to lose strength, and with care an old well-used dry plank is possibly more trustworthy than an untried, new one. Moist

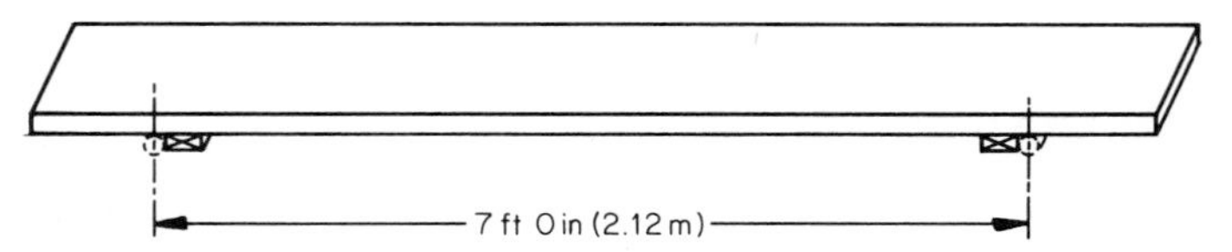

FIG. 16.21 Cleated single-span planks.

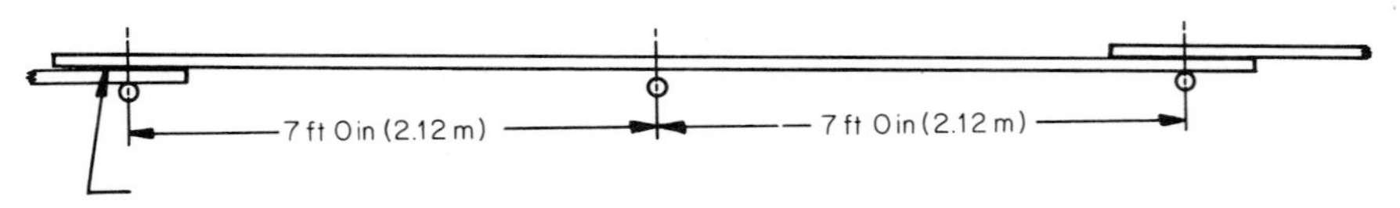

FIG. 16.22 Overlapped planks.

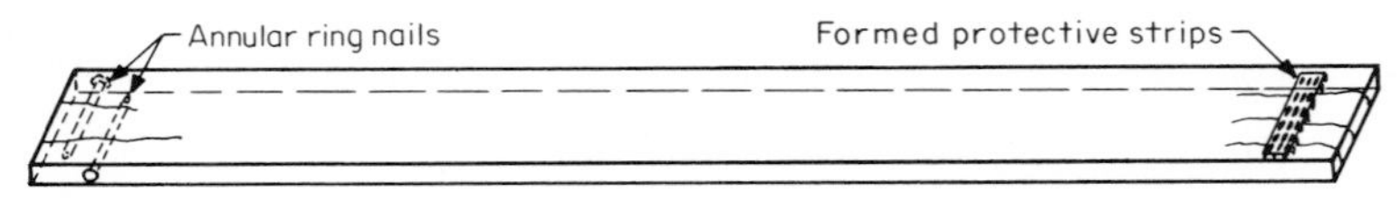

FIG. 16.23 End-protected plank.

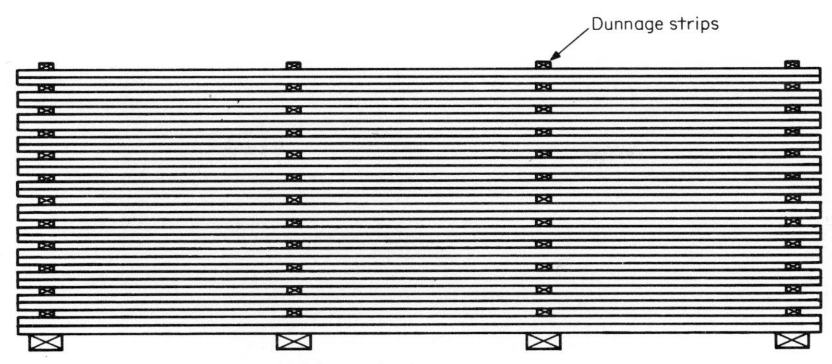

FIG. 16.24 Stacked scaffolding planks.

conditions over a period of time can reactivate growth of incipient decay fungi already present; moisture does not initiate decay and wood does not itself decay. Decay is a parasitic fungus often present in wood grown in moist, warm conditions, which after drying or seasoning becomes dormant. Prolonged reexposure to moisture can cause these fungi to reactivate, and they do so by eating through the cell walls to eat the contents of the cells. This is why decayed wood is lighter than comparable wood of the same species, so planks which appear lighter than their associated ones should be suspected of being decayed. ("Light" planks may also be of a lighter density, weaker subspecies as previously discussed.) Nevertheless, the lightness indicates one or the other condition. Once decay has started, it is almost impossible to stop. However, it is practical to treat well-dried and seasoned wood with a moisture-resistant antifungicide, such as pentacloro-phenol which, when applied in accordance with the manufacturer's instructions, can prevent propagation of decay fungi from without or within. Decay is often called "dry rot" incorrectly.

GENERAL APPLICATIONS OF SCAFFOLDS

Wall Scaffolds

The use of horizontal diagonal bracing is recommended on high scaffolds; it should normally be positioned close to the same height increments and levels as the building ties. Although not always used, horizontal bracing does impart additional rigidity to the scaffold. To facilitate installation, braces should be positioned also at the lowest frame level in order to assure squareness of frames in relation to each other; all brace connections have sufficient tolerance of fit to allow a nonsquare scaffold.

Figure 16.25 shows a scaffold using 5-ft (1.5-m)-high mason-type frames, which, because of their inability to provide intermediate platform egress, are generally used for work at the top platform level only, which is progressively raised as frames are added to the height and planks relocated upwards. These frames are generally 4 ft 6 in, 5 ft, and 6 ft 6 in (140, 150, and 200 cm) high.

FIG. 16.25 Masons' scaffold, using 5-ft (1.5-m)-wide and 5-ft (1.5-m)-high mason frames, working from the upper west level. *(Courtesy of The Patent Scaffolding Co., Inc.)*

Figure 16.26 shows an adaptation of 6-ft 6-in (7.98-m)-high walk-through-type frame construction adapted for use by masons. This method involves the use of inset sidewall brackets (also known as outriggers—wrongly), extensions, and lookouts. Used for masonry, they give the facility of providing alternating work-level positions at intermediate heights of the frames, allowing easier reach to the work and more convenient height locations than offered by frame support bars. For masons' work above 18 ft (5.5 m) or so, the use of the 6-ft 6-in (200-cm)-high frame with brackets provides substantial labor economy over top-platform-level 4-ft 6-in or 5-ft (140- or 150-cm)-high frames. For any given height, there is a 25% or more savings of the total quantities of frames and braces required and 25% fewer plank relocations; the additional expense involved in the brackets and planks for them is minimal.

Figure 16.27 shows a sidewalk scaffold for exterior renovation or similar work. Such scaffolds can be constructed in accordance with manufacturers' instructions to provide appreciable pound per square foot (newton per square meter) loadings on the overhead planked level in addition to support of the platform load(s) from the upper working level(s) and the weight of the superimposed scaffold.

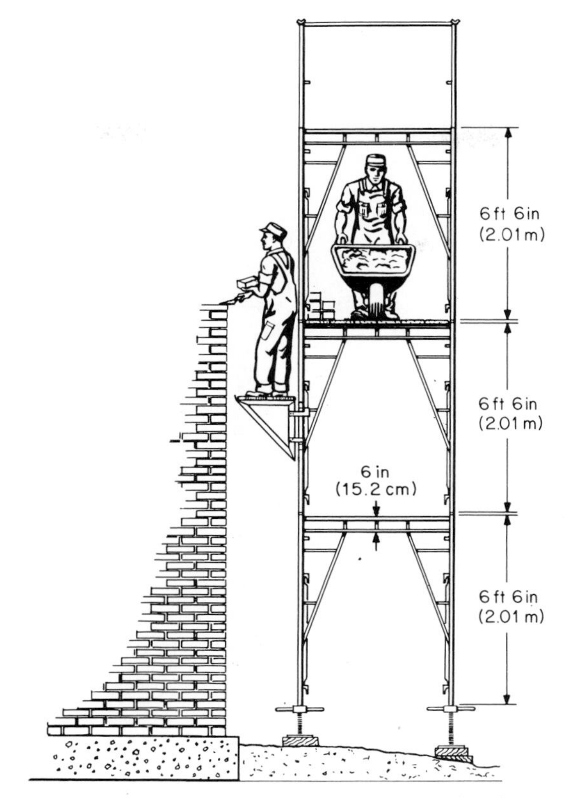

FIG. 16.26 Walk-through-type frames 6 ft 6 in (1.98 m) high used by masons.

Rolling Towers (Mobile Scaffolds)

With suitable bracing and accessories, these are adaptable to almost any work requirement. Ranging from the single width, single bay tower, they can be expanded in both frame width and brace length directions.(See Fig. 16.7.)

Large rolling towers should always use ties connecting the lower legs, so that there is multiplied resistance against excessive deflection caused by one leg or caster hitting a hole or obstruction. All frames in rolling towers should be connected to each other in the vertical tiers by means of positive-locking connection pins, as mentioned previously, since uplift is always a hazard more prevalent to this type of construction than most others. Casters should always be pinned or otherwise securely connected to the bottom frames.

Horizontal diagonal bracing should always be installed on rolling towers to avoid racking; OSHA regulations require that it be positioned at levels (1) as close to the rolling surface as possible and (2) at least at successive height intervals of not more than 20 ft (6 m). Free-standing towers must not be used when

FIG. 16.27 Sidewalk scaffold. (*Courtesy of The Patent Scaffolding Co., Inc.*)

the top working platform height exceeds 4 times the minimum base dimension—which in most instances is the frame width—*unless* they are securely tied off to a building structure and/or the base is widened by adding more frame and brace elements to widen the base dimensions. Some manufacturers have special frame-type accessories, often called "outriggers," to assist accomplishing this. Some states may have more restrictive dimensional requirements than the mentioned 4:1 height-to-base ratio (3:1 or 3½:1). Rolling towers frequently need to be equipped with adjustable screw legs to compensate for slightly sloping floors and for irregularities in the rolling surface; few surfaces are truly flat or level and a minor 1-in (25-mm) deviation could cause a caster to lose contact with the surface transferring the weight of the scaffold leg it supports to the interconnecting braces. This could result in the overload of casters adjacent to it. Frequent readjustment of screw legs can overcome this problem.

To refer back to the OSHA construction regulations, subsection 1926.451(e) allows personnel to ride scaffolds while they are being moved, provided the floor is within 3° of level and is free from pits, holes, or obstructions; provided the scaffold has a height-to-base ratio of 2:1 (instead of the normal 4:1); and provided all tools and materials are secured or moved from the platform. As a safe practice, riding of towers is not recommended since obstructions may not be obvious. Columns may not be noticed if the scaffold rolls out of line, and what may appear to be an obstruction-free surface could suddenly become obstructed by an unrelated worker at the job placing some material in the path of the scaffold. The resulting shock as a caster hits such an obstruction could cause a rider to fall on the platform. When moving all towers, whether using outriggers or other means of lateral support or not, the scaffold should preferably be moved in the direction of its longest base dimension and the motive force applied as close to the base as possible. Casters must always be locked when the scaffold is in use or being climbed.

Interior Scaffolds

The application of large-area interior scaffolds is limitless; access to almost any wall or ceiling surface can be achieved using ingenuity *and* preplanning the use of the correct various accessories. Typical scaffolds often use spanning trusses (to minimize the amount of vertical support components), cantilevered short trusses, brackets, and tube and couplers to reach the various surfaces. All open sides of such scaffolds require guardrails to be installed in accordance with the criteria mentioned earlier. Availability of variously sized cross braces facilitates dimensional adaptability to existing seat row spacings and similar obstructions. However, depending on the availability of special accessories and the experience of available erectors, such interiors may well be just as conveniently scaffolded using tube and coupler scaffolds. Sometimes building configurations are such that, without expert help to lay out the work, the time needed to make sectional scaffolding fit is greater than the normal time saved when using sectional scaffolds instead of simple scaffolds; thereby both methods may be equally feasible.

There are occasions when economies are achievable by spanning prefabricated platforms between stationary rows of sectional scaffold. Such platforms as used for two-point suspended swinging scaffolds are sometimes used for this purpose. They are normally available in sizes up to 28 in (71 cm) wide by 32 to 36 ft (9.75 to 11.0 m) long. If use of this method is proposed, extra precautions must be taken. It is not safe to rest the end of such a platform (usually aluminum) on one or two wood planks running atop the sectional scaffold; the point load reactions are excessive, and safely securing the platforms is difficult. This method is favored when only a portion of the area needs to be worked on at one time, necessitating frequent relocation of the (heavy) aluminum platforms. Depending on the work (i.e., spot work such as electrical or decorative work requiring a reasonably large contiguous work area), guardrails must be used along the sides of the platforms when the platforms are not directly adjacent to one another. A major difference between using this method and using rows of scaffolds connected by trusses is that the latter provides bracing rigidity between the scaffold rows; the U-bolted connections used with the platforms do not provide this feature so that on scaffolds over 20 ft (6 m) high, alternative, separate means of stabilizing the scaffold rows is required.

Outrigger Scaffolds

These consist of a series of outriggers, or thrust-outs, projecting from the face of a structure. The outboard cantilevered ends support a planked platform or a built-up scaffold. The inboard end must be anchored by securely strutting it to a floor slab or beam above it or restrained from movement by a positive means such as U bolts passing over the outrigger and (1) passing through the floor or (2) connecting to looped steel inserts previously set in a concrete floor. The outriggers must be securely braced against lateral overturning, and the point of bearing at the fulcrum point of the beam is required (by OSHA) "to extend at least 6 inches in each horizontal dimension." This last statement is not clear since it could mean either a 6-in (15-cm)-square wood block or other surface, or an extension of 6 in (15 cm) in all horizontal directions from the fulcrum center point. Although this type of scaffold originated with wood beam outriggers, current practice is to use steel beams or sectional steel trusses. From the steel outriggers, built-up sectional scaffold can be erected on the cantilevered end or, when necessary, built down, or hung, from the outrigger. Figure 16.28 shows a scaffold such as used for surface finishing (such as thin stripes of brick veneer, special mortar course pointing, flashing, caulking) where the work is occasional and not sufficiently continuous to require a two-point suspension scaffold. Figure 16.29 shows a scaffold used to reach *down* from the outrigger level using a truss as an outrigger. In calculation of uplift restraint forces at the inboard end, the previously mentioned safety factor of 4 is applicable using 4 times the live and dead loads of the workers, material, and equipment supported by the cantilevered part of the beam.

Shipbuilding Scaffolds (Steel)

These are generally specialized semifabricated steel components, having evolved over the years from a combination of tube and coupler scaffolding and sectional frame scaffolding. Special OSHA maritime regulations cover shipbuilding and repairs: subparts 1915, 1916, 1917, and 1918.

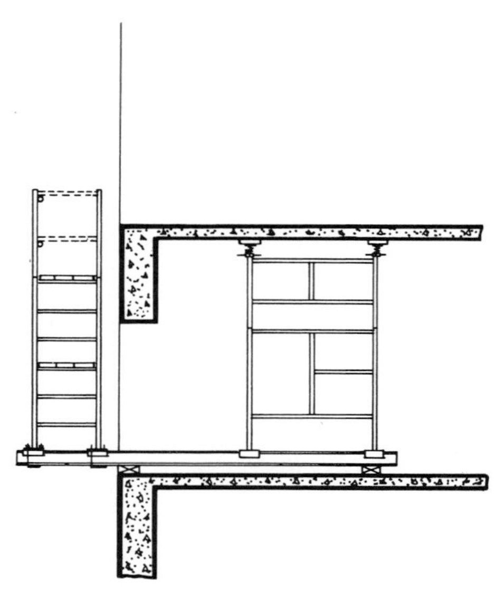

FIG. 16.28 Outrigger scaffold braced against floors inside the building.

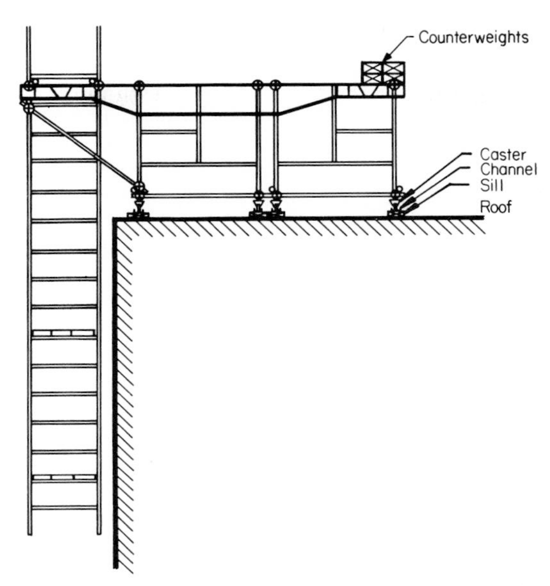

FIG. 16.29 Outrigger rolling scaffold for work below the roofline.

Wood Built-up Scaffolds

It is not intended to describe wood scaffolds in any detail. The OSHA and ANSI regulations give more complete and comprehensive guidelines for wood scaffolds than for any other type. One seldom sees wood scaffolds in any quantity, their use being generally restricted to low-rise housing. Many states and cities still give extensive construction details in regulatory bylaws. In principle, the construction of wood scaffolds is essentially the same as that of tube and coupler scaffolds; however, since the common fastening is nails, which have little moment rigidity, wood scaffolds require substantial amounts of diagonal bracing in both transverse and longitudinal planes. Construction labor costs are high in comparison with those for metal scaffolds. Carpenters' bracket scaffolds, bricklayers' square scaffolds, and horse scaffolds are prefabricated wood modules used essentially on small housing projects. They fall into the same categories as built-up wood scaffolds.

There are also a number of wood and steel bracket scaffolds with restricted use for formwork platforms, steelworkers platforms, and the like. These are quite specialized in nature and of little value for generalized use.

SUSPENDED (HANGING) SCAFFOLDS

Multiple-Point Suspension Scaffolds

The first mechanization of equipment specifically designed for use as scaffolding occurred between 1870 and 1880. The concept involved simple, manually operated, exposed-gear winches, used in pairs at regular intervals along the length of a building face and developed for high-rise masonry work. Each pair of winches supported a steel "putlog" which served as a bearer to support longitudinal planking. Each winch had a toothed drum upon which ½-in (12.7-mm) steel wire rope was wound by a simple ratchet lever mechanism. The wire ropes were suspended in pairs from I-beam outriggers cantilevered from a floor or roof of the building. The inboard ends of the outriggers were anchored by U bolts to the concrete slab or structural steel beams of the building. Cranking the ratchet lever of each pair of winches simultaneously wound the wire rope on the drums and raised the putlog at that position. Successive cranking of each pair of drums was repeated along the length of the scaffold in approximately 9- to 12-in (23- to 30-cm) increments so that the whole scaffold platform was constantly raised to keep the masons at an optimum working height in relation to the top course of masonry.

This scaffolding continues to be popular 100 years later, having been subjected to minor refinements over the years (Fig. 16.30). It enjoys rather unique terminology; two winch mechanisms, or drums, plus their connecting putlog (bearer), plus the outrigger I beam and its anchorage bolts were collectively known as a "machine" and continue to be termed thus. Originally, the platforms were 5 ft (150 cm) wide; 6-ft 6-in and 8-ft (200- and 240-cm) platforms became available, so as to give more access room for the passage of materials. Putlogs were developed which gave one-plank or two-plank insets between the inside scaffold winch and the masonry wall so that masons would work uninterrupted by passage of materials. Overhead canopy attachments were added to provide the masons with some protection from hazards caused by other work being done overhead, and this is now an OSHA requirement.

The inset plank methods were later refined to be positioned 16 to 20 in (40 to 50 cm) below the material platform. This lessened fatigue resulting from the masons needing to constantly bend over to reach their brick and mortar. The wider 6-ft 6-in and 8-ft (200- and 240-cm) width platforms required stacked pairs of outriggers to obtain the necessary cantilever strength. The outriggers used were once a standard size popularly used as intermediate joists between larger structural beams, being 7-in by 15.3-lb (18-cm by 7-kg) beams, 15 ft (4.5 m) long.

FIG. 16.30 "Gold Medal" suspended scaffolding for masonry work all around the building. *(Courtesy of The Patent Scaffolding Co., Inc.)*

This is no longer a popular size, but they are still available in large quantities from suppliers of this equipment.

Multiply stacked 7-in (18-cm) beams become logically disfavored because of inconvenience. In some areas of the country 8-in (20-cm), 20-ft (6-m) long steel and aluminum beams are preferred but of limited availability.

For use with stone masonry, methods were developed to use the 7-in, 15.3-lb (18-cm, 7-kg) beams as monorails to facilitate movement of stone blocks along the scaffold from an end lifting location, using 1- or 2-ton trolleys and chain hoists. The use of this type of equipment on even taller skyscrapers involved a special procedure known as "jumping," since the machines were, and still are, typically equipped with 100- or 150-ft (30- or 45-m) lengths of steel wire rope. The method involves use of extension cables attached to the outriggers at the top and to the drum cables at their lower levels. When the scaffold rises to the joint between the cables, a tying-off procedure is carried out and a level of extension cables removed. This is repeated until the scaffold is high enough so that the drum cables can be directly attached to the outriggers. The determination whether to use extension cables or outrigger "jumps" is generally based upon the rate of rise of the building frame and the timing of the beginning of the masonry construction. Unless the machines are removed by derrick or crane at the top of the building, extension cables are often used in reverse sequence to wind down the scaffold.

As with all other scaffolding components, the universal safety factor of 4 is used except for the steel wire ropes, which OSHA requires to have a safety factor of 6. The labor required to install and remove suspended scaffolds of this type is high. Steel frame scaffolds are cheaper overall, but costs rise geometrically with height so that for any given project, there is a break-even height above which the suspended scaffold is more economical overall for masonry. Approximately, this break-even height varies between 80 and 120 ft (24 and 36 m). Naturally, high buildings tend to use masonry less and less because of the availability of a multiplicity of other exterior cladding materials, often installed from the building floors and requiring no external scaffolding. Much institutional work using federal and state funding continues to specify brick exteriors, which are preferred for their superior insulation qualities.

Two-Point Suspension Scaffolds (Swing Stages)

Originally conceived for shipbuilding and repairs and later adapted for construction work, this type of scaffold began as a planked platform supported at both ends by ropes suspended from an overhead anchorage. To obtain mechanical advantage, multipart blocks were used at the platform ends to make it easier for people to lift their own weight (see Fig. 16.31). Such platforms are still used on ships, by some light-work building trades, and for some window cleaning and painting operations.

To provide additional safety and security, manually operated winches were developed using steel wire ropes wound on drums. These winches were more sophisticated mechanisms than those used for multiple-point suspension scaffolds. Raising was accomplished by a long ratchet lever direct drive to a toothed drum, lowering being accomplished by means of a handle crank driving through a worm reduction gear. Subsequently, electric or air-powered winches were

FIG. 16.31 Typical two-point swinging scaffold. This type of scaffold consists of a platform suspended by two manually operated Gold Medal Junior machines. Manually operated scaffolds such as this are ideal for tuck-pointing, caulking, and similar light-duty work which does not require the higher speed obtainable with electric or air-powered machines. (*Courtesy of The Patent Scaffolding Co., Inc.*)

developed to replace the manual winches to achieve faster vertical relocation of the platform and elimination of physical effort for raising and lowering.

Platforms

Early platforms were merely planks spanning between the winches, which limited platform length. These evolved into use of planked ladders lying flat, which were relatively unsafe, to extend the span and give increased lateral working area for efficiency. Eventually, the ladder type platform was developed using stronger siderails and supporting light planks or wooden slats on top of the rungs.

The demands by industry and construction trades for lighter, easier handling equipment spurred the development of aluminum platforms, generally designed with tubular or extruded I sections or C sections for siderails and with aluminum rungs to support either plywood or extruded aluminum slats; the latter is the most popular form.

Metal platforms are available up to 40 ft (12 m) long. However, the norm is limited to approximately 32 ft (10 m) for handling reasons. The design of power winches became more sophisticated, embodying safety devices to sense and prevent (1) too high a rate of speed when lowering, (2) slippage of wire rope around the drum, (3) rope slippage when power is off, and (4) overloading. Powered winches are of two major types: drum wound and traction. "Drum wound" is self-explanatory; the "traction" type is designed to have the wire rope reeved around grooved drums for between three and four wraps, one of the drums being driven by the electric or air motor. The rope passes through the winch and exits at the bottom; thus the winch and supported platform effectively *climb* the rope. The traction type is generally preferred over the drum type, since the latter is hampered by the weight of the wire rope wound and stored on the drum. It limits practical application to low heights compared to the traction type which can be operated for any height over which the wire rope is long enough to extend. Heights of 1200 ft (365 m) and greater have been scaffolded with these machines. The wire rope, of course, needs to be a single length; it is not practical or safe to use with tying-off "jumping" procedures as for multiple-point suspended scaffolds.

Metal platforms for swinging scaffolds are required by OSHA to be "approved" by Underwriters Laboratories in accordance with "rated" safe load capacity, which must be indelibly marked on the platform and/or have a load rating affixed to it in a manner and format prescribed by UL. The generally used load ratings are for: (1) two workers and hand tools not exceeding 500 lb (226 kg) and (2) three workers and hand tools not exceeding 750 lb (340 kg). These loads are applied within 3 ft (90 cm) of the center of the platform span. The maximum rated load is multiplied by 4 for testing in accordance with the standard requirement for a safety factor of 4.

Similarly, all winches for swinging scaffolds must be of a design tested and approved by UL per OSHA regulations. UL labels are now restricted to the wording "UL-tested for load capacity." The wording or connotation of "approval" is not implied by them for scaffolding devices. In some sections OSHA calls for "testing and approval"—in others "testing and listing" as well as "meeting the requirements of Underwriters Laboratories." All such statements call for such testing by Underwriters Laboratories or Factory Mutual Engineering Corp. The latter does not test or certify any scaffolding components.

Rigging

Anchorage of wire ropes for scaffolding winches is accomplished by three basic methods: (1) positive attachment to solid and secure structural elements of the building, (2) attachment to "cornice hooks," also known as "S hooks" used over a building cornice or a parapet wall, (3) attachment to the ends of cantilevered outriggers. When using outriggers, the inboard ends must be positively secured against uplift by means of positive anchorage to the building or by counterweighting. Counterweighting methods used to be extremely hazardous since field personnel were often uninformed of the requirements of using a safety factor of 4 against overturning so that in many cases far too little counterweight was applied.

Methods used often consisted of bags of flowable material such as sand, which obviously exposed a hazard of the bags splitting and losing their contents. Also, heavy concrete blocks were sometimes used atop planking or plywood. Both methods seldom included provision for positive attachment of the counterweight material to the outrigger beam creating obvious hazards in accidental removal.

The current and preferred counterweighting method is to use steel or cast-iron counterweight blocking which can be positively attached to special brackets bolted to the outrigger beams. For handling ease, these blocks are generally made to weigh 50 lb (22.6 kg) each. The original conception of multiple, attachable counterweights used specially made blocks of poured concrete. An attendant problem with these is the hazard that with rough usage over a period of time they can be chipped and thereby become progressively lighter over the years. Strict inspection and control to maintain the original poured concrete weight is necessary.

A development of the I-beam outrigger method is to mount pairs of them upon interconnected rolling towers, often known as "benchwork," so that the total rigging and counterweighting can be quickly relocated at successive lateral positions as required by the work. Such an application is shown in Fig. 16.19. It is important to follow the proper method of calculating the necessary counterweights which can use the self-weight of the benchwork to reduce the amount of separate counterweights to be placed at the rear of the towers. Seating of the outriggers can be done in two ways: U-bolted to the top head bars of the scaffold frames located in U heads or bolted concentrically over a frame leg. The latter must be carefully calculated to avoid overload of the caster supporting that leg. The head-bar method is generally preferred since the reaction loads at the fulcrum are shared by *two* legs and *two* casters if the outrigger is centrally placed on the frame.

Single-Point Suspension Bosun Chairs

For occasional spot work, a single-point suspension scaffold for use by one worker is often desirable. These generally contain a single powered winch mounted in a "work cage." These are often extended by adding additional components, often called "extension baskets" to both sides of the work cage for use by two workers—one in each extension. The weights of each worker plus tools should be carefully arranged to be approximately equal; otherwise imbalance will be experienced causing the assembly to hang at an angle other than horizontal since the point of lifting of the winch is at the center. When raising or lowering the winch, one worker has to operate the controls at the winch center. To avoid an imbalance hazard this means that both workers should relocate to the center section while raising or lowering the winch. There is economy of equipment involved here, to the detriment of safety. Realistically, if two workers are required, it could be done more safely by using a two-point suspension scaffold with a short platform. Some one-person cages are available using manually operated winches.

Although it is possibly a misprint or unfortunate word choice, the OSHA requirement for single-point scaffolds (as differentiated from a bosun chair) is that the *complete scaffold* must be tested by UL, whereas only the *winches* are required to be UL-tested with all other types of suspended scaffolds, plus the platforms for two-point suspended scaffolds.

The original derivation of a one-person scaffold is a bosun chair, consisting of a wood seat, with rope slings attached to a multipart rope block and rope falls. This was used originally for ship work. Current technology generally uses a metal or plastic formed seat attached to a powered winch. Rigging for both types of single point should follow the same criteria as for two-point suspended scaffolds (swing stages).

Stone Setters' Adjustable Multiple-Point Suspension Scaffolds

These are used now only on rare occasions. They generally utilize a combination of two side-by-side winch drums using a common lifting-lowering ratchet mechanism. The cable from one winch half is connected vertically to the end of an outrigger beam above the outside face of the scaffold; the other cable passes along a cantilevered arm under the scaffold platform, over a pulley at the inboard, and thence vertically connected to an inner position on the same outrigger beam. Two-point suspension-type platforms are placed between pairs of the machines. The platforms are positioned so that the end of each closely abuts the end of similar platform(s) adjacent to it. Having the winch mechanism on the outside face of the platform enables clear egress from one platform to the next. It was originally used for stonemasons in circumstances ensuring that material was not stored on the platform—i.e., it was handled by monorail, crane, or derrick. Generally, this type of scaffold is known as a "twin-drum" scaffold for obvious reasons; it is essentially a marriage between a multiple-point suspension masons' scaffold and a two-point suspension scaffold. Having two cables supporting the inner and outer sides of the platform gives enhanced stability over that of a two-point scaffold which has a tendency to tilt transversely as workers move between the inner and outer edges of the platform. Normally, these scaffolds are used with overhead canopy attachments which support planking to guard against overhead hazards. Also, when not mated to an adjoining platform, chain guardrails are used across the open end(s) of the scaffold.

Safety Considerations for Suspended Scaffolds

Multiple-point suspended scaffolds are usually constructed using wire mesh between the guardrail and the toe board; OSHA and ANSI call for this to be of ½-in (12.7-mm) mesh and made from no. 18 gauge U.S. standard wire. This should be no. 19 gauge since no. 18 has not been commercially available as mesh for many years. Safety belts and lifelines are not required for workers on the scaffold. During erection of outriggers, safety belts and lanyards should be used in accordance with prevailing standard practice for structural steel work. During jumping and tying-off operations, it is recommended that belts and lanyards be tied off to the building or some part of the scaffold.

Safety belts, lanyards, and lifelines (safety lines) should be used at all times when working on two-point suspension scaffolds and bosun chairs. Multitiered two-point platform scaffolds present a problem; use of a safety belt and separate lifeline could result in an upper platform tier striking someone, should one end of the scaffold give way when a worker is being held at a relatively fixed height on the lifeline. For this use, the safety line is often allowed to be attached along the length of the scaffold, on the supposition that it would be rare that *both* end cables or overhead anchorage points would give way.

Similarly, safety lines and belts are not required by OSHA for use with single-point suspension scaffolds since a like chain of events could occur as with multitiered two-point scaffolds. The overhead fairlead entry at the top of the cage could strike someone who has been arrested by a lifeline safety device.

Other Applications

Many scaffolding applications involve the use of two or more different types of scaffold. Tubes and couplers can be used to greatly enhance the flexibility of sectional scaffolding when used in conjunction with them to provide lacing and ties between disconnected sections, to brace a high tower to a spread base, and to provide additional work levels not achievable with standard sectional scaffolding accessories.

Scaffolding winches suspended from roof steel or similar anchorage can be used to provide a suspended working platform with no support being required from the ground. The winches support bearers and runners of tubes and couplers or steel I-beams which in turn support a planked platform. OSHA refers to this type of scaffold as an interior hung scaffold.

The use of rolling bench works to support swinging scaffolds has already been discussed. Human ingenuity is the only limiting factor in providing solutions to scaffolding problems, and it is important to keep in mind the facility of combining two or more disparate scaffolding products into a safe access or support system.

SAFE PRACTICES

We cannot cover all safety aspects of scaffolding, but some highlights are given in the following comments. The majority of scaffolding manufacturers through their direct operations and through their agents and distributors normally provide free safety rules and instructions with all shipments of equipment. These safety rules contain detailed instructions on safe use and proper construction. The Scaffolding, Shoring and Forming Institute, Inc. (SSFI), besides providing safety rules through its industry members, also has available booklets on safe erection of certain products. The user should ensure that copies of such publications are obtained from a member supplier or directly from the institute and should most importantly ensure that they reach the field personnel who will be building and *using* the scaffolding.

An important reference for the industry is the *Safety Requirements for Scaffolding,* ANSI A10.8-1977, by the American National Standards Institute, Inc., 1430 Broadway, New York, NY 10018.

Guardrails

As mentioned, for all scaffolds OSHA and ANSI call for use of 2 × 4-in (5 × 10 cm) guardrails and 1 × 6-in (2.5 × 15-cm) midrails. Attachment of wood guardrails to steel components should not be considered normal. In most cases installation can be achieved only by using (often inadequate) wire binding. Use the metal or other guardrails made for the particular brand of equipment being used.

Guardrailing can be relocated continuously to different working levels, provided that further work will not be necessary at a vacated level; if the planking is left in place, there is no collateral need for guardrails unless the platform is to be *reused* at a later time.

In the absence of guardrailing designed for use in conjunction with any particular type of scaffolding, expedient measures can be taken to assure safe conditions and compliance. Tube and coupler scaffolds use tubes and couplers for guardrailing working levels; tubing can also be used for guardrailing of sectional steel scaffolding, being attached to frame legs with 1⅝- × 2-in (41- × 50.8-mm) standard couplers. Similarly, manila, synthetic, and steel wire ropes can also be used for guardrails and midrails; ropes of any kind are easy to attach by simply making a loop or a half-hitch around an upright frame leg or post. To comply with the requirement for "minimum deflection," they must be strung tautly along the scaffold. Rope guardrails are particularly useful in guardrailing intermediate platform levels which may be occupied infrequently for uses such as inspection or light repairs. Most sectional scaffolding frames do not have built-in locking devices at levels suitable for attaching guardrailing for a simple reason: most scaffold frames are connected with cross bracing positioned in the vertical plane immediately *inside* the frame legs, thereby occupying the physical space in which a leg-attached guardrail would have to be located. However, separate, attachable devices are available for such use.

There exists an uncertainty in current regulations about guardrails being furnished and installed to comply with the *intent* of the regulations; minor differences of dimensions in specific instances should not be the subject of jobsite controversy. Citations should not be issued for regulation infringement on minor points. The lack of *any* guardrailing on scaffolds is frequently observed all around the country, and *this omission* should be the prime target of safety enforcement.

Toeboards

Toeboards have been discussed. They are required wherever guardrails are installed. Many prefabricated planks and platforms have hinged toeboards or "drop-in" receptacles to receive them. For use with normal planking, various devices are available to facilitate toeboard positioning.

Planking

Planking has been covered; however, it is a regulation that, when overlapped on the scaffold support member, the overlap is required to be a minimum of 6 in (15 cm) and a maximum of 12 in (30 cm). Variously, some sections of OSHA require that planking be laid "tight," implying that there should be no openings through which tools and rubble could fall. Using five full or nominal 10-in (25-cm) planks with 5-ft (150-cm)-wide frames results in a *tightly* planked width between 51 and 46½ in (129 and 118 cm) so that to plank the full width of the scaffold frame [58⅜ in (148 cm) between frame legs] there is between 7⅜ and 11⅞ in (18.7 and 30.1 cm) unplanked space. Where there is no work in progress beneath the platform, distributing this excess space by leaving gaps in the planking is reasonable; where there is need to protect persons or property under the platform, plywood sheets should be laid atop the planking. Prefabricated planks are designed to fill the whole 58⅜-in (148-cm) interior frame width, being from 19 to 19⅜ in (48 to 49 cm) in width.

To provide continuous planking around corners, planks must be placed so as to avoid tipping; to avoid overloading the side planks of a platform, overlapped planks at right angles to each other should extend across the whole platform so that the load at the side is distributed over other planks when the most heavily loaded planks deflect. Where possible to lay planks diagonally across a corner, the diagonal planks should be placed first, running from bearer to bearer; thence the planking from the two adjoining elevations of scaffold are placed atop the diagonal planks. Under *these* conditions, the previous 6 to 12 in (15 to 30 cm) overlap does not apply and the diagonal planks must overlap their bearers by *at least* 12 in (30 cm).

If planks are to be butted to provide a smooth walking surface, ends of butted planks should be seated on separate bearers or equivalent solid seating. This is achieved simply by using additional bearers with tube and coupler scaffolds attached to runners. With sectional frame scaffolding various means of lumber packing can be used, or tube and coupler runners can be placed under the frame head bars. These, in turn, can support the additional required bearers. Where dislodgement of planks is possible, they should be cleated, nailed down, and/or prevented from uplift by wiring or some combination of all three. Where necessary to place planks at an angle other than horizontal, as for ramps or elevation changes of the platform for work on variable-height ceilings, any slope greater than 1:10 should have cleats nailed to the tops of the planks 12 in (30 cm) apart. Debris and rubble should not be allowed to accumulate on planking and should be removed as quickly as possible. Platforms should be cleared of ice or snow before being used for work.

Ladders

Ladders have been discussed previously; some additional comments are pertinent. Unless manufactured portable wood or metal ladders or job-built cleated ladders are used, there is no specific requirement that rungs be spaced 12 in (30 cm) apart. Variations from 12-in (30-cm) rung spacing are necessary with scaffolds for very simple reasons: frames may not be of even 1-ft (30.48-cm) heights, and where they are, the additional 1-in (2.54-cm) collar of the typical coupling pin adds 1 in (2.54 cm) to each frame tier. Therefore, three 5-ft (150-cm)-high frames give a height of 15 ft 2 in (4.62 m) to the top bearer; one 6-ft 6-in (1.98-m)-high frame plus two 5-ft (1.52-m)-high frames give a top bearer height of 16 ft 8 in (5.08 m). Common sense requires that rungs be relatively *evenly* spaced, but not necessarily exactly spaced, and in most cases they cannot be at an even 12 in (30 cm) unless some nonstandard increments are used from the ground to the lowest rung and from the uppermost rung to the platform. Pitched ladders are generally required to be placed with an overrun of 36 in (91 cm) above the top bearing support of the ladder; used at the side or end of a scaffold, this requires a person to have to jump down from the guardrail level from the platform or, as is unfortunately prevalent, the guardrails are *removed* at that point. The most important approach in applying finite restricting dimensions to scaffold access and guardrailing is to remember that by nature, all scaffolds are variable structures, with built-in flexibility for nonstandard and vastly varying circumstances and conditions. Common sense, practical measures, and broad interpretations of dimensional regulations should prevail. An excellent example is the 12-in (30-cm) ladder spacing mentioned before; there are no standards or studies which can show that 12-in (30-cm)-rung spacing is inherently more or less safe than, say, 10, 11½, or 13½ in (25, 29, or 34 cm). The 12-in (30-cm) spacing is convenient, traditional, and standard for building portable or fixed industrial ladders, which does not mean that scaffolds must be designed to comply with ladder regulations; rather they are products which must be evaluated on what they can achieve as a building tool having completely individual and separate form from any other product since there are no other comparable products which fulfill a similar function.

17

Temporary Bracing and Guying

by Wheeler H. Rucker

The design of structures is generally premised on load distributions which are applicable only after completion of the total structure. During construction, gravity loads are often supported by temporary shoring and lateral loads by use of temporary braces or guys. The considerations and methods of support of gravity loads are discussed elsewhere in this handbook. In this section, we shall review generally the sources of the lateral loads to be restrained as well as the principles to be considered in the development of systems of braces and/or guys to provide adequate lateral restraint. Although development, analysis, and detailing of an effective and economical lateral bracing system for a specific application may require a very sophisticated approach, the principles to be followed for all systems are very simple. It is these principles that will be described and illustrated in this chapter.

First and by way of definition, the term "guy" as used herein refers to a tension member normally installed at some angle other than vertical or horizontal, and the term "brace" refers to a compression member likewise installed at some angle other than vertical or horizontal.

Temporary bracing and guying may be installed either externally to or internally within the permanent structure under construction, and in either event they will function until the integrity of the permanent structure is such that it will withstand the lateral loads independently of the bracing and guying, at which time the braces and guys are removed.

Temporary bracing is generally designed as pin-connected, axially loaded members having both vertical and horizontal loads or reactions at each end. Although it is the horizontal component of this load or reaction that we are primarily concerned with, we must not lose sight of the effect of the vertical component. Because of the temporary and removable purpose of bracing and guying, attention must also be given to the details of their attachment to the structure and to the effect of any elastic deformation of the structure during erection. In many instances, it is desirable to include adjustable elements such as turnbuckles in cables or jacks in struts and braces for the facilitation of both installation and removal of the guy or brace.

In the development of a system of bracing and guying to be used in the construction or erection of a structure, it is most important that the system be consistent with the procedure actually to be followed in the pursuance of the work. This seems like a very simple criterion, and yet it is the one which, when ignored, most frequently results in failure of the structure. If the design assumptions are not consistent with the actual work sequence or procedure, the lateral restraint provided through the bracing and guying system will very possibly not be consistent with the needs of the structure under construction at certain times.

In this chapter we shall discuss the general philosophy and procedure to be followed in the design of bracing and guying, erection considerations, and coordination of the total bracing system into the overall program.

DESIGN LOADS

The normal code-specified design wind and seismic loadings are as applicable to the design of temporary bracing and guying systems as they are to the design of the permanent structure. In addition to these loads, however, are various loadings which are unique to the structure under construction. Typical of these are loads applied through acceleration or deceleration of equipment operating on the structure during construction, soils loadings which may be different during construction than after completion of the structure, and other loads applied by the structure which may be different during construction than after completion, due to various intermediate stages of geometry of the structure as it is being constructed.

Other chapters of this handbook discuss construction loadings. It is left to the designer of the lateral bracing system to identify which of these construction loads are applicable and the extent to which they are applicable at each of the various stages of construction. Once identified, it is merely a matter of form to accommodate these loads and forces in the bracing and guying system to be employed.

In addition to the design load, the designer must consider the allowable stresses to be used in the design. In general, and notwithstanding the "temporary" nature of the bracing and guying system, it is the generally accepted practice to use the same allowable stresses as for permanent use in the applicable code.

GUYS

Guys are tension members, normally comprised of one or more cables (usually wire rope) tensioned by turnbuckles or other means, depending on design requirements. Although a number of different wire rope fittings are available on the market, by far the most common means of attachment of wire rope guys is by means of wire rope clips, as shown in Figs. 17.1 and 17.2. This type of clip, if properly installed, can be expected to develop up to 80% of the nominal strength of the rope. In the United States the safety factor used on the breaking strength of wire rope guy cables is between 3 and 7.

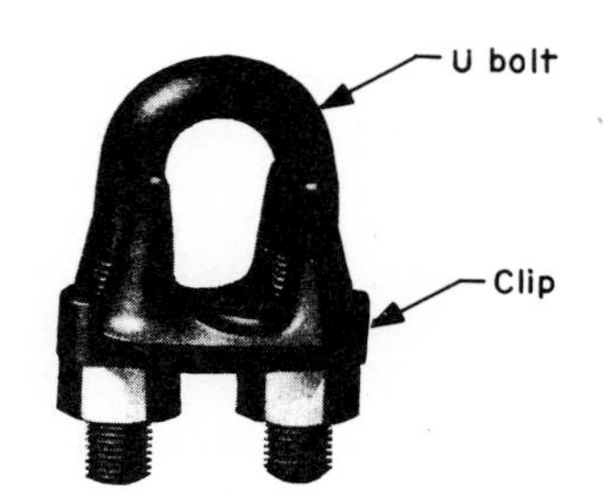

FIG. 17.1 Crosby-type wire rope clip.

Guys are normally installed to function as opposing pairs to provide the required lateral restraint. It should be noted, however, that temporary guys, even though they are initially tensioned somewhat, are generally regarded as passive members which are not loaded until a load or force is applied for them to withstand. Figures 17.3 and 17.4 illustrate two common systems of guying a building structure. In Fig. 17.3, compression is being introduced into the columns and into each of the two floors. In Fig. 17.4, tension is being introduced into the members of the upper-floor system, compression is introduced into the columns, and this compressive force is carried down to whatever supports the columns, and deadmen are required. Each of the two systems has its own application, depending on availability of space, equipment being used, the sequence of the operation, and other factors dictated by conditions at the jobsite.

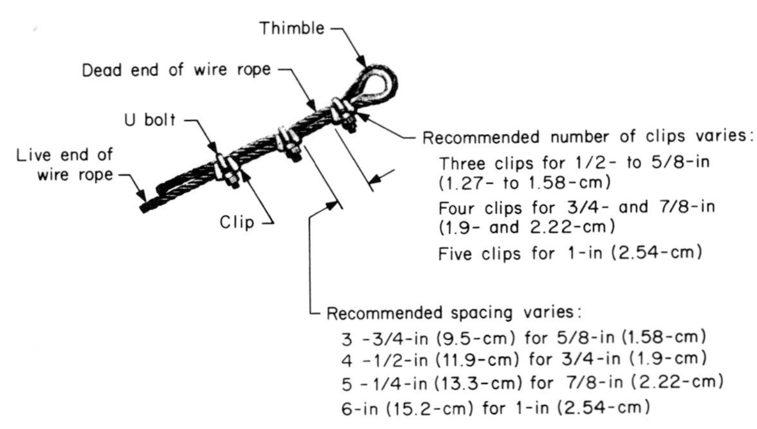

FIG. 17.2 Typical installation of wire rope clips. (Note: Consult manufacturer of wire rope and fittings for specific recommendations and allowable loads.)

In the design of either system, the lateral forces to be resisted are first determined. The vertical component of load in the guy is a function of the slope of the guy, and the effects of the various temporary loads from the guys on the permanent structure are a matter for analysis for the designer. If it is desired to preload the guys, this must be done on an equalized basis between opposing pairs so that net lateral forces are not introduced into the structure prior to its having the ability to withstand these.

It goes without saying that the guy lines must be attached to anchors which have the ability to withstand the load in the line. These anchors may be other portions of the structure, concrete foundations, or deadmen designed specifically for the purpose; in some instances, heavy equipment may be used as anchorage.

In the design of a deadman, a miminum safety factor of 1.33 should be observed for each of the following design considerations:

1. The weight of the deadman plus the weight of any earth on top of it must exceed the vertical component of load in the guy line.
2. The face of the deadman against the earth must have sufficient area against the passive pressure of the earth to exceed the horizontal component of load in the guy line.

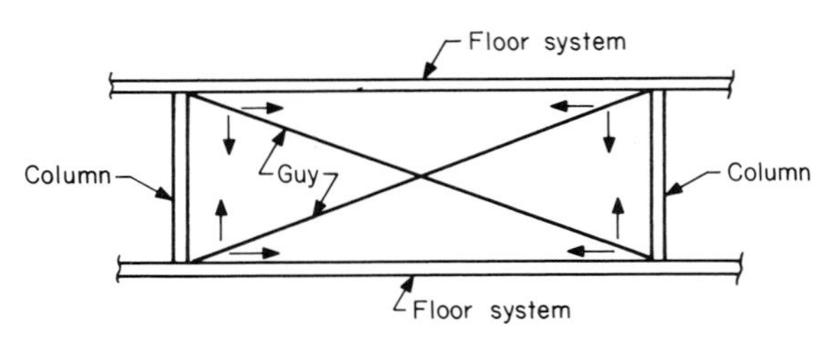

FIG. 17.3 Internal guy system.

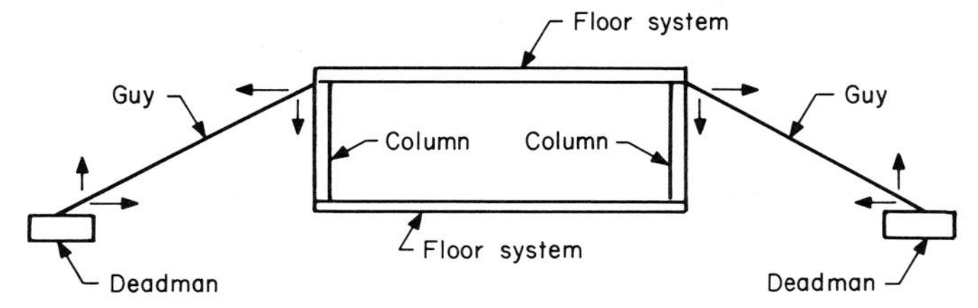

FIG. 17.4 External guy system.

3. The connection of the guy line to the deadman must be located so as to not cause overturning of the deadman.

Many methods have been developed through the years by engineers and riggers for attaching and tensioning wire rope guy lines. The method used for any particular application will vary depending on the personal preference of the personnel involved and the equipment available. However, whatever the method, it is imperative that the connections, linkages, shackles, sheaves, clips, and other hardware that are used be compatible with the wire rope used for the guy line. Design information on this hardware, as well as on the wire rope itself, is best obtained from the rope manufacturers, most of which produce excellent catalogs containing this information. For unusual or highly sensitive applications, it is recommended that the designer consult with one or more experienced riggers in the developmental stages of the design.

FIG. 17.5 Attachment of internal guy to structure.

FIG. 17.6 Internal system of guys.

FIG. 17.7 Attachment of guy to structure.

FIG. 17.8 Attachment of internal guy to structure.

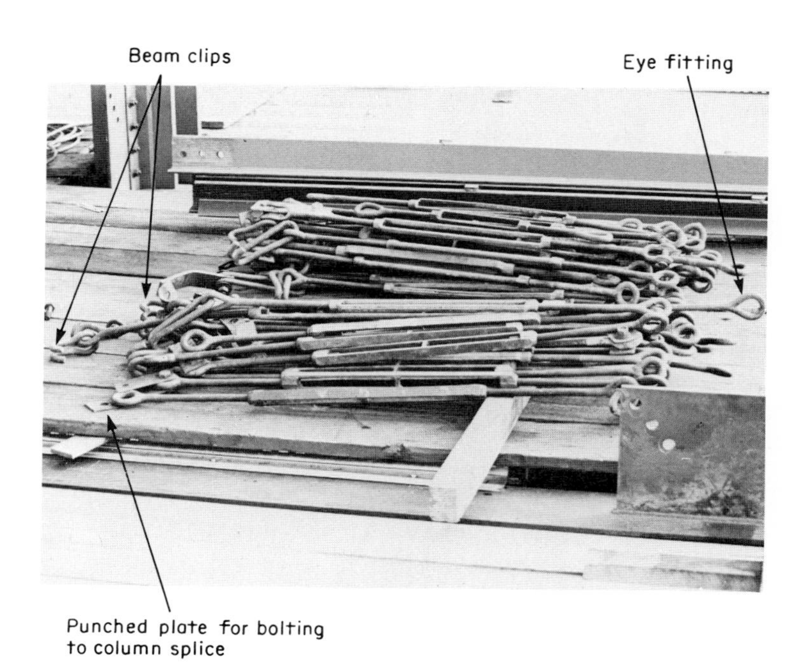

FIG. 17.9 Job stockpile of turnbuckles.

FIG. 17.10 External guy attached to deadman.

FIG. 17.11 External guys on a crane tower structure.

FIG. 17.12 Internal guying system.

Thus far in our comments about guys, we have limited our discussion to wire ropes used for this purpose. Although wire ropes are the most frequently used material for guying, they are by no means the only material. Steel shapes, timbers, and other materials used in construction will also function very well so long as the general constraints set forth herein are attended to in the design.

FIG. 17.13 Temporary lateral column brace.

FIG. 17.14 Temporary brace.

Figures 17.5 through 17.14 are photographs of a number of guying and bracing applications and materials representative of current practices.

BRACES

Braces are compression members usually made of structural-steel shapes or heavy timbers. They are inclined members, usually designed for axial load only. Although in some applications loads are introduced to them by means of calibrated hydraulic jacks, they are normally designed as passive members, carrying no load until it is applied through the structure which they are bracing.

Being inclined members designed for axial load, there are horizontal and vertical forces to be taken into account in both the design of the support for the brace and in the analysis of the structure or system which is being braced. Figure 17.15 illustrates a typical bracing installation where sheet piles have been driven around an area to be excavated and then bracing of the sheet-pile wall is installed concurrently with the excavation.

It will be noted from looking at Fig. 17.15 that vertical forces, as well as horizontal, must be attended to at the connection between the brace and the sheet-pile wall; and since the brace will tend to lift the sheet piling upward, the piles

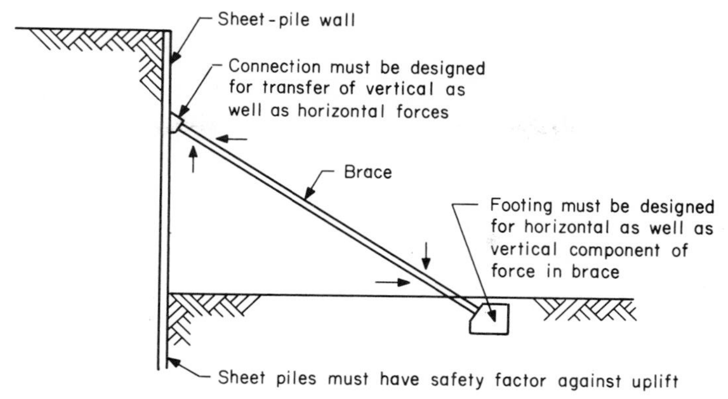

FIG. 17.15 Braced sheet-pile wall.

will have to be driven to a point where they have an adequate safety factor against uplift. In the absence of more stringent code requirements, the normal practice is to use a minimum safety factor of 1.33 for this application. Looking at the footing which supports the brace, we see that it must be designed for a vertical bearing load, that the face of the footing opposite the brace must have sufficient surface area against the passive earth pressure to prevent sliding, and that the brace must be supported by the footing at a location which will not cause overturning of the footing.

Figure 17.16 illustrates externally installed braces to provide lateral restraint to a floor system. With this application, it can be seen that the braces, in providing the lateral restraint, will introduce compression into the floor system and tension into the columns, and that further, they will tend to lift the columns from whatever supports them. The external footings for this application may be either cast-in-place or precast concrete or timber or steel grillage, so long as the footing is designed to have adequate vertical-load-bearing capacity, adequate bearing area to preclude lateral displacement, and geometry of the connection such that it will not overturn.

Internally opposed braces, as shown in Fig. 17.17, will introduce tension into the floor system, or compression into the floor system of the adjacent span, and tension into the column. They will also tend to lift the column from whatever

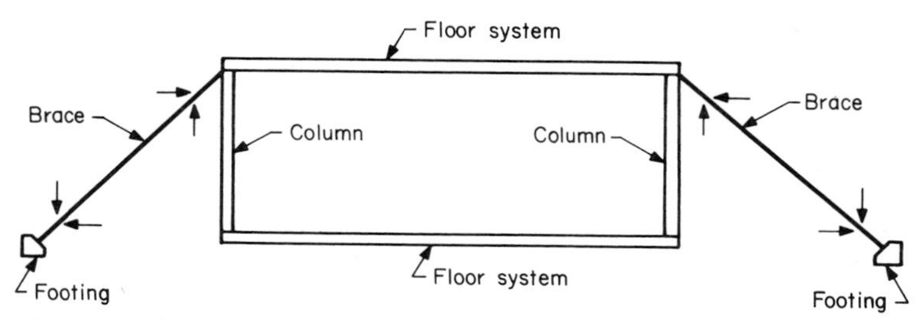

FIG. 17.16 Externally opposed braces.

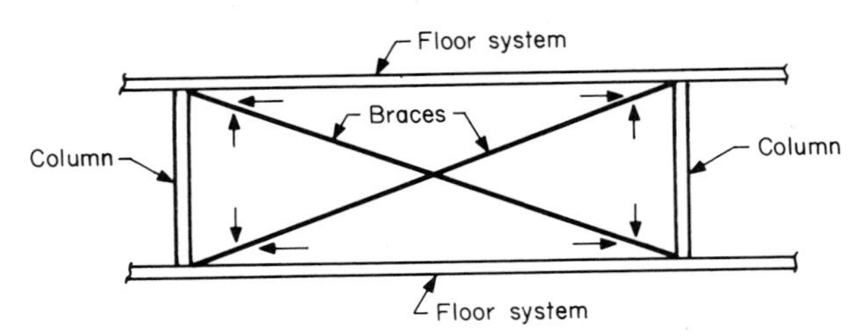

FIG. 17.17 Internally opposed braces.

supports it. It should be recognized that only one of the braces in each opposing pair will be loaded at any one time.

In the development of a system of bracing, it should be recognized that at some point it will be necessary to remove the system. To facilitate this, special attention should be given to the details of the connections, as the brace which was installed under a no-load condition may, in fact, be subjected to a load at the time it is to be removed. Provision must therefore be made, in some instances, for transfer of the load from the brace to the permanent structure prior to removal of the brace.

ERECTION TOLERANCES

One of the functions of a lateral-bracing system, in addition to providing temporary resistance to lateral loads which may be applied during construction or erection, is to provide the ability to laterally adjust the position of elements of the structure during construction so as to maintain the dimensions of the completed structure within the tolerances developed in the original design. For this reason, it is incumbent on the designer of the temporary bracing and guying system to become acquainted with the allowable tolerances of the original design and to make provisions in the design for the necessary adjustment by means of turnbuckles. However, many other systems and procedures are in popular use, ranging from the simplicity of wedges to highly complex equalized and calibrated hydraulic jacking systems.

ERECTION MANUALS

In many instances, failure of a temporary lateral-bracing system is due to the construction of the permanent structure "getting ahead" of the temporary bracing system. To preclude this possibility and to assure that all elements are braced as their needs develop, it is most important that an erection manual be prepared to illustrate the sequence of construction or erection; the arrangement of the temporary bracing or guying system during each phase of the construction; the method of installation and removal of the temporary system; the equipment to be used in the construction; the required tolerances and controlling dimensions; and all other factors which will have a bearing on the organization of the work of the contract. The erection manual should be keyed to the construction schedule.

DESIGN CALCULATIONS

Prior to commencing the design calculations for the temporary bracing and guying system, it is essential that three elements of information be obtained:

1. The complete drawings of the permanent structure to be constructed.
2. An understanding of the lateral forces which the permanent structure is designed to withstand, together with an understanding of the function and details of the system within the permanent structure which will act to withstand these lateral forces.
3. The proposed method and sequence of construction, including all equipment which will be used, where the equipment will be placed, and all other related information which would assist in identification of the individual elements to be braced and the lateral loads to which these will be subjected.

As calculations for the temporary lateral-bracing system proceed, depending on the complexity of the project, it will further be necessary to coordinate these with preparation of the erection manuals and with the schedule, and if necessary, to revise the method and sequence of erection.

In development of the detailed design, the next step is to outline the general geometry of the proposed lateral-bracing system for each stage of the construction program. The criteria for determination of this overall geometry should include consideration of several factors:

1. It should have the capability of withstanding the applied lateral loads, or combinations of loads, at each of the successive stages of construction.
2. It should not unduly interfere with required access to the construction, nor should it interfere with the construction itself.
3. It should facilitate lateral positioning of the structure as required to comply with design requirements.
4. It should be readily removable.

After the general geometry of the temporary lateral-bracing system has been developed and reconciled with all relevant parties involved in the construction program, it becomes time to proceed with the detailed design of the system. The classic steps of this phase are as follows:

1. Identify and define the intensity and manner of application of the loads and load combinations to be provided for in the design.
2. Identify the load paths for transfer of these loads and load combinations through the elements of both the permanent structure and the temporary lateral-bracing system.
3. Identify any special needs, such as adjustment capability, preloading capability, or specialized handling, and provide for these in the design.

4. Size the members, including, in addition to the temporary braces and guys, stress checks of the members of the permanent structure which may be loaded by the temporary system.
5. Detail the connections.

It is of utmost importance that the design and implementation of the temporary lateral-bracing system be coordinated with the field force that will accomplish the construction. For this reason, and prior to proceeding with the drawings and companion erection manuals, the entire program to date should at this stage be reviewed again with the relevant parties involved in the construction program to assure that all understand and support the program and system being pursued.

DRAWINGS

Drawings should be prepared reflecting the sequence of installation and removal, in addition to the details of the temporary bracing and guying system. The drawings should clearly define the status of the temporary system at each stage of construction and should be such that in addition to illustrating the properties and details of the construction elements, they also have an operational usefulness in planning of the procedure of construction around the installation and removal of the temporary bracing and guying system. The drawings, in addition to being necessary for the construction and removal of the temporary system, will serve as companion documents for the erection manuals and for the erection check sheets.

The drawings when completed shall illustrate the following:

1. Materials and geometry of the installation at each phase.
2. Details of connection.
3. Instructions for special procedures for the installation and/or removal.
4. Details and instructions related to provisions required for preloading of the temporary system or for lateral positioning of the permanent structure.
5. Illustrations of the extent and status of the temporary bracing and guying system at each of the various stages or phases of construction of the permanent structure.
6. Operational plans to illustrate equipment access, material flow, and the general method of procedure of the work during each of the various stages of utilization of the temporary bracing and guying system.
7. Any other considerations which may be unique to the project in hand and which will be useful in utilization of the temporary system.

INSPECTION

Effective inspection is a vital element in providing assurance of the continuous integrity of the temporary lateral-bracing system throughout the entire program of construction of the permanent structure. The person responsible for this inspection should be thoroughly conversant with the erection manuals and with the drawings and should use these to prepare a comprehensive checklist for use in observation and reporting of the status, procedures, and workmanship related to the temporary bracing and guying system and to the relationship of this temporary system to the permanent structure.

SUMMARY

In this chapter we have touched on some of the elemental aspects of temporary bracing and guying systems. By way of review, the following checklist is presented as a guide to the development and implementation of the total temporary lateral-bracing requirements for a single structure.

1. Identify and understand the lateral-bracing system of the completed structure.
2. Develop and define the method and procedure of construction.
3. Define the lateral forces and loads to be withstood by the temporary system, and design the bracing and guying system for these as defined herein.
4. Prepare erection manuals and drawings for the guidance of those responsible for the various elements of the lateral-bracing system.
5. Inspect the pursuance of the temporary bracing and guying system to assure that it is being accomplished in accordance with all of the criteria set forth in the erection manuals and drawings.

18

Temporary Structures in Concrete-Bridge Construction

by Man-Chung Tang

A bridge can be built in many different ways. For different construction methods, different temporary structures are required. As new materials and equipment are introduced into the market and new ideas are developed and improved, construction techniques become more and more variable. Only very few types of temporary structures can be standardized for general uses; most of them are modified or custom-made for the specific construction method of a specific bridge. In order to understand the function of a temporary structure, it is necessary first to understand what it is used for.

In each of the following sections a construction method will be described first and then the temporary structures for that construction method will be discussed. Some general remarks on special considerations in the design of temporary structures are given at the end of the chapter.

FALSEWORK MATERIALS AND DESIGN

Using falsework is the oldest way of bridge construction. It is still the most commonly used method for concrete bridges, especially for short-span bridges such as grade-separation overpasses, for structures with complicated configurations, and for low-level viaducts where soil conditions are good and the site is easily accessible.

Falsework is used to provide supports for the concrete formwork and the required working platform. The most common materials for falsework are steel and lumber. There are many patented steel falsework systems available on the market. The advantage of a patented system is that it can be assembled and reassembled easily to satisfy various height and width requirements. Most of the patented systems consist of four vertical posts (or legs) which are braced against each other by cross bracings to increase the buckling load. These posts and bracings are connected by field bolting to form modules. The common dimensions of these modules are 10 ft x 10 ft x 10 ft (3 m x 3 m x 3 m) or 8 ft x 8 ft x 10 ft (2.5 m x 2.5 m x 2.5 m), but other dimensions are also available. Some systems have triangular shapes with three posts; however, they are less stable and require more cross bracing. After the modules are assembled, they are combined together to provide the height, width, and length of the required support. Patented falsework systems are available in different load capacities: 20-, 40-, 50-, 80-, or 100-kip (89-, 178-, 222-, 356-, or 445-kN) working load per leg. Thus a heavy-duty falsework tower with 100-kip (445-kN) legs could support a 400-kip (1780-kN) load.

Many falsework systems come with handy accessories and hardware. Telescopic support at the top and bottom of a falsework tower can eliminate a lot of shimming for height adjustment. Spreader plates or grillage can distribute the load to a larger bearing area. Cross-beam assemblage on top of a falsework tower can distribute the load evenly to the legs.

If the bridge is wide and falsework towers are placed side by side to provide the required width of the support, they should be braced to each other to increase their stability against lateral loads. The actual loading transferred to each leg of the assemblage depends on the framing detail used and should be carefully investigated. Thus a tower with four 100-kip (445-kN) legs does not necessarily provide sufficient capacity for a 400-kip (1780-kN) reaction unless the cross beams on top of the towers are arranged in such a way as to transfer equal load to each leg.

Falsework can also be fabricated using steel profiles. Very often, contractors may have available H piles, wide flanges, or other profiles which can be welded or bolted together to form a falsework support. Sometimes overlength piles are driven and used simply as falsework columns.

Lumber is popular for low-profile falsework. The initial cost of lumber is lower than steel, which is advantageous if the material is not planned for reuse. It is also easier to use lumber for more complicated configurations. The main disadvantage of lumber is that it is not economical for tall supports and is not suitable for repeated use.

An important consideration in the design of falsework is the compressibility of its members and the settlement of its foundations. Large falsework settlement may cause unacceptable stresses or deformations in the bridge superstructure. Other points that should be considered and investigated are listed:

- Compatibility of deflections between falsework and adjacent piers.
- Joint take-up and uneven bearing at module connections.
- Temperature and shrinkage movements of the superstructure, which may induce horizontal forces and tilting of the falsework.
- Differential volume changes which may affect the distribution of vertical forces.
- Framing details above falsework towers should be designed to distribute the loads uniformly between the tower legs.

If the superstructure is cast in segments, the resultant reaction on the falsework along the bridge length may not be uniformly distributed. Depending upon segment length and pouring sequence, the variation in falsework reaction can be significant.

FREE-CANTILEVER CONSTRUCTION, CAST IN PLACE

For bridges spanning over navigation channels where traffic may not be interrupted, or spanning over terrain where falsework is not economical or not permitted, the free-cantilever construction method can be used advantageously. This method, since its introduction by U. Finsterwalder in 1952, has been used to build hundreds of concrete bridges around the world. Because of its high labor efficiency, it has become competitive in many cases even where local falsework is possible. Using this method, the construction starts by casting a short portion of the superstructure, usually between 35 and 50 ft (10.5 and 15 m) on top of a completed pier. This portion of the superstructure is referred to as the pier table or hammerhead and is cast by using local formwork supported either by brackets which are attached to the pier or on falsework which is supported from the ground. Two form travelers are then erected on top of this pier table. By successive casting of segments and advancing of form travelers in opposite directions, a pair of cantilevers is constructed. The superstructure of the bridge is then completed by closure pours connecting these cantilevers approximately at the middle of each span. Thus, except for the pier table, no falsework support is required.

The length of segments which are cast by means of form travelers may vary between 7 and 20 ft (2 and 6 m). The choice of segment length depends on the capacity of the form travelers and optimal use of the labor force. The work involved in completing one segment is called a work cycle. Each work cycle consists of the following steps: (1) placing reinforcement and tendons, (2) placing concrete, (3) curing concrete, (4) prestressing tendons, and (5) advancing form travelers.

The basic components for a free-cantilever construction are the form travelers, the brackets for the pier table, and local bracings for the closure pours. Sometimes an auxiliary bridge is used to transport personnel and material.

Form Travelers

There are various types of form travelers available from different suppliers. The main features, however, are practically the same; so only one type will be described (Fig. 18.1).

The basic elements of a form traveler are the main frame, the guide rails, and

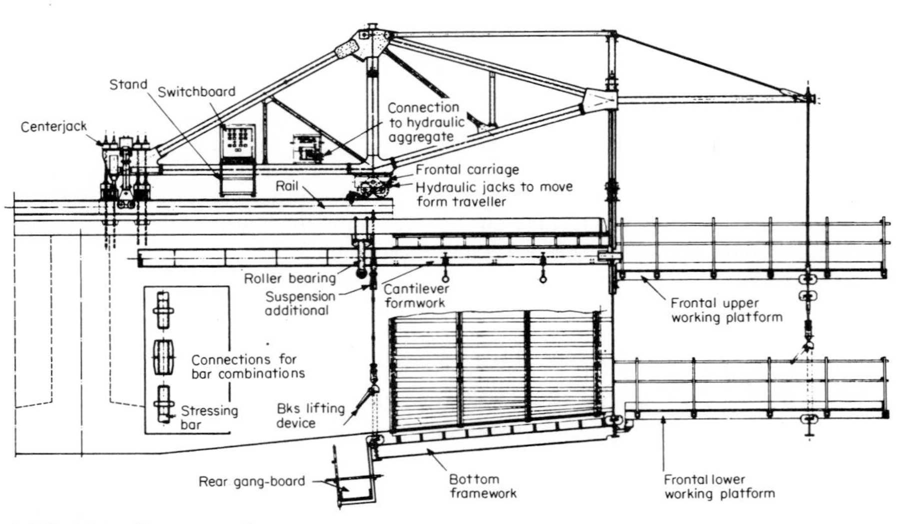

FIG. 18.1 Form traveler.

FIG. 18.2 Pine Valley Creek Bridge.

the suspended platforms. The main frames run on the upper guide rails and both the frames and the rails are attached to the finished portion of the bridge by means of tie-down anchors. All forms for the superstructure are suspended from the main frames. The suspension system should be designed in such a way that stripping can be done in a single operation by lowering the main frames. In a well-designed form traveler all operations, such as advancing, adjusting of elevations, and stripping, should be carried out either hydraulically or electrically and controlled by a single central panel. The efficient and smooth operation of form travelers is essential for successful free-cantilever construction because the form traveler is always on the critical path of the construction.

Figure 18.2 shows form travelers in operation in the Pine Valley Creek Bridge, California, which was the first major cast-in-place segmental cantilever bridge in the United States, with 450-ft (137.2-m) main spans. The superstructure is supported by 350-ft (106.7-m)-high piers. Figures 18.3 and 18.4 show form travelers used to construct other bridges.

Pier Brackets

Pier brackets are used to provide support for the formwork of pier tables. There are many types of pier brackets depending upon the local conditions. If a pier is high, the pier brackets are usually built out from the pier cap and pier shafts,

FIG. 18.3 Parrotts Ferry Bridge.

such as in the Pine Valley Creek Bridge and the Kipapa Stream Bridge. If a pier is low, the bracket may be supported on the footing of the pier (Figs. 18.4 and 18.5) or directly on the ground. If a pier bracket is supported directly on the ground, the cost of the bracket is proportional to the length of the pier table. However, the cost of a pier bracket increases very rapidly with length of pier table if it is built out from the pier cap and pier shaft—an important factor to be considered during design.

FIG. 18.4 Knight Street Bridge.

FIG. 18.5 Black Gore Bridge (Vail Pass).

Auxiliary Bridges

For bridges with high piers or those over difficult terrain, it is often preferable to transport personnel and materials at the level of the superstructure. Auxiliary bridges spanning from the finished portion of the bridge to the next pier table can be used. An auxiliary bridge should be light and mobile. Figure 18.6 shows the auxiliary bridge used for the Pine Valley Creek Bridge, which was specially made for this purpose. In many other cases patented trusses such as the Bailey system may also be used.

FIG. 18.6 Auxiliary bridge, Pine Valley Creek Bridge.

MODIFIED CANTILEVER CONSTRUCTION

The repetitive operation using form travelers in free-cantilever construction is very efficient. This advantage, however, can be achieved only if the cantilevering operation can be continuous for many segments. If the bridge spans are short and there are only a few segments to be cast in each cantilever, the cost of setting up and disassembling of the form travelers on pier tables could be disproportionally high, which may make the free-cantilever construction unattractive. Various modifications to the free-cantilever method have been developed to avoid this shortcoming.

Overhead Gantry

An overhead gantry can be used to carry the form travelers from pier to pier, to avoid dismantling and reassembling of the form travelers after each pair of cantilevers is completed. In most cases the gantry is also used for personnel and material transport. If the superstructure is supported by bearings, the overhead gantry may be designed to carry the unbalanced moments resulting during cantilever construction.

Commonly the gantries are composed of steel trusses, but steel-box and plate girders have been used for some bridges. A typical gantry has one movable support and two fixed supports, one at the middle and one at the rear end, as shown in Fig. 18.7. The fixed supports are equipped with either steel rollers or rubber tires for moving and with hydraulic jacks for height adjustments. Gantries are usually moved forward either by cable winches or by means of hydraulic pushers.

The form travelers required for this technique are slightly different from those used for free-cantilever construction, because the bottom slab forms must be released open to clear the pier columns during the moving operation (Fig. 18.8). Some travelers are designed in such a way that they can be combined together to serve as formwork for the pier table. This is an important advantage because the construction of the pier table is usually a costly and time-consuming operation in the free-cantilever construction.

FIG. 18.7 Overhead gantry, cast in place. *(Dyckerhoff & Widmann.)*

FIG. 18.8 Modified form traveler. *(Dyckerhoff & Wydmann.)*

Under-Deck Gantry

The principle of this scheme is very similar to the use of the overhead gantry, except that the truss which supports the form travelers is placed underneath or alongside the superstructure and is supported by brackets at the pier columns. This scheme offers the advantage of having the form travelers supported directly on the truss, which simplifies the construction of the form traveler. The gantry is easy to move ahead. However, the disadvantage is the requirement of pier brackets which have to be dismantled after the truss has been advanced. The mechanics of moving the form travelers past the pier column is also more difficult here than when using the overhead gantry.

It is important that the truss is tied properly to the superstructure to avoid differential movement when the new segment is cast.

Temporary Cable Stays

In the classic free-cantilever method, construction always starts at a pier and the cantilevers extend approximately to the center of each span. To extend the cantilevers beyond the midspan will require significant increase of prestressing for

FIG. 18.9 Temporary cable stays. *(Dyckerhoff & Wydmann.)*

the construction stages because the bending moments increase approximately proportional to the square of the cantilever length. An alternative to the strengthening of the girder by additional reinforcement is to use staying cables to provide elastic supports to the bridge girder (Fig. 18.9). At the lower end the cables are usually anchored directly to the concrete superstructure. At the upper end, the cables are anchored to a temporary tower placed over the pier. The tower is usually made up of precast concrete sections which are vertically stressed together. This stressing is required to reduce the effects of bending moments that usually develop in the towers.

By using cable stays, the length of the cantilevers can be extended beyond the half-span length. If the cantilevers are extended to reach the adjacent piers, it is then possible to build the bridge in successive operations without pier tables at the subsequent piers (Fig. 18.10).

Cables may be placed at each segment or at every second segment. Dywidag high-strength threadbars are most commonly used for the cables. The larger size and lower stress level of these threadbars make them less susceptible to corrosion. The continuous thread along the entire length of the threadbar offers great convenience for anchorage and coupling. They are also reusable many times.

Temporary Supports

Instead of the staying cables, temporary supports from the ground may be used to carry the superstructure during construction. An example of this type of support is shown in Fig. 18.11. The feasibility and cost of these supports depend largely on the soil conditions and the height of the bridge.

FIG. 18.10 Temporary cable stays—continuous operation. *(Dyckerhoff & Widmann.)*

FIG. 18.11 Temporary support.

Because the loads are high, temporary supports of this type are usually constructed with either built-up steel frames or reinforced-concrete columns. If footing settlement might be significant, hydraulic jacks are required to adjust the elevation of the support as settlement increases. These hydraulic jacks can also be used to facilitate dismantling of the supports. If footing settlements are negligible, sand jacks may be used more economically for the release of the supports.

SPANWISE CONSTRUCTION

In long viaducts with short spans, the material quantities involved in each span are relatively small. Therefore, it is often economical to build these structures one span at a time successively. There are several methods available for this type of construction.

A simple method is to use falsework supports for the superstructure. This method will not be discussed here. Description will be mainly on the use of form carriers and trusses.

Form Carrier

The form carrier is a large steel structure that is designed to carry the weight of an entire bridge span. Figure 18.12 shows a typical form carrier, which usually consists of a center girder and a series of steel claws. The formwork is either suspended from the top girder or supported by the bottom claws. The original form carriers were designed for the so-called mushroom-type bridge superstructures which require only one single bottom formwork. However, more complicated cross sections such as girder bridges with longitudinal and transverse ribs have also been built by form carriers.

In the concreting position the girder is supported at the rear on the completed

FIG. 18.12 Form carrier. *(Dyckerhoff & Wydmann.)*

bridge deck and at the front on the pier column, and the formwork is suspended from the upper arm of the claws. After the concrete has hardened and the required tendons are posttensioned, the formwork is stripped by lowering the whole carrier. The formwork at the center of the cross section is then pulled back to both sides to clear the pier column when the form carrier is launched ahead.

The launching of the form carrier starts with the advancing of the center girder to the next pier. After it is properly supported at the top of the pier column, the rest of the form carrier is then pushed ahead using the advanced center girder for support.

FIG. 18.13 Form truss. *(Dyckerhoff & Wydmann.)*

A form carrier is a heavy piece of structure which weighs hundreds of tons, and therefore its use is only economical for long viaducts with many approximately equal spans. It may also be competitive for projects where other methods become too expensive.

Form Truss

Instead of a form carrier, a truss spanning under the superstructure can be used for the same purpose. Figure 18.13 shows the use of a truss for a bridge supported at the piers by means of steel brackets. Because the viaduct is not high, pier brackets of this type are not expensive and are easy to install. Steel rollers are often used at the brackets to facilitate launching (Fig. 18.14). The truss must be tied to the end of the finished portion of the bridge during concrete operation to avoid differential displacement in the superstructure at the construction joint.

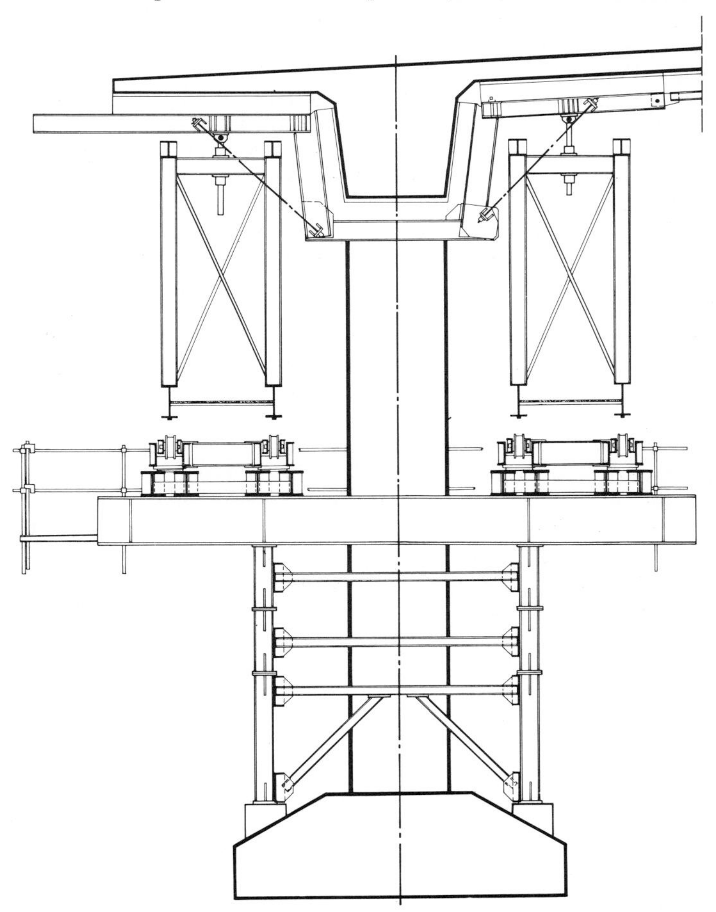

FIG. 18.14 Pier bracket.

The truss is moved forward by means of hydraulic winches which may be installed stationary at the rear end of the truss, at the next pier, or at the finished portion of the bridge. These winches should be powerful enough to overcome all friction forces at the supports and the additional load from the horizontal component of the weight of the truss and formwork if the bridge is sloping. The truss usually transports with it the external formwork of the bridge. Intermediate supports for the truss may be used if the spans are too long or the load too heavy. However, these supports are not usually required during launching.

INCREMENTAL-LAUNCHING CONSTRUCTION

In the incremental-launching method the formwork is placed stationary on a specially designed foundation at one end of the bridge. The bridge superstructure is pushed or pulled forward after each segment is completed. Depending on the lengths of the spans, temporary intermediate support between the final piers may be used (Fig. 18.15).

The length of formwork used is usually equal to the length of two segments. The formwork for the first segment consists of the form for the bottom slab and lower part of the webs, while the formwork of the second segment consists of the forms for the rest of the cross section: top slab and webs. The construction starts by first casting the bottom slab of a segment, which is then advanced to the next position after the concrete has attained the required strength and has been partially posttensioned. As the top slab and the rest of the cross section are poured, the bottom slab of the next segment will be cast at the same time. This work cycle is repeated until the whole bridge is cast and pushed to its final position.

At the front end of the superstructure, a launching nose is usually attached to minimize the cantilever bending moment in the bridge before it reaches the next vertical support. The launching nose may either be a truss or, if the bridge is small, a steel wide-flange section.

At the piers and intermediate supports, special sliding blocks are used to reduce friction during launching. A typical sliding block consists of a concrete block covered with a thin stainless-steel sheet. Teflon sheets are fed into the sliding block between the soffit of the bridge and the top of the stainless-steel sheet. This arrangement reduces the horizontal friction force to about 2% of the vertical loading. Before each advance, a vertical hydraulic jack which is padded with a teflon sheet and slid on top of stainless-steel base plates will lift the bridge slightly off the support at the abutment. The advancing of the superstructure is done by means of either pulling or pushing jacks located underneath the soffit of the bridge at the abutment.

Intermediate Supports

As a rule of thumb, the depth of bridge girder to be launched incrementally should have a span-to-depth ratio of 17:1 or smaller. If the depth of the bridge superstructure is too shallow, intermediate supports are used to reduce the bending moments during launching.

Intermediate supports can be either concrete columns or steel frames. Because the superstructure slides over the bearings on top of the supports, the

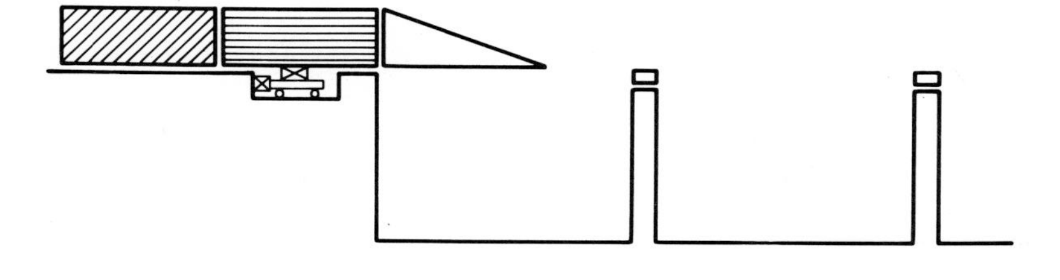

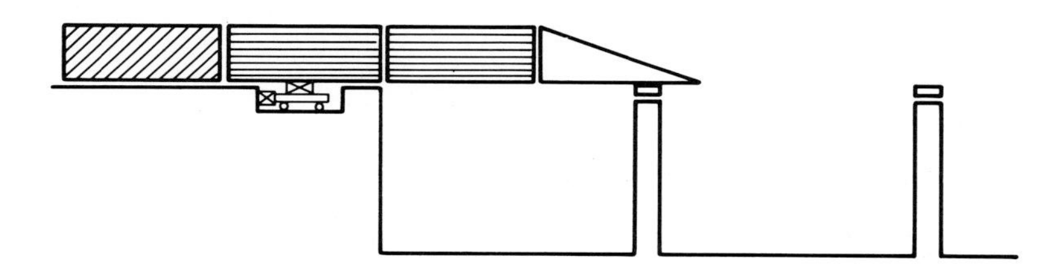

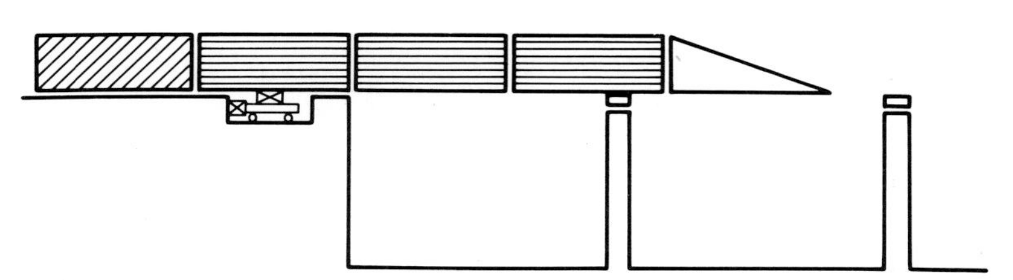

FIG. 18.15 Incremental launching.

horizontal force caused by friction at the sliding blocks must be considered in the design.

A simple method to resist these horizontal forces is to tie the temporary support to the pier column with tie-rods if the pier columns are strong enough. Another way is to connect the intermediate supports by tie-rods through the piers back to the abutment where the launching jacks are connected. However, if the bridge is long, these tie-rods will not be effective until large displacement of the intermediate supports takes place. In this case, local bracings must be used instead.

Launching Nose

The launching nose is a lightweight-steel girder which is tied to the front end of the bridge girder to reduce the cantilever moment during launching operations. If the bridge is straight or only slightly curved, the launching nose may be either a steel truss or a plate girder. If the bridge alignment is on a sharp curve, the launching nose should then preferably be a steel-box girder which provides sufficient torsional stiffness.

The bottom flange of the launching nose should be smooth and free of attachments, because it is advanced on the sliding blocks. To compensate for the deflection of the nose before it reaches a pier or intermediate support, the front end of a launching nose is usually tapered upward to facilitate easy engagement with the sliding blocks. Prestressed ties anchoring well inside the concrete girder are used to connect the nose to the front end of the concrete superstructure.

PRECAST CANTILEVER BRIDGES

Precast cantilever bridges can be erected in a manner similar to cast-in-place cantilever bridges, except that no form traveler is required because the segments have already been precast. The principles of overhead gantry, falsework supports, underlying truss, temporary cable stays, etc. which have been described previously can all be applied to the erection of precast bridges. However, the loadings may be slightly different, as discussed below.

Overhead Gantry

The overhead gantry in a cast-in-place bridge is used to transport form travelers, materials, and personnel. For precast bridges, it is mainly used to transport precast segments. Thus the size of the segment depends on the capacity of the gantry.

Figure 18.16 shows the gantry for the Kishwaukee River Bridge, which is the first precast bridge to be erected by a gantry in the United States. The gantry supports a hoist which travels along the entire length of the truss. The hoist picks up the precast segment at the rear end of the truss and transports it to its proper location. During transport the segments are usually oriented so that they can pass between the legs of a truss, which are normally placed over the webs. The segments are rotated to their correct orientation before they are lowered into position and prestressed to the end of the cantilevers.

Pier Brackets

Pier brackets for precast cantilever bridges are very similar to those of cast-in-place cantilever bridges, except that they are usually heavier. In order to simplify the formwork, the precast segments are usually exactly symmetrical about the pier axis so that the bracket will have to carry the unbalanced bending moment of at least one segment; in cast-in-place construction, the segments are usually arranged in such a way that there will always be only half a segment out of balance. Furthermore, in many cases the brackets are designed to carry the entire

FIG. 18.16 Overhead gantry, precast.

bridge span (two cantilevers together) to allow adjustment of the bridge elevations before final setting of the bearings.

Temporary Cable Stays

They are the same as those in cast-in-place cantilever construction.

Cableways

A cableway can be used instead of an overhead gantry to erect the precast segments. A cableway consists basically of two towers, the head tower at one end and the tail tower at the other; a hoist; a cable carriage; and a slack-rope carriage. They are available in various capacities and configurations.

Because cableways usually span from one end of the bridge to the other, it is possible to transport and erect segments at various spans of the bridge simultaneously. The segments are transported to the abutment where they can be picked up by the cableway and transferred to any location of the bridge.

Cableways have not been common for precast cantilever bridges. The main reasons are probably that overhead gantries are more convenient to operate and

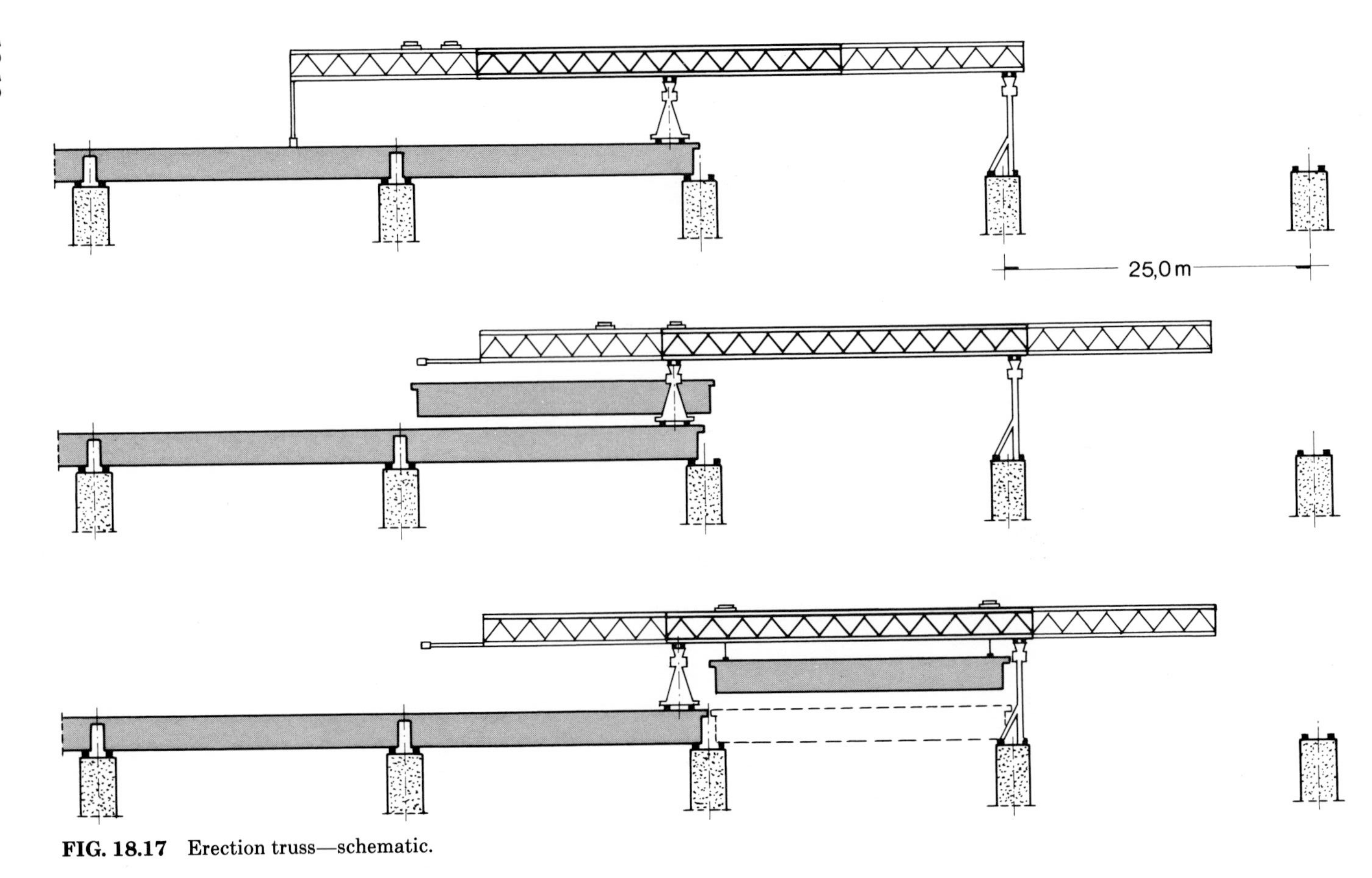

FIG. 18.17 Erection truss—schematic.

the weight of normal precast segments are quite heavy so that cableways become uneconomical.

PRECAST GIRDERS

One of the most commonly used construction methods for short-span viaducts is the combination of a cast-in-place deck slab on prestressed, precast I girders. The I girders are usually erected using mobile cranes. However, if erection positions are not accessible or the terrain is not suitable for crane operations, then a mobile overhead gantry may be used to erect the precast girders.

Figure 18.17 shows schematically the working of such an overhead gantry. Figure 18.18 shows the actual gantry at site.

SPECIAL CONSIDERATIONS

Following are some items which are important in the design of temporary structures for bridge construction.

Design Loadings

In the design of temporary structures, loading assumptions should be reasonable and clearly specified. Most specifications give the minimum loads that should be considered, which include the weight of concrete, reinforcement, forms, construction, equipment, wind, earthquake, temperature, and live load, as well as the dead load of the temporary structure itself.

The gravity loadings can be calculated by adding up the weight of structural and nonstructural elements. Equipment loading should be based on the weight of the actual equipment selected for the job.

FIG. 18.18 Erection truss at work. *(Dyckerhoff & Wydmann.)*

Wind loadings could be one of the major loadings for the design. The magnitude of wind pressure, wind area, and direction should be carefully considered and combined with other loadings. The wind loading on equipment and other auxiliary structures should not be neglected.

Temperature is not a loading per se. However, additional stresses caused by temperature movement of various parts of the structure can not be overemphasized. Many failures of structures have been directly or indirectly attributable to temperature changes. The rise and fall in temperature specified in most codes should be used as a guideline for design purposes. Differential temperature between various portions of the structure should be considered, especially in regions where direct exposure to the sun may raise the temperature of some members of a structure and cause unacceptable deformations and stresses.

Earthquake loads are usually specified in the codes; however, in cases when these loads are unreasonably high, the engineer may want to assess the risk and consequence of using less severe design loads. Because earthquake intensities used for the design of permanent structures are unlikely to occur during the construction period, it is more economical to use a smaller earthquake load and risk some minor damage.

Construction live load should be considered carefully, especially for new construction methods. In many cases the actual weight of construction can be estimated and is considered in the design. The construction live load then consists merely of small tools, stored material, and other minor items. In some cases of long-span bridges, a uniform live load as low as 8 lb/ft^2 (383 N/m^2) has been found to be reasonable.

There are other items which may require careful consideration. Vibrations and resonance due to wind or equipment could be serious. The slope of the bridge structure and slope of bearings may cause additional horizontal reactions. Deformation of forms and falsework may cause secondary stresses that are not negligible. Whenever tolerance is allowed or is unavoidable, due consideration shall be given to its consequences. Tolerance in formwork dimensions, for instance, could cause additional unbalanced moments in free-cantilever construction. Buckling and stability problems are also of major importance.

To avoid confusion or error, the time of application and removal of equipment loading should always be indicated on the appropriate drawings.

Formwork

The trend of modern construction methods is to build a bridge in many repetitive working cycles. This usually means that formwork is also used repeatedly. A well-designed and sturdy formwork is a prerequisite for this type of construction. Steel forms are usually durable, but relatively heavy for handling. In most cases, special wood forms are used. Some structures require a good finish of concrete surfaces. Plastic-coated plywood forms should meet this requirement, and can often be reused up to 30 times. The slightly worn outside form skin can be used to line the inside formwork.

If the bridge has variable depths, timber forms are usually advantageous because cutting of the form to fit the shape of the bridge is necessary.

Gantry Design

The design of gantries is relatively simple. However, special attention should be given to the required safety factor for various loading combinations and their durations. In most cases, some load combinations are present only for a very short period of time. For instance, launching may take only 1 or 2 h. It is usually possible to specify that the gantry shall be launched only when wind velocity is below a certain value (specify wind velocity instead of intensity). Live load or equipment can also be limited during many critical loading periods. This may allow a truss to be designed for a smaller loading than the maximum possible load combinations.

On the other hand, it is important to note that construction loads are all actual loadings which unfortunately are usually underestimated instead of overestimated as in the case of live loads in most final structures. The engineer must carefully weigh the available safety factor , which should be appropriate for each specific loading combination. Underdesigned structures are dangerous; overdesigned structures are not only uneconomical, but could often totally eliminate a special construction method from consideration. Realistic and good engineering judgment is the basic rule for all structures.

19

Demolition Operations and Equipment

by J. Mark Loizeaux

In this age of high technology and product liability, most manufacturers tell contractors how to do almost everything with their products. Major construction material categories are backed by industry-funded institutes, and handbooks advise contractors how to bid, order, ship, store, and install materials. After these data are combined with the design directives of architects and engineers, a contract is usually won or lost on the basis of who can complete the work while tolerating the lowest profit margin.

When contractors are faced with the need to modify or demolish a structure, however, they find that there are no industry standards to go by. Nor are there any institutes or manuals available with guidelines to help the contractor make even basic decisions in getting the project underway.

This chapter will enlighten contractors in what must be a subjective approach to the most varied and unpredictable field in the construction trades: demolition.

PROJECT ANALYSIS

Demolition consultants or contractors often find that their clients simply do not know where to begin with demolition operations. A good place to start is with a generic definition: *Demolition* can be described as the breakdown and handling of materials scheduled for removal in consideration of the work environment and available time, labor, and equipment.

The first step is to simplify the project with a basic problem-solving outline. Realistic projection based on adequate historical data must be made in this initial operation if the project goals are to be effectively and economically reached.

- Establish the ultimate goal for the project.
- Gather all relevant data, and break the project down into independently definable subproblems.
- List possible solutions for each subproblem based on technological, labor, economic, time, and site-condition factors.
- Evaluate possible and probable side effects of each of the alternative solutions.
- Rate each of the subproblems in terms of importance, and rearrange as required in consideration of time and cost-weighting factors.
- Start with the last phase of the project and work backward, test-interfacing possible solutions against problems in reverse order until the first phase of the on-site work is reached. Eighty percent of the decisions which originally confused you will be eliminated by the natural functions which must be followed in order to reach the project goal.
- Formulate a "critical path" in consideration of the above.
- Test the estimated cost of the work against the available budget.
- Effect on-site operations with whatever is required to meet the schedule.
- Use critical-path management (CPM) to push the project to completion.

The preceding steps can be applied to any type of project, regardless of size or complexity. Time taken in this planning phase will pay for itself many times over.

DEMOLITION TECHNIQUES AND EQUIPMENT

Hand Demolition

With the rapidly changing economic status of labor and escalating pay scales, any type of hand operation is probably the most expensive means of demolition on a unit-production basis. Large machines might easily do 100 times the work an individual can with hand tools in the same time period. From this cost standpoint alone, manual demolition operations requiring the use of hand-held and -operated tools should be reserved for work that simply cannot be accomplished mechanically with larger equipment.

This technique usually applies to very small projects, material salvage operations, partial-removal strip-out projects, or demolition work immediately adjacent to or adjoining exposures which are to remain. On small projects, the budget

FIG. 19.1 Manual and light mechanical demolition being used to dismantle a multistory structure. In this case, partition and curtain walls have been removed first to provide easy access for segmentation of the reinforced-concrete framing. Note the pointing-up and weatherproofing being performed on the party wall on the right side.

will not allow for the mobilization of large pieces of equipment. The other categories mentioned generically preclude the use of high-production mechanical or explosives demolition methods due to the sensitivity of the work being performed. Manual demolition operations usually begin at the top of the project, with the roof being removed first. Internal partitions, columns, and perimeter wall-and-column panels are pulled in and down on the adjacent floor slab to be fragmented on a level-by-level basis. Debris is then either dropped, chuted, or lowered to ground elevation for removal from the demolition site. Where pedestrian or vehicular rights-of-way are immediately adjacent to the work in progress, standard "wing" scaffolding is erected to prevent unintentional dropping of debris. On this type of operation, scaffolding sections are removed from the exterior of the building or wings (which are extended from windows one floor below the level where demolition is taking place) are lowered as demolition progresses to lower elevations.

Pneumatic jackhammers, chipping hammers, and impact rams are the workhorses of this technique. Rams are for higher-production involvements and are particularly effective when mounted on either track drill carriages (called hobknockers) or small hydrostatic loaders. When the structure being demolished will carry the load and the volume of demolition to be performed is sufficient, small cranes are mounted on skid-sleds for the use of small headache balls in the fragmentation or segmentation of slabs and structural components. These small cranes can literally let themselves down on a floor-by-floor basis as demolition proceeds to lower levels.

Demolition debris can be handled in a variety of ways. Chuting down open elevator shafts or special dustproof metal chutes can put lighter materials

FIG. 19.2 Full-scale mechanical demolition being performed by cranes from ground elevation on an 11-story steel-framed structure. Debris is being balled down to the ground on a bay-by-bay basis, with equipment (lower right) waiting to load into trucks.

directly into trucks. Heavier elements such as columns and beams are dropped to impact areas for rehandling with equipment at grade. The use of fast-moving hydrostatic loaders for the handling of light debris at each floor level will tremendously improve the speed of demolition operations. Small track loaders can be used for heavier elements if the floor will carry the load on a square-foot basis. Either of these types of loaders can be used to load debris into chutes, or even into skip boxes for lowering to removal areas by cranes on tightly confined projects where chuting may not be feasible or legal.

Mechanical Operations

Where the nature of the structure being demolished and adjacent exposures permit, work performed through the application of mechanical demolition techniques can range from piecemeal cut-and-pick crane operations generating a few hundred cubic feet of demolition per hour, to high-yield wrecking operations

FIG. 19.3 Standard scaffolding used to the full height of a structure on "an exposed side" where mechanical operations cannot be utilized due to the overhanging new structure. Note that windows and window framing have been preremoved.

FIG. 19.4 Hand demolition operations with pneumatic chipping hammer being used to demolish a reinforced-concrete chimney. The chimney is being "segmented" into panels which are chuted down the center of the chimney for removal at the bottom.

FIG. 19.5 Mechanical demolition of a small masonry and frame structure. The extension attached to the track loader being used to pull the structure down is called a "rake." Where permitted by local codes, such techniques are used to demolish structures up to five stories high.

Fig. 19.6 Fire hoses being used to control dust generated by mechanical demolition of reinforced-concrete industrial buildings. A 40-ton (36-t) crane could be expected to generate from 70,000 to 100,000 ft^3 (1,980 to 2,830 m^3) of building volume per shift on this type of project.

producing thousands of cubic feet in an equal time period. On any project, the sensitivity of the work being performed, the sensitivity of adjacent properties, the type of structure being demolished, the capacity of debris-handling equipment, and the expertise of the equipment operators all have a dramatic effect on actual production levels.

Although some of the more productive methods covered under the manual techniques above are actually light mechanical operations, they are usually associated with hand demolition activities when there is a limited amount of room for the overall demolition project.

Equipment-oriented demolition operations are usually carried out on structures up to 12 stories in height with truck- or track-mounted cranes. The efficiency of crane demolition operations is greatly reduced above this elevation, and the expense of moving in and setting up larger rigs is frequently prohibitive. Many contractors use a combination of light mechanical and hand operations down to a level where a 40- or 50-ton (36- or 47-t) mobile crane can be employed to complete the demolition. Such procedures are frequently utilized on projects with limited but adequate distance to adjacent exposures. Debris can usually be balled or dropped into a free staging area at the base of the structure for handling with track loaders. If the free area at the base of the structure is equivalent to 50% of the structure's height, this method can generally be very effective up to within one bay (the distance between two columns) of adjacent properties. In this case, crane operations should not proceed too far ahead of manual removal operations on the perimeter bay, as hazardous, unstable, and unsupported conditions might result.

FIG. 19.7 Rigid steel truss being floated off of piers by crib-and-barge method. Span can be lifted by the use of normal tidal changes or by pumping water out of partially filled barges shimmed into place under the span. Span will be floated to a work area alongside the bank for torching and removal.

TRITON/COI
FANTA

FIG. 19.8 CDI demolished this 30-story reinforced-concrete structure in Sao Paulo, Brazil after only 40 days of on-site preparations. Mechanical or hand demolition with available technology in Brazil would have taken over 12 months to complete the operation.

Explosives Operations

Explosives are chemical-energy tools which can readily be converted into mechanical energy through the use of gravity. The advantage of explosives is the virtually limitless amount of work which can be performed in a continuous, controlled sequence of operations. The application of explosives demolition has been perfected over a surprising range of operations: from work inside delicate nuclear reactors to the felling of 30-story buildings in congested urban areas. Its viability on any given project, however, is a function of the environment, the material being removed, and most importantly, the skill and experience of the explosives technician.

Over about four decades of commercial explosives applications in this country, a few select explosives contractors have removed over 1000 major structures and completed over 10,000 explosive applications in specific service to the construction and demolition trades. Because of the relatively subjective nature of the work, it has taken almost that long for these contractors' clients to understand the type of structure which can be realistically considered for "implosion."

As long as adjacent exposures are farther away than the equivalent of 150% of the height of the structure to be demolished, there should be no problem in the use of explosives demolition. It is assumed, in this case, that the permit-letting agency, property owner, and contractor have ascertained that the explosives technician has sufficient credentials and experience to perform the project. Explosives methods have successfully felled structures over 30 stories tall in heavily congested urban areas, pulled 14-story structures away from major buildings only inches away, and fragmented heavily reinforced 400-yd^3 (300-m^3) concrete foundations right alongside other in-plant units operating "on line" without tripping delicate mercury control switches. These projects are, however, the exception rather than the rule. Decisions to use explosives on such delicate operations are based on detailed inspection, computerized structural analysis, and careful reflection using decades of hands-on explosives experience. It is a field not to be entered into lightly.

There are three words which accurately describe the advantages of explosives demolition on certain projects: speed, safety, and economy.

Speed

Where time is of the essence, explosives are the most desirable means of demolition as they reduce the entire structure to a pile of rubble in a matter of seconds. Instead of waiting with trucks and a single loader for debris to be generated by hand or mechanical operations, a fleet of loaders and backhoes can be applied to the debris pile for accelerated removal operations.

Safety

This is the first consideration on all projects. The unique advantage of explosives with respect to safety is that no one is near the structure when actual demolition takes place. The exposure to employees working in high places is eliminated, and weeks or months of mechanical operations and their conflict with adjacent properties and the general public can be compressed into a few seconds early on a

FIG. 19.9 CDI felled this 605-ft (184-m) tall reinforced-concrete chimney after only 4 days of preparation. Manual demolition of the chimney had been projected at almost 90 working days.

FIG. 19.10 CDI used linear-shaped charges to cut this 451-ft (137-m) bridge truss into sections to facilitate removal. The main channel of the Mississippi River was cleared in less than 4 h.

Sunday morning. The incidence of bodily injury during explosives operations in this country is virtually nil. It should be mentioned, however, that a major problem with injuries and fatalities on explosives projects arises when unqualified contractors pre-weaken and overprepare a building for implosion, resulting in premature progressive collapse.

The importance of engineering analysis and precaution is emphasized here in the hope that prime contractors, property owners, and permit-letting agencies will recognize the need for greater control of contractors who wish to perform explosives demolition projects. Federal regulations specifically call for engineering analysis of structures prior to and during demolition operations. Past experience has taught, however, that only the industry can effectively enforce such requirements in the interest of overall project safety.

The principle of explosives demolition is not unlike underpinning operations as practiced on wide open, conventional demolition sites. Selected supports are knocked out, creating cantilever loading on what were originally designed as "rest beams." Failure and rotation of the cantilever portion will occur either instantaneously or as a time function of the relationship between the length and weight of the cantilever and the resistance of the structural elements. In order to take maximum advantage of the weight of the structure, explosives operations usually call for the elimination of support columns or walls in the basement or first-floor areas. If the structure being felled is particularly resistant, additional explosives weakening can be performed on vertical supports at upper elevations to ensure the continued vertical failure of the structure.

The key to explosives handling as a demolition mode is directly tied to the efficient application of explosives on each individual support member and to the timing of support weakening or elimination in order to precipitate a controlled progressive collapse. By selectively accelerating certain portions of a structure during collapse while slowing others, utilizing natural stiffeners in buildings such as elevator shafts, and adding additional integrity (through the use of shoring and cross cabling), an experienced explosives technician can choreograph a building into a neat pile of debris. It should be stressed, however, that the possession of a federal or state explosives license does not qualify an individual as an explosives expert in the demolition field. Consistent performance does.

Economy

The economy of explosives operations is obvious, in that it is far less labor-intensive than manual operations. Working time is shortened (sometimes as much as 60%) on the project, reducing the probability of interference from outside factors such as weather, legal involvements, and injunctions from well-meaning historical-preservation societies wishing to save the structure. Reduced demolition time frames allow for the rapid removal of structures from tax roles; late evacuation of tenants, to maximize rental revenues; early initiation of construction operations, with the reduction of construction financing which follows; reduction of union contract and material cost increases for new construction; and early completion of the new structure, for positive cash flow from the sale or rental of facilities.

DEMOLITION DECISION MAKING

Contractors and architects must recognize that demolition is only one step in a balanced series of events designed to change the utility of a given area. Understanding the effects of time and sequence in site rehabilitation is necessary in order to develop an intelligent solution to any demolition problem. Each step of the project must mesh with other on-site activities. For this reason, the basic demolition program should always be presented to the prime contractor and property owner for their early assessment of its impact on other activities. During this preliminary "diagnostic" phase, necessary changes in operations or schedules can be negotiated, and a totally workable critical path for the entire project can be established, be implemented, and succeed through critical-path management.

Examination of Material to Be Removed

Look at the demolition problem in consideration of your in-house experience in the skills required for this type of work, available debris-handling equipment, and exposure to adjacent properties or facilities. Be certain that you have the technical expertise to identify and analyze the structural aspects of the project and how they compare with your demolition experience. Never attempt operations of which you are uncertain. Virtually all injuries, fatalities, and accidents in demolition operations occur when the contractors either overestimate their own abilities or underestimate the structures they are dealing with.

Debris Handling and Disposal

It is often asked why contractors need worry about debris handling and disposal before a decision has been made on the type of demolition technique to be used. Here lies the great mistake of most contractors getting into demolition. What point is there in breaking structures down to a given size unless the removal equipment and disposal area are capable of handling that size material?

On the average, debris handling and disposal represents anywhere from 25 to 40% of the total cost of demolition projects. Go slowly here, as dump site conditions and requirements will occasionally dictate the type of demolition operation to pursue. Disposal areas should be contacted early in your review of the project with respect to their requirements, and equipment manufacturers should be asked to furnish guidelines as to the capacity and performance of their products under field conditions you anticipate. Again, if you are uncomfortable in this area, contact a demolition consultant as to the probable productivity of different pieces of equipment in handling the type of debris the project will generate.

Escalating diesel fuel costs and unit dumping charges are playing a more critical role in the expense of debris disposal. As a result of the environmental regulations (EPA) of the 1960s and 1970s, many bids will be won or lost on the basis of who finds the closest legal dump to the demolition site. Whenever possible, take an option on the use of a landfill facility, as it will improve your odds of winning the contract. A few extra days spent on researching disposal areas are usually well worth the effort.

Demolition-Technique Selection

In consideration of engineering requirements dictated by the type of structure being demolished, debris handling, and the project critical path, select one or a combination of demolition modes to perform the project. Continue on to select alternate techniques of demolition in the event that restrictions or developments will not allow the total application of your initial program. Plan ahead by expecting the unexpected in all demolition operations.

Temporary Supports

Structures occasionally have a direct effect on the stability of adjacent facilities to remain. If you do not have the in-house engineering capabilities to recognize and define these relationships, retain an independent consultant to review the situation. The prime contractor and/or the property owner will often become involved in this area, as local regulations usually put the onus on them to stabilize site conditions before allowing demolition to proceed.

Protection of Adjacent Exposures

Protective measures are usually the application of good common sense, and the adage "an ounce of prevention is worth a pound of cure" is most applicable on demolition projects. Although simplistic, the easiest means of avoiding problems with adjacent properties is to utilize a demolition technique which will not affect them. Where the application of the selected technique is likely to create undue risk, temporary barriers can be erected ranging from light plastic sheeting or tarpaulins to prevent dust damage, to heavy plywood mounted on support scaffolding. Remember, however, that virtually nothing can protect an adjacent property from damage if an incorrect demolition technique is selected or if equipment operators and technicians are inattentive or incompetent. The best insurance available is to hire the most experienced, reputable demolition superintendent one can find for the project. Whether an in-house employee or a subcontractor on that single project, the reputable demolition superintendent will save time, money, and legal problems by making the right decision on technique and field personnel the first time. Although there are only a handful of true demolition consultants in the country, their advice can be invaluable. It is often more economical for a contractor to seek their input at a reasonable fee before starting the project rather than calling a consultant once a project is already in trouble and the profit margin is lost.

BIDDING

No two structures are exactly alike. They vary due to design, structural materials, workmanship during construction, alterations, weathering, and fatigue. When available, design drawings of a structure are frequently helpful, and as-built drawings are somewhat credible. Usually, however, only a detailed inspection of the structure is totally useful. Any demolition expert with solid field experience (and no ego problems) will readily admit that the only way to fully understand a specific structure is to take it apart. Unfortunately, that ultimate data are not available during the bidding phase, so another means of planning and pricing a project must be used. There is no substitute for historical data when bidding in any of the construction trades. The utilization of personal experience or reliable in-house information is certainly the best way to approach the pricing of any prospective project. By breaking historical data down into units (such as cost per cubic or square foot, in-place yards of material, and truck debris-yards generated), one can come up with surprisingly well-defined parameters for planning and pricing new projects on a similar unit basis. It is critical that one consider differentials in the following areas when projecting costs of completed projects against future work being bid:

Effects of inflation on labor, materials, supplies, equipment maintenance. overhead, bonding, and insurance.

Productivity of labor in different countries, states, cities, and even job locations in the same city.

Difference in type of construction and size of equipment required or available.

The physical and political nature of adjacent exposures. (Adjacent property owners with the wrong attitude can literally shut a job down.)

The attitude of the contract-letting agent, engineers, architects, and public or private agencies who have a hand in administration or inspection of the project.

The effect of seasonal changes on productivity levels.

Specification requirements and details, as a single word can totally change a contractor's obligations on a project.

Availability and capability of subcontractors and labor required for the new project.

Changes in dumping locations, dumping fees, and related regulations.

Fluctuating values of salvageable materials, contents, and construction materials.

Changes in local, state, and federal regulations governing the performance of the work.

Policy changes in one's own company, such as different financial ratio requirements, procedures, or goals.

While the above list does not cover all considerations, it gives the contractor a checklist for review.

STRUCTURAL COMPARISONS

A comment is necessary here about the comparison of types of construction on completed projects with new work under consideration. The following represent a few factors in design and construction where differences can be critical. These variations have a direct effect on demolition, job safety, productivity, and the nature of debris generated.

Concrete: Unreinforced, reinforced, cast in place, prestressed, posttensioned, flat slab, shear wall, specialty concrete (unique aggregate, strength, special additives, or types of reinforcement.)

Masonry: Brick, block, tile, or stone units. Is the mortar harder than the unit itself? Is there internal reinforcing?

Wood: Type of wood. Has it been treated with material which will affect handling or disposal, and which way does the grain run in the timbers? Is it bolted, strapped, nailed, pegged, or screwed in place? Is it salvageable?

Structural Steel: The grade of steel and its working characteristics in structural members; the design of the members (built-up, hot-rolled, H, I, channel, tubular, or box). Are the members welded, bolted, or riveted? Are the structural relationships between the members altered by the addition of gussets, other stiffeners, or types of encasement?

Also consider the present condition of the structural materials, as weathering and fatigue have an effect on their ability to do the work for which they were originally intended. Any combination of these materials or components complicates the contractor's task in defining and handling the demolition project.

For example, demolition of a structural-steel-framed building requires certain expertise and equipment. A great deal of historical experience in this type of structure would generate very accurate unit costs for the handling of such a project. A simple change, such as the use of cast-in-place structural-concrete encasements around the steel columns, could almost double the cost of breaking down the structure for removal. The further addition of vertical rebar and spiral wrapping to the concrete encasement could add as much again to the dismantling expense and preclude salvage of the steel.

All construction materials can be handled quite effectively in their pure state. When different materials are combined, however, the weaknesses of some materials are bolstered by the strengths of others. Only on-site activities or a great deal of prior experience will provide an accurate picture of the time and expense required for the demolition. Obviously, one who does not have historical information to use in bidding should get the help of a qualified consultant, engage a subcontractor, or work on a cost-plus basis.

Experience in the demolition of eight-story column-and-beam cast-in-place reinforced-concrete structures does not necessarily qualify a contractor for the demolition of 20-story structures of the same construction configuration and material. The engineering considerations which change drastically in the design of these taller structures will certainly affect such structures' reactions to demolition operations. The contractor must be prepared to handle these variations, and cannot always apply previous experience on smaller but "similar" structures.

For example, many pre-1950 cast-in-place reinforced-concrete structures cannot be compared to more recent construction. The introduction of flat-slab and shear-wall designs have changed the dead- and live-load relationships between components in these complex force systems. Demolition techniques which were safely applied in older, more conservatively designed structures would precipate catastrophic progressive collapse in today's highly interdependent computer-designed structures.

Contractor experience qualifications for structural demolition should be divided into categories of magnitude, construction material, and construction design. It is generally recommended that contractors compare demolition methods and historic costs on structure height categories of up to 9 stories, 10 to 15 stories, 16 to 22 stories, and over 22 stories in each construction design and material category. Contractors must have adequate data and experience in each area to properly utilize their historical productivity and cost factors against future work. Experience with one or two projects of a specific magnitude in a given material and design category is hardly sufficient to be considered "averaged data." When more data are required, check to see how similar projects have been recently priced and performed on a unit-and-time basis by competition in the area. If that is still insufficient to make you comfortable with your approach to the project, retain a consultant who keeps current, regularly averaged data.

INSURANCE

Even in instances of minimal exposure, it is recommended that demolition projects carry higher insurance limits than those required for construction operations of similar magnitude. This is primarily due to the relative unpredictability of demolition operations as compared to construction involvements, and the absolute liability nature of demolition under the law. It is suggested that the contractor contact either a demolition consultant or insurance broker as to the levels of insurance currently being recommended for such operations. In any case, the type of coverage required should be bodily injury (BI), property damage (PD), and combined single limit (CSL) coverage including explosion, collapse, and underground (XCU). It is not recommended that a contractor follow the lead of permit-letting agencies or contract specifications in determining the amount of coverage to carry on a project. Most architects, engineers, and agencies are notorious for being slow to respond to precedent in insurance settlements. Probably the most qualified person to advise the contractor on insurance limits to be carried in consideration of the project, adjacent exposures, and legal precedents is the demolition consultant.

PUBLICATIONS AND ORGANIZATIONS

For decades, wrecking has been an unsavory word in the construction trades. The industry's reputation and all of the connotations that go with it are not totally deserved, however, and there are highly qualified demolition contractors available across the country to perform even the most complex project with flawless professionalism. Through the efforts of publications like the *Wrecking and Salvage Journal,* local contractor groups, and the National Association of Demolition Contractors (NADC), qualified demolition contractors have begun to work together to upgrade the entire spectrum of demolition activities and services. Their efforts, combined with positive changes of state and federal regulatory requirements, have served to bring demolition into focus as a professional industry which rightfully deserves a place alongside the other construction trades.

Improved technology, greater contractor-governmental cooperation on regulations, and clear communication between contractors, project managers, and property owners in the future should clear the way for even greater advances for the demolition industry.

20

Protection of Site, Adjacent Areas, and Utilities

by Richard C. Mugler, Jr.

GENERAL PRACTICES, LIABILITIES

The problems of protection at construction sites are immense. The one statistic, provided by the National Safety Council, that most clearly gives the problem its dimension is the fatality rate at construction sites. An average of eight people are killed every single day in the construction industry. The insurance companies feel that at least 60% of these accidents are avoidable. This figure suggests that at least 60% of all accidents are due to negligence or improper protection at the construction site.

The word "protection" clearly implies a liability, and the bulk of litigation derived from construction sites alleges improper protection of lives and property. Over a lifetime in construction one is likely to spend many days at examination before trial and then in the courtroom defending the adequacy of the protection provided at the construction site. Keep in mind that most construction damage claims involve more than one contractor; therefore, if an award is made, it is often apportioned among the defendants on the basis of degree of liability. Since this is often a very subjective area, you will find that a claim will turn against you on seemingly unimportant evidence such as having used the wrong color danger sign. A clever lawyer working with a minor technicality will have a jury thinking you regularly worked with total disregard for safety. The main point to remember is that form may be as important as substance in a legal sense. In other words, it is not enough to put up a required danger sign; it must be the prescribed color and size. It is on such seemingly trivial details that major damage suits are won and lost.

With this legalistic point of view in mind, we will review the various considerations in the protection of a construction site.

OSHA AND OTHER SAFETY REGULATIONS

Every contractor must retain a copy of the OSHA manual for construction titled *Construction Safety and Health Regulations, Part 1926.* Copies may be ordered by requesting stock no. 2915-00034 (price $1.55) from: Superintendent of Documents, U.S. Government Printing Office, Washington, DC 20402.

Many of the OSHA requirements, as will be seen in the following pages, are nothing more than good housekeeping and require little or no expense for compliance. Experienced contractors treat all OSHA citations seriously due in part to the fact that each citation becomes part of the contractors' permanent records regardless of where they work in the United States. These records become especially important in the event of serious accidents. Many prior OSHA citations will imply a pattern of disregard for safety which can prejudice the contractor's position in defenses against liability and property damage claims, plus the ever-present threat of criminal negligence indictments. In the 1978 collapse of the cooling-tower scaffold in West Virginia that resulted in the death of 51 men, it was immediately reported in the newspapers that the *site* had received numerous OSHA citations many months before the accident. This clearly implied to the public that there was some connection between the collapse and bad safety procedures. In reality the citations were against other contractors not connected with the scaffold collapse, but it serves to show how a bad record will compound the problems when a serious accident occurs.

Aside from the OSHA regulations which apply uniformly across the country, each state and locality may have differing building codes and labor laws that will affect construction in their jurisdictions. Most seasoned contractors when working in new localities will make personal contact with the building departments and agencies that will oversee construction at their sites. This is the kind of courtesy and diplomacy that can turn a potential adversary into a valuable ally. The personnel of the various building departments and agencies tend to be quite protective of their powers and responsibilities. A contractor who ignores, snubs, or bruises these sensibilities may be in for a rude awakening.

SIGNS

Proper signs are an important part of a contractor's legal obligation to protect the public and workers from the hazards that may exist around the construction site. Signs scribbled on plywood and boards are an invitation to allegations of negligent protection of the site. OSHA outlines very simple sign requirements in part 1926.200. Approved signs are available at most contractor supply houses. The two basic signs are "Danger" and "Caution," as shown in Fig. 20.1.

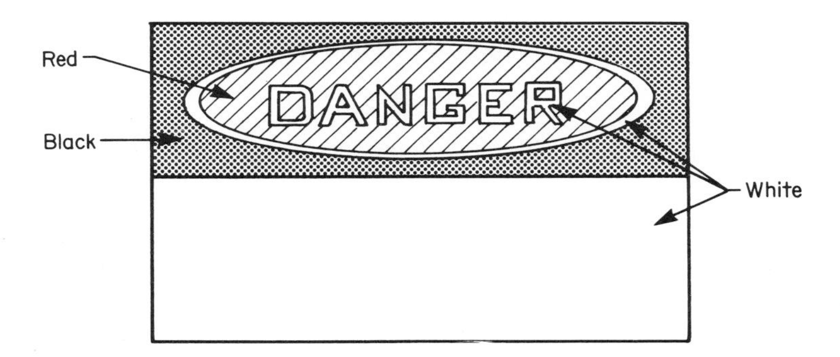

FIG. 20.1 OSHA-approved basic warning signs.

Traffic control signs become more involved, and because moving vehicles are involved, it is very important to comply with regulations for directing traffic. Traffic control signs have been standardized nationally and are detailed in American National Standards Institute D6.1-1971 *Manual on Uniform Traffic Control Devices for Streets and Highways.* Local highway departments usually have copies of this manual.

When conditions are such that a flagman is needed to direct traffic, the degree of liability is escalating. The contractor's employee is actually directing traffic, and if an accident occurs, it will usually be alleged to have been caused by improper traffic direction. For this reason it is especially important that the flagman be properly equipped, using a red flag of at least 18 in (45 cm) square as required by OSHA, using red lights at night and wearing a red or orange warning vest by day and a reflectorized vest at night.

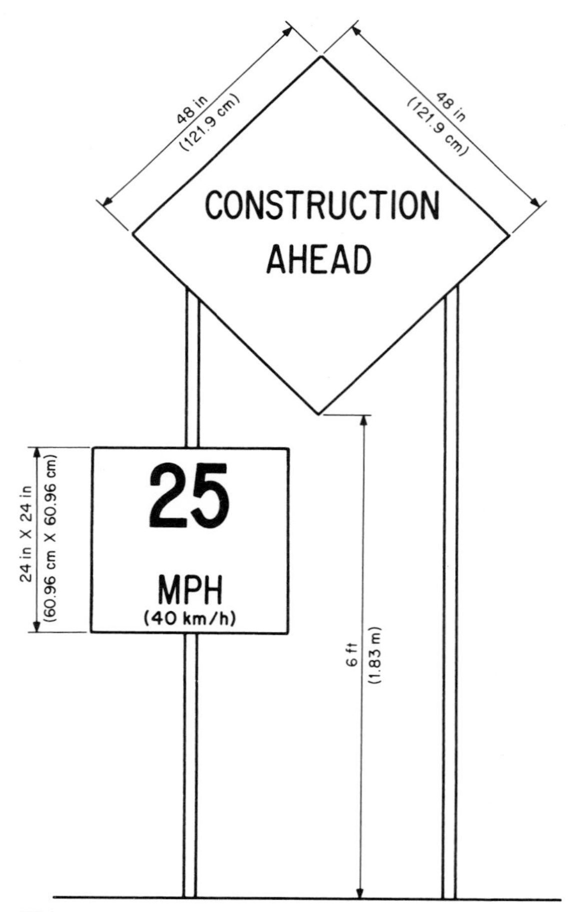

FIG. 20.2 Basic traffic-control sign.

A few general rules should be observed in the use of traffic control signs. Traffic signs should be mounted at right angles to traffic and about 2 ft (60 cm) from the roadway. Advance warning signs should begin at least 1500 ft (450 m) before the construction site and then be followed by repeat signs at 1000 and 500 ft (300 and 150 m) from the site. A typical sign would be diamond-shaped with an orange background and black lettering, as shown in Fig. 20.2.

Suggested speed limits, when posted by the contractor in the prescribed orange background and black lettering, are not intended as enforceable speed limits. The suggested speed limits should be reasonable for existing conditions. If they are too low, they will be ignored. The contractor should use only the orange and black signs which are now used universally for construction warning signs. The use of standardized regulatory signs of the red-and-yellow and black-and-white types normally placed and enforced by police and highway departments should be avoided. If they are to be used, it should be with the approval and supervision of the public agency having jurisdiction. Such signs are not just warning signs, as are the typical construction site signs; they impose legal obligations on the public. Unauthorized use of regulatory signs will cause unnecessary legal complication in the event of an accident.

BARRICADES AND FENCES

Barricades take many forms and are fully covered in the American National Standards Institute D6.1-1971 *Manual on Uniform Traffic Control Devices for Streets and Highways.* Figure 20.3 illustrates a typical heavy-duty street barricade that is both strong and movable. The type illustrated is approved for use on Port Authority of New York construction sites. Barricades have been generally standardized into three types as shown in Fig. 20.4.

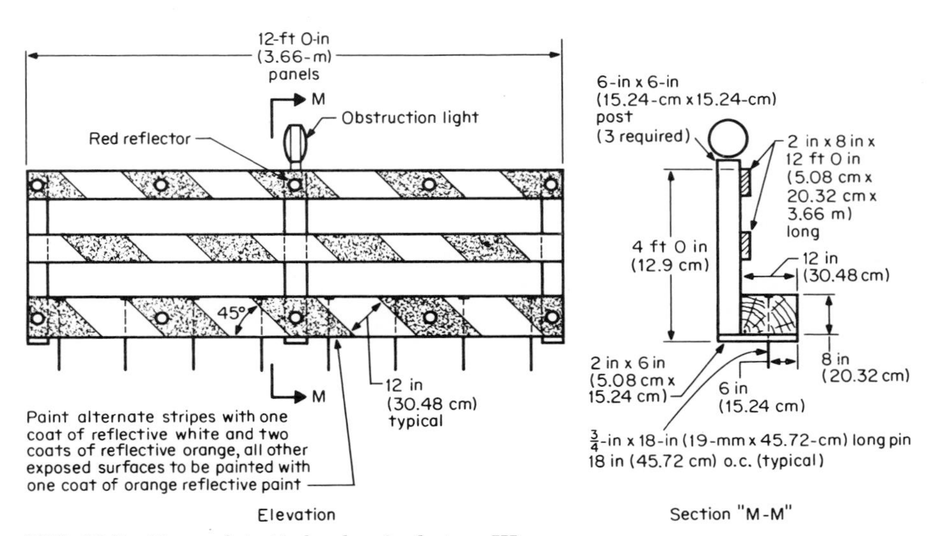

FIG. 20.3 Heavy-duty timber barricade type III.

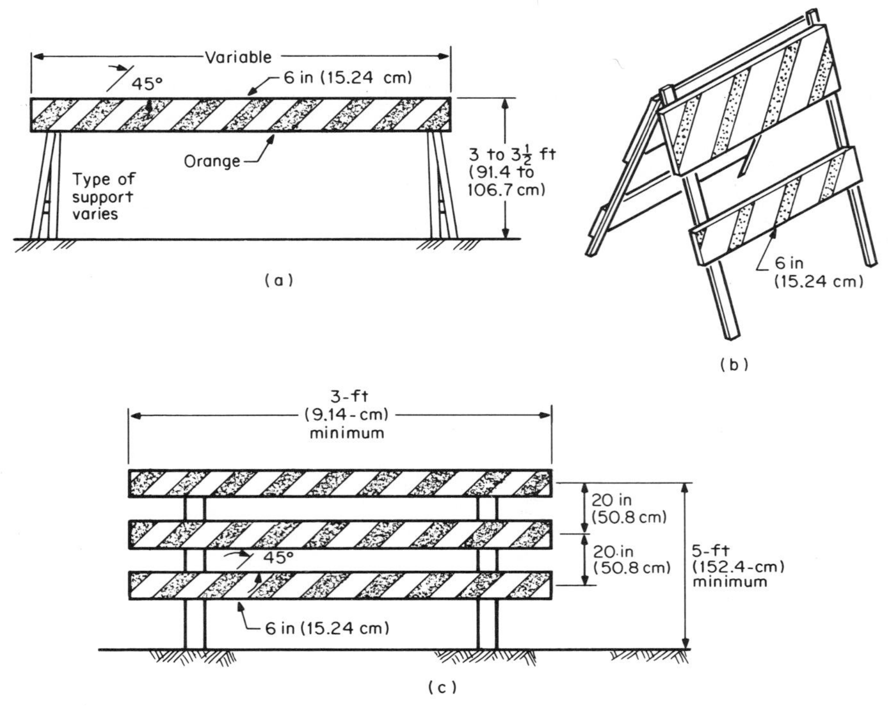

FIG. 20.4 Three basic barricades: (*a*) Type I; (*b*) type II; (*c*) type III.

Drums of the 30-gal (0.11-m^3) and 55-gal (0.21-m^3) size are very good traffic guides when they are closely spaced. They should be striped with 6-in (15-cm) orange stripes and 4-in (10-cm) white stripes using reflectorized paint. Plastic drum shapes are now sold for traffic control by building supply houses.

Cones are another good traffic control device, especially for short-duration jobs of less than a day. They should be at least 18 in (45 cm) high and preferably orange in color.

Drums and cones should never be used in the road without advance-warning signs as described in the previous section.

Figure 20.5 is a sample of a good jobsite-condition report used by many contractors for closing down the job each day. When faithfully used it will make a real contribution to safety. In the event of accident, it becomes a powerful piece of evidence to refute claims of negligence.

Fences are important not only to protect equipment and materials, but also to reduce the contractor's exposure to public-liability claims. Construction sites have repeatedly been held by the courts to be attractive nuisances for children. The primary burden seems to be on the contractor to keep them out by virtually impenetrable fences.

Chain link is in common use on large, open sites, but it is not suitable for confined urban sites. A plywood fence is the most practical for city construction.

JOB SITE CONDITION REPORT ____________

____________ for signs, barricades, and lighting

Date ________ Time ________ M Weather ____________

Job Location ____________________

	Yes	No
1. Are signs posted in advance of job limits and hazards		
2. Are flares and flashers lit		
3. Are signs clean and in good repair		
4. Is job site kept free of holes and ruts		
5. Are tools, equipment, etc. off roadway		
6. Any signs no longer needed		
7. Are detours well identified and lighted		
8. Is road kept free of rocks and spillage		
9. Are shoulders well identified		
10. Dust control precautions taken		

*Note: All unsafe conditions are to be corrected before leaving job site.

Unsafe conditions corrected:

1. ____________________
2. ____________________
3. ____________________
4. ____________________
5. ____________________

Signed ____________________

Reviewed by ____________________ Title ____________

Date ____________________

Instructions:

This report is to be kept with our permanent records

Standards for signs and barricades, along with proper methods for constructing detours are available at the field office.

FIG. 20.5 Typical jobsite condition report.

It reduces the invitation to thefts and offers better protection to pedestrians. This is especially important if blasting will occur at the site. A typical fence would be constructed of ½-in (12-mm) plywood mounted on an 8-ft x 8-ft (2.5-m x 2.5-m) frame. The plywood should be mounted on frames to facilitate alterations to the fence line, which will certainly occur on confined urban sites. The fence posts should be a minimum of 4-in x 4-in (10-cm x 10-cm) posts set at every 8 ft (2.5 m) or less in 2-ft (60-cm)-deep holes. It is a good practice to brace the fence with 2-in x 4-in (5-cm x 10-cm) braces back to stakes driven into the ground at every 8 ft (2.5 m) to stabilize the fence against wind. A poorly built fence will blow down without warning and can cause severe pedestrian injuries.

Gates for truck entry are often a neglected item at construction sites. A typical truck entrance is 16 ft (5 m) wide. Swinging gates are not suitable for an opening of this size, due to the wind hazard. They also pose a maintenance problem and are difficult to secure well enough to prevent entry by children. The safest and most secure gate is a chain-link sliding gate, which can be built inexpensively using low-cost tracks and rollers available at most building supply houses.

The use of spectator viewing openings in the fence should be considered carefully. On major projects the owners often specify that the fence will have a certain number of viewing locations on the theory that it is good public relations. This may be good for the owner, but it creates an added liability for the builder and subcontractors. The fewer people attracted to the site, the better, especially if blasting is in progress. The average pedestrian does not respond to the blasting whistle. If viewing windows are unavoidable, they should be covered with heavy shatterproof plastic and located away from compressor hoses. The compressor hoses in any case should be kept as far from the fence as possible, since they do occasionally rupture and thrash violently.

One final admonition in the maintenance of good fences and barricades: a persistent cause of pedestrian and vehicular accidents at construction sites is the failure to properly close fences and place barricades at the end of the day. Everything is there to make the job safe for the night, but someone neglects to place them properly. This is more likely to happen on evenings when a crew that does not normally close down the job is working late. In any case, the safe closing of a job is the direct responsibility of the construction superintendent.

HEAT PROTECTION

The proper use of temporary heat at the construction site is regulated by a great deal of state and local law as well as by detailed OSHA regulations. Strict compliance is important for safety reasons as well as legal reasons. Serious fires due in part to violation of fire regulations fall in the criminal negligence area. Inspections for compliance with fire regulations by various agencies are quite frequent, and noncompliance tends to be quite obvious.

Almost every construction site will store a certain amount of flammable or combustible liquids and liquid petroleum gas (LP-Gas). Flammable liquids are those that will ignite below 140°F (60°C), and combustible liquids are those that will ignite above 140°F (60°C). Flammable and combustible liquids must never be stored in hallways, on stairs, or at exits. OSHA regulations permit storage of

TABLE 20.1 Storage Regulations for LP Gas in the Vicinity of Buildings

Quantity of LP gas stored,		Distance from nearest building,	
lb	kg	ft	m
500 or less	228	0	0
501 to 6000	229 to 2736	10	3
6001 to 10,000	2737 to 4560	20	6
Over 10,000	Over 4560	25	7.6

up to 25 gal (0.1 m^3) of flammable or combustible liquids within a building without special storage cabinets. Quantities above 25 gal (0.1 m^3) must be stored in an approved metal cabinet. A given storage cabinet cannot have over 60 gal (0.22 m^3) of flammable liquid or 120 gal (0.45 m^3) of combustible liquids. Liquefied petroleum gases (LP gas) such as butane and propane are widely used on construction sites for temporary heat. The fuel containers or cylinders must never be stored inside the building. This is a serious and dangerous violation in every jurisdiction due to the explosive nature of these gases. The OSHA regulation for storage of containers outside of buildings is shown in Table 20.1.

Approved fire extinguishers must always be available within 50 ft (15 m) of stored flammables. Regulations vary widely as to the approved number, type and size of extinguishers for a given condition. Local fire departments are the most readily available source for such information. Water-type extinguishers are never to be used on flammable or combustible liquids and electrical fires; only foam, carbon dioxide, and dry chemicals are suitable and effective.

Temporary heaters used to protect concrete pours during freezing weather are usually LP gas or solid fuel salamanders. LP gas heaters should not be used without first consulting with the local fire department. There are many regulations controlling the use of these heaters due to the ever-present danger of explosion. One of the basic OSHA rules requires that the heater be at least 6 ft (1.8 m) from the LP gas container. Individual heaters and containers should be kept at least 20 ft (6 m) away from other heater units.

Solid fuel salamanders using coke or coal only are acceptable by OSHA standards if properly ventilated. This is important due to the danger of carbon monoxide buildup.

All heaters must be kept at least 10 ft (3 m) from tarpaulins to avoid being turned over by a thrashing tarpaulin in high winds.

NOISE CONTROL

The problems of noise control have received a great deal of attention in recent years from the public and various governmental agencies. The recognition of the need for more control came very slowly, probably for two reasons. First, the equipment in use gradually became more powerful over a long span of time. Second, the damage to the human ear is itself a function of time. Hearing damage tends to be gradual and the person affected may not be aware of a change until an appreciable loss has occurred. This would be typical of heavy-equipment operators over many years of operation without protection.

This new awareness has resulted in a steep rise in compensation claims for loss of hearing. Awards in most states are for "loss of function," meaning loss of hearing, even though there may be no loss of earning power. The contractor therefore has a strong financial as well as legal obligation to enforce noise-control regulations.

Noise-related ear damage is almost always irreversible. For this reason noise-control programs must be effective and rigorously enforced. Noise or sound is measured in decibels (dB). OSHA has established that for noise levels above 90 dB for an 8-h working day, ear protection must be provided. A worker may be exposed to higher decibels without protection, providing it is for a limited period of time, as shown in Table 20.2 reprinted from the *Construction Safety and Health Regulations.*

Table 20.3 shows the noise levels for various types of construction equipment and will help alert the reader to the danger areas.

The types of recognized ear protection fall into two categories: earplugs and ear covers or muffs. Cotton is not an acceptable earplug; it will give some comfort but very little protection. Earplugs of the soft rubber type are effective when properly fitted. They do cause irritation to some people and are not convenient for continuous removal and reinsertion; they must also be kept clean and cannot be used by people with ear infections or drainage. The earmuffs which cover the entire ear have none of the earplug disadvantages and are readily removable for conversations. In extreme situations plugs and muffs can be used together for maximum protection. Listed below are four manufacturers of earmuffs.

Hear-Guard Hearing Protector, American Optical Company, Safety Products Division, Southbridge, MA 01550; Quiet Ear Protector, Bausch and Lomb, Rochester. NY 14602; Straightaway Ear Protectors, David Clark Company, Inc., 360 Franklin St., Worcester, MA 01604; and MSA Noise Foes, Mine Safety Appliances Company, 201 North Braddock Avenue, Pittsburgh, PA 15208.

TABLE 20.2 OSHA Table D-2 Permissible Noise Exposures

Duration per day, h	Sound level slow response, dBA
8	90
6	92
4	95
3	97
2	100
1½	102
1	105
½	110
¼ or less	115

SOURCE: OSHA, *Construction Safety and Health Regulations,* part 1926, U.S. Dept. of Labor, Occupational Safety and Health Administration, Washington, D.C., 1974, p. 22809.

TABLE 20.3 Construction Equipment Noise Levels

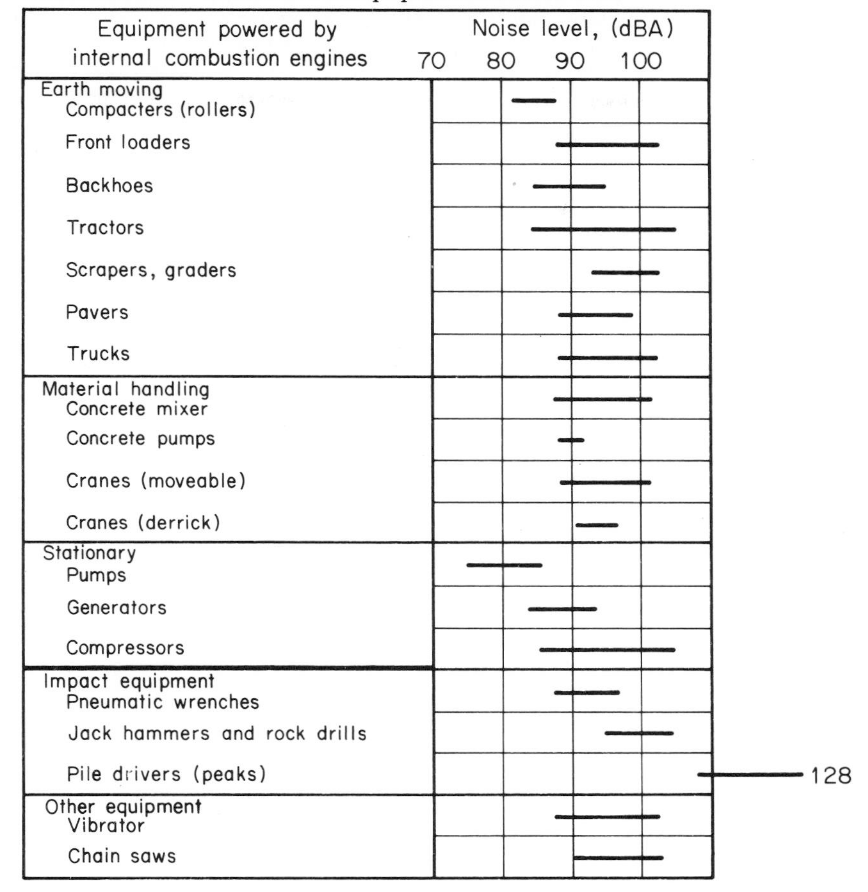

There are several steps that can be taken at the construction site to reduce the level of noise emanating from the site. The construction fence made with plywood is the single most effective noise barrier that is readily available to the contractor. Wood is an excellent noise attenuator and is used extensively in the control of sound. Noise attenuation is a function of material and mass; therefore, the heavier the plywood, the more effective the noise barrier. The height of the fence is also important in reducing the amount of noise that leaves the site. As a practical matter, fences over 12 ft (3.6 m) pose stability and wind problems, although with proper engineering anything is possible in this area.

Special enclosures around excessively noisy machines are often appropriate. This option is particularly important during alterations inside existing buildings. Noise levels from machinery within a building can reach dangerous levels very quickly. The use of plywood panels sandwiched with insulation is an effective do-it-yourself enclosure; however, the fact that plywood is combustible may make its use dangerous or illegal within buildings where fireproof materials are required. There are available today a variety of prefabricated fireproof shelters and enclosures that will provide a noise reduction of 25 to 40 dB. The basic component is a galvanized metal panel with either double or triple walls packed with a high-density mineral fiber. These panels are interlocking tongue-and-groove design and self-supporting. A typical panel might be 2 ft x 8 ft (60 cm x 240 cm) and weigh 6 to 7 lb/ft^2 (30 to 35 kg/m^2). Window and door panels are also available.

Sound absorbers may be useful to temporarily reduce the construction noise leaving an enclosed area. There are two main types available. One is a cylinder of absorbent material the size of a 50-gal (0.2-m^3) drum designed to be hung from the ceiling, and the other is a sound absorbent panel for wall and ceiling mounting.

A leading manufacturer of noise control products is: Gale Corporation, P. O. Box 183, North Brunswick, NJ 08902

SIDEWALK SHEDS

Sidewalk sheds as typically used in all major cities are primarily for protection of the public from falling construction debris and secondarily for the storage of construction materials. The standard heavy-duty bridges as required in cities such as New York are sturdy structures capable of withstanding great impact and sustaining loads of 300 lb/ft^2 (14 kN/m^2) (see Fig. 20.6). For construction more than six stories above the sidewalk, anything less than the heavy-duty bridge does not offer good pedestrian protection.

For low-rise construction sites, a light-duty sidewalk bridge is used which is essentially a heavy-duty scaffold. This is available from many equipment-rental companies and scaffolding contractors. The sidewalk bridge should be built with the following minimum dimensions: clear height for pedestrians, 7 ft (2.1 m); width of walkway, 5 ft (1.5 m); distance from curb, 2 ft (0.6 m) but not over 4 ft (1.2 m); protected lights every 8 ft (2.4 m).

Fences should not be attached to the bridge on the curb side. This creates a concealed tunnel which is attractive to muggers.

There are conditions under which a sidewalk bridge should not be built. If a

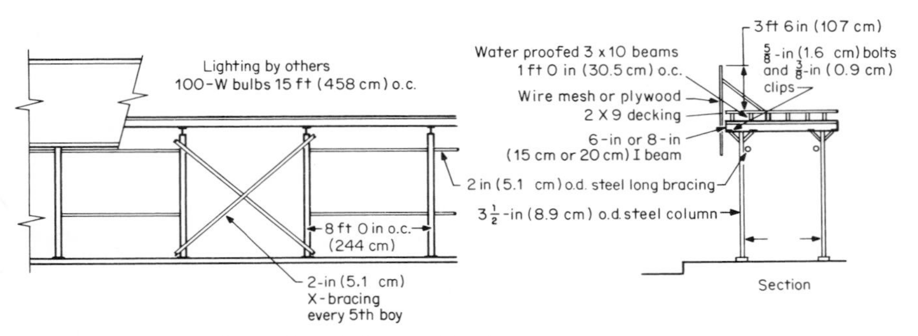

FIG. 20.6 Heavy-duty sidewalk bridge (shed).

building develops structural problems such as a buckling curtain wall or failing cornices or parapets where a major collapse of material is possible, the only safe procedure is to close the sidewalk and even the street if necessary. The use of a sidewalk bridge under such conditions would be false security. There is an element of judgment involved, but the decision is rarely the contractor's alone, since the local building department would certainly be on the scene.

The cost of a typical heavy-duty loading sidewalk bridge, assuming lengths of at least 100 ft (30 m), is approximately $50/ft ($165/m). The cost of a light-duty nonloading sidewalk bridge runs approximately $35/ft ($115/m).

CATCH PLATFORMS AND SCAFFOLDS

Catch Platforms

Catch platforms—or outriggers, as they are commonly called—are discussed in section 1926.451 of the OSHA construction manual.

A sturdy, well-built catch platform can be a very effective form of protection in high-rise construction. A great deal of material inadvertently falls from a skyscraper during construction. Quite naturally, the higher the building, the less of this material falls conveniently on the sidewalk bridge. The only way to overcome this problem is to follow below the construction deck with a catch platform. The first catch platform should be built at about the tenth floor. To be effective, it should be raised as construction progresses so as not to be more than three or four floors below the construction deck.

Because a catch platform is a cantilevered structure, it is subject to great stress and must be built of sound design. In addition, it must be able to withstand impact and severe winds. OSHA permits the use of 2-in x 10-in (5-cm x 25-cm) lumber for light-duty catch platforms, but it is good practice not to use less than 3-in x 10-in (7.5-cm x 25-cm) lumber for the outrigger beams. This will reduce the risk of a failure during the construction of the platform when the carpenters must be on the platform. Safety belts should be a requirement for those building the platform. Figure 20.7 details a heavy-duty catch platform that

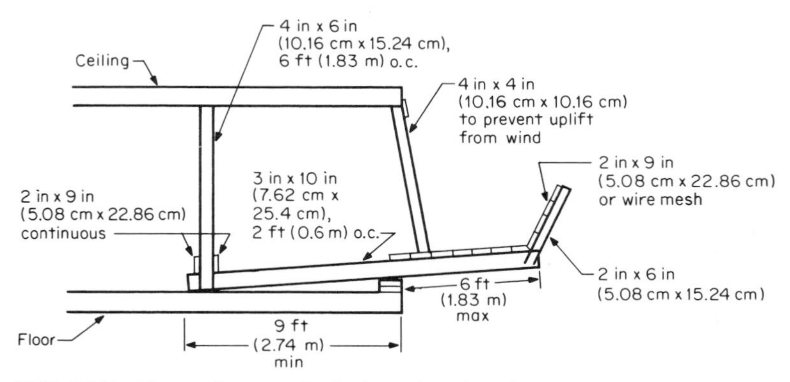

FIG. 20.7 Heavy-duty catch platform (outrigger).

will cost approximately $100/ft ($330/m) to construct. Due to the extreme hazard of an improperly designed catch platform, all catch-platform designs should be reviewed by a professional engineer.

Scaffolds

Scaffolds are one of the primary causes of injury to workers and the public. Scaffolds are constantly being built and used by workers not necessarily versed in the basic safety principles of scaffold construction. OSHA covers scaffold requirements in section 1926.451, and this is must reading for every contractor and superintendent.

There are several fundamental precautions that must be reviewed constantly to reduce the likelihood of scaffold accidents. Toe boards and handrails must always be in place. Cover the scaffold with wire mesh when working over the sidewalk. Quantities of materials, tools, and hardware fall through the scaffold planking. For this reason it is also a good rule not to allow one trade to work above another on a scaffold even with overhead protection. Improper planking is another area of great abuse. Planking should be scaffold-grade full-thickness undressed lumber usually 2 in x 9 in (5 cm x 22 cm). Spans over 10 ft (3 m) are not permitted for any scaffold plank. A typical scaffold on the side of a building should be secured to the building at every 30 ft (9 m) horizontally and 26 ft (8 m) vertically. These ties must be checked daily due to the problem of careless workers removing ties to facilitate their work and then not replacing them. Many scaffolds have blown off buildings for this very reason.

Observe the height rules for a given make scaffold. Most steel-frame scaffolds are designed for heights up to 125 ft (38 m), but not all. Ask your contractor for a copy of the manufacturer's specifications or a state approval. If the planned height of the scaffold exceeds either state or manufacturer's specifications, do not allow the scaffold to be built without modifications approved by a professional engineer.

Scaffold accidents that can be traced to a violation of any one or more of the basic safety rules are a particular threat to builders and/or their superintendents. When serious injury or death is involved, there is a strong possibility that criminal indictments will follow for those responsible for having permitted or overlooked the unsafe use of a scaffold.

DAMAGE DUE TO CONSTRUCTION OPERATIONS

Blasting

The largest single risk associated with blasting is the accidental discharge of electric blasting caps. Current can be induced in blasting caps by radar, microwaves, radio transmitters, lightning, adjacent power lines, dust storms, or any other source of electricity.

Due to the ever-present risk of accidental detonation, it is essential that OSHA-required warning signs be prominently displayed as illustrated in Fig. 20.8.

The most likely source of extraneous electricity is radio transmission, which has greatly increased in recent years with the widespread use of CB radios. The risk of picking up energy from this source can be considerably reduced by making certain that all detonating wires are kept on the ground and are not allowed to go over fences or materials. An elevated wire has a much greater ability to receive radio signals.

During the preconstruction survey of the jobsite where blasting is planned, it is essential to determine if any adjacent buildings house radio transmitters. Arrangements can usually be made with amateur radio operators, but commercial users may present very difficult and costly legal problems. This kind of problem would have to be resolved even before an excavation contract is let. To proceed without a resolution will ultimately lead to either the involuntary shutting down of the commercial transmitter or the cessation of the blasting operation. Either one can be very costly.

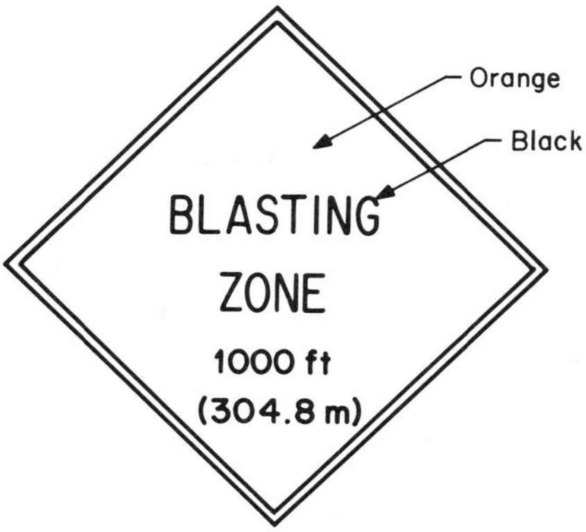

FIG. 20.8 OSHA-required warning signs for blasting: (*a*) 48 in × 48 in ± (122 cm × 122 cm ±) and (*b*) 42 in × 36 in ± (107 cm × 91 cm ±).

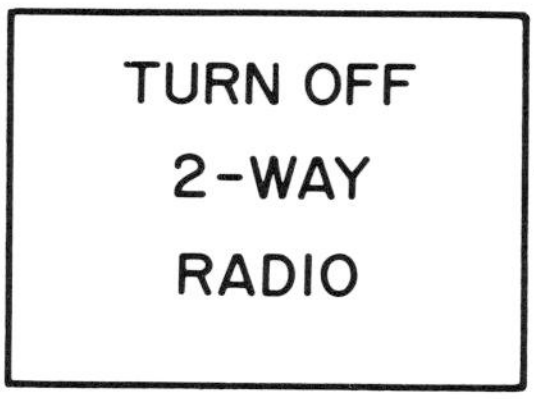

Careful planning of the blasting operation can reduce the possibility of damage to adjacent structures to a minimum. This can be accomplished by first obtaining a designed blasting pattern that will assure that the shock waves transmitted to adjacent buildings will be of low enough magnitude not to cause damage. Specialists in blasting techniques can make on-site field tests to determine the transmission factor of the rock at the site. This test will determine the energy ratio being transmitted through the rock. From this data a blasting pattern can be designed that will assure a safe blasting operation.

It is worth mentioning that the key to designing safe but effective blasting is the use of proper delays in detonation. The U.S. Bureau of Mines estimates that an instantaneous detonation of a given amount of dynamite will produce an energy ratio 11 times greater than a timed delay detonation. The work produced in breaking the rock is the same in either case, but the power of the shock wave jumps dramatically in an instantaneous blast.

In supervising a blasting operation there are several precautions to monitor on a day-to-day basis. Check the blasting mats to be certain the blaster is using the proper number of mats and is not using torn mats. Observe the strike of the rock to determine the possibility of a rock slide. Folded rock, common in the eastern United States, can be extremely dangerous under the condition illustrated in Fig. 20.9 and may require pinning or bracing.

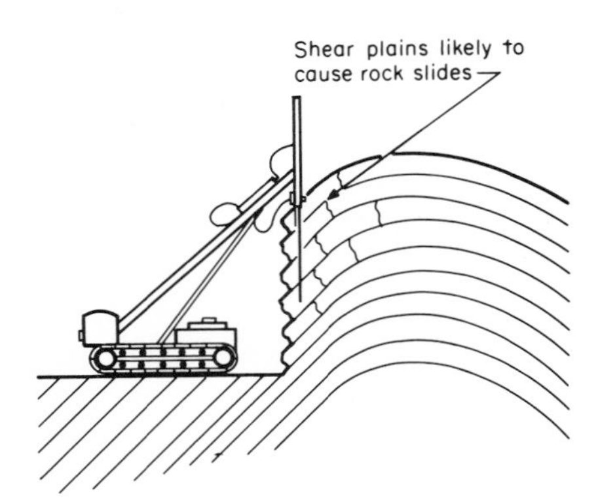

FIG. 20.9 Rock drilling under conditions likely to cause rock slides.

Drilling should not be permitted within 50 ft (15 m) of a loaded hole in the event of accidental detonation.

As the drilling and blasting approaches adjacent buildings, recheck elevations to be certain the cellar of the adjacent building is not behind the rock that will be blasted. As obvious as this may sound, careless excavators are occasionally blasting in foundation walls.

The storage of explosives and related hardware is strictly regulated by federal and state law. Violations in this area will usually involve criminal law; therefore, strict adherence to the law is essential.

Every construction site using explosives is required to have two very secure magazines, one for explosives and one for blasting caps, detonating primers, and primed cartridges. The owner of the explosives must keep an accurate inventory and use record of all explosives. The quantity of explosives stored at the construction site should be kept to a minimum for safety and theft reasons. Terrorists and criminal elements have always been tempted by construction site magazines. Any thefts, discrepancies in inventory, or attempted break-ins must be reported immediately to the local police and the FBI.

In the event of a fire close enough to the magazines to possibly cause detonation, the generally recommended procedure is to evacuate personnel rather than attempt to fight the fire.

Pile Driving

Damages due to pile driving are, on average, not difficult to control. The operation itself is noisy and conspicuous, thus attracting the attention of the public and adjacent owners. Human sensitivity to vibration is very acute and even more so when accompanied by noise. People are aware of vibrations of an energy level only 1/100 of that required to do damage to a structure. Even vibrations perceived to be severe by people are only one-fifth of the vibrations necessary to damage a sound structure.

The only professional way to control pile-driving vibrations is to retain a consulting engineer to monitor the site with an accelerograph. This instrument is a velocity seismograph which measures vibrations in three directions (one vertical and two horizontal) and records the maximum disturbance that was measured in any one of the three directions in inches per second of particle velocity—more commonly referred to as the energy ratio. For energy ratios up to 3, damage is unlikely to occur. For energy ratios between 3 and 6, slight damage is possible. In general, for a distance of 20 ft (6 m) from a pile driver, energy ratios of 1.0 or less will develop in most soils. This level of vibration would be comparable to a train passing at the same distance. Soil type has a direct bearing on the transmission of vibratory energy. In general, the energy ratio in sand will drop to almost zero at a distance from the pile equal to the length of the piles. The energy ratio increases somewhat as the sand grades off to gravel. In clay soils measurable energy will be transmitted a distance of 2 or 3 times the pile length. The above generalizations assume dry conditions. A high water table may cause a much higher energy ratio for a given distance from the pile. Backfilled areas, rubbish, and swamps are special conditions where high energy ratios may be encountered. Vibration recordings are an invaluable tool under these conditions.

The above generalizations are only a guide and an aid in avoiding problems. Pile driving is fraught with dangers due to the wide range of variables. Vigilance is the best protection against damages from the unpredictable.

A word of caution in using energy ratios: The safe ranges assume sound structures. In most of our urban centers, this is often not the case. In the eastern United States it is very likely that some adjacent building will be 100 years old and constructed of lime mortar that has been thoroughly leeched out. Walls, chimneys, and parapets will often be found out of plumb, bulging, and in various stages of incipient collapse. Even the slightest vibration can be damaging to such buildings. A pre-job survey should reveal such conditions and must be done well in advance of construction to allow time for remedies. Many owners of such rundown buildings are often unaware of their true condition. The local building department will usually issue a violation to the owner when defects are pointed out. This removes some of the responsibility from the shoulders of the builder and makes it a matter of public record that an adjacent building is unsound. The problem of making it safe enough for the builder to begin work can become very complicated legally. Many owners are absentee or destitute or both. In any case, they often cannot be induced to remedy their own problems. Builders are then forced to brace, tie, or in some manner shore up the buildings at their own expense. Most state and city codes allow for this problem to the extent that the project cannot be stalemated by an uncooperative adjacent owner. Legal counsel is essential in these matters to avoid problems of trespass and alleged damages.

The most hazardous part of pile driving is to the pile-driving crew. The equipment is heavy and there is a great deal of hoisting and moving about. The builder can contribute little to improving the safety of the crew, since the equipment, crew, and operations are under the control of a pile-driving contractor. The builder can, however, monitor and control two important aspects of the operation. The site itself must be made safe for a pile rig to move about. The rig must not be required to work too close to the top of excavated banks or under banks that do not have a safe angle of repose. The second area of responsibility for the builder is to be certain that the rig does not have to work within 10 ft (3 m) of a power line carrying up to 50 kV. For power lines over 50 kV, greater distances are required. Consultation with the power company is essential. It is possible to eliminate all hazard by temporarily shutting down power.

Dewatering

The subject of dewatering is covered in depth in Chapter 6. This section will be confined to a few remarks on the potential hazards of dewatering.

The average dewatering problems encountered will usually be overcome without any detectable damage to surrounding ground and structures. When possible, it is best to pump continuously with automatic sumps rather than allowing the water to rise and fall. Not allowing the water table to fluctuate up and down reduces the degree of consolidation of the soil and also reduces the flow of fine-grained material to the pumps. There are, however, conditions under which an improperly conceived dewatering plan may cause massive damage to surrounding buildings, roads, and utilities. Compounding the problem is the likelihood that a site requiring heavy dewatering will also require piles. The net effect is that the water is withdrawn from the soil, permitting consolidation, and the vibration of pile-driving further encourages the consolidation. The use of vibratory pile-driving hammers can aggravate the consolidation process to an even greater degree.

The best course to follow when heavy dewatering is encountered is to put the problem in the hands of a soils engineer and dewatering expert.

Excavation

Safe excavation is a matter of constant vigilance and adherence to several elementary rules. Most problems that develop during excavation are not really accidents, but would more properly be called gambles taken and lost. The temptation to dig a little further without sheet piling or closer to a building than is safe seems irresistible, but the consequences can be costly.

The most elementary rule to observe in excavation is to maintain 30° slopes when the depth of the excavation exceeds 5 ft (1.5 m). If these slopes cannot be maintained, sheet piling should be used for earth banks and underpinning in the case of adjacent buildings. As far as safe slopes are concerned, OSHA does allow variations in the 30° rule if the soil type is clearly a more compacted material. Table 20.4 is taken from the OSHA *Construction Safety and Health Regulations, Part 1926.*

TABLE 20.4 Approximate Angle of Repose for Sloping of Sides of Excavations

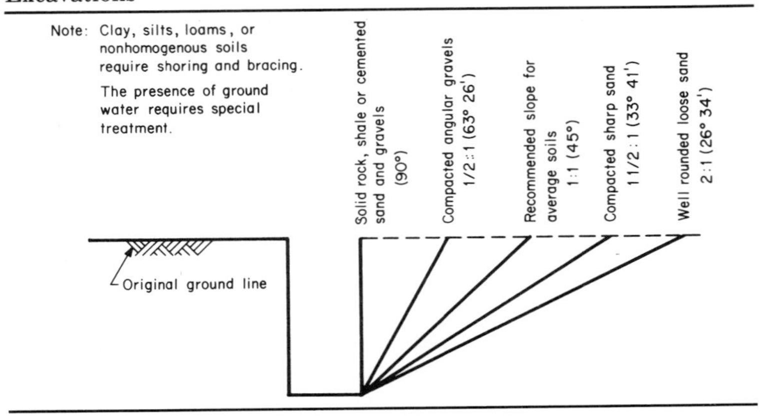

Wood sheet piling should never be used in lieu of underpinning for supporting adjacent buildings. The 30° rule should always be observed regardless of soil type when excavating toward adjacent buildings. The exact method of underpinning is a matter to be determined by a professional engineer.

If groundwater is encountered, the OSHA table does not apply and excavation near adjacent buildings becomes a special condition requiring the judgment of a professional engineer.

A problem frequently encountered when excavating is a change in soil type. This can be dangerous when the excavation begins in a compacted material and then abruptly changes to a loose sandy material overlain by a compacted material. The loose sand will have a much lower angle of repose and result in undermining and collapsing of the compacted material above, as shown in Fig. 20.10. If this condition is encountered, the bank above must be cut back or sheet piling must be installed.

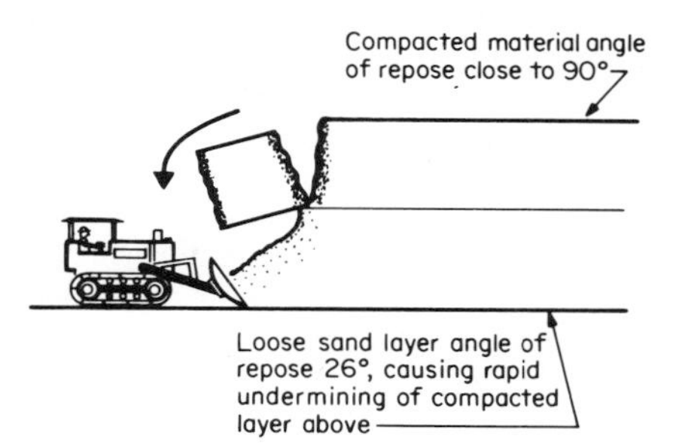

FIG. 20.10 Dangers inherent in excavation of materials of sharply varying compactness.

Special attention must be given to the operation of heavy equipment above the banks next to the excavation. The OSHA table for safe slopes does not take into account superimposed loads of equipment or materials. This becomes especially critical if sheet piling is supporting the banks. The load of equipment can easily exceed the soil pressures for which the sheeting was originally designed. Unless the sheeting is designed to support the additional load of equipment, it is good practice to fence or barricade the sheeting for a distance equal to approximately 45° back from the base of the sheeting, as shown in Fig. 20.11.

The excavation of pits poses special hazards mainly to those doing the work in the pits. Many needless fatalities occur every year from careless pit operations. The basic rule is to sheet-pile pits over 5 ft (1.5 m) deep, regardless of how com-

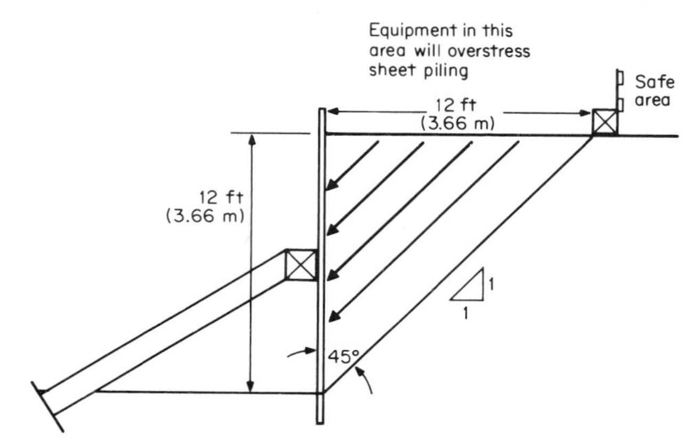

FIG. 20.11 Area of potential danger adjacent to sheet piling.

pact the soil seems. Pits that might otherwise seem safe are always subject to the movement of heavy equipment close to them, causing a sudden surcharge that will shear the soil without warning and bury the workers below. For this reason pits should always be barricaded as well. For small pits up to 4 ft x 4 ft (1.2 m x 1.2 m) square, 2-in x 9-in (5-cm x 22-cm) undressed lumber is usually adequate for sheet-piling a pit. For larger pits the variations in conditions are so numerous that generalizations are dangerous; therefore, the sheeting should be designed by a professional engineer. For pits over 10 ft (3 m) deep, special care must be taken to assure constant clean air in the pit. Carbon monoxide from nearby gasoline engines is the greatest danger. Smoking should not be allowed in pits, due to the possibility of trapped gases being encountered.

PROTECTION OF NEARBY STRUCTURES

The best tool for protecting nearby structures is the pre-job survey. When this job is done by an engineering firm specializing in surveys, very little will be left to chance. The physical integrity of nearby structures can directly affect such operations as pile driving and blasting. When adjacent buildings are sound, these operations can be carried on with little danger of damage, providing energy levels are controlled and monitored with accelographs as discussed earlier. If, on the other hand, walls, parapets, and chimneys are out of plumb, bulging, and in various stages of collapse, extensive shoring and reinforcement will be required.

The most direct technique available for supporting weak or failing walls is the common brace. To be effective, the brace should not be steeper than 2½:1 or about 22° above horizontal. Figure 20.12 shows a typical bracing system using a heavy timber on the wall to distribute load at floor level and a solid heel to develop sufficient bearing value in the soil. This bracing system is not suitable for all conditions. The type, size, and extent of bracing for each condition should be determined by a professional engineer.

Some of the problems of protecting nearby structures develop during the demolition of adjacent buildings. Party walls left standing on adjacent buildings, while quite safe when supported from both sides, become unstable when required to stand alone with support from only one side. A typical procedure to reinforce such walls is to attach channels to the floor beams as shown in Fig. 20.13. Con-

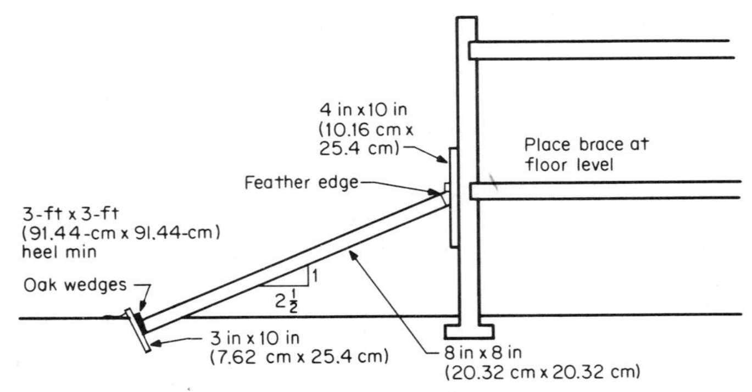

FIG. 20.12 Bracing of unstable buildings adjacent to construction sites.

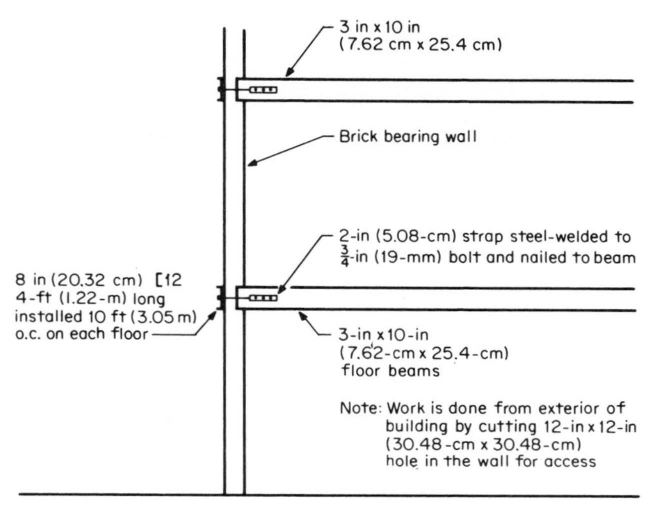

FIG. 20.13 Reinforcing old brick walls using channels and rod ties.

ditions under which this type of reinforcing is done vary so widely that it is best to have the system designed by a professional engineer.

Old foundation walls left standing on property lines with earth loads behind them usually become very unstable when the floor is removed that retained the top of the wall. Here again, conditions vary widely and a professional engineer should be consulted to design a safe bracing system.

An old free-standing building adjacent to an excavation often requires reinforcement of all four walls. A system of channels and rods as shown in Fig. 20.14 is a typical solution, but again the system requires careful engineering.

The picture in Fig. 20.15*a* is graphic evidence of the failure to properly tie-in an old wall. The building on the left lost its wall when settlements began during excavation for a new office building on the right. Note that complete collapse of the building was avoided by timely shoring of all floors with 8-in x 8-in (20-cm x 20-cm) post and headers.

Figure 20.15*b* is a photograph of an old brick building that has lost its front and side wall simultaneously during excavation for a new building on the left, behind the plywood fence. The three-story building is balanced precariously on its own partitions. The building was too dangerous to save and was demolished. A properly designed tie-in system as shown in Fig. 20.14 would have prevented collapse of the walls.

After the new building comes out of the ground, there are still precautions that must be taken with adjacent buildings. Roof damage is a primary concern when the new building rises any distance above the adjacent property. The most severe claims actually come from water damage due to clogged drains caused by debris from the new construction. Sawdust, paper, and wood chips can clog a roof drain in minutes. A good precaution is to circle the roof drains with ½-in (12-mm) wire mesh at least 2 ft (60 cm) from the drain. The protection of the roof itself will vary with the height of the work above the roof. Certainly ½-in (12-mm) plywood placed on the adjacent roof would be a minimum protection. For

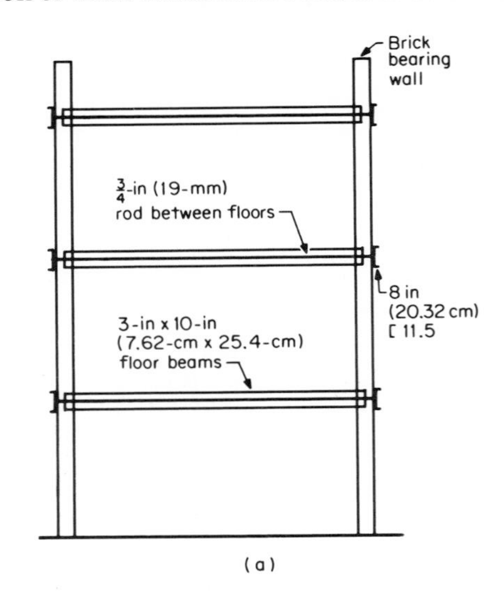

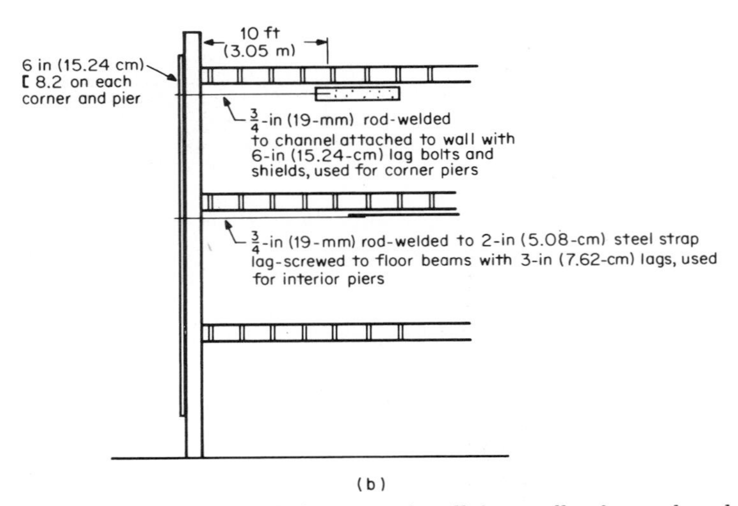

FIG. 20.14 A strong tie-in system for all four walls of a weakened building: (*a*) Side bearing walls and (*b*) front and rear walls.

five floors or more above adjacent roofs, serious consideration has to be given to impact from falling debris. The kind of roof construction will have a bearing on the amount of protection. Obviously, a concrete roof is better protection than a typical wood beam and board construction. The occupancy of the structure to be protected is also a factor. A building used for dead storage is clearly less of a concern than a crowded apartment building.

When heavy protection is required, a typical design would be 3-in (7.5-cm) lumber on the flat in a direction of right angles to the existing roof beams, fol-

(a)

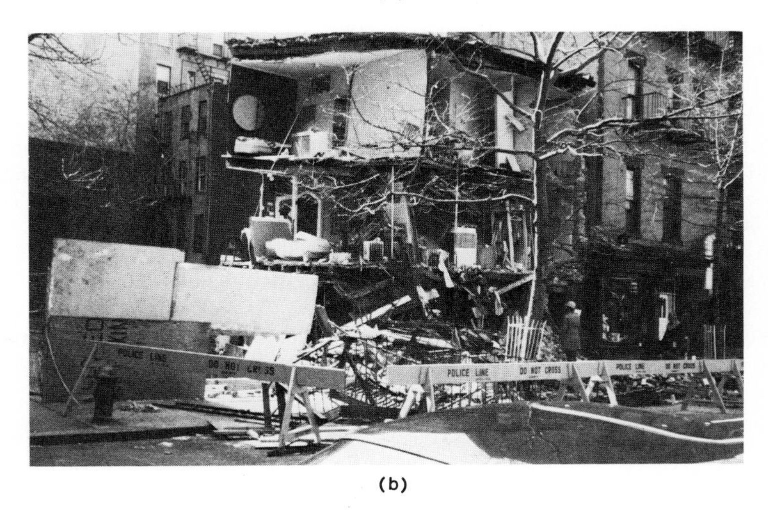

(b)

FIG. 20.15 (*a*) Failure to reinforce the bearing wall resulted in total collapse of the wall; (*b*) a three-story brick building after collapse of the front and side walls.

lowed by a 3 in x 10 in (7.5 cm x 25 cm) on edge covered with at least 2-in (5-cm) lumber. This pattern could be repeated again for added protection (see Fig. 20.16.)

Shoring of the roof from below may be required due to the weight of the protection itself and to assure its ability to withstand impact.

The photograph in Fig. 20.17 shows a heavy cribbing system that has just withstood the impact of a 10-ft x 15-ft (3-m x 4.5-m) section of brick wall that fell 12 stories onto the cribbed roof. There was no roof damage. The cribbing had been placed there when a buckle developed in the wall of a completed office building. The wall was too dangerous to work on and was allowed to collapse.

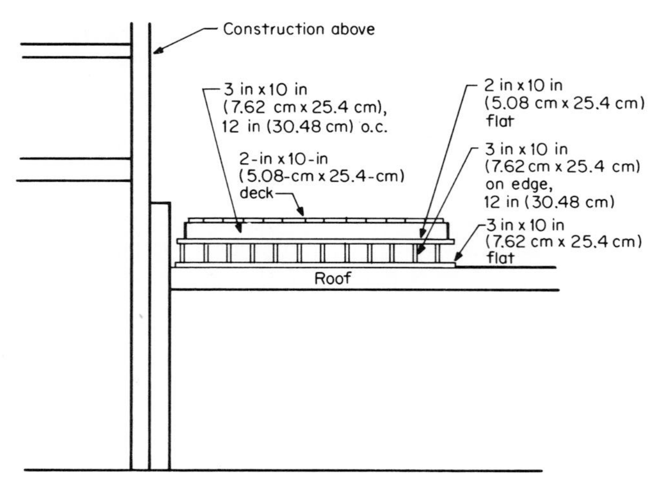

FIG. 20.16 Heavy-duty protection of roofs exposed to falling material.

The distance to carry protection out from the new building is a matter of judgment based on the relative height of the new construction over adjacent buildings. Certainly 10 ft (3 m) out would be a minimum for any condition. Where the new building is hundreds of feet higher, protection is often carried out 50 ft (15 m) and more.

After completion of the new building, it is a good practice to have another survey of adjacent buildings. Damage, if any, should be carefully documented and certified by a professional engineer. The after-job survey is particularly important in protecting the contractor from questionable or outright fraudulent claims. Many an owner sees adjacent construction as an opportunity to get a new

FIG. 20.17 Heavy-duty roof protection after collapse of a brick wall above.

roof, make plaster repairs, and repaint it at the builder's expense. The survey reports, if professionally done, will carry great weight in any future litigation for damages.

PROTECTION OF UTILITIES

Temporary support of utilities must be approached with extreme care. The utility company or agency controlling the utility in question should always be consulted. All temporary supporting systems should be designed by a professional engineer and approved by the controlling agency. The location of various shutoffs for the utilities in the vicinity of the jobsite should be marked in the field and posted in the construction office to assist emergency crews in the event of an accident.

The first step in protecting utilities is to mark their approximate location, keeping in mind that utility drawings are notoriously inaccurate.

The utilities of concern to the builder are those just outside the foundation walls and utilities that may cross the construction site. The utilities just outside the excavation but within the angle of repose will have to be supported by sheet piling. The strength and type of sheeting will depend on the proximity of the utilities, the depth of excavation, and soil material. There is always reconsolidation of material behind the sheeting; therefore, some movement of the ground is to be expected. Solid-steel sheeting, where conditions permit its use, offers the most positive retention of the soil. If H beams are to be driven for the sheeting, numerous test holes should be dug to positively locate the utilities. Blind driving, depending on utility drawings, is an invitation to disaster.

If plans call for the utilities to cross the construction site and remain in place, a careful plan of support and/or protection must be developed. Utilities that cross the site at the bottom of the excavation are relatively easy to protect. Barricades are necessary to keep machinery away. Hoisting over the utilities should be avoided. Bridges over the utilities for the movement of equipment are usually necessary. Earth bridges are very risky since the utility must withstand a good percentage of the equipment load that passes over it. The best bridge is one that permits no load on the utility, as illustrated in Fig. 20.18.

Utilities that must be undermined present a more difficult problem of support. The system used will depend on the weight of the utility, its own structural strength, and its joint interval. For lighter utilities a single H beam driven next to it at appropriate intervals with a carrying beam is usually adequate. The pho-

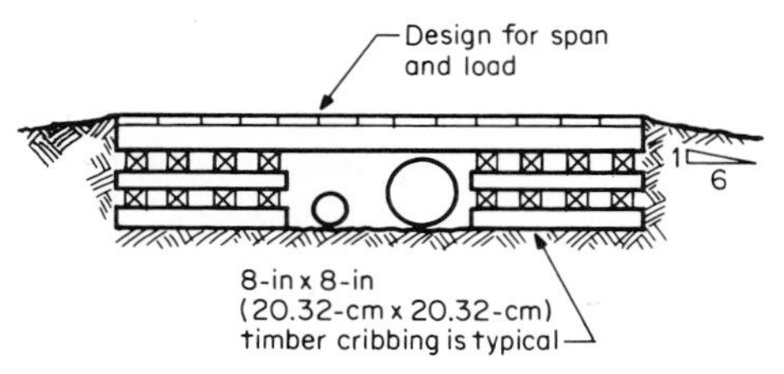

FIG. 20.18 Construction ramp over utilities.

FIG. 20.19 Temporary hanging of concrete-encased conduit at an excavation.

tograph (Fig. 20.19) shows a concrete-encased conduit hung from 12-in (30-cm) HP 53s at 20 ft (6 m) above an excavation.

For heavier utilities a frame must be built, again using H beams driven on both sides, as shown in Fig. 20.20.

On some sites utilities are relocated and temporarily hung from brackets on the sheet piling. This may be convenient, but it is a dangerous practice. Sheet piling is under great stress and subject to sudden movements due either to a failure or to equipment knocking down a brace.

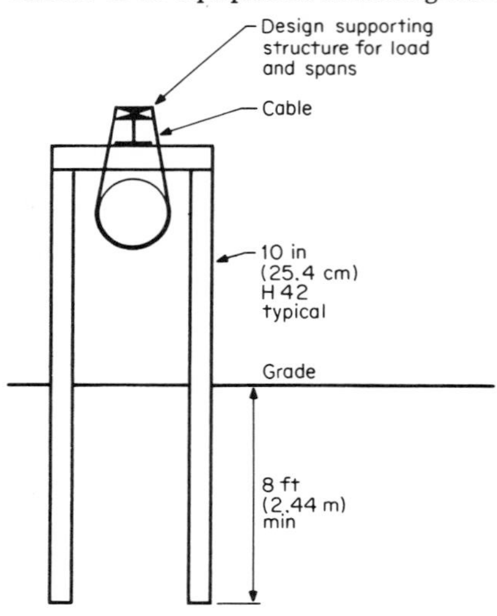

FIG. 20.20 Hanging heavy utilities during excavation.

Utilities that come through foundation walls into the excavation are of particular concern. This is most common where a new wing is being added to an existing building. If the utility is temporarily supported and later backfilled, again great care must be taken to prevent a shear at the foundation wall. The ground below the utility must be fully compacted, or the backfill will move down the wall with glacierlike force and take the utility with it. The photograph in Fig. 20.21 shows sheet piling for an addition to a hospital. The gas, water, and sewer pipes are all hung from the sheet piling as they leave the existing building. The new wall for the addition went up about 4 ft (1.2 m) from the sheet piling. When the space between the sheeting and the new wall was back-

FIG. 20.21 A potentially dangerous method of supporting utilities.

filled, all three utilities were ripped from the sheeting and sheared off at the wall of the existing building.

Utility poles are often found along the curb just outside foundation walls. Excavation for a new building will likely disturb the anchors, requiring temporary guying of the pole. A suitable anchor for the guy is often a problem. If H beams are being driven for sheet piling, a cable can be attached to a beam with ease. Since the sheet piling is usually adjacent to the poles, it is a good practice to drive one or two long H beams for cabling to the pole, as shown in Fig. 20.22.

MONITORING SURROUNDING GROUND AND NEARBY STRUCTURES' CONDITIONS

This section deals mainly with spotting the problems that develop during the excavation phase of a building project.

Probably the best early warning of impending trouble is simply constant visual inspection. Many of the points made here will seem simplistic; yet, over and over again, perfectly obvious signs are overlooked in their early stages.

A well-designed sheet-piling system will not show any visible signs of movement. If, however, a failure is developing, the first signs of movement will usually show behind the sheet piling—in the earth which is actually held in place by the sheet piling. Cracks will start to form a few feet back from the sheeting and parallel to the line of sheeting. These cracks will grow measurably each day if movement is taking place, and new lines will form farther back from the sheeting.

If these cracks are along a curb line of a roadway, they must be filled daily until movement stops. Water from heavy rains follows the curbs and will run down these cracks, causing a dramatic increase in sheeting load and probable loss of material from behind the sheeting. Soil and road cracks can in some cases result from reconsolidation of the soil, due possibly to loss of material while

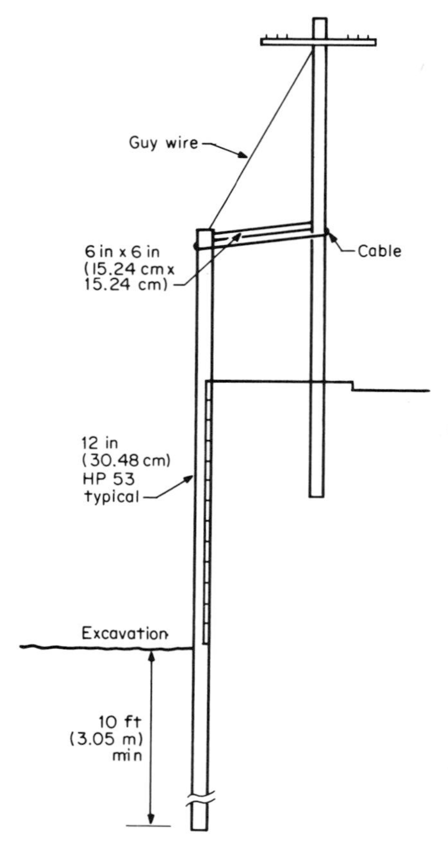

FIG. 20.22 Cabling utility poles adjacent to excavations.

installing the sheeting. If this is so, the movement will stop, but in any case the appearance of such cracks requires thorough examination of the sheeting itself. Check for bending boards, braces, or walers. The heels for the braces may be failing or may have been disturbed by machinery. All these signs call for attention and review by a professional engineer. Sheet-piling failures usually have a domino effect in that they trigger water, gas, and sewer main breaks and the loss of job momentum.

Buildings are relatively easy to monitor visually if done daily by the same person. If a building starts to settle or a wall moves laterally, it will show almost immediately on the inside by cracks in the corners and separation of the floors from the walls. Inside inspection is vastly superior to external inspection. A crack developing in plaster on the inside may not show in the brick on the outside until considerably more movement has taken place.

The use of the level and transit to monitor roads, buildings, and utilities is the best legally acceptable record of job conditions when kept in log form. Most of the larger projects shoot levels and lines on a daily basis and record them in a log in order to detect vertical or lateral movements in ground and structures. Many projects seem to put too much reliance in shooting levels, probably because it is easier to do. Detecting lateral movement can in some cases be much more significant than detecting vertical movement. Sheet piling and supported utilities almost always fail laterally.

The older, structurally unsound masonry buildings will often begin failure by the walls bulging out in the middle. The bulge could grow several inches before the level readings would pick it up, if at all. Lateral readings will pick up the movement immediately and allow valuable time for bracing or tying in of the walls.

Concrete underpinning requires special attention to prevent and detect lateral movement. It is the nature of the underpinning operation that the underpinning must for a period act as a retaining wall until the adjacent walls of the new building are installed against the underpinning. Most underpinning is designed to only take the vertical load of the structure above. It has basically no lateral stability; therefore, if the earth loads behind it are significant, it must be braced until the new foundation walls are built. During this period while waiting for the new walls to be built, the entire stability of the building is dependent on these braces. Lateral readings are especially important during this period to assure that the bracing system is adequate and holding. There have been many dramatic collapses in the past due to failure to properly brace the underpinning. Underpinning when it collapses appears to be instantaneous, but it is almost always preceded by several days of lateral creep which would easily be detected by lateral readings.

Under certain conditions the monitoring of interior columns of adjacent buildings can take on great significance. The operations that can affect interior columns and footings are pile driving, blasting, dewatering, and underpinning. The detection of movement by level readings is the same regardless of the cause. The control of settlements due to excessive vibration was discussed earlier in this chapter in reference to the use of seismograph readings and proper design of blasting patterns.

The chapter on construction dewatering in this book details the implementation of a safe dewatering system that will greatly reduce the risk of settlements from improper dewatering.

Underpinning is a special condition under which one can inadvertently undermine interior footings. If foundation plans are not available, which is the usual case, the pre-job survey will indicate whether the interior footings will be undermined. If there is doubt as to the depth of interior footings, arrangements must be made for a test pit to determine exact footing elevations. Problems with adjacent owners can complicate work inside the building, but there is no way for the builder to avoid the responsibility of underpinning interior footings if required. The general rule governing the slope between underpinning and interior columns is 2:1, which is the same as in most footing relationships. Figure 20.23 illustrates the application of the 2:1 rule.

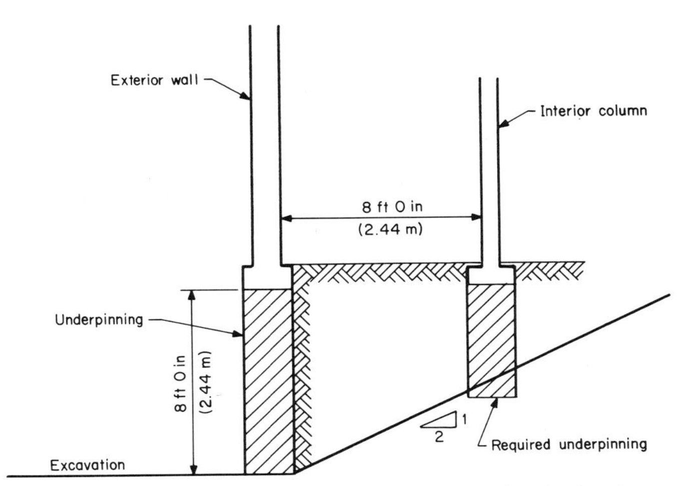

FIG. 20.23 The often-neglected problem of underpinning interior columns.

BIBLIOGRAPHY

Blasting Vibrations and Their Effects on Structures, Bulletin 656, U.S. Bureau of Mines, Washington, D.C., 1971.

Commerce in Explosives, IRS publication 26 CFR 181, Internal Revenue Service, Washington, D.C.

Construction Safety and Health Regulations, part 1926, U.S. Dept. of Labor, Occupational Safety and Health Administration, Washington, D.C., 1974.

Construction Safety Notes, Liberty Mutual Insurance Co. Loss Prevention Service, Boston, Mass., (undated).

Crandell, F. J.: "Ground Vibration Due to Blasting and Its Effect Upon Structures," reprinted from the *Journal of the Boston Society of Civil Engineers,* April 1949.

Crandell, F. J.: *Evaluation of Vibration Waves Due to Blasting,* National Safety Council, 425 North Michigan Ave, Chicago, Ill., 1954.

Crandell, F. J.: "Transmission Coefficient for Ground Vibrations Due to Blasting," reprinted from the *Journal of the Boston Society of Civil Engineers,* vol. XLVII, no. 2, April 1960.

Flammable and Combustible Liquids Code, NFPA 30-1969, National Fire Prevention Association, Boston, Mass., 1969.

Foundation Facts, vol. II, no. 2, Raymond Concrete Pile Division of Raymond International, Inc., New York, N.Y., 1967.

Maintenance and Use of Portable Fire Extinguishers, NFPA no. 10A-1970, National Fire Protection Association, Boston, Mass., 1970.

Manual on Uniform Traffic Control Devices for Streets and Highways, ANSI D6.1-1971, American National Standards Institute, New York, 1971.

Seelye, Elwyn E.: *Foundations Design and Practice,* John Wiley & Sons, New York, 1956.

Uniform Traffic Control Devices for Streets and Highways, U.S. Dept. of Transportation, Federal Highway Administration, Washington. D.C., 1968.

Index

About the Editor

Robert T. Ratay, P.E., is a highly respected Consulting Engineer and Professor of Civil/Structural Engineering. He has more than 20 years experience in the design, construction, and investigation of structural and civil engineering projects in the USA and abroad, as well as university teaching of structural design and foundation engineering. He is the author of papers in the journals of the American Society of Civil Engineers and other technical publications. Dr. Ratay was educated at the University of Massachusetts where he earned B.S., M.S., and Ph.D. degrees in Civil/Structural Engineering.

DEAN C. BERNETT
STATE OF WASHINGTON
REGISTERED
PROFESSIONAL ENGINEER